CAPILLARY FLOWS WITH FORMING INTERFACES

CAPILLARY FLOWS WITH FORMING INTERFACES

Yulii D. Shikhmurzaev

Professor of Applied Mathematics
University of Birmingham, UK

CRC Press

Taylor & Francis Group

Boca Raton London New York

CRC Press is an imprint of the
Taylor & Francis Group, an **informa** business

A CHAPMAN & HALL BOOK

Cover photographs of a spreading drop courtesy of Dr. R. Rioboo.

CRC Press
Taylor & Francis Group
6000 Broken Sound Parkway NW, Suite 300
Boca Raton, FL 33487-2742

First issued in paperback 2019

© 2008 by Taylor & Francis Group, LLC
CRC Press is an imprint of Taylor & Francis Group, an Informa business

No claim to original U.S. Government works

ISBN-13: 978-1-58488-748-5 (hbk)
ISBN-13: 978-0-367-38865-2 (pbk)

Library of Congress Cataloging-in-Publication Data

Shikhmurzaev, Y. D.
  Capillary flows with forming interfaces / Yulii Damir Shikhmurzaev.
    p. cm.
  Includes bibliographical references and indexes.
  ISBN-13: 978-1-58488-748-5 (hardcover : alk. paper)
  ISBN-10: 1-58488-748-6 (hardcover : alk. paper)
    1. Fluid mechanics--Mathematics. 2. Capillarity--Mathematical models. 3. Solid-liquid interfaces. I. Title.

QA911.S299 2008
530.4'27--dc22   2007021669

Visit the Taylor & Francis Web site at
http://www.taylorandfrancis.com

and the CRC Press Web site at
http://www.crcpress.com

Contents

Preface

> How wonderful that we have met with a paradox.
> Now we have some hope of making progress.
>
> Niels Bohr

Capillarity plays an important role in numerous natural phenomena and various technological processes. The latter range from film flows in coating devices and dynamics of drops in ink-jet printing to biological and biomedical applications of fluid mechanics, bioengineering and microfluidics. The complexity of effects associated with capillarity and a continuously widening spectrum of applications have kept capillary flows in the focus of considerable research effort for more than two centuries. This effort brought in many remarkable successes in the understanding of the fundamentals and the quantitative modelling of capillary phenomena in different situations.

However, research also highlighted a number of intriguing paradoxes where an attempt to describe mathematically some, often quite ordinary, capillary flows by combining well-tested elements from the arsenal of classical fluid mechanics leads to solutions with manifestly unphysical properties. Furthermore, in some situations there are no solutions at all! These 'paradoxical flows' include the spreading of liquids on solid surfaces, coalescence and breakup of drops and bubbles, breakup of jets, disintegration of liquid films on a solid substrate, and the formation of two-dimensional singularities in the free-surface curvature, to mention but a few examples.

In each case when a paradox is encountered, it becomes a *problem*. Such *problems* mark the boundary that the standard computer-aided applied research cannot cross without descending to a semi-empirical level. Ironically, many applications, especially in the emerging technologies, lie on the other side of this boundary and cannot be 'scaled up' to be dealt with on a semi-empirical basis. As a result of this situation, each of these *problems* becomes the subject of intensive research in its own right, with its own 'club' of investigators, ad hoc 'laws' and all other attributes of a long siege.

The key idea of the present monograph is a realization that many seemingly very different 'paradoxical flows' are in fact particular cases from a general class of fluid motion — capillary flows with *formation* (and/or disappearance) of interfaces. For example, the famous 'moving contact-line problem' arises in the continuum description of the process known as 'dynamic wetting', and the very name 'dynamic wetting' actually means a process by which an initially dry solid surface becomes 'wet', i.e., a process of *formation* of a new liquid-solid interface (a 'freshly wetted' solid surface). Then it becomes clear why classical fluid mechanics fails to adequately describe dynamic wetting as well as other flows where the formation of interfaces is essential: this process is simply not accounted for in the model. It is also clear why treating 'paradoxical flows' separately on an ad hoc basic, as purely mathematical *problems*, leads to less than satisfactory results: there is very little chance that one will accidentally come up with an adequate mathematical description of a phenomenon without realizing what physics controls it and the place of this physics in a broader physical context.

The objectives of this book are described in a section of the first chapter entitled "Scope of the book". Here it is appropriate to outline the book's style, methodology and structure.

In this respect, the main overall goal was to achieve textbook clarity in the exposition of both the physical ideas and the mathematics they are wrapped in. To bring the readers with different educational backgrounds to the same starting level, assumed to be the level of the final-year undergraduate students, the first two chapters give a research-oriented exposition of the fundamentals of the fluid mechanics. This part can be seen as a concise 'textbook for future researchers'. A particular emphasis is on the topics that are rarely covered in standard texts, like the Lagrangian representation of the equations of motion, the use of the Airy stress function for the Stokes flows, assumptions behind boundary conditions, especially on the solid surface, and their limits of applicability. To make the reading easier, appendices recapitulate the necessary elements of mathematics in a self-sufficient user-friendly way. Since the clause "as is known" often becomes a barrier separating knowledge from familiarity, it is completely avoided in the first two chapters and wherever possible in what follows.[1] Chapter 2 ends with a discussion of basic criteria that must be satisfied by a physically meaningful solution of a mathematical problem formulated in the framework of fluid mechanics. These minimal criteria of physical meaningfulness and self-consistency can then be used to assess theories proposed to remove the paradoxes.

The subsequent chapters are devoted to the 'paradoxical flows'. In each case, first, we discuss the essence of the problem inherent in the classical fluid mechanics formulation. This is followed by a detailed review and analysis of relevant experiments and, where available, theories developed alongside them to remedy the problem. In this analysis, the experiments are looked at through the eyes of a theoretician and the conceptual frame of fluid mechanics since, as Einstein put it, "it is the theory that decides what we can observe". This remark is particularly relevant to the phenomena that we consider.

---

[1] Aristotle's observation that "what is known is known to a few" seems to be as true now as it was in his time.

The theories are examined against the following:

(i) The basic criteria of physical relevance,

(ii) The underlying assumptions and hence the limits of applicability of continuum mechanics within which these theories have been formulated as well as assumptions behind the theories themselves,

(iii) Their ability to describe experimental observations and provide a conceptual framework which would allow one, if necessary, to incorporate other physical effects.

This analysis allows the reader to have a broad picture of the situation with the flow in question and see advantages as well as disadvantages of different theoretical approaches to its modelling.

Then, we consider how the common root of the problems with 'paradoxical flows' — the missing physics of the interface formation in the mathematical model — emerges from different guises. Once the essence of the difficulty is identified for the first time (Chapter 3), we develop the simplest (irreducible) ready-to-use model addressing the difficulty (Chapter 4) which we then test, without any ad hoc adjustments, for other flows against the *same criteria* as the theories reviewed earlier. This approach ensures consistency of the analysis and indicates how the model's parameters can be determined from independent experiments.

The book is intended for the following categories of readers:

- Graduate and final-year undergraduate students in applied mathematics, physics and engineering whose programs of research or study includes fluid mechanics,

- Chemical, mechanical, design and bioengineers working with fluids, in particular, in the areas of microfluidics and coating technologies,

- Researchers in applied mathematics and physics interested in capillary flows, interfacial phenomena and related aspects of fluid mechanics.

Throughout the book cross-referencing is used to highlight logical connections between different phenomena and the unified way in which they are addressed. In order to counterbalance this interweaving of material, the text is punctuated with summaries that outline the main points and are intended to facilitate practical use of the book. More than 130 figures provide visual support and illustration of the main concepts. The author-plus-year referencing system together with the Index of Authors make it easy to find references to particular works. The references include all main publications directly related to the problems considered and the physics of interface formation; references to the works that deal primarily with influences *additional* to the physics of interface formation (effects of electromagnetic field, surfactants, chemical reactions, evaporation/condensation, etc.) have not been included and can be found in the cited papers and reviews.

Acknowledgements. I wish to thank my long-time collaborators Terry Blake, Andrew Clarke and Mark Wilson for stimulating discussions that contributed, directly and indirectly, to this book; Osman Basaran, Alvin Chen, Ilker Bayer and Constantine Megaridis for providing original figures from their works; Warren Smith, Eamonn Gaffney and my students Paul Suckling and James Sprittles for their feedback on some parts of the manuscript at different stages of its preparation, and Mike Finney for technical assistance.

Birmingham Y.D.S.

List of Figures

List of Tables

List of Symbols

Ca — capillary number

$\mathbf{E}$ — rate-of-strain tensor

$\mathbf{e}_i$ — basis vector

e_{ij}, e^{ij} — components of the rate-of-strain tensor

$\mathbf{F}$ — body force

F_m, F_v — free energy per unit mass and per unit volume

g_{ij}, g^{ij} — components of the metric tensor

$\mathbf{I}$ — metric tensor

L, ℓ — macroscopic and molecular length scales

$\mathbf{n}$ — unit normal to an interface

$\mathbf{P}$ — stress tensor

p_{ij}, p^{ij} — components of stress tensor

p — pressure

$\mathbf{q}^{\pm}$ — heat fluxes in the bulk on the opposite sides of an interface

$\mathbf{q}^s$ — heat flux along an interface

Re — Reynolds number

S_m, S_v — entropy per unit mass and per unit volume

S^s — surface entropy

T — temperature

$\mathbf{U}$ — velocity of a solid

U_m, U_v — internal energy per unit mass and per unit volume

$\mathbf{u}$ — fluid's velocity

$\mathbf{u}^{\pm}$ — velocities of fluids on the opposite sides of an interface

u, v, w — velocity components in an orthogonal coordinate system

$\mathbf{v}^s$ — surface velocity

W — Airy stress function

$\mathbf{w}$ — displacement vector

x^i — coordinates in Eulerian description of motion

α, β, γ — (dimensional) material parameters of an interface

$\Delta \equiv \nabla^2$ — Laplacian

ε_{ij}, ε^{ij} — components of the deformation tensor

θ_{app} — apparent contact angle

θ_d — dynamic contact angle

θ_s — static contact angle

μ — dynamic viscosity

ν — kinematic viscosity

ξ^i — coordinates in Lagrangian description of motion

$\mathbf{\Pi}$ — momentum flux tensor

ρ — density

$\rho^{\pm}$ — densities of fluids on the opposite sides of a an interface

ρ^s — surface density

ρ_e^s, ρ_G^s — dimensional and dimensionless equilibrium surface densities of a liquid-gas interface

ρ_{1e}^s, ρ_S^s — dimensional and dimensionless equilibrium surface densities of a liquid-solid interface

$\rho_{(0)}^s$ — surface density corresponding to zero surface tension

$\boldsymbol{\sigma}$ — tensor of viscous stresses

σ_{ij}, σ^{ij} — components of tensor of viscous stresses

σ — surface tension

σ_e	equilibrium surface tension of a free surface	τ	surface-tension-relaxation time
$\sigma_{(2)}$, σ_{sg}	dimensional and dimensionless surface tension of liquid-gas interface	Ψ_m	chemical potential
		ψ	stream function
		$\boldsymbol{\omega}$	vorticity

Chapter 1

Introduction

1.1 Free-surface flows in nature and industry

Free surface flows are ubiquitous in nature and industry. One can say without exaggeration that free surfaces are present, or can appear, in almost every technological process or natural phenomenon where liquids are involved. The characteristic length and time scales associated with these flows vary in a very wide range, from nanometers and nanoseconds in, as one might expect, nanotechnologies to kilometers and hours typical for tidal waves and lava flows.

In order to understand and predict natural phenomena, design new technologies and improve the existing ones, we need to be able to describe, both qualitatively and quantitatively, the behaviour of free-surface flows in different situations. From the quantitative point of view, this means to know the distributions of the flow parameters with the required resolution and accuracy. The ultimate way of obtaining this information is, of course, a high-accuracy direct experiment. However, in many cases this is difficult or impossible for one or both of the following two main reasons.

The first reason is that often the effect to be understood or predicted is also the one to be avoided. A typical example of this is a critical situation at nuclear power stations where one would prefer theory (and recommendations based on it) to full-scale experiment. Large-scale natural phenomena, such as the interaction of a tsunami wave with the coastline, is another example of flow where experiments, though technically feasible, are undesirable and quantitative predictions have to be based on the results of mathematical modelling.

The second reason that often makes one prefer theory to experiment is a very high cost of performing experiments with the accuracy and resolution that would allow the experimenter to draw unambiguous conclusions from the data and make reliable quantitative predictions. From another perspective, this can be seen as the limited accessibility of the required information to existing experimental techniques. This source of difficulties is usually associated with very small length and time scales. The area of fluid mechanics dealing with such scales is called 'microhydrodynamics', and its branch addressing flows where interfacial effects are important is known as mechanics

FIGURE 1.1: Impact of a microscopic drop on a solid surface. The initial stage of impact, when the point of contact grows into an area, is the most important one and it is not accessible to experimental observation. (Courtesy of P.M. Suckling.)

of capillary flows.

An important aspect of experimentation in the area of capillary flows is that often it is necessary to investigate the role of material parameters almost invariably involving properties of interfaces. This brings in a multidimensional parameter space which, in most cases, cannot be investigated experimentally since, in order to vary a dimensionless similarity parameter formed by material constants, one has to change the composition/conditions of the medium thus causing variation in other similarity parameters. As a result, an experiment goes along a line in the parameter space. Drawing conclusions from the data obtained along such a line or a set of lines of different shapes in a multi-dimensional parameter space becomes a problem difficult even from a methodological, leave alone practical, point of view.

An example of flow where all these difficulties are wrapped in one is the impact of a microscopic droplet on a solid surface (Fig. 1.1). This is an element of ink-jet printing as well as of some bioengineering technologies. In order to understand the processes involved in the initial stage of the impact, one would require, ideally, an experiment capable of resolving the flow down to molecular length and time scales. Although this is feasible in principle, the cost of a systematic study of this phenomenon covering the whole parameter space would be prohibitively high. The process of spreading of the liquid over the solid substrate that follows the initial stage of the impact is determined by both hydrodynamic factors and material properties of the gas/liquid/solid system. Experimental investigation of the former in the immediate vicinity of the three-phase-contact line requires a high degree of spatial resolution, whereas the latter are inseparable in experiments and hence their role cannot

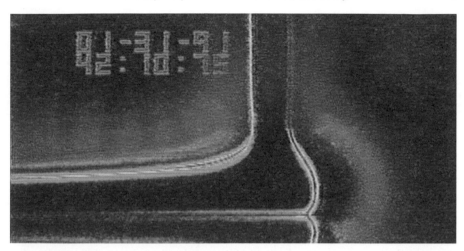

FIGURE 1.2: Curtain coating is one of the most flexible methods of depositing liquid films on solid substrates. The quality of coating is largely determined by the processes near the moving three-phase-contact line which are virtually impossible to monitor experimentally. (Courtesy of A. Clarke.)

be studied independently.

All the previously mentioned factors make it desirable to develop methods capable of modelling capillary flows mathematically in different situations that occur in applications and natural processes. Adequate mathematical models would make it possible to obtain full information with the required accuracy and, if a complex experiment is performed, correctly interpret its results and extend them to the areas which are not accessible to measurements. An important aspect is that the theory would also make it possible to study separately the role of different factors inseparable in a real-life experiment and suggest how the composition of materials could be modified to achieve the set technological objectives. Below, we will briefly describe two broad classes of capillary flows where the issues of modelling are of prime importance and which are the main subject of this book.

A wide class of capillary flows have at their core the process of 'dynamic wetting', that is, the spreading of a liquid over a solid substrate. This process is the key element of so-called 'coating flows' (Kistler & Schweitzer 1997, Weinstein & Ruschak 2004) defined by the technological need to 'coat' the solid surface with a film of liquid. The ultimate goal is either to modify properties of the solid surface, like in painting, lamination, etc., or, by using the solid as a temporary support, to produce a film of solidified liquid (e.g., polymer films, metal sheets, etc.). The flow configurations employed in coating technologies include curtain coating (Fig. 1.2), forward and reverse roll coating, slot and knife coating, spin coating, to mention but a few. A relatively new area of application is micro- and nanotechnology, where it is important to describe

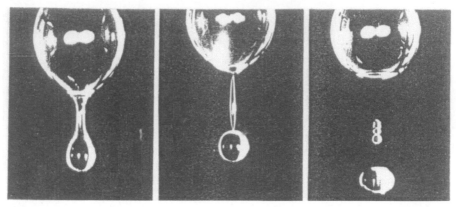

FIGURE 1.3: What happens during the final stage of breakup of a liquid thread is not accessible to experimental observations and has to be understood from analysing the pre-breakup evolution of the thread and the post-breakup recoil of its pieces. The latter can be seen as a macroscopic 'summary' of how the breakup has taken place.

how microscopic films spread over topography of various types.

In addition to the general area of coating flows, dynamic wetting is also central to a number of other processes and industrial applications. They include various capillary flows in biological systems, interaction of drops and bubbles with particles (flotation, wetting of powders, etc.), flows of emulsions in capillaries and porous media (chemical technologies, oil recovery), flows of foams, hydrodynamics of boiling (heat exchangers, nuclear power stations), etching and electro-chemical treatment of surfaces, flows in liquid bridges (crystal growth in microgravity) and many others. In all these processes, one has to understand the governing physical mechanisms and formulate a quantitative model which would not only address a particular problem but also provide a conceptual framework for incorporating, if necessary, other physical factors.

Another wide class of capillary flows where the issues of modelling are of primary importance can be generally labelled as flows with topological transitions in the flow domain. They include coalescence of drops and bubbles, breakup of liquid threads (Fig. 1.3) and jets, rupture of free films, disintegration of films on a solid substrate, evolution of foams, nucleation of bubbles and many others. From the point of view of application, topological transitions can be either the required outcome (e.g., drop formation in ink-jet printing or spray painting, sintering, targeted deposition of fluid on a patterned substrate, etc.) or a cause of defects (e.g., disintegration of coating films, rupture of a tear film in the dry eye syndrome, etc.). In both cases, one has to know what factors control the process and the role played by material parameters. It is in the flows with topological transitions where the free surface manifests itself in the most spectacular way, making both experimental studies and the mathematical description of fluid motion so notoriously difficult. At the same time, due to the complex nature of these flows there are still many amazing effects waiting to be discovered and understood even in seemingly simple phenomena we see every day (e.g., Thoroddsen & Takehara 2000, see Fig. 6.9).

In our brief outline of capillary flows we grouped them into broad classes which comprise what are usually regarded as different types of fluid motion. Traditionally, free surface flows are classified by the object of study (e.g., thin film flows, dynamics of drops and bubbles, flows in pipes and channels, etc.) and/or by simplifications one can make in their mathematical description (e.g., axisymmetric flows, creeping flows, boundary layer theory, etc.). The reason for our slightly unorthodox classification is explained below.

1.2 Scope of the book

Theoretical fluid mechanics has always been at the intersection of applied mathematics, physics and engineering. Since the conception of modern sciences in the early seventeenth century, this triad in fluid mechanics has been in what can be described as the state of a dynamic equilibrium. The classical models of an ideal and later a viscous fluid provided a common foundation to three main trends: (a) mathematical research focussed on idealized model problems and general features of fluid motion, (b) the modelling side aimed at incorporating various physical and/or chemical factors into the given conceptual framework, and (c) engineering applications dealing with real problems with rather crude mathematical means. The 'common denominator' to these trends was the fact that even the simplest set of equations at the core of fluid mechanics are extremely difficult to solve analytically. For example, only a handful of nontrivial analytical solutions of the Navier-Stokes equations are known (Drazin & Riley 2006) and even simplified equations, for example those for creeping, potential or boundary-layer flows, are tractable analytically only for very simple flow configurations and relatively simple boundary conditions. Perhaps, one could define classical fluid mechanics as an art of finding approximate analytical solutions to the known equations of fluid motion. Given the undisputed complexity of the equations and the lack of essentially new and much more powerful analytical methods, this activity is more and more turning into that of finding the problem for the method rather than vice versa.

The situation started to change radically a few decades ago when the advent of computer-aided research began to redefine the very essence of fluid mechanics. Indeed, not only the simple idealized problems but also much more complex ones, dealing with realistic situations and involving elaborate models, which are far beyond the reach of even the most sophisticated analytical techniques, can be now routinely solved by standard numerical methods. Computational fluid dynamics (CFD) with its emerging multi-purpose codes no longer needs separation of flows and associated problems into classes on the basis of possible simplifications in their analytical treatment. The simple 'classical' model problems are turning into mere exercises seen now either as building blocks for computer algorithms (e.g., in the theory of multiphase flows) or, more often, as routine tests for numerical codes. Finding a solution describing a regular laminar flow of a simple fluid is not a challenging fluid-mechanical problem any more; it is just a matter of diligence, time and computer resources.

At the same time, fast progress of computational fluid dynamics highlighted some fundamental issues of modelling which stand in the way and make numerical algorithms powerless in dealing with a number of problems of real importance. It is these issues which are now becoming the new focus of research in theoretical fluid mechanics. Surprisingly, even for simple Newtonian fluids there are entire *classes* of flows which cannot be described in the standard way, that is, as solutions of the Navier-Stokes equations subject to classical boundary conditions. The standard formulation will either lead to physically unacceptable singularities in the solutions or even to the situations where no solutions exist. The two broad classes considered in the previous section belong to this category. Thus, in order to be able to solve practical engineering tasks, incorporate additional physical effects and apply powerful computational methods, one has first to resolve the problems lying at a deeper level in a conceptually consistent and experimentally verifiable way.

The scope of this book is to address the modelling issues at the core of the problems where standard fluid mechanics fails to satisfactorily describe capillary flows. We will analyze the roots of the singularities arising in solutions and identify the 'extra' physics whose absence in the problem formulation is responsible for them. This will make it possible to put the problems in question into a more general physical context, where the 'singular free-surface flows' become particular cases of more general physical phenomena. The ways of incorporating this 'extra' physics into mathematical models will then be examined and the results illustrated by considering a number of particular problems. We will also highlight a few outstanding unresolved issues and discuss possible approaches to their analysis.

This book is addressed to the following partially overlapping groups of readers.

The first one comprises researchers and students, primarily in the areas of applied mathematics, physics and engineering, who are interested in the modelling side of continuum mechanics. The selected material covers not only the achievements of the last few decades but also a number of unresolved theoretical and important practical problems for which the book sets the background and provides the conceptual framework. As far as the students are concerned, the goal of the book is to give a 'flying start' for their research in one of the most interesting areas of fluid mechanics. This is especially the case for research in the emerging areas of application, such as biofluid mechanics and microfluidics, where an open-ended conceptual framework is vital for further progress leading to new reliable self-consistent mathematical models.

The second group of readers to whom the book might be of interest includes 'practitioners', that is, computational fluid dynamicists, chemical, mechanical, design and bioengineers who work on particular problems involving capillary flows and are interested in how to adequately describe them mathematically and/or numerically in the most efficient way. In many of their problems, the singularities become difficulties of principle which cannot be dealt with in crude 'practical' ways, like those built in many commercial numerical codes. Such an approach would reduce the modelling to a semi-empirical level and hence have almost no predictive power. In many situations, it would also introduce uncontrollable errors which make quantitative predictions misleading. In this respect, the first goal of the present book is to explain the essence of the problems and highlight the issues involved in the modelling of the 'singular flows', thus allowing the reader to make an informed choice of strategy regarding available codes or their possible development 'in-house'. Secondly, the book provides a set of simple ready-to-use experimentally-verified models (or, more precisely, variants of the same model) together with a number of solutions illustrating applications of these models to particular problems. A detailed discussion of alternative approaches to these problems known in the literature, as well as a review of relevant experiments, are intended to help the reader see these problems in a general context.

Scientific interactions in the framework of multidisciplinary projects often reveals that collaborators with different backgrounds can understand and interpret very differently even the very basics of what is supposed to be their common area of expertise, such as, in our case, fluid mechanics. In order to have a common starting point for analysis of more complex issues, we will begin with the fundamentals of fluid mechanics, review its main concepts and assumptions behind them, consider the standard ways in which its problems are formulated, and discuss the basic criteria one has to apply as the first test of physical meaningfulness of the solutions.

Chapter 2

Fundamentals of fluid mechanics

In this chapter, we give an overview of the main concepts of fluid mechanics in their methodological context. Our emphasis is on the assumptions made in the classical fluid-mechanical model with the main attention focussed on those associated with the boundary conditions. A detailed exposition of particular problems analysed in classical fluid mechanics can be found in a succession of textbooks (Lamb 1932, Kochin, Kibel & Roze 1964, Sedov 1965a, 1997, Batchelor 1967, Landau & Lifshitz 1987, Panton 2005) and is beyond our goal.

2.1   Main concepts

2.1.1   Physical properties of liquids and gases

The defining property of 'fluids', which embrace both liquids and gases, is that, unlike solids, they can 'flow' easily changing their shape. This pictorial definition goes back to the times of the ancients, who derived it from their observation of properties of water and air, and it implies that 'fluidity' manifests itself on the time and length scale of the observation. The latter is important for our perception and description of the process. Indeed, on a geological time and length scale the mountains are quite 'fluid',[1] whilst a microscopic droplet of water on a microscopic time scale can behave like a solid particle. Another property of common fluids, which is perhaps less palpable but more useful for their quantitative description, is that, unlike solids, their resistance to external forces is proportional primarily to the rate of deformation rather than to its degree. The key word here is 'primarily' since in reality there is almost a continuous spectrum of materials which exhibit fluid-like and solid-like features in different proportion. For example, polymer solutions acquire more features of a solid as the concentration of polymer increases, and the rheological properties of some paints strongly depend on their temperature and the degree of deformation to which they are subjected. Thus, from a

---

[1]This observation, made quite a while ago, later resulted in the notion of the 'Deborah number' (Reiner 1964).

macroscopic point of view, the distinction between fluids and solids is not sharp and determined by how we see the correlation between their dynamic and kinematic properties on the length and time scale we are interested in. In other words,'fluid' and 'solid' are models we use to describe real media.

At one end of the fluid-solid spectrum, there is the so-called 'simple' fluid. In all types of flow, it exhibits resistance proportional to the rate of externally caused deformation, so that a finite degree of deformation can result from the action of infinitesimal forces. This idealized model provides an accurate description for water, air, oils and their derivatives, liquid metals and many other fluids. It is a remarkable and surprising fact that, although from the mechanical point of view these real fluids are systems with very many degrees of freedom, one can very accurately describe their behaviour in *all types* of flow using a closed set of equations involving only a few macroscopic parameters and the spatial coordinates and time as independent variables. This success, which was not guaranteed by any known 'law of nature', forms the foundation of the whole of fluid mechanics as a science. Despite ongoing effort, we do not have the same degree of universality for complex fluids, such as, for example, polymer solutions or melts, where one has models only for particular classes of flow.

The arrival of statistical mechanics and later of molecular dynamics simulations stimulated attempts to understand physical mechanisms behind the success of fluid-mechanical description of simple fluids. As far as their molecular structure is concerned, liquids and gases are very different. In a gas, for most of the time each molecule moves independently of its neighbours and occasionally interacts with them through 'collisions', when two or more molecules come sufficiently close together to feel each other's presence through repulsive intermolecular forces. For a gas at temperature 0°C and a pressure of one atmosphere, the concentration of molecules is 2.687×10^{19} molecules per cubic centimeter so that the average distance between them is of order 3×10^{-7} cm, which is ten times greater than the range of the repulsive forces for simple molecules. In these conditions, a molecule of oxygen, for example, travels on average 6.5×10^{-6} cm between collisions, which is more than two orders of magnitude greater than its size, and, having a velocity of the order of 4×10^4 cm s^{-1}, experiences 7×10^9 collisions per second.

Unlike the gas, the liquid is a *condensed* phase. Its molecules are packed together so closely that each of them constantly feels its neighbours through both attractive and repulsive intermolecular forces. The macroscopic motion of a liquid is a result of continuous microscopic rearrangements of its molecules, which include formation and splitting of molecular groups of different kind.

The reason why the fluids which are organised so differently behave qualitatively in the same way when observed on a macroscopic length and time scale is still not quite clear. It is even less clear why the chain of causal links between processes on different length and time scales, from the macroscopic down to the molecular level, can be broken to form a closed set of equations involving a small number of parameters on the macroscopic level with the microscopic 'tail' replaced by a few, in many cases constant, coefficients (e.g., viscosity, thermal conductivity, etc.). These and some other questions are being intensively studied by statistical physics and molecular dynamics simulations, which in the future might provide a better understanding of the nature of liquids as well as a 'bridge' between macroscopic fluid mechanics and real-life experiments. At the moment, however, fluid mechanics remains being fed primarily by empirical, heuristic and phenomenological inputs.

From a macroscopic point of view, the principal distinction between gases and liquids following from their molecular structure is that, whilst a gas in equilibrium occupies the whole volume available to it, the volume of a given mass of liquid remains constant. As a result, when a given amount of liquid is poured into an open vessel, the liquid will have a free surface. For most free-surface flows, compressibility of the liquid can be neglected so that it can be treated as an incompressible medium that can change its shape while its volume remains fixed.

The molecules of a liquid in a thin layer on the free surface, which can be modelled as a 'surface phase', experience nonsymmetric action of intermolecular forces from the two bulk phases, and this asymmetry gives rise to the property of the free surface known as the surface tension. From the macroscopic mechanical point of view, the surface tension is a force per unit length acting on every line drawn on the free surface in the direction normal to this line and tangential to the free surface. As a result, the free surface behaves like a stretched membrane trying to minimize its area. For most processes, at a given temperature the surface tension is a constant characterizing the pair of contacting media. A nonuniform distribution along the free surface of temperature, surfactant concentration or the intensity of electric field can cause gradients of the surface tension and generate motion. This effect, first discovered for temperature (Marangoni & Stefanelli 1872, 1873), is known as the Marangoni effect. There is also a wide class of flows where the surface tension gradients can be caused by the flow itself, and these gradients in their turn will have a reverse influence on the flow that generated them. Surprisingly, this *flow-induced* Marangoni effect has received relatively little attention from the fluid-mechanical community. We will return to this effect later in the book.

2.1.2 The continuum approximation: density and velocity

In the mathematical modelling of mechanical properties of the fluid one can choose between different levels of description, ranging from direct simulations of molecular motion to the field description of continuum mechanics. Each of these levels is associated with its own conceptual framework, advantages and limitations. The approach of molecular dynamics simulations based on treating the fluid as a discrete system of interacting molecules is conceptually rather simple since the problem is essentially reduced to that of considering ordinary differential equations describing the motion of individual molecules. The most attractive feature of this approach is that, potentially, it provides complete information about the fluid. However, this advantage is moderated in three ways.

First, the resulting set of ordinary differential equations is not only absolutely untractable analytically; it is also extremely demanding computationally. To have a realistic quantitative description of a liquid, one has to consider about 10^{23} molecules per cm^3, which is far beyond the computational power available in the foreseeable future. Even in doing computations with the number of molecules below the required one by many orders of magnitude, it is necessary to resort to simplifying assumptions to be able to carry out the computations in a reasonable time.

Secondly, the very approach involves fundamental assumptions associated with the modelling of intermolecular interactions. While in gases one can often consider only pair interactions of molecules and model them in a simple way as 'collisions', in liquids the picture is much more complex and the interactions include both classical and quantum components as well as multiparticle effects.

Thirdly, the interpretation of results in macroscopic terms is a problem in its own right. Its solution requires 'macroscopic guidance' telling one what averaged macroscopic quantities are to be calculated from the masses of numerical data and how to do this. To some extent, this guidance is provided by statistical physics which, in the present context, can be seen as a bridge between molecular dynamics simulations and macroscopic fluid mechanics: it accounts for the discrete nature of matter but operates with distribution functions associated with its various properties rather than the primitive dynamic variables.

Instead of considering an extremely complex (and still not entirely realistic) dynamical system of interacting molecules and then face a problem of macroscopic interpretation of the results it provides, it would be desirable to describe the fluid using a relatively small number of macroscopic parameters defined and continuous at every point in the bulk. These parameters can be chosen to be those measured in macroscopic experiments and, assuming that their distributions are sufficiently smooth, one can use differential calculus and express relationships between them in terms of partial differential equations. This gives a relatively simple model to analyze and the results can be interpreted in a straightforward way, without the necessity to invoke sophisticated constructions linking the quantities featuring in the model with what is actually measured in experiments. These advantages motivate fluid mechanics as an approach to the mathematical modeling of fluid flow and come as a result of a conceptually nontrivial representation of a discrete molecular system in terms of smooth fields describing averaged macroscopic properties of this system. This representation is known as the continuum approximation.

On a descriptive level, the essence of the continuum approximation is that the actual fluid flow is modelled as the motion of a hypothetical 'moving continuum'.[2] In specifying this picture mathematically, it is important to understand how this 'moving continuum' results from the description of fluid as a discrete system, the assumptions behind the conceptual framework of fluid mechanics, and the limits of applicability of the approach they bring with them.

From a molecular viewpoint, fluid-mechanical variables represent some averages of molecular properties. These averages can be divided into (a) variables, like density and velocity, which are associated with kinematics of molecules characterized by their positions and velocities, i.e., in terms of statistical physics, single-molecule distribution function, and (b) variables describing intermolecular interactions. The former can be introduced in a straightforward way as spatial and/or temporal averages of the corresponding molecular properties, which is consistent with the way these parameters are measured experimentally.

Spatial averaging defines the density ρ and velocity $\mathbf{u}$ of the fluid as a continuous medium via

$$\rho = \frac{1}{V} \int_V m \, dV, \qquad \rho\mathbf{u} = \frac{1}{V} \int_V m\boldsymbol{v} \, dV,$$

where m and $\boldsymbol{v}$ are masses and velocities of molecules, and V is the volume of averaging. For ρ and $\mathbf{u}$ to be independent of the shape and size of V, the latter must contain a very large (in the limit, infinite) number of molecules, i.e., the size of the volume of averaging l_{av} must be much greater than a length scale ℓ_1 characterizing intermolecular distances. Secondly, the spatial variation of the molecular distribution across the volume of averaging must be negligibly small. In other words, l_{av} must be small compared with a macroscopic length scale L on which the model is applied and the macroscopic (averaged) quantities vary. Mathematically, the limits

$$\frac{\ell_1}{l_{av}} \to 0, \qquad \frac{l_{av}}{L} \to 0 \tag{2.1}$$

ensure both independence of the averages of the shape and size of the volume of averaging and their continuity as functions of the position of this volume. The second limit also implies that an averaged quantity provides a *local* characteristic of the fluid corresponding to a point in space. Thus, conceptually the continuum approximation is essentially the 0th-order approximation in the limit $\ell_1/L \to 0$, which follows from (2.1) and ensures the existence of l_{av}. Qualitative arguments related to time averaging can be considered in a similar way.

---

[2]Hence the often used term 'continuum hypothesis' implying not a hypothesis about any of physical properties of the fluid, which could turn out to be true or false; it refers to the hypothetical 'continuous medium' used as a model of the real fluid.

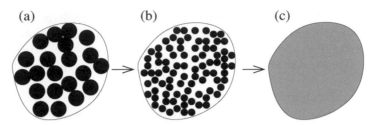

FIGURE 2.1: A sketch illustrating how a 'continuum medium' results from the thermodynamic limit (a)→(b)→(c).

For gases, a full fluid-mechanical model can be derived from statistical physics, where the macroscopic quantities are introduced via a more flexible technique of ensemble averaging which considers possible microscopic 'realisations' of a macroscopic state of the fluid. Although this derivation involves a number of assumptions and hence, strictly speaking, is not mathematically rigorous, it is useful as an illustration of the limit in which the molecular description turns into the field description of continuum mechanics. The continuum appoximation results from the so-called 'thermodynamic limit' as the number of molecules in the system N and the size of the volume V that contains them both tend to infinity with their ratio remaining finite. The thermodynamic limit can be redefined by scaling the molecules with the size of the volume, as schematically illustrated in Fig. 2.1, which is equivalent to the limit $\ell_1/L \to 0$ we have just considered. The charateristic time scale associated with molecular effects is proportional to ℓ_1 (and inversely proportional to the characteristic velocity of molecular motion) so that the thermodynamic limit is also associated with the 0th order in the ratio of molecular-to-macroscopic time scales.

The limits (2.1) for the spatial averaging imply that the macroscopic 'length scale of interest' L characterizes all three dimensions of the domain where the fluid parameters are modelled. One can relax this requirement and introduce averages applicable to the situations where there is anisotropy in the molecular distribution. (We are still dealing with variables representing single-molecule distributions and properties.) This is required, for example, to describe in continuum terms the structure of interfacial layers between different fluids (van der Waals 1893; see Chapter 4). The required macroscopic parameters can then be introduced via surface averaging (or the corresponding modification of ensemble averaging) so that (2.1) now applies to the dimensions along the surface of averaging. Then, given that no restrictions are imposed on the scales in the direction normal to the surface of averaging, one can consider variation of the macroscopic quantities introduced in this way on a length scale comparable with ℓ_1, that is having in this direction $L \sim \ell_1$. Obviously, the averaged quantities specified via spatial and surface averaging coincide in the situations where both definitions are applicable.

One can relax the definition even further by introducing line averaging so that the constraints (2.1) to be satisfied will be associated with the scale along the line of averaging.[3]

It should be emphasized here that the *actual* averaging of molecular properties is required only when one is examining the links between the fluid-mechanical and molecular-dynamic description of some phenomenon. In fluid mechanics, macroscopic variable are introduced heuristically on the basis of qualitative arguments relating them to the microscopic picture and the properties they are supposed to represent. These arguments help to ensure that the assumptions behind different macroscopic variables are compatible and indicate the expected limits of applicability of macroscopic models and solutions obtained in their framework.

An important point to note is that, if in describing certain physical characteristics of the fluid, one needs only the concepts of density and velocity, i.e., the macroscopic variable associated with the single-molecule distribution function, these can be defined in a meaningful way not only via spatial averaging but also via surface and line averaging (or the appropriate ensemble averaging) to accommodate anisotropy of the molecular distribution thus allowing for $L \sim \ell_1$ in the direction normal to the surface or line of averaging.

2.1.3   The continuum approximation: forces and stresses

Every molecule in a fluid experiences a range of different external forces, from gravity to intermolecular forces from its neighbours, and moves in accordance with Newton's second law, where all these forces, external to the molecule, appear in the same way. When we employ the continuum approximation and consider the fluid as a continuous medium, it becomes necessary, first, to formulate the conservation laws, in particular, the momentum balance, in an integral form for a control volume of the fluid modelled as a moving continuum. At this stage, we can qualitatively distinguish between, on the one hand, long-range forces, like gravity, acting on all molecules in this control volume and, on the other hand, intermolecular interactions involving only molecules near the boundary of this volume and the momentum fluxes associated with chaotic molecular motion across it. As a result we come to the notions of body forces and internal stresses.

Long-range forces include gravity, electromagnetic forces and, if the motion is considered in a noninertial coordinate frame, the forces of inertia. These forces vary on a macroscopic length scale so that, if $\mathbf{F}(\mathbf{r}, t)$ is a time-dependent

---

[3]An illustration of line averaging is a geological probe going through a rock. The probe collects information on its way, i.e., along a line, and this gives an accurate average for the whole rock provided that the distribution of minerals is spatially uniform. The line average no longer represents the volume average if the variations of the mineral concentrations are steep in the directions normal to the path of the probe, but even then it remains a meaningful local characteristic and can be useful for specific purposes.

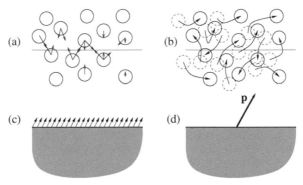

FIGURE 2.2: In continuum mechanics, the intermolecular forces acting across the surface of a control volume (a) and the momentum fluxes associated with the molecular motion through it (b) are represented as a distributed force acting on this surface (c). This force, known as stress, is characterised by its density $\mathbf{p}$, shown in (d), so that the stress acting on an elementary surface dS can be expressed as $\mathbf{p}\,dS$. Thus, the introduction of stresses implies that the spatial range of intermolecular forces and the time scale associated with molecular motion are negligible compared with the corresponding scales characterizing the motion of the continuum to be described using the notion of stress. An explicit inclusion of intermolecular forces into a continuum mechanics model in addition to the stress would mean double-counting of the physical effects already taken into account in the stress.

distribution in space of the mass density of these forces (defined as force per unit mass), then, to leading order, the force acting on an elementary volume dV will be given by

$$\rho(\mathbf{r}, t)\mathbf{F}(\mathbf{r}, t)\, dV,$$

where ρ is the fluid's density defined as mass per unit volume. In particular, for the gravity force acting on an elementary volume of incompressible liquid we have

$$\rho\mathbf{g}\, dV,$$

where $\mathbf{g}$ is the acceleration of gravity. The forces acting on a mass of fluid, like gravity or inertial forces, are called mass forces or body forces. In some cases, body forces can be used as a modelling tool, when they are introduced artificially, for example in aerodynamics, where an airfoil can be replaced by a distribution of body forces applied externally to the liquid.

Compared to body forces, intermolecular forces act (and, in terms of statistical physics, influence molecular distributions) on a much shorter length scale ℓ_2. Physically, one can have both situations, $\ell_1 \gg \ell_2$ (for rarified gases) and $\ell_1 \ll \ell_2$ (for condensed matter). When we consider a control volume, only molecules at distances on the scale of ℓ_2 from the surface confining the control volume exert forces on each other across this surface.

Another contribution to the momentum flux across the surface of the control volume comes from the transport of momentum across the boundary by migrating molecules. This process is associated with the molecular length scale, i.e., the distance a molecule travels between two collisions or the amplitude of its oscillations, as well as with a molecular time scale. This component of the momentum flux plays the main role in gases, where intermolecular interactions can be modelled in a classical way and even a simple statistical model of colliding rigid balls produces adequate results. In solids, where interatomic forces (of quantal nature) have the range much larger than interatomic distances and there is no free motion of atoms, the momentum flux is due to interatomic forces. In liquids, one has both classical and quantal components to intermolecular forces and both contributions to the momentum flux, due to intermolecular forces and molecular motion, are comparable.

As we introduce the continuum approximation, the momentum fluxes due to intermolecular interactions and molecular motion across the surface of a control volume are now modelled as a distributed force per unit area acting on the *surface* of this volume (Fig. 2.2). This force density is known as *stress*.

This notion is a conceptual leap from the molecular description: on the molecular level one has only masses of molecules, their velocities and the (external) forces acting upon them, whereas on the continuum level we now have densities, velocities, body forces *and* stresses. It is this distinction between the types of forces arising on the macroscopic level as the result of the continuum approximation which is at the core of continuum-mechanics modelling. One may say that this distinction defines the continuum approximation in its application to dynamic properties.

The introduction of stresses, i.e., distributed forces acting on surfaces, imposes a more restrictive requirement on the ratio of scales associated with molecular and macroscopic processes. One has that the continuum modelling of fluids is actually the 0th-order approximation in the limit

$$\frac{\ell}{L} \to 0, \qquad \frac{t_{\mathrm{mol}}}{t_{\mathrm{macro}}} \to 0, \tag{2.2}$$

where $\ell = \max(\ell_1, \ell_2)$, t_{mol} is the time scale of molecular processes and t_{macro} is characteristic time scale for the macroscopic motion to be described. Thus, once the notion of stress is introduced into the continuum description, one has the following implications:

- Intermolecular forces do not appear explicitly in the continuum mechanics description. All such forces are accounted for in the macroscopic notion of stress and their inclusion in addition to it would mean double counting of the same physical effects. Intermolecular forces, as well as characteristics of molecular motion, manifest themselves only via macroscopic quantities (such as transport coefficients, parameters constitutive equations, etc.).[4]

- The length and time scales associated with molecular processes are both negligible (i.e., equal to zero) in the continuum approximation. In particular, interfacial layers between immiscible fluids, whose thickness is determined by the range of intermolecular interactions, must be modelled as mathematical surfaces separating two bulk phases.[5] In other words, once the forces are separated into body forces and stresses, this distinction immediately brings in the notions of 'bulk' phases and sharp 'interfaces' separating them.[6] The latter can be regarded as two-dimensional 'surface phases' whose averaged properties per unit area are analogous the averaged properties used to describe the bulk phases.

- All length and time scales featuring explicitly in continuum mechanics models are, by definition, macroscopic, not molecular.

From what is said it is obvious that 'material points' (also referred to as 'fluid particles') forming the 'moving continuum' should not be confused with molecules, in particular in comparing the results with experiments or molecular dynamics simulations: according to the continuum limit, material points represent the averaged behaviour of an infinite number of molecules.

---

[4] The implications of relaxing condition (2.2) will be discussed in §7.7.2.

[5] The interfacial layers whose thickness is determined by macroscopic mechanisms (e.g. diffusion) can be modelled as having a finite thickness. Such layers appear, for example, in the process of mixing of mutually soluble fluids, in the liquid-vapour systems near the critical point and some other situations.

[6] It has been shown that there is no unique way of defining stress in the interfacial layer (Schofield & Henderson 1982, see also Rowlinson 2002, pp. 295–298).

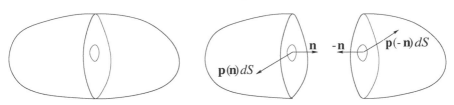

FIGURE 2.3: Internal stresses acting on each of two parts of a material volume in (a) are shown in (b).

As a result of the continuum approximation, intermolecular forces and momentum fluxes associated with molecular motion transform into distributed forces acting on the boundary of a volume of fluid (Fig. 2.2). In general, these forces will depend on the orientation of the surface element they are acting upon. We will characterise this orientation by a unit vector $\mathbf{n}$ normal to the surface and pointing from the side of the fluid acted upon towards the fluid exerting the force. Then, the surface force acting on an elementary area dS located at $\mathbf{r}$ at time t can be written as a vector: $\mathbf{p}(\mathbf{n}, \mathbf{r}, t)\, dS$, where $\mathbf{p}$ is the force per unit area. Clearly, since two portions of fluid having a common boundary act on each other with the surface forces equal in magnitude and opposite in direction (Fig. 2.3), we have

$$\mathbf{p}(-\mathbf{n}, \mathbf{r}, t) = -\mathbf{p}(\mathbf{n}, \mathbf{r}, t). \tag{2.3}$$

In order to incorporate internal stresses, i.e., forces depending on the orientation of the surface they act upon, into the differential equations of motion we will later introduce the stress tensor.

It should be noted that the connection between the field description of continuum mechanics and the molecular-level description of statistical physics, as well as the resulting justification of continuum models via those used in statistical physics, can be regarded only as conceptual. In practice a number of assumptions inherent in all statistical models are neither more obvious nor more experimentally verifiable than those of phenomenological fluid mechanics, and there is no formal and rigorous 'derivation' of fluid-mechanical models, especially for liquids (as opposed to gases) where statistical physics encounters some formidable difficulties of principle associated with the necessity to take into account multiple interactions of molecules. However, the notion of the continuum approximation as the 0th order in the ratio of molecular-to-macroscopic length and time scale is very helpful in several respects. Importantly, it points out connections between different elements of continuum mechanics modelling, e.g., the intimate interweaving of the notions of stresses and sharp interfaces.

Another side of this connection is that it becomes clear that, to be self-consistent, a step from the 0th order, given by the continuum approximation, towards accounting for the discrete structure of the matter would have to be highly nontrivial: it will have to include thermal fluctuations of macroscopic properties, 'traces' of intermolecular forces and many other factors. The on-going research in this direction has produced some results though, at present, they cannot be regarded as entirely satisfactory (see Burnett 1935a,b and Agarwal, Yun & Balakrishnan 2001, 2002).

Since we do not have a rigorous way of deriving fluid-mechanical models from the statistical ones, it is not possible to estimate the accuracy of the continuum approximation *a priori*. On the theoretical side, we can only make sure that the models are self-consistent in the sense discussed above, and the results they produce, when applied to particular problems, remain consistent with their underlying assumptions. Then, the results can be compared with experiments for an *a posteriori* assessment of their accuracy. In most cases, the measured quantities can be interpreted directly as the macroscopic parameters introduced by the continuum approximation. However, there are some situations where the problem of interpretation of experiments becomes more subtle. We will discuss some of them in §3.2.

Finally, we should also mention that the idea of describing the fluid as a medium filling the space continuously is fruitful in one more respect. It provides a natural conceptual framework for generalisations of the simplest models to multicomponent and, furthermore, multiphase systems, which can then be described as composed of mutually penetrating continua filling the same space (e.g., Nigmatulin 1991). This of course raises the question of how to model interactions between components or phases, which is far from being trivial, but it does not outweigh the advantages offered by the continuum approximation.

2.1.4 Eulerian and Lagrangian description of motion

The notion of motion is always determined with respect to a system of coordinates of the 'observer', which we will denote by x^1, x^2, x^3 and call the Eulerian coordinates. In classical mechanics, we say that we know the motion of a material point if we have a set of algebraic equations

$$x^i = f^i(t) \qquad (i = 1, 2, 3), \tag{2.4}$$

where f^i are known functions of time. In order to extend this notion to the motion of a fluid modelled as a continuous medium, we need to 'individu-alize' the material points which form the continuum. For example, we can characterize every point by its Eulerian coordinates at some initial moment $t = t_0$ and these three numbers will label it throughout its motion. This actually means that we introduce a coordinate system "frozen" into the fluid, which will move and deform with it (Fig. 2.4) and in which individual particles will have the same coordinates at all times. These coordinates are called Lagrangian, and we will denote them by ξ^1, ξ^2, ξ^3. Clearly, the use of the initial positions of material points as their Lagrangian coordinates is just one possible way of introducing the latter.

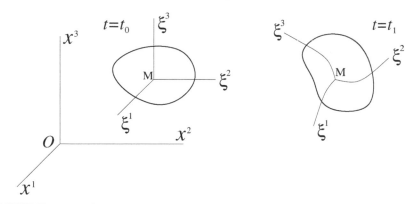

FIGURE 2.4: A definition sketch for Lagrangian coordinates.

Now, if (ξ^1, ξ^2, ξ^3) are the Lagrangian coordinates of an arbitrary point in the continuum, then, instead of (2.4), the full description of the moving fluid will be given by a set of equations

$$x^i = f^i(\xi^1, \xi^2, \xi^3, t) \qquad (i = 1, 2, 3), \qquad (2.5)$$

where f^i are known functions of all four variables. For a fixed t, equations (2.5) will give locations of individual points of the continuum; if we fix (ξ^1, ξ^2, ξ^3), these equations will describe the motion of every material particle.

An approach to the description of fluid motion aimed at finding (2.5) is known as Lagrangian. It uses ξ^1, ξ^2, ξ^3 and t as independent variables and can be regarded as a direct generalisation of the methods used in classical mechanics. The main advantage of the Lagrangian approach is the simplicity with which one can describe kinematics of the flow. Indeed, in a complete analogy with classical mechanics, the velocity $\mathbf{u}(\xi^1, \xi^2, \xi^3, t)$ and acceleration $\mathbf{a}(\xi^1, \xi^2, \xi^3, t)$ of every material particle are expressed as the time derivatives of its position vector with respect to the observer's reference frame:

$$\mathbf{u} = \frac{\partial \mathbf{r}}{\partial t}(\xi^1, \xi^2, \xi^3, t), \qquad \mathbf{a} = \frac{\partial \mathbf{u}}{\partial t} = \frac{\partial^2 \mathbf{r}}{\partial t^2}(\xi^1, \xi^2, \xi^3, t). \qquad (2.6)$$

This advantage of the Lagrangian approach is more than counterbalanced by its disadvantages associated with the necessity to consider interactions of material particles whose positions with respect to each other change continuously during flow. As a result, even for the simplest case of an incompressible fluid with no internal friction the condition that the fluid remains continuous and Newton's laws of motion are expressed as nonlinear partial differential equations which are very difficult to solve. We will briefly consider these equations in §§2.2.1, 2.2.4 and 2.2.6.

An important aspect of the Lagrangian approach is the notion of material volumes, surfaces and lines as consisting of the same fluid particles throughout the motion. These notions allow one to formulate physical laws and, in some cases, reach general conclusions irrespective of the way in which the motion is described.

In the Eulerian approach, we are interested not in what happens to a given material particle but in what is going on at a given point in space, and hence use as independent variables x^1, x^2, x^3 and t. The problem is, however, that Newton's laws of motion and all other laws of nature are formulated for material substances and not for geometrical points in space.[7] In order to formulate the equations governing the fluid motion in the observer's coordinate frame, we have to introduce the notion of a material derivative and consider how to differentiate with respect to time an integral taken over a moving material volume.

When physical quantities are specified in the Lagrangian variables the 'material' derivative is simply a partial derivative with respect to time, as in (2.6). Now, let $f(\mathbf{r}, t)$ be a physical quantity, scalar, vector or tensor, associated with moving material particles but specified in Eulerian variables, that is as a time-dependent field in space. Then, expressing the Eulerian variables through the Lagrangian ones by (2.5), we can differentiate as in (2.6)

$$\left(\frac{\partial f}{\partial t}\right)_{\boldsymbol{\xi}} \equiv \frac{\partial f}{\partial t}(\mathbf{r}(\boldsymbol{\xi}, t), t) = \frac{\partial f}{\partial t}(\mathbf{r}, t) + \frac{\partial f}{\partial x^i}(\mathbf{r}, t)\frac{\partial x^i}{\partial t}(\boldsymbol{\xi}, t)$$

$$= \frac{\partial f}{\partial t}(\mathbf{r}, t) + \mathbf{u}(\mathbf{r}, t) \cdot \nabla f(\mathbf{r}, t).$$

Here, in addition to the summation convention to be used throughout the book (see Appendix A), we have used that in Euler's description $\mathbf{u}(\mathbf{r}, t)$ is the velocity of the material particle whose position vector at moment t is $\mathbf{r}$, that is the same quantity as $\partial \mathbf{r}/\partial t$ in the Lagrangian description.

In particular, the acceleration of a fluid particle can now be expressed in the Eulerian variables as

$$\mathbf{a}(\mathbf{r}, t) = \frac{\partial \mathbf{u}}{\partial t} + \mathbf{u} \cdot \nabla \mathbf{u},$$

which includes the local variation $\partial \mathbf{u}/\partial t$ and the convective term $\mathbf{u} \cdot \nabla \mathbf{u}$.

For the material derivative in Euler's variables we will use the notation

$$\frac{D}{Dt} \equiv \frac{\partial}{\partial t} + \mathbf{u} \cdot \nabla. \tag{2.7}$$

Let us now consider differentiation with respect to time of an integral over a material volume. If a given mass of fluid occupies volume V at moment t and V' at moment $t + \delta t$ in the observer's reference frame, then the volume $V' - V$ is formed due to the motion of each surface element dS of the surface S, which confines V, in the normal direction by $\mathbf{u} \cdot \mathbf{n}\,\delta t$ (Fig. 2.5). Hence $dV = \mathbf{u} \cdot \mathbf{n}\,dS\,\delta t$ and, by using the definition of a derivative, we have

---

[7]We are not talking here about relativistic issues where both space and time themselves become subjects of the 'laws of nature'.

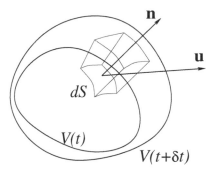

FIGURE 2.5: Moving material volume.

$$\frac{d}{dt} \int_{V(t)} f(\mathbf{r},t)\,dV \equiv \lim_{\delta t \to 0} \frac{1}{\delta t} \left[\int_{V'} f(\mathbf{r},t+\delta t)\,dV - \int_{V} f(\mathbf{r},t)\,dV \right]$$

$$= \lim_{\delta t \to 0} \frac{1}{\delta t} \left[\int_{V} (f(\mathbf{r},t+\delta t) - f(\mathbf{r},t))\,dV + \int_{V'-V} f(\mathbf{r},t+\delta t)\,dV \right]$$

$$= \int_{V} \frac{\partial f}{\partial t}(\mathbf{r},t)\,dV + \int_{S} f\mathbf{u} \cdot \mathbf{n}\,dS.$$

After applying Gauss' divergence theorem to the last integral, one finally arrives at

$$\frac{d}{dt} \int_{V(t)} f\,dV = \int_{V(t)} \left[\frac{\partial f}{\partial t} + \nabla \cdot (f\mathbf{u}) \right] dV. \tag{2.8}$$

Alternatively, (2.8) can be derived in the following more formal way. If we take an arbitrary material volume, its points will have the same coordinates in the Lagrangian space throughout its motion, while their Eulerian coordinates will change with time. For simplicity and without loss of generality, we can choose Cartesian coordinates x, y, z as the observer's reference frame. Let u, v and w be, respectively, the x, y and z components of velocity, and

$$\frac{\partial(x,y,z)}{\partial(\xi^1,\xi^2,\xi^3)} \quad \text{and} \quad \frac{\partial(\xi^1,\xi^2,\xi^3)}{\partial(x,y,z)}$$

denote the (time-dependent) Jacobians of transformations from the Eulerian to Lagrangian coordinates and back. Then,

$$\frac{d}{dt} \int_{V(t)} f(x,y,z,t)\,dx\,dy\,dz = \int_{V_\xi} \frac{\partial}{\partial t} \left[f(\mathbf{r}(\boldsymbol{\xi},t),t) \frac{\partial(x,y,z)}{\partial(\xi^1,\xi^2,\xi^3)} \right] d\xi^1\,d\xi^2\,d\xi^3$$

$$= \int_{V(t)} \frac{Df}{Dt} \, dx \, dy \, dz + \int_{V_\xi} f \frac{\partial}{\partial t} \left[\frac{\partial(x, y, z)}{\partial(\xi^1, \xi^2, \xi^3)} \right] d\xi^1 \, d\xi^2 \, d\xi^3 = \int_{V(t)} \frac{Df}{Dt} \, dx \, dy \, dz$$

$$+ \int_{V(t)} f \left[\frac{\partial(u, y, z)}{\partial(\xi^1, \xi^2, \xi^3)} + \frac{\partial(x, v, z)}{\partial(\xi^1, \xi^2, \xi^3)} + \frac{\partial(x, y, w)}{\partial(\xi^1, \xi^2, \xi^3)} \right] \frac{\partial(\xi^1, \xi^2, \xi^3)}{\partial(x, y, z)} \, dx \, dy \, dz$$

$$= \int_{V(t)} \frac{Df}{Dt} \, dx \, dy \, dz + \int_{V(t)} f \left(\frac{\partial u}{\partial x} + \frac{\partial v}{\partial y} + \frac{\partial w}{\partial z} \right) dx \, dy \, dz$$

$$= \int_{V(t)} \left(\frac{Df}{Dt} + f \nabla \cdot \mathbf{u} \right) dx \, dy \, dz.$$

Then, using the definition of material derivative (2.7) we arrive at (2.8).

If the Eulerian description of motion is known, the corresponding Lagrangian one can be obtained in the following way. By integrating a set of ordinary differential equations

$$\frac{dx}{u(x, y, z, t)} = \frac{dy}{v(x, y, z, t)} = \frac{dz}{w(x, y, z, t)} = dt, \tag{2.9}$$

one can find trajectories of fluid particles, where the three constants of integration can be interpreted as the Lagrangian coordinates individualizing the trajectories. The result will have the form of (2.5). Given that all parameters of the fluid are known as functions of the Eulerian variables, one can use (2.5) to express them in terms of the Lagrangian ones. It should be noted that in a general case trajectories of fluid particles differ from the streamlines, which are a set of lines tangential to the velocity field at a given moment. The latter are determined by the set of equations

$$\frac{dx}{u(x, y, z, t)} = \frac{dy}{v(x, y, z, t)} = \frac{dz}{w(x, y, z, t)} = d\lambda, \tag{2.10}$$

where λ is the independent variable, and t is a parameter. For steady flows, streamlines and trajectories obviously coincide.

If the flow is specified in the Lagrangian variables, one can also obtain its description in the Eulerian ones. Adding the velocity field found from (2.6) to other parameters of fluid known as functions of ξ^1, ξ^2, ξ^3 and t we can then find ξ^1, ξ^2, ξ^3 as functions of x^1, x^2, x^3 and t from the set of algebraic equations (2.5).

2.2 Governing equations

The set of equations of fluid mechanics comprise two subsets: the equations expressing fundamental conservation laws (conservation of mass, momentum, angular momentum, energy) and supplementary equations, which include constitutive equations and equations of state. Conservation laws are essentially the same for all media, and their mathematical representation defines and quantifies the basic characteristics to be used to describe the fluid. The set of equations expressing conservation laws is not closed, and one needs additional constitutive equations and equations of state to specify a particular type of fluid or class of flows. These equations are ultimately rooted in experimental observations and include material constants specific to particular fluids. Many relevant observations are accumulated in the laws of classical thermodynamics and therefore sometimes the supplementary equations are said to be of a thermodynamic origin.

In many important cases, additional assumptions about the flow, for example the assumptions that compressibility of the fluid and/or its viscosity can be neglected, make it possible to consider a closed set of equations comprising some of the conservation laws and some of constitutive equations. This set of equations can be studied independently of the rest, and the remaining equations examined afterwards will, if necessary, provide additional information about the parameters they describe. We will consider some of possible simplifications later but, first, we have to formulate the conservation laws.

2.2.1 Conservation of mass

Consider an arbitrary volume V whose position is fixed relative to the observer's reference frame, and let S be a surface confining this volume and $\mathbf{n}$ be a unit outward normal to S. Then, if volume V is entirely occupied by fluid of possibly variable density ρ, then

$$\int_V \rho \, dV$$

is the mass of fluid inside V at any instant. The net rate at which the mass inside the volume changes due to the mass flux across its boundary is given by

$$\int_S \rho \mathbf{u} \cdot \mathbf{n} \, dS.$$

If there are no sources and sinks of mass inside V, we have an equation

$$\frac{d}{dt} \int_V \rho \, dV = - \int_S \rho \mathbf{u} \cdot \mathbf{n} \, dS, \tag{2.11}$$

which describes conservation of mass in an integral form. This equation does not require that ρ and $\mathbf{u}$ are smooth functions inside V and can be applied to flows with surfaces across which the flow parameters are discontinuous. If ρ and $\mathbf{u}$ are smooth, then, since volume V is fixed in space, differentiation can

be taken under the integral sign and, after using Gauss' theorem to transform
the integral on the right-hand side of (2.11) into that over V, we arrive at

$$\int_V \left[\frac{\partial \rho}{\partial t} + \nabla \cdot (\rho \mathbf{u}) \right] dV = 0. \tag{2.12}$$

Since this equation is valid for all volumes lying entirely in the fluid, the
integrand has to be identically zero, and we arrive at the mass conservation
law in a differential form

$$\frac{\partial \rho}{\partial t} + \nabla \cdot (\rho \mathbf{u}) = 0. \tag{2.13}$$

This equation is also known as the continuity equation.

An alternative way of deriving equation (2.13) is by considering a moving
material volume which consists of the same fluid particles and hence contains
the same mass of fluid throughout its motion. Then, in the observer's reference
frame, we have

$$\frac{d}{dt} \int_{V(t)} \rho \, dV = 0, \tag{2.14}$$

and, by using (2.8), arrive at (2.12) and hence (2.13).

In many flows of liquids and gases, the changes in their density are negligible, and they can be accurately described by the model of an incompressible
fluid. Then, the density of each fluid particle remains the same,

$$\frac{D\rho}{Dt} = 0,$$

and the continuity equation (2.13) becomes

$$\nabla \cdot \mathbf{u} = 0. \tag{2.15}$$

Vector fields satisfying this equation are called solenoidal. The continuity
equation in different coordinate systems can be found in Appendix B.

In the cases of two-dimensional and axisymmetric flows of incompressible
fluid and steady flows of compressible fluid, the continuity equation, (2.15) or
(2.13), will contain only two derivatives. Then, instead of looking for a vector
field $\mathbf{u}$ or $\rho\mathbf{u}$, one can reduce the problem to that of finding a scalar function
which will make the continuity equation satisfied identically and allow one to
obtain $\mathbf{u}$ or $\rho\mathbf{u}$ by differentiation. We will illustrate how this function can be
introduced in the case of flow of an incompressible fluid.

Let $Oxyz$ be a Cartesian coordinate system in the observer's reference frame
in which the fluid's velocity is independent of z. Without loss of generality
we can assume that the z-component of velocity is zero and the flow is in the
Oxy plane. If u_x and u_y are the x and y components of velocity, equation
(2.15) will take the form

$$\frac{\partial u_x}{\partial x} + \frac{\partial u_y}{\partial y} = 0, \tag{2.16}$$

and hence $u_x\,dy - u_y\,dx$ is a full differential of a function which we will denote as $\psi(x,y,t)$ and call the *stream function*. Then,

$$\psi(x,y,t) - \psi_0 = \int (u_x\,dy - u_y\,dx) \tag{2.17}$$

and

$$u_x = \frac{\partial\psi}{\partial y}, \qquad u_y = -\frac{\partial\psi}{\partial x}. \tag{2.18}$$

Equation $d\psi = u_x\,dy - u_y\,dx = 0$ is a two-dimensional case of (2.10), and hence ψ is constant along every streamline. It is easy to show that the difference between the values of ψ on two stream lines represents the flux of volume of fluid (per unit length in the z-direction) across any curve connecting them.

If r, θ are polar coordinates in the plane Oxy ($x = r\cos\theta$, $y = r\sin\theta$) and u_r, u_θ are the radial and transversal components of velocity, the continuity equation written down in these variables (Appendix B) is satisfied identically for

$$u_r = \frac{1}{r}\frac{\partial\psi}{\partial\theta}, \qquad u_\theta = -\frac{\partial\psi}{\partial r}.$$

If the flow is axisymmetric and r,θ,z are the appropriate cylindrical coordinates, the stream function can be introduced in a similar way and

$$u_r = \frac{1}{r}\frac{\partial\psi}{\partial z}, \qquad u_z = -\frac{1}{r}\frac{\partial\psi}{\partial r}.$$

Finally, for the axisymmetric flow specified in spherical coordinates r, φ, θ, where the velocity is independent of φ, one has

$$u_r = \frac{1}{r^2\sin\theta}\frac{\partial\psi}{\partial\theta}, \qquad u_\theta = -\frac{1}{r\sin\theta}\frac{\partial\psi}{\partial r}.$$

In the case of a steady compressible flow, the vector $\mathbf{u}$ in the above expressions must be replaced by $\rho\mathbf{u}$.

2.2.1.1 Conservation of mass in the Lagrangian description

Let us also derive the continuity equation for the Lagrangian specification of motion. If $Ox^1x^2x^3$ is the Cartesian coordinate frame of the observer, then (2.14) can be written down as

$$0 = \frac{d}{dt}\int_{V(t)} \rho\,dx^1\,dx^2\,dx^3 = \frac{d}{dt}\int_{V_\xi} \rho\frac{\partial(x^1,x^2,x^3)}{\partial(\xi^1,\xi^2,\xi^3)}\,d\xi^1\,d\xi^2\,d\xi^3$$
$$= \int_{V_\xi} \frac{\partial}{\partial t}\left[\rho\frac{\partial(x^1,x^2,x^3)}{\partial(\xi^1,\xi^2,\xi^3)}\right]\,d\xi^1\,d\xi^2\,d\xi^3.$$

Given that the material volume V_ξ is arbitrary, we obtain a differential equation

$$\frac{\partial}{\partial t}\left[\rho\,\frac{\partial(x^1,x^2,x^3)}{\partial(\xi^1,\xi^2,\xi^3)}\right] = 0.$$

After integration over a time interval from t_0, when the density distribution $\rho(\xi^1,\xi^2,\xi^3,t_0)$ and the Eulerian coordinates of individual particles, $x_0^i = x^i(\xi^1,\xi^2,\xi^3,t_0)$ were known, to the current moment t, this equation takes the form

$$\rho(\xi^1,\xi^2,\xi^3,t)\frac{\partial(x^1,x^2,x^3)}{\partial(\xi^1,\xi^2,\xi^3)} = \rho(\xi^1,\xi^2,\xi^3,t_0)\frac{\partial(x_0^1,x_0^2,x_0^3)}{\partial(\xi^1,\xi^2,\xi^3)}.$$

If the Lagrangian coordinates individualizing material particles are chosen to be the Eulerian coordinates of these particles at $t = t_0$, then the Jacobian on the right-hand side of the above equation is equal to 1, and the continuity equation becomes

$$\rho(\xi^1,\xi^2,\xi^3,t)\frac{\partial(x^1,x^2,x^3)}{\partial(\xi^1,\xi^2,\xi^3)} = \rho(\xi^1,\xi^2,\xi^3,t_0). \tag{2.19}$$

Finally, if the fluid is incompressible, then $\rho \equiv \text{const}$ and (2.19) simplifies even further:

$$\frac{\partial(x^1,x^2,x^3)}{\partial(\xi^1,\xi^2,\xi^3)} = 1. \tag{2.20}$$

If the observer's coordinate frame is not Cartesian, the equations (2.19) or (2.20) will have to be transformed with the help of the appropriate Jacobian.

It should be noted that, unlike (2.15), equation (2.20) is nonlinear. This reflects the fact that (2.20) imposes a constraint on trajectories of material particles which have to preserve continuity of the fluid, whereas (2.15) is associated with the instantaneous velocity distribution. If one decides to switch from the Eulerian to Lagrangian specification of motion, the nonlinearity will be recovered in (2.9).

2.2.2   Equation of motion. Stress tensor

In classical mechanics, the momentum of a system of N material points of masses m_i moving with velocities $\mathbf{u}_i$ is defined as

$$\sum_{i=1}^{N} m_i\mathbf{u}_i, \tag{2.21}$$

which results in Newton's second law being formulated for the system of material points in the same way as for one such point. A natural generalization of (2.21) for a material volume containing a continuum of material particles is

$$\int_{V(t)} \rho\mathbf{u}\,dV.$$

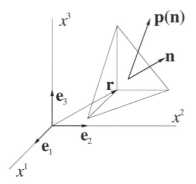

FIGURE 2.6: A definition sketch for the introduction of the stress tensor.

It is assumed that for a material volume the rate of change of momentum is equal to the sum of external forces (body forces and stresses) acting on it:

$$\frac{d}{dt} \int_{V(t)} \rho \mathbf{u}\, dV = \int_{V(t)} \rho \mathbf{F}\, dV + \int_{S(t)} \mathbf{p}(\mathbf{n})\, dS, \qquad (2.22)$$

where $\rho \mathbf{F}$ is the sum of external body forces, $\mathbf{p}$ are stresses, acting on the surface S confining volume V, and $\mathbf{n}$ is a unit normal vector to S pointing outwards. Equation (2.22) is the main dynamic relationship postulated in continuum mechanics. This equation requires only that the functions involved are integrable and can be applied to volumes containing surfaces of discontinuity.

Using (2.8) and (2.13), and assuming that ρ and $\mathbf{u}$ are smooth in V, we can write down the term on the left-hand side of (2.22) as

$$\frac{d}{dt} \int_{V(t)} \rho \mathbf{u}\, dV = \int_{V(t)} \left[\frac{\partial(\rho\mathbf{u})}{\partial t} + \nabla \cdot (\rho\mathbf{u}\mathbf{u}) \right] dV$$

$$= \int_{V(t)} \left[\mathbf{u} \left(\frac{\partial \rho}{\partial t} + \nabla \cdot (\rho\mathbf{u}) \right) + \rho \frac{D\mathbf{u}}{Dt} \right] dV = \int_{V(t)} \rho \frac{D\mathbf{u}}{Dt}\, dV$$

and hence equation (2.22) will take the form

$$\int_{V(t)} \rho \frac{D\mathbf{u}}{Dt}\, dV = \int_{V(t)} \rho \mathbf{F}\, dV + \int_{S(t)} \mathbf{p}(\mathbf{n})\, dS. \qquad (2.23)$$

In order to derive a differential form of the momentum balance equation, we have to understand how $\mathbf{p}$ depends on the orientation of $\mathbf{n}$. To do this, let us take an elementary volume in the shape of a tetrahedron with three ribs parallel to the axes of a Cartesian coordinate system, which we will use

here as the observer's reference frame (Fig. 2.6). Let h be the distance from
$\mathbf{r}$ to the opposite face of the tetrahedron, and S_1, S_2, S_3 and S_n denote the
areas of the tetrahedron's sides normal to $\mathbf{e}_1$, $\mathbf{e}_2$, $\mathbf{e}_3$ and $\mathbf{n}$, respectively. As
$h \to 0$, for the stresses on all four sides of the tetrahedron we can use their
Taylor expansions about $\mathbf{r}$. It is also clear that, as $h \to 0$, $S_i = O(h^2)$ for
$i = 1, 2, 3, n$. Thus, as $h \to 0$, equation (2.23) takes the form

$$\rho(\mathbf{r}, t)\frac{D\mathbf{u}}{Dt}(\mathbf{r}, t)\tfrac{1}{3}S_n h = \rho(\mathbf{r}, t)\mathbf{F}(\mathbf{r}, t)\tfrac{1}{3}S_n h + \mathbf{p}(\mathbf{n}, \mathbf{r}, t)\, S_n$$

$$+\mathbf{p}(-\mathbf{e}_1, \mathbf{r}, t)\, S_1 + \mathbf{p}(-\mathbf{e}_2, \mathbf{r}, t)\, S_2 + \mathbf{p}(-\mathbf{e}_3, \mathbf{r}, t)\, S_3 + o(h^2). \qquad (2.24)$$

After eliminating S_i $(i = 1, 2, 3)$ using

$$S_i = S_n \mathbf{n} \cdot \mathbf{e}_i = S_n n_i \qquad (i = 1, 2, 3).$$

we can divide (2.24) by S_n and take the limit $h \to 0$. As a result, after taking
into account the general property of stresses (2.3) and introducing a notation
$\mathbf{p}^i = \mathbf{p}(\mathbf{e}_i)$, we arrive at

$$\mathbf{p}(\mathbf{n}) = \mathbf{p}^i n_i. \qquad (2.25)$$

This equation expresses $\mathbf{p}(\mathbf{n})$ in terms of the stresses acting on the surface
elements orthogonal to the basis vectors and components of the normal vector
$\mathbf{n}$, which determines the (arbitrary) orientation of a surface element the stress
$\mathbf{p}$ is acting upon.

Now, Gauss' divergence theorem makes it possible to turn the last integral
in (2.23) into an integral over $V(t)$,

$$\int_{S(t)} \mathbf{p}(\mathbf{n})\, dS = \int_{S(t)} \mathbf{p}^i n_i\, dS = \int_{V(t)} \nabla_i \mathbf{p}^i\, dV,$$

so that equation (2.23) becomes

$$\int_{V(t)} \left[\rho\frac{D\mathbf{u}}{Dt} - \rho\mathbf{F} - \nabla_i\mathbf{p}^i\right] dV = 0.$$

Given that the material volume $V(t)$ is arbitrary, we obtain

$$\rho\left(\frac{\partial\mathbf{u}}{\partial t} + \mathbf{u} \cdot \nabla\mathbf{u}\right) = \rho\mathbf{F} + \nabla_i\mathbf{p}^i. \qquad (2.26)$$

This differential equation of motion, as well as equation (2.25), explicitly
relies on the stresses acting upon surfaces normal to the axes of an orthogonal
coordinate system, which are an auxiliary tool we used in the derivation. To
make the description of stresses more general, let us introduce the components
of $\mathbf{p}^i$ by

$$\mathbf{p}^i = p^{ik}\mathbf{e}_k \qquad (i = 1, 2, 3) \qquad (2.27)$$

and rewrite (2.25) in the form

$$\mathbf{p} = p^{ik}\mathbf{e}_k\mathbf{e}_i \cdot \mathbf{n}. \qquad (2.28)$$

This equation relates vectors $\mathbf{p}$ and $\mathbf{n}$, which, as all vectors, are objects invariant with respect to transformations of coordinates, and the basis vectors $\mathbf{e}_i$ $(i = 1, 2, 3)$, which are specific to the coordinate system we use. If we change the coordinate system $(x^1, x^2, x^3) \mapsto (y^1, y^2, y^3)$, then the new basis vectors $\mathbf{e}'_j$ $(j = 1, 2, 3)$ will be related to the old ones by (see Appendix A)

$$\mathbf{e}_i = \mathbf{e}'_j \frac{\partial y^j}{\partial x^i},$$

and (2.28) will take the form

$$\mathbf{p} = p^{ik}\mathbf{e}_k\mathbf{e}_i \cdot \mathbf{n} = p^{ik}\left(\mathbf{e}'_j \frac{\partial y^j}{\partial x^k}\right)\left(\mathbf{e}'_n \frac{\partial y^n}{\partial x^i}\right)\cdot \mathbf{n} = p'^{ik}\mathbf{e}'_k\mathbf{e}'_i \cdot \mathbf{n},$$

where

$$p'^{ik} = p^{mn}\frac{\partial y^i}{\partial x^m}\frac{\partial y^k}{\partial x^n}.$$

Thus, coefficients $\{p^{ik}\}$ and $\{p'^{ik}\}$ in the two coordinate systems are related as contravariant components of a tensor

$$\mathbf{P} = p^{ik}\mathbf{e}_i\mathbf{e}_k, \qquad (2.29)$$

which is known as the *stress tensor*. As one can see from (2.28), components of the stress tensor have a clear physical meaning: p^{ik} is the kth component of the stress acting on a unit area normal to the ith axis of a given coordinate system. For the stress acting on a unit area with the outward (i.e., pointing into the fluid) normal $\mathbf{n}$ we obviously have

$$\mathbf{p}(\mathbf{n}) = \mathbf{n} \cdot \mathbf{P}. \qquad (2.30)$$

Now, using (2.27) and (2.29), we can finally write down (2.26) as

$$\rho\left(\frac{\partial \mathbf{u}}{\partial t} + \mathbf{u}\cdot\nabla\mathbf{u}\right) = \rho\mathbf{F} + \nabla\cdot\mathbf{P}. \qquad (2.31)$$

This equation is invariant with respect to the coordinate system and expresses the law of motion in a differential form. Given that $\nabla_i\mathbf{e}_k = 0$ for all i and k, equation (2.31) can be also written down in components as

$$\rho\left(\frac{\partial u^i}{\partial t} + u^k\nabla_k u^i\right) = \rho F^i + \nabla_k p^{ki}, \qquad (i = 1, 2, 3). \qquad (2.32)$$

It is worth mentioning that we have derived the notion of the stress tensor from the fact that to leading order as $h \to 0$, the stresses acting on an elementary volume are in balance, that is from (2.25). The motion is driven by the imbalance at the next order and as a result equations of motion (2.31) and (2.32) include *gradients* of the stress tensor.

The conservation laws for mass and momentum considered so far give us four scalar equations, (2.13) and (2.31), for 13 unknown scalar functions: ρ, three components of $\mathbf{u}$ and nine components of $\mathbf{P}$. If the fluid is assumed to be incompressible so that ρ becomes a known constant, then still there will be four equations, (2.15) and (2.31), for twelve unknown components of $\mathbf{u}$ and $\mathbf{P}$. To add more equations, or, as it will turn out, to reduce the number of unknowns, we have to consider the balance of the angular momentum.

2.2.3 Conservation of angular momentum. Symmetry of stress tensor

In classical mechanics, for the angular momentum of a system of N material points of mass m_i located at $\mathbf{r}_i$ $(i = 1, 2, \ldots, N)$ one has an equation

$$\frac{d\mathbf{K}}{dt} = \sum_{i=1}^{N} \left(\mathbf{r}_i \times \mathbf{F}_i^{(e)} \right), \tag{2.33}$$

where $\mathbf{F}_i^{(e)}$ $(i = 1, 2, \ldots, N)$ are external forces acting on the system and

$$\mathbf{K} = \sum_{i=1}^{N} (\mathbf{r}_i \times m_i \mathbf{u}_i) \tag{2.34}$$

is, by definition, the angular momentum. The angular momentum can be also written down in the form

$$\mathbf{K} = \mathbf{r}_* \times M\mathbf{u}_* + \sum_{i=1}^{N} (\mathbf{r}_i' \times m_i \mathbf{u}_i'),$$

where $M = \sum_{i=1}^{N} m_i$ is the total mass; $\mathbf{r}_*$ and $\mathbf{u}_*$ are the position and velocity of the center of mass; $\mathbf{r}_i'$ and $\mathbf{u}_i'$ are positions and velocities of points in a moving coordinate frame with the origin at the center of mass.

In classical mechanics, equation (2.33) follows from the equations of motion for material points forming the system and one has only to define the angular momentum $\mathbf{K}$ by (2.34). In fluid mechanics, the situation is more complex. The difficulty comes from the fact that, besides the motion of a fluid volume as a whole and the motion of fluid particles composing it with respect to the center of mass of this volume, there can be an internal angular momentum, for example, due to rotation of fluid particles caused by external torques. Another possible source of internal angular momentum is orientation of molecules as a result of an external magnetic field. An example of a medium with an internal angular momentum is the so-called magnetic fluid (Shliomis 1971, Gogosov, Naletova & Shaposhnikova 1981), which is a suspension of small magnetic particles ($\sim 100 \, \mathring{A}$ in diameter) in a nonmagnetic liquid. An external rotating magnetic field acting on the particles can make their orientation anisotropic and generate rotation which has to be accounted for in the angular momentum balance equation.

Therefore, the angular momentum balance equation in fluid mechanics has to be postulated independently of the equation of motion. In a general case, it has the form

$$\frac{d}{dt}\left[\int_{V(t)} (\mathbf{r} \times \rho\mathbf{u}) \, dV + \int_{V(t)} \rho\mathbf{k} \, dV\right] = \int_{V} (\mathbf{r} \times \mathbf{F}\rho) \, dV + \int_{S} (\mathbf{r} \times \mathbf{p(n)}) \, dS$$

$$+ \int_{V} \rho\mathbf{h} \, dV + \int_{S} \mathbf{Q(n)} \, dS, \tag{2.35}$$

where $\mathbf{k}$ is the density of internal angular momentum associated with rotation of fluid particles, $\mathbf{h}$ is the density of external body torques, and $\mathbf{Q(n)}$ is the surface density of external torques acting on the surface S confining V.

Below, we will consider the classical case and assume that $\mathbf{k} = \mathbf{h} = \mathbf{Q} = 0$ so that equation (2.35) becomes simply

$$\frac{d}{dt}\int_{V(t)} (\mathbf{r} \times \rho\mathbf{u}) \, dV = \int_{V} (\mathbf{r} \times \mathbf{F}\rho) \, dV + \int_{S} (\mathbf{r} \times \mathbf{p(n)}) \, dS. \tag{2.36}$$

The second integral on the right-hand side can be transformed as

$$\int_{S} (\mathbf{r} \times \mathbf{p(n)}) \, dS = \int_{S} (\mathbf{r} \times \mathbf{p}^i)n_i \, dS = \int_{V} \nabla_i(\mathbf{r} \times \mathbf{p}^i) \, dV$$

$$= \int_{V} (\mathbf{r} \times \nabla_i\mathbf{p}^i) \, dV + \int_{V} (\nabla_i\mathbf{r} \times \mathbf{p}^i) \, dV$$

$$= \int_{V} (\mathbf{r} \times \nabla_i\mathbf{p}^i) \, dV + \int_{V} (\mathbf{e}_i \times \mathbf{e}_k)p^{ik} \, dV. \tag{2.37}$$

Here we have used equation (2.25) for $\mathbf{p}$, Gauss' divergence theorem, the definition of the basis vector, $\mathbf{e}_i = \nabla_i\mathbf{r}$ (see Appendix A), and expressions (2.27) for $\mathbf{p}^i$. Now, after using (2.8) and (2.37), we can rewrite equation (2.36) in the form

$$\int \left[\mathbf{r} \times \left(\frac{D\mathbf{u}}{Dt} - \mathbf{F} - \frac{1}{\rho}\nabla_i\mathbf{p}^i\right)\right] dV = \int_{V} (\mathbf{e}_i \times \mathbf{e}_k)p^{ik} \, dV.$$

The expression in parentheses on the left-hand side is identically zero due to (2.26) so that, since V is arbitrary, we arrive at

$$(\mathbf{e}_i \times \mathbf{e}_k)p^{ik} = 0. \tag{2.38}$$

Thus, a convolution of $\mathbf{e}_i \times \mathbf{e}_k$, which is antisymmetric with respect to i and k, with a second rank tensor $\mathbf{P}$ is identically zero, and hence $\mathbf{P}$ is symmetric. Indeed, the double sum in (2.38) can be split in two

$$\underbrace{(\mathbf{e}_i \times \mathbf{e}_k)p^{ik}}_{i<k} + \underbrace{(\mathbf{e}_i \times \mathbf{e}_k)p^{ik}}_{k<i} = 0,$$

and, after changing the order of multiplication in the second term and renaming the dummy indices $i \to k$, $k \to i$, we obtain that

$$(\mathbf{e}_i \times \mathbf{e}_k)(p^{ik} - p^{ki}) = 0 \qquad \text{for } i < k,$$

and hence $p^{ik} = p^{ki}$ for $i, k = 1, 2, 3$.

Thus, after showing that the stress tensor is symmetric and hence can at most have only 6 different components, we have 4 scalar equations (2.13), (2.31) involving 10 unknowns: ρ, $\mathbf{u}$ and 6 components of $\mathbf{P}$.

2.2.4 Ideal fluid

Intuitively, it is clear that the stress acting upon an elementary surface should depend on the local rate of deformation of fluid particles. Then, one has to quantify the notion of deformation, its rate and consider how the stress tensor is related to it. However, we can formulate a class of models useful in describing a number of flows without explicitly invoking these issues. The idea comes from an observation that in many situations the force acting on a surface positioned in the fluid can be very accurately described in terms of pressure, that is a force per unit area directed against the outward normal to the surface. If the tangential component of this force, which in a general case of a real fluid is nonzero, is not essential to the flow to be described, then one can consider an idealized model assuming that $\mathbf{p} = -p\mathbf{n}$, where the pressure p is a scalar function of parameters determining the local state of the fluid. Since obviously $\mathbf{n} = \mathbf{I} \cdot \mathbf{n}$, where $\mathbf{I}$ is the metric tensor (see Appendix A), the stress tensor will be given by

$$\mathbf{P} = -p\mathbf{I}. \tag{2.39}$$

Fluids (or, strictly speaking, their mathematical models) with the stress tensor of this form are referred to as ideal or inviscid. The stress tensor given by (2.39) implies that there is no internal friction between fluid layers moving parallel to each other and, if a solid is immersed into the fluid, there will be no force tangential to its surface.

Now the divergence of $\mathbf{P}$ is equal to the pressure gradient, $\nabla \cdot \mathbf{P} = -\nabla p$, so that equation (2.31) becomes

$$\frac{\partial \mathbf{u}}{\partial t} + \mathbf{u} \cdot \nabla \mathbf{u} = -\frac{1}{\rho}\nabla p + \mathbf{F}, \tag{2.40}$$

which is known as Euler's equation.

If the fluid is assumed to be incompressible, then (2.40) together with the corresponding mass balance equation (2.15), i.e.,

$$\nabla \cdot \mathbf{u} = 0 \tag{2.41}$$

form a closed set of equations describing the fluid as a purely mechanical system with no dissipation of mechanical energy, and one can find p and $\mathbf{u}$ as functions of x^1, x^2, x^3 and t irrespective of thermodynamic properties of the medium.

If compressibility of the fluid is essential to the flow to be described and hence the distribution of ρ has to be found, then to close the set of equations

$$\frac{\partial \rho}{\partial t} + \nabla \cdot (\rho \mathbf{u}) = 0 \tag{2.42}$$

and (2.40) it becomes necessary to consider the fluid also as a thermodynamic system. In a general case, one has to add to (2.42), (2.40) the energy conservation law and formulate the equations of state, which would specify thermodynamic properties of a particular medium. These equations will be considered in §2.3.2. A simpler closure can be employed for a particular class of flows, known as barotropic, where pressure and density are related by an algebraic equation, $p = p(\rho)$ or $\rho = \rho(p)$. This equation added to (2.13), (2.40) closes the system. In this context, the incompressible fluid is just a particular case, where $\rho = \text{const}$. An important subclass of barotropic flows is known as polytropic flows, for which $p = C\rho^n$.

The analysis of barotropic flows of ideal fluids can be considerably simplified in the case where the body forces are potential, so that $\mathbf{F} = -\nabla\Pi$. Then, Thomson's theorem (Thomson 1869, Lamb 1932, Kochin, Kibel & Roze 1964) gives that the circulation of velocity over a closed material contour confining a material surface remains constant during flow. A corollary of this general result is Lagrange's theorem, which states that if initially there were no vortices in a certain mass of an ideal barotropic fluid, they will not appear in the subsequent motion. In particular, if the flow starts from rest, it will always be irrotational.

Then, after introducing the velocity potential,

$$\mathbf{u} = \nabla\phi, \tag{2.43}$$

one can rewrite (2.15) as Laplace's equation for ϕ:

$$\Delta\phi = 0. \tag{2.44}$$

In many particular problems, this equation can be analyzed independently of the pressure distribution which then can be obtained *a posteriori*, either from (2.40) or, more conveniently, from the Cauchy-Lagrange integral

$$\frac{\partial \phi}{\partial t} + \frac{|\nabla\phi|^2}{2} + \int \frac{dp}{\rho} + \Pi = f(t), \tag{2.45}$$

which follows from (2.40) after substituting (2.43) into it and integrating with respect to the spatial coordinates. The function $f(t)$ resulting from this integration can be incorporated into the velocity potential by replacing ϕ with

$$\bar{\phi} = \phi + \int\limits_0^t f(t)\,dt.$$

In the case of free surface flows, equations (2.44) and (2.45) become interrelated through the boundary conditions and have to be analyzed simultaneously.

The theory of potential flows is well developed and one can find a detailed exposition of obtained results in a number of books and review articles (e.g., Lamb 1932, Kochin, Kibel & Roze 1964, Sedov 1965b).

2.2.4.1 Ideal fluid in the Lagrangian description

Let us give an equivalent of (2.40) for the Lagrangian specification of motion. For simplicity we will use Cartesian coordinates x^1, x^2, x^3 as the observer's reference frame. Components of the body force **F** will be denoted by F^i ($i = 1, 2, 3$). Using the expression (2.6) for the acceleration in the Lagrangian coordinates, we can rewrite (2.40) as

$$F^i - \frac{\partial^2 x^i}{\partial t^2} = \frac{1}{\rho}\frac{\partial p}{\partial x^i}, \qquad (i = 1, 2, 3).$$

Then, multiplying the ith equation by $\partial x^i/\partial \xi^k$, adding them up, and taking into account that

$$\frac{\partial p}{\partial x^i}\frac{\partial x^i}{\partial \xi^k} = \frac{\partial p}{\partial \xi^k},$$

we arrive at the equation of motion in the form

$$\sum_{i=1}^3 \left(F^i - \frac{\partial^2 x^i}{\partial t^2} \right) \frac{\partial x^i}{\partial \xi^k} = \frac{1}{\rho}\frac{\partial p}{\partial \xi^k}, \qquad (k = 1, 2, 3). \tag{2.46}$$

Equations (2.20), (2.46) describe flows of ideal incompressible fluids and can be referred to as Euler's equations in the Lagrangian specification of motion.

2.2.5 Elements of theory of deformation

2.2.5.1 Tensors of deformation

In order to describe fluid motion and internal stresses more realistically, we need to consider deformation of the fluid continuum during flow. The essence of deformation is the displacement of individual material points forming a moving continuum with respect to one another which leads to changes in distances between them. To quantify this notion, let us consider the same material volume at two moments, $t = t_1$ and $t = t_2$, and look at how an elementary segment $d\mathbf{r}'$ connecting two infinitesimally close material points M_0 and M_1 evolves during flow (Fig. 2.7).

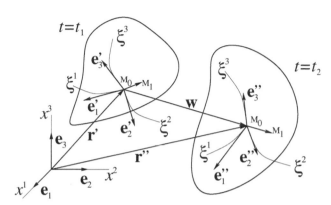

FIGURE 2.7: Deformation of a fluid particle during flow. At $t = t_1$ and $t = t_2$ the Lagrangian coordinates generate two different coordinate systems in the observer's space with different metrics.

Clearly, the Lagrangian coordinates at $t = t_1$ and $t = t_2$ can also be regarded as two different coordinates systems in the *observer's* space. For these systems we will obviously have two different metric tensors with covariant components $\{g'_{ij}\}$ and $\{g''_{ij}\}$, respectively. By definition,

$$g'_{ij} = \mathbf{e}'_i \cdot \mathbf{e}'_j \quad \text{and} \quad g''_{ij} = \mathbf{e}''_i \cdot \mathbf{e}''_j \qquad (i, j = 1, 2, 3), \qquad (2.47)$$

where $\mathbf{e}'_i$ and $\mathbf{e}''_i$ are the basis vectors in the coordinate system (ξ^1, ξ^2, ξ^3) at $t = t_1$ and $t = t_2$, respectively (Fig. 2.7). The material points M_0 and M_1 by definition have the same Lagrangian coordinates throughout their motion and the elementary vectors $d\mathbf{r}'$ and $d\mathbf{r}''$ both have the same coordinates $(d\xi^1, d\xi^2, d\xi^3)$. Then, if $|d\mathbf{r}'| = ds'$ and $|d\mathbf{r}''| = ds''$, one has (Appendix A)

$$(ds')^2 = g'_{ij} d\xi^i d\xi^j \quad \text{and} \quad (ds'')^2 = g''_{ij} d\xi^i d\xi^j$$

so that

$$(ds'')^2 - (ds')^2 = 2\varepsilon_{ij} d\xi^i d\xi^j, \qquad (2.48)$$

where

$$\varepsilon_{ij} = \tfrac{1}{2}(g''_{ij} - g'_{ij}). \qquad (2.49)$$

Since in the Lagrangian space the left-hand side of (2.48) will obviously remain invariant if we change the Lagrangian coordinate system, say, from (ξ^1, ξ^2, ξ^3) to (η^1, η^2, η^3), the coefficients $\{\varepsilon_{ij}\}$ defined by (2.49) will have to transform as covariant components of a tensor. Thus, unsurprisingly, variations of lengths during flow can be associated with the corresponding variations of the metric tensor generated in the observer's space by the Lagrangian coordinates.

The tensor has to be an object invariant with respect to transformations of the coordinate system. In our case, we can achieve this invariance in two ways, that is by assigning $\{\varepsilon_{ij}\}$ to two different coordinate systems. We can multiply the covariant components ε_{ij} by polyadic products either formed by contravariant basis vectors $\mathbf{e}'^i$ corresponding to $t = t_1$ or by $\mathbf{e}'''^i$ corresponding to $t = t_2$. As a result we will end up with two tensors

$$\mathcal{E}' = \varepsilon_{ij}\mathbf{e}'^i\mathbf{e}'^j \quad \text{and} \quad \mathcal{E}'' = \varepsilon_{ij}\mathbf{e}''^i\mathbf{e}''^j, \tag{2.50}$$

both having the same covariant components but in different coordinate frames. These tensors are called the tensors of deformation or the *strain tensors*. In a sense, the very concept of the deformation tensor implies that deformation is an *act* relating two distinct states: 'before' and 'after' deformation. Hence one ends up with two tensors (2.50) associated with these two states. In order to describe the flow, we need a concept which would convey the idea that deformation is a *process* and hence would be related with the velocity field. To do this, we must, first, consider a preliminary step and relate ε_{ij} with displacements of material points.

If $\mathbf{w}$ is a vector characterizing the displacement of M_0 (Fig. 2.7), then

$$\mathbf{w} = \mathbf{r}'' - \mathbf{r}', \tag{2.51}$$

where $\mathbf{r}'$ and $\mathbf{r}''$ are the position vectors of M_0 in the observer's reference frame at $t = t_1$ and $t = t_2$, respectively. As we have already noted, the Lagrangian coordinates at $t = t_1$ and $t = t_2$ generate different coordinate systems in the observer's space, and the corresponding basis vectors defined by (Appendix A)

$$\mathbf{e}'_i = \frac{\partial \mathbf{r}'}{\partial \xi^i} \quad \text{and} \quad \mathbf{e}''_i = \frac{\partial \mathbf{r}''}{\partial \xi^i}$$

are different. Thus, differentiating (2.51) with respect to ξ^i, we obtain

$$\frac{\partial \mathbf{w}}{\partial \xi^i} = \frac{\partial \mathbf{r}''}{\partial \xi^i} - \frac{\partial \mathbf{r}'}{\partial \xi^i} = \mathbf{e}''_i - \mathbf{e}'_i \quad (i = 1, 2, 3)$$

and hence

$$\mathbf{e}''_i = \mathbf{e}'_i + \frac{\partial \mathbf{w}}{\partial \xi^i} \quad (i = 1, 2, 3). \tag{2.52}$$

Now, using the definitions (2.49), (2.47) and expressing $\partial \mathbf{w}/\partial \xi^i$ via covariant derivatives of the components of $\mathbf{w}$,

$$\frac{\partial \mathbf{w}}{\partial \xi^i} = (\nabla'_i w_k)\mathbf{e}'^k = (\nabla'_i w^k)\mathbf{e}'_k,$$

we obtain

$$\varepsilon_{ij} = \tfrac{1}{2}(g''_{ij} - g'_{ij}) = \tfrac{1}{2}(\mathbf{e}''_i \cdot \mathbf{e}''_j - \mathbf{e}'_i \cdot \mathbf{e}'_j)$$

$$= \frac{1}{2}\left[\left(\mathbf{e}'_i + \frac{\partial \mathbf{w}}{\partial \xi^i}\right) \cdot \left(\mathbf{e}'_j + \frac{\partial \mathbf{w}}{\partial \xi^j}\right) - \mathbf{e}'_i \cdot \mathbf{e}'_j\right]$$

$$= \frac{1}{2}\left(\frac{\partial \mathbf{w}}{\partial \xi^i} \cdot \mathbf{e}'_j + \mathbf{e}'_i \cdot \frac{\partial \mathbf{w}}{\partial \xi^j} + \frac{\partial \mathbf{w}}{\partial \xi^i} \cdot \frac{\partial \mathbf{w}}{\partial \xi^j}\right)$$

$$= \tfrac{1}{2}[\nabla'_i w_j + \nabla'_j w_i + (\nabla'_i w^k)(\nabla'_j w_k)].$$

Thus,

$$\varepsilon_{ij} = \tfrac{1}{2}[\nabla'_i w_j + \nabla'_j w_i + (\nabla'_i w^k)(\nabla'_j w_k)], \tag{2.53}$$

where the prime indicates that the covariant derivatives are taken with respect to the coordinates generated in the observer's space by the Lagrangian coordinates at $t = t_1$. Similarly, one has

$$\varepsilon_{ij} = \tfrac{1}{2}[\nabla''_i w_j + \nabla''_j w_i - (\nabla''_i w^k)(\nabla''_j w_k)], \tag{2.54}$$

where the double prime corresponds to the coordinates in the observer's space coinciding with the Lagrangian coordinates at $t = t_2$.

It should be noted that equations (2.53), (2.54) are valid for *finite* deformations.

2.2.5.2 The rate-of-strain tensor

Now we can introduce the rate-of-strain tensor which will describe the process of deformation. Let moments $t = t_1$ and $t = t_2$ be infinitesimally close to each other so that, as $t_2 - t_1 \equiv \delta t \to 0$, to leading order the field of displacements $\mathbf{w}$ is related with the velocity field $\mathbf{u}$ by $\mathbf{w} = \mathbf{u}\,\delta t$. Then, after using $w_i = u_i\,\delta t$ and $w^i = u^i\,\delta t$ in (2.53), where we will replace ε_{ij} with $\delta\varepsilon_{ij}$ to emphasize that deformations are infinitesimal, and taking the limit

$$\lim_{\delta t \to 0} \frac{\delta\varepsilon_{ij}}{\delta t} = \tfrac{1}{2}(\nabla'_i u_j + \nabla'_j u_i) \equiv e'_{ij},$$

we introduce covariant components e'_{ij} of a symmetric tensor, which is called the *rate-of-strain tensor*. In the coordinate system of the observer, covariant components of the rate-of-strain tensor are obviously given by

$$e_{ij} = \tfrac{1}{2}(\nabla_i u_j + \nabla_j u_i), \tag{2.55}$$

where the covariant derivatives are taken with respect to x^i instead of ξ^i. In particular, if the Cartesian coordinates are used as the observer's reference frame, we have simply

$$e_{ij} = \frac{1}{2}\left(\frac{\partial u_j}{\partial x^i} + \frac{\partial u_i}{\partial x^j}\right).$$

Covariant components of the rate-of-strain tensor in some orthogonal curvilinear systems are given in Appendix B.

In order to present the rate-of-strain tensor in an invariant form we have to use the convolution of e_{ij} with the corresponding basis vectors:

$$\mathbf{E} = e_{ij}\mathbf{e}^i\mathbf{e}^j = \tfrac{1}{2}(\nabla_i u_j + \nabla_j u_i)\mathbf{e}^i\mathbf{e}^j = e_{ij}g^{ik}g^{jm}\mathbf{e}_k\mathbf{e}_m$$

$$= \tfrac{1}{2}(\nabla^i u^j + \nabla^j u^i)\mathbf{e}_i\mathbf{e}_j.$$

A convenient notation for the rate-of-strain tensor is

$$\mathbf{E} = \tfrac{1}{2}[\nabla\mathbf{u} + (\nabla\mathbf{u})^*], \tag{2.56}$$

where $\nabla\mathbf{u} = \nabla^i u^j \mathbf{e}_i\mathbf{e}_j$ and $(\nabla\mathbf{u})^* = \nabla^j u^i \mathbf{e}_i\mathbf{e}_j$.

Intuitively, it was clear from the very beginning that the rate of deformation has to be somehow related with the velocity gradient. However, it is necessary to distinguish between the components of the velocity gradient associated with the rigid-body rotation of the fluid and those responsible for its genuine deformation. Let us show that it is the rate-of-strain tensor (2.55) which describes (the rate of) genuine deformation of a fluid element. Indeed, the Taylor expansion of the velocity field about an arbitrarily chosen point $\mathbf{r}_0$ in the moving continuum gives that for an infinitesimally close point $\mathbf{r}_1$ we have

$$\mathbf{u}_1 = \mathbf{u}_0 + (\nabla_i u_j)_0 x^i \mathbf{e}^j + O(|\mathbf{r}|^2)$$

$$= \mathbf{u}_0 + \tfrac{1}{2}(\nabla_i u_j + \nabla_j u_i)_0 x^i \mathbf{e}^j + \tfrac{1}{2}(\nabla_i u_j - \nabla_j u_i)_0 x^i \mathbf{e}^j + O(|\mathbf{r}|^2)$$

$$= \mathbf{u}_0 + (e_{ij})_0 x^i \mathbf{e}^j + (\omega_{ij})_0 x^i \mathbf{e}^j + O(|\mathbf{r}|^2) \qquad \text{as } |\mathbf{r}| \to 0, \tag{2.57}$$

where $\mathbf{r} \equiv \mathbf{r}_1 - \mathbf{r}_0 = x^i \mathbf{e}_i$. Then, after introducing a scalar function $\Phi_0 = \tfrac{1}{2}(e_{ij})_0 x^i x^j$ and a vector $\boldsymbol{\omega}_0 = \tfrac{1}{2}(\nabla \times \mathbf{u})_0$, to leading order as $|\mathbf{r}| \to 0$, we can rewrite (2.57) as

$$\mathbf{u}_1 = \mathbf{u}_0 + \nabla\Phi_0 + (\boldsymbol{\omega}_0 \times \mathbf{r}). \tag{2.58}$$

This equation differs from Euler's formula for the distribution of velocities in a rigid body known in classical mechanics,

$$\mathbf{u}_1 = \mathbf{u}_0 + (\boldsymbol{\omega}_0 \times \mathbf{r}), \tag{2.59}$$

by the second term on the right-hand side, which therefore describes the departure from rigidity.

Using (2.58), one can show that for an elementary material segment $\mathbf{r} = x^i \mathbf{e}_i$ the rate of relative elongation is given by

$$e_r \equiv \frac{1}{|\mathbf{r}|}\frac{d|\mathbf{r}|}{dt} = \frac{1}{2|\mathbf{r}|^2}\frac{d|\mathbf{r}|^2}{dt} = \frac{1}{|\mathbf{r}|^2}\left(\mathbf{r}\cdot\frac{d\mathbf{r}}{dt}\right) = \frac{1}{|\mathbf{r}|^2}(\mathbf{r}\cdot\nabla\Phi_0)$$

$$= \frac{2\Phi_0}{|\mathbf{r}|^2} = e_{ij}\frac{x^i}{|\mathbf{r}|}\frac{x^j}{|\mathbf{r}|}.$$

In particular, in a Cartesian coordinate frame of the observer the diagonal components of $\{e_{ij}\}$ describe the rate of relative elongation of elementary material segments initially parallel to coordinate axes.

Similarly, it can be shown that the rate of relative change of an elementary material volume is equal to the first invariant of the rate-of-strain tensor,

$$\lim_{\substack{\delta t \to 0 \\ V(t_0) \to 0}} \frac{V(t_0 + \delta t) - V(t_0)}{V(t_0)\delta t} = e_i^i = \nabla_i u^i \equiv \nabla \cdot \mathbf{u},$$

and the condition of incompressibility (2.15) in terms of the theory of deformation means that the first invariant of the rate-of-strain tensor, $I_1 = \mathbf{I} : \mathbf{E} = e_i^i$, is identically zero.

2.2.6 Viscous incompressible liquid

In real fluids, the momentum transport due to molecular motion leads to what can be macroscopically described as internal friction between fluid elements moving with respect to one another. In order to take this effect into account in a mathematical model, it is convenient to present the stress tensor as a sum of two components

$$\mathbf{P} = -p\mathbf{I} + \boldsymbol{\sigma}, \tag{2.60}$$

where $\boldsymbol{\sigma}$ describes internal friction. The assumptions one has to make about $\boldsymbol{\sigma}$, which is known as the viscous stress, will specify the fluid's rheology, the order and type of the resulting equations of motion and ultimately the way in which we formulate problems to describe the fluid's motion and dynamic state.

In a general case, $\boldsymbol{\sigma}$ can depend on $\mathcal{E}'$ (or $\mathcal{E}''$), $\mathbf{E}$, the history of deformation of fluid elements as well as on scalar, vector and tensor characteristics specifying physical properties of the fluid to be described. Then, one will have to address the fundamental issues associated with the very possibility of obtaining a closed set of equations describing macroscopic behaviour of the fluid without invoking its dynamics on a microscopic level. We will leave all these issues aside and consider the simplest case assuming that $\boldsymbol{\sigma}$ at every moment t depends only on the local rate of deformation of fluid elements at this moment. The quantitative measure of the rate of deformation is the rate-of-strain tensor $\mathbf{E}$, and among different possible dependences of $\boldsymbol{\sigma}$ on $\mathbf{E}$ (and obviously the metric tensor $\mathbf{I}$) we will again look at the simplest case assuming that $\boldsymbol{\sigma}$ is a linear function of $\mathbf{E}$:

$$\sigma^{ij} = A^{ij\alpha\beta} e_{\alpha\beta}. \tag{2.61}$$

The coefficients $\{A^{ij\alpha\beta}\}$, which are obviously components of a tensor of the fourth rank, depend on material properties of the fluid and, possibly, its local state, but do not explicitly depend on the velocity distribution. The right-hand side of (2.61) can be seen as the first term of a Taylor expansion of

σ where all components of the rate-of-strain tensor are small in some sense. Qualitatively, the meaning of this argument becomes clear if we look at the motion at a microscopic level. The dimension of the rate of strain is $1/\text{time}$ and the macroscopic motion in the bulk can be associated with a certain characteristic time scale, τ_M. On the other hand, the molecular motion is also associated with some characteristic time scale, τ_m, which can be estimated as the time required for molecular interactions to establish statistical equilibrium at the microscopic level. If $\tau_m/\tau_M \ll 1$, the macroscopic motion is only a small perturbation of the microscopic state and, to leading order, the viscous stress generated by this motion is proportional to the rate of strain.[8] However, in order to make a quantitative theoretical assessment of the limitations of macroscopic models which follow from (2.61) when this equation is seen as the leading term in a Taylor expansion, one has to examine the underlying microscopic (statistical) model. This is possible, though also with a number of additional assumptions, only in a few limiting cases. Therefore, here we will treat (2.61) as a macroscopic hypothesis whose limits of applicability have to be determined experimentally.

Due to symmetry of σ^{ij} and $e_{\alpha\beta}$, the tensor $A^{ij\alpha\beta}$ is also symmetric in the indices i, j and α, β, and hence among its 81 components at most only 36 are independent. To further reduce the number of independent components of $A^{ij\alpha\beta}$ or, more precisely, to establish correlations between them, we have to consider symmetries inherent in the physical properties of the fluid. Mathematically, symmetries of physical properties mean invariance of components of the tensors describing these properties with respect to certain groups of transformations of the coordinate system. We will consider the case of an isotropic fluid and define isotropy as invariance of the physical properties with respect to translations, rotations and mirror reflections of the coordinate system (orthogonal group).

To consider the implications for $A^{ij\alpha\beta}$ in the case of an isotropic fluid, it is convenient to use a Cartesian coordinate system with the axes directed along the principal axes of $e_{\alpha\beta}$. In these coordinates, $e_{\alpha\beta} = 0$ for $\alpha \neq \beta$ and hence in (2.61) one has to consider only the coefficients $A^{ij\alpha\alpha}$. After a mirror reflection of the coordinate system with respect to a coordinate plane which changes the direction of the ith axes to the opposite, that is $(x^1, x^2, x^3) \mapsto (y^1, y^2, y^3)$, where $y^j = x^i$ for $j \neq i$ and $y^i = -x^i$, we have

$$A'^{ij\alpha\alpha} = A^{klmn} \frac{\partial y^i}{\partial x^k} \frac{\partial y^j}{\partial x^l} \frac{\partial y^\alpha}{\partial x^m} \frac{\partial y^\alpha}{\partial x^n} = -A^{ij\alpha\alpha} \qquad \text{for } \alpha, j = 1, 2, 3, \ j \neq i,$$

which for an isotropic fluid means that $A^{ij\alpha\alpha} = 0$ for $i \neq j$. Thus, in the

---

[8]This argument also gives the key to understanding why for simple fluids the 7-dimensional space used in the kinetic theory can be successfully replaced by a 4-dimensional space of continuum mechanics, whilst for complex fluids continuum mechanics fails to produce a universal mathematical model applicable to all types of flow of a given medium and therefore one has to consider different models for different types of motion instead.

Cartesian frame coinciding with the principal axes of the rate-of-strain tensor only the coefficients A^{iijj} and A^{iiii} can be nonzero. Due to isotropy, each of these two types of coefficients has the same value, and for them we will use the notations $A^{iijj} = \lambda$ and $A^{iiii} = 2\mu + \lambda$. Now, equations (2.61) become

$$\sigma^{ii} = \lambda(e_{11} + e_{22} + e_{33}) + 2\mu e_{ii} \qquad \text{for } i = 1, 2, 3. \qquad (2.62)$$

Given that in Cartesian coordinates $g^{ij} = g_{ij} = \delta^i_j$, where δ^i_j is the Kronecker symbol, and noticing that the sum in brackets is the first invariant of the rate-of-strain tensor, $I_1 = e^\alpha_\alpha = \nabla \cdot \mathbf{u}$, we can now rewrite (2.62) in an invariant form valid for all coordinate systems:

$$\sigma^{ij} = \lambda e^\alpha_\alpha g^{ij} + 2\mu g^{i\alpha} g^{j\beta} e_{\alpha\beta}, \qquad \sigma_{ij} = \lambda e^\alpha_\alpha g_{ij} + 2\mu e_{ij}. \qquad (2.63)$$

The first of these equations immediately gives an expression for the coefficients $A^{ij\alpha\beta}$, which, as one should expect, in the case of an isotropic fluid are functions of components of the metric tensor only:

$$A^{ij\alpha\beta} = \lambda g^{ij} g^{\alpha\beta} + \mu(g^{i\alpha} g^{j\beta} + g^{i\beta} g^{j\alpha}).$$

Thus, equation (2.60) now takes a specific form

$$\mathbf{P} = -p\mathbf{I} + \lambda e^\alpha_\alpha \mathbf{I} + 2\mu \mathbf{E},$$

or

$$\mathbf{P} = -p\mathbf{I} + \lambda \mathbf{I}(\nabla \cdot \mathbf{u}) + \mu[\nabla\mathbf{u} + (\nabla\mathbf{u})^*]. \qquad (2.64)$$

Fluids with rheological behaviour described by this equation are referred to as Newtonian.

The coefficient μ is called dynamic viscosity and instead of λ it is convenient to introduce the second viscosity ζ by

$$\zeta = \lambda + \tfrac{2}{3}\mu.$$

Then,

$$\boldsymbol{\sigma} = (\zeta - \tfrac{2}{3}\mu)e^\alpha_\alpha \mathbf{I} + 2\mu \mathbf{E} = \zeta\mathbf{I}(\nabla \cdot \mathbf{u}) + \mu[\nabla\mathbf{u} + (\nabla\mathbf{u})^* - \tfrac{2}{3}\mathbf{I}(\nabla \cdot \mathbf{u})] \qquad (2.65)$$

and the first invariant of the viscous stress will now depend only on ζ:

$$\sigma^\alpha_\alpha = 3\zeta(\nabla \cdot \mathbf{u}),$$

while μ will be associated with the deviatoric stress. As will be shown in §2.3.2, the second law of thermodynamics requires that $\zeta > 0$, $\mu > 0$.

Thus, we finally have the stress tensor in the form

$$\mathbf{P} = -p\mathbf{I} + \zeta\mathbf{I}(\nabla \cdot \mathbf{u}) + \mu[\nabla\mathbf{u} + (\nabla\mathbf{u})^* - \tfrac{2}{3}\mathbf{I}(\nabla \cdot \mathbf{u})] \qquad (2.66)$$

and for an incompressible liquid it becomes simply

$$\mathbf{P} = -p\mathbf{I} + \mu[\nabla\mathbf{u} + (\nabla\mathbf{u})^*]. \tag{2.67}$$

In particular, in a Cartesian coordinate frame one has

$$p_{ij} = -p\delta_{ij} + \mu\left(\frac{\partial u_i}{\partial x^j} + \frac{\partial u_j}{\partial x^i}\right).$$

For $i \neq j$ this gives a relationship for the tangential stress known as the Navier law (Navier 1823).

Coefficients μ and ζ are material parameters characterizing particular Newtonian fluids and have to be determined experimentally. They are independent of velocity but, in a general case, can be functions of the parameters determining the local state of the fluid, such as pressure and temperature. Therefore, after substituting (2.64) into (2.31), we have

$$\rho\left(\frac{\partial\mathbf{u}}{\partial t} + \mathbf{u}\cdot\nabla\mathbf{u}\right) = -\nabla p + \nabla[(\zeta - \tfrac{2}{3}\mu)(\nabla\cdot\mathbf{u})] + \nabla\cdot\{\mu[\nabla\mathbf{u} + (\nabla\mathbf{u})^*]\} + \rho\mathbf{F},$$
$$\tag{2.68}$$

where the coefficients cannot be taken out of the sign of covariant differentiation. In most cases, however, variations in μ and ζ caused by the flow and external influences are negligible, and they can be regarded as material constants. Then, equation (2.68) becomes

$$\rho\left(\frac{\partial\mathbf{u}}{\partial t} + \mathbf{u}\cdot\nabla\mathbf{u}\right) = -\nabla p + (\zeta + \tfrac{1}{3}\mu)\nabla(\nabla\cdot\mathbf{u}) + \mu\Delta\mathbf{u} + \rho\mathbf{F}, \tag{2.69}$$

which is known as the Navier-Stokes equation. As in the case of an ideal compressible fluid, in order to close the set of equations (2.13), (2.69) one has either to consider the conservation of energy equation and related thermodynamic issues, or, if the process is known (or assumed) to be barotropic, use the appropriate relationship between p and ρ.

For an incompressible fluid, the mass and momentum balance equations form a closed set of equations

$$\nabla\cdot\mathbf{u} = 0, \tag{2.70}$$

$$\frac{\partial\mathbf{u}}{\partial t} + \mathbf{u}\cdot\nabla\mathbf{u} = -\frac{1}{\rho}\nabla p + \nu\Delta\mathbf{u} + \mathbf{F}, \tag{2.71}$$

where $\nu = \mu/\rho$ is known as the kinematic viscosity. These equations describe the fluid as a mechanical system and allow one to determine p and $\mathbf{u}$ independently of thermodynamic parameters (such as temperature) which can be found, if necessary, *a posteriori* with the known dynamics of the liquid as a background. Equations (2.70), (2.71) will be central to our consideration of free-surface flows. They are often referred to collectively as the Navier-Stokes equations for viscous incompressible fluid. The form these equations take in some orthogonal curvilinear coordinate systems is given in Appendix B. Values of μ and ν for some liquids are given in Appendix D. More information about the viscosity of liquids can be found in Viswanath et al. (2007).

Equations (2.70), (2.71) for an incompressible liquid has been first published by Navier (1823), who in his derivation used some assumptions about the microscopic structure of the fluid. The corresponding equations for a compressible fluid were derived six years later by Poisson (1829). These equations were derived again in a different way by Saint-Venant (1843) and, finally, by Stokes (1845) essentially in the way we derive them today.

2.2.6.1 Alternative forms of the Navier-Stokes equation

In some cases, it is convenient to use alternative forms of (2.70), (2.71) which we will now consider.

Using (2.70), one can rewrite (2.71) in the form of a conservation law:

$$\frac{\partial(\rho\mathbf{u})}{\partial t} + \nabla \cdot (\rho\mathbf{uu} - \mathbf{P}) = \rho\mathbf{F}. \tag{2.72}$$

This equation states that the local variation of momentum per unit volume results from the divergence of the momentum flux and the body force acting on the volume. The tensor

$$\mathbf{\Pi} = \rho\mathbf{uu} - \mathbf{P} \tag{2.73}$$

is known as the *momentum flux tensor*. Its physical meaning is as follows: if $\mathbf{n}$ is a unit vector, then the vector $\mathbf{n} \cdot \mathbf{\Pi}$ is the flux of momentum across a unit area of the surface normal to $\mathbf{n}$.

In the case of a viscous incompressible liquid, equation (2.72) becomes

$$\frac{\partial(\rho\mathbf{u})}{\partial t} + \nabla \cdot \{p\mathbf{I} + \rho\mathbf{uu} - \mu[\nabla\mathbf{u} + (\nabla\mathbf{u})^*]\} = \rho\mathbf{F}, \tag{2.74}$$

and the momentum flux tensor has the form

$$\mathbf{\Pi} = p\mathbf{I} + \rho\mathbf{uu} - \mu[\nabla\mathbf{u} + (\nabla\mathbf{u})^*]. \tag{2.75}$$

Equations (2.72), (2.74) are useful in the analysis of conditions on surfaces of discontinuity and in designing numerical algorithms.

Using that

$$\mathbf{u} \cdot \nabla\mathbf{u} = \nabla\frac{|\mathbf{u}|^2}{2} + 2(\boldsymbol{\omega} \times \mathbf{u}), \qquad \boldsymbol{\omega} = \tfrac{1}{2}\nabla \times \mathbf{u},$$

representing the Laplacian of velocity as

$$\Delta\mathbf{u} = \nabla(\nabla \cdot \mathbf{u}) - 2\nabla \times \boldsymbol{\omega},$$

and taking into account the continuity equation $\nabla \cdot \mathbf{u} = 0$, one can rewrite (2.71) in Gromeka-Lamb's form

$$\frac{\partial \mathbf{u}}{\partial t} + \nabla \frac{|\mathbf{u}|^2}{2} + 2(\boldsymbol{\omega} \times \mathbf{u}) = \mathbf{F} - \frac{1}{\rho}\nabla p - 2\nu(\nabla \times \boldsymbol{\omega}).$$

If the body forces are potential, $\mathbf{F} = -\nabla\Pi$, equations (2.71), (2.70) take the form

$$\frac{\partial \mathbf{u}}{\partial t} + 2(\boldsymbol{\omega} \times \mathbf{u}) + \nabla H + 2\nu(\nabla \times \boldsymbol{\omega}) = 0, \tag{2.76}$$

$$\nabla \cdot \mathbf{u} = 0,$$

$$\nabla \times \mathbf{u} = 2\boldsymbol{\omega},$$

$$H = \frac{|\mathbf{u}|^2}{2} + \frac{p}{\rho} + \Pi.$$

It may seem surprising that the viscous term in (2.76) is associated with vorticity, whilst it has been shown by comparing (2.58) and (2.59) that vorticity is not what deforms a fluid particle and hence is not related with the stress tensor. The explanation, however, is that the equation of motion (2.32) includes *gradients* of the stress tensor, that is, in the case of a Newtonian fluid, gradients of the rate-of-strain tensor which are also gradients of the vorticity involved in (2.76).

2.2.6.2   Viscous incompressible fluid in the Lagrangian description

For completeness, consider also the form which the Navier-Stokes equations take if we use the Lagrangian specification of motion (Monin 1962). Choosing the Cartesian coordinates x, y, z as the observer's reference frame and the coordinates of material point at some initial moment as the Lagrangian variables ξ^1, ξ^2, ξ^3, we can write down the continuity equation (2.70) in the form of (2.20), that is

$$\frac{\partial(x, y, z)}{\partial(\xi^1, \xi^2, \xi^3)} = 1. \tag{2.77}$$

The acceleration of a material particle on the left-hand side of (2.71) and its velocity on the right-hand side are given in the Lagrangian description by (2.6), that is

$$\frac{\partial^2 x}{\partial t^2}, \quad \frac{\partial^2 y}{\partial t^2}, \quad \frac{\partial^2 z}{\partial t^2} \quad \text{and} \quad \frac{\partial x}{\partial t}, \quad \frac{\partial y}{\partial t}, \quad \frac{\partial z}{\partial t},$$

respectively. To express components of the gradient and Laplacian in the equation of motion (2.71) in the Lagrangian variables, we will, first, notice that for an arbitrary function f the partial derivatives with respect to the Eulerian coordinates can be written down in terms of the corresponding Jacobians:

$$\frac{\partial f}{\partial x} = \frac{\partial(f, y, z)}{\partial(x, y, z)}, \quad \frac{\partial f}{\partial y} = \frac{\partial(x, f, z)}{\partial(x, y, z)}, \quad \frac{\partial f}{\partial z} = \frac{\partial(x, y, f)}{\partial(x, y, z)}.$$

Then, using properties of Jacobians and equation (2.77), one has

$$\frac{\partial(A, B, C)}{\partial(x, y, z)} = \frac{\partial(A, B, C)}{\partial(\xi^1, \xi^2, \xi^3)} \frac{\partial(\xi^1, \xi^2, \xi^3)}{\partial(x, y, z)} = \frac{\partial(A, B, C)}{\partial(\xi^1, \xi^2, \xi^3)}.$$

Introducing the notation

$$\frac{\partial(A, B, C)}{\partial(\xi^1, \xi^2, \xi^3)} = [A, B, C],$$

we now have

$$\frac{\partial f}{\partial x} = [f, y, z], \quad \frac{\partial f}{\partial y} = [x, f, z], \quad \frac{\partial f}{\partial z} = [x, y, f],$$

$$\Delta f = [[f, y, z], y, z] + [x, [x, f, z], z] + [x, y, [x, y, f]], \tag{2.78}$$

so that the projections of equation (2.71) on the coordinate axes can be written as

$$\frac{\partial^2 x}{\partial t^2} = -\frac{1}{\rho}[p, y, z] + \nu\Delta\frac{\partial x}{\partial t} + F_x, \tag{2.79}$$

$$\frac{\partial^2 y}{\partial t^2} = -\frac{1}{\rho}[x, p, z] + \nu\Delta\frac{\partial y}{\partial t} + F_y, \tag{2.80}$$

$$\frac{\partial^2 z}{\partial t^2} = -\frac{1}{\rho}[x, y, p] + \nu\Delta\frac{\partial z}{\partial t} + F_z, \tag{2.81}$$

where the operator Δ is given by (2.78). If $\nu = 0$, we will have another form of equations of motion for an ideal incompressible fluid in the Lagrangian variables. A review of solutions for the Euler equations and another form of the Navier-Stokes equations in the Lagrangian description can be found in Abrashkin & Yakubovich (2006).

2.2.6.3   Slow viscous flows: The Stokes equations

Consider a particular class of flows where the Navier-Stokes equations (2.70), (2.71) can be considerably simplified. For brevity we will neglect the body forces: the corresponding criterion can be easily formulated and, if it is not satisfied, the body forces can be brought back into the picture in a straightforward way.

Let L, U, $T = L/U$ and $\mu U/L$ be characteristic scales for length, velocity, time and pressure. Then the dimensionless form of (2.70), (2.71) is given by

$$\nabla \cdot \mathbf{u} = 0, \qquad Re\left(\frac{\partial \mathbf{u}}{\partial t} + \mathbf{u} \cdot \nabla \mathbf{u}\right) = -\nabla p + \Delta \mathbf{u},$$

where $Re = \rho L U/\mu$ is the Reynolds number.[9] For fluid motion where $Re \ll 1$, e.g., for microhydrodynamic flows associated with small characteristic length

---

[9]The rather curious history of this number is described by Rott (1990).

scale, one may consider an asymptotic limit $Re \to 0$ and in this limit, to leading order in Re, the flow parameters satisfy a set of equations

$$\nabla \cdot \mathbf{u} = 0, \qquad \nabla p = \Delta \mathbf{u}, \tag{2.82}$$

known as the Stokes equations. The corresponding fluid motion is usually referred to as the Stokes flow or creeping flow. Equations (2.82) play an important role in a great number of applications and, due to their linearity and the parametric time-dependence, are much easier to handle than (2.70), (2.71). A detailed exposition of the main results obtained for (2.82) can be found, for example, in Langlois (1964) and Happel & Brenner (1965).

Elimination of $\mathbf{u}$ from (2.82) yields

$$\Delta p = 0, \tag{2.83}$$

and by applying $\nabla \times$ to the second equation of (2.82) one obtains

$$\Delta \boldsymbol{\omega} = 0, \tag{2.84}$$

where, as before,

$$\boldsymbol{\omega} = \tfrac{1}{2} \nabla \times \mathbf{u}. \tag{2.85}$$

For plane two-dimensional flows the Stokes equations admit a number of useful representations where one can draw parallels with, and use results from, the linear theory of elasticity. Below, we consider some of them.

Let $Oxyz$ be the Cartesian coordinate frame with the xy-plane parallel to the plane of flow. Then, the continuity equation in (2.82) takes the form (2.16), i.e.,

$$\frac{\partial u_x}{\partial x} + \frac{\partial u_y}{\partial y} = 0, \tag{2.86}$$

and hence one can introduce the stream function (2.17) so that (2.86) is satisfied identically and the velocity components are given by (2.18), i.e.,

$$u_x = \frac{\partial \psi}{\partial y}, \qquad u_y = -\frac{\partial \psi}{\partial x}. \tag{2.87}$$

The vorticity vector $\boldsymbol{\omega}$ now has only the component normal to the plane of flow, $(0, 0, \omega)$, and (2.84) becomes the scalar Laplace equation

$$\Delta \omega = 0 \tag{2.88}$$

(hereafter in this section $\Delta = \partial^2/\partial x^2 + \partial/\partial y^2$). The expression (2.85) for ω now having the form

$$2\omega = \frac{\partial u_y}{\partial x} - \frac{\partial u_x}{\partial y},$$

can be combined with (2.87) to give

$$\Delta \psi = -2\omega. \tag{2.89}$$

In many situations, the (ψ, ω)-representation (2.88), (2.89) is more convenient than equations (2.82) in the primitive variables. In particular, it is often used in numerical methods, such as the boundary integral equation method.

Finally, elimination of ω from (2.88), (2.89) reduces the Stokes equations to one biharmonic equation for the stream function

$$\Delta^2 \psi = 0. \tag{2.90}$$

The absence of inertial terms in (2.82) allows one to use another convenient representation for the plane two-dimensional Stokes equations originating from the linear theory of elasticity (Muskhelishvili 1963). In the Cartesian coordinates, the second equation in (2.82) can be written down as

$$0 = \frac{\partial p_{xx}}{\partial x} + \frac{\partial p_{xy}}{\partial y}, \quad 0 = \frac{\partial p_{yx}}{\partial x} + \frac{\partial p_{yy}}{\partial y}, \tag{2.91}$$

where

$$p_{xx} = -p + 2\frac{\partial u_x}{\partial x}, \quad p_{yy} = -p + 2\frac{\partial u_y}{\partial y}, \tag{2.92}$$

$$p_{xy} = p_{yx} = \frac{\partial u_x}{\partial y} + \frac{\partial u_y}{\partial x}.$$

Each of equations (2.91) allows one to introduce a function that makes this equation satisfied identically, similar to how the stream function ψ was introduced to satisfy (2.86). Let F_1 and F_2 be defined by

$$p_{xx} = \frac{\partial F_1}{\partial y}, \quad p_{xy} = -\frac{\partial F_1}{\partial x}; \quad p_{yy} = \frac{\partial F_2}{\partial x}, \quad p_{yx} = -\frac{\partial F_2}{\partial y}.$$

Then, given that $p_{xy} = p_{yx}$, one has

$$\frac{\partial F_1}{\partial x} = \frac{\partial F_2}{\partial y},$$

which is the necessary and sufficient condition for the existence of a function W defined by

$$F_1 = \frac{\partial W}{\partial y}, \quad F_2 = \frac{\partial W}{\partial x}.$$

Then, components of the stress tensor can be expressed as

$$p_{xx} = \frac{\partial^2 W}{\partial y^2}, \quad p_{xy} = -\frac{\partial^2 W}{\partial x \partial y}, \quad p_{yy} = \frac{\partial^2 W}{\partial x^2}, \tag{2.93}$$

and equations (2.91) become identically satisfied.

The function W is known as the *Airy stress function* (Airy 1862, 1863). By combining (2.86), (2.92) and (2.93) p can be expressed in terms of W,

$$p = -\tfrac{1}{2}(p_{xx} + p_{yy}) = -\tfrac{1}{2}\Delta W,$$

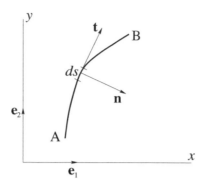

FIGURE 2.8:   Sketch establishing notation for the derivation of the expression for the stress acting on a boundary in plane two-dimensional Stokes flows. The unit normal **n** points into the fluid exerting the stress and the tangent **t** indicates the direction of parametrisation of AB.

and, given that, according to (2.83), p is harmonic, one arrives at the biharmonic equation for the Airy function

$$\Delta^2 W = 0. \tag{2.94}$$

The Airy function is widely used in the linear theory of elasticity (Muskhelishvili 1963) where formulation of problems in terms of stresses is one of the key elements of modelling. As we will see in Chapter 6, the Airy function can also be applied to the dynamics of free-surface flows where its use allows one to obtain some general analytic results.

A useful analytic tool in dealing with plane two-dimensional Stokes flows is the theory of analytic functions of the complex variable which uses the linearity of these flows and biharmonicity of ψ and W. Consider the complex representation of the stress acting on a line in the plane of flow in terms of the Airy stress function (Muskhelishvili 1963, p. 113). Let $z = x + iy$ be the complex variable defining the plane of flow and AB be an oriented line in this plane parametrised by the arclength s with the tangent vector $\mathbf{t} = (t_x, t_y)$ showing the direction of parametrisation and the normal $\mathbf{n} = (n_x, n_y)$ pointing into the fluid exerting the stress (Fig. 2.8):

$$z = z(s) \equiv x(s) + iy(s); \qquad t_x = x'(s) = -n_y, \quad t_y = y'(s) = n_x. \tag{2.95}$$

The stress per unit length of the line is given by (2.30), i.e., in our case

$$\mathbf{p}(\mathbf{n}) = (\mathbf{n} \cdot \mathbf{P} \cdot \mathbf{e}_1)\mathbf{e}_1 + (\mathbf{n} \cdot \mathbf{P} \cdot \mathbf{e}_2)\mathbf{e}_2,$$

and for the two projections one has

$$\mathbf{n} \cdot \mathbf{P} \cdot \mathbf{e}_1 \equiv X_n = p_{xx}n_x + p_{xy}n_y, \quad \mathbf{n} \cdot \mathbf{P} \cdot \mathbf{e}_2 \equiv Y_n = p_{xy}n_x + p_{yy}n_y.$$

Using (2.93) and (2.95) the above expressions can be written down as

$$X_n = \frac{\partial^2 W}{\partial y^2}\frac{dy}{ds} + \frac{\partial^2 W}{\partial x \partial y}\frac{dx}{ds} = \frac{d}{ds}\left(\frac{\partial W}{\partial y}\right),$$

$$Y_n = -\frac{\partial^2 W}{\partial x \partial y}\frac{y}{ds} - \frac{\partial^2 W}{\partial x^2}\frac{dx}{ds} = -\frac{d}{ds}\left(\frac{\partial W}{\partial x}\right),$$

or in a complex form

$$X_n + iY_n = \frac{d}{ds}\left(\frac{\partial W}{\partial y} - i\frac{\partial W}{\partial x}\right) = -i\frac{d}{ds}\left(\frac{\partial W}{\partial x} + i\frac{\partial W}{\partial y}\right).$$

The stress acting on an element ds of the line is then given by

$$(X_n + iY_n)ds = -id\left(\frac{\partial W}{\partial x} + i\frac{\partial W}{\partial y}\right). \qquad (2.96)$$

It can be shown (see Appendix C) that every biharmonic function, in our case the stream function ψ and the Airy stress function W, can be represented either as the real or the imaginary part of an expression

$$\bar{z}A_1(z) + A_2(z),$$

where A_1 and A_2 are analytic functions and the overbar denotes the complex conjugate.

Let

$$W = \text{Re}\{\bar{z}A_1(z) + A_2(z)\}$$

or, equivalently,

$$W = \tfrac{1}{2}[\bar{z}A_1(z) + z\overline{A_1(z)} + A_2(z) + \overline{A_2(z)}].$$

Then, one can verify that the representation (2.96) takes the form

$$(X_n + iY_n)\, ds = -id\{A_1(z) + z\overline{A_1'(z)} + \overline{A_2'(z)}\}. \qquad (2.97)$$

Similarly, if W is represented as

$$W = \text{Im}\{\bar{z}A_1(z) + A_2(z)\},$$

then

$$(X_n + iY_n)\, ds = -d\{A_1(z) - z\overline{A_1'(z)} - \overline{A_2'(z)}\}. \qquad (2.98)$$

The expressions (2.97) and (2.98) are useful in formulating boundary conditions at free surfaces.

2.3   Elements of thermodynamics

2.3.1   General concepts

Equations (2.70), (2.71), which we will use to describe free surface flow, were formulated without specifying thermodynamic properties of the fluid. However, in our subsequent analysis of interfaces we will need to make use of thermodynamics of irreversible processes, and it is appropriate to recapitulate its basic elements and their application to moving fluids. A detailed exposition of various aspects of irreversible thermodynamics, including its application to multicomponent systems, can be found in a number of monographs (e.g., de Groot & Mazur 1962, Prigogine 1967) and remains beyond our goal here.

If, as was once remarked,[10] "analogy and experience are the two crutches we rely on on the road of reasoning", then classical thermodynamics is perhaps the science which makes the most of these two elements by accumulating and formalizing experimental observations into a few universal laws and general concepts that provide a reliable framework for modelling even when applied to a completely new and untested thermodynamic system. The subtleties are in the detail, that is in the way thermodynamic concepts and relationships are specified for this particular system.

The starting point of a thermodynamic analysis is a postulate that every closed thermodynamic system will evolve towards the state of a thermodynamic equilibrium. In equilibrium, the state of the system can be completely characterized by a finite number of independent parameters, called the parameters of state, with other parameters related to them through a set of equations called the equations of state. In thermodynamics of irreversible processes, the notion of parameters of state was extended to systems which are not in equilibrium and whose state can be specified by a greater number of parameters than in equilibrium. Once the equations of state and constitutive equations describing how the parameters relax towards their equilibrium values are known, one has a complete macroscopic description of the system. Thermodynamics suggests how to introduce parameters which can be used as parameters of state and its laws impose constraints on the form these equations may take. However, when applied to a new system, these are only plausible hypotheses which have to be verified experimentally. We will, first, describe the general notions of thermodynamics and then specify them for a particular case of a two-parametric fluid. The subscript m will refer to a system with a fixed mass or, in the case of a fluid, to parameters specified per unit mass; subscript v will mark quantities specified per unit volume.

---

[10]By Frederick II of Prussia (Khoromin 1918).

The first law of thermodynamics postulates conservation of energy.[11] Using slightly more precise terms, it states that for every thermodynamic system there exists a single-valued function E_m of parameters of state and velocity, called the total energy, whose variation is the result of work done on the system by external forces $\delta W^{(e)}$ and heat $\delta Q^{(e)}$ added to it from outside[12]

$$dE_m = \delta W^{(e)} + \delta Q^{(e)}. \tag{2.99}$$

It is important to emphasize that on the left-hand side of this equation we have a differential of a (single-valued) function whilst the two terms on the right-hand side are just infinitesimal quantities. By definition, the total energy is the sum

$$E_m = K_m + U_m \tag{2.100}$$

of kinetic energy K_m, which accounts for macroscopic motion, and internal energy U_m, which characterizes a thermodynamic state of the system and depends on the parameters of state. The kinetic energy is introduced via the equation of motion and, as in classical mechanics, its variation is equal to work done by both external and internal forces

$$dK_m = \delta W^{(e)} + \delta W^{(i)}. \tag{2.101}$$

Subtracting this equation from (2.99) and using (2.100), one arrives at

$$dU_m = \delta Q^{(e)} - \delta W^{(i)}, \tag{2.102}$$

which shows that the internal energy can be increased by adding heat to the system, and it decreases due to work done by internal forces. For slow processes where the kinetic energy variation can be neglected, $dK_m = 0$, one has from (2.101) that

$$\delta W^{(i)} = -\delta W^{(e)}$$

and hence

$$dU_m = \delta Q^{(e)} + \delta W^{(e)}.$$

This equation provides the framework for determining U_m from experiments where one can (slowly) vary external influences and monitor the response of the system.

Equation (2.99) gives the energy balance but does not indicate in which direction the system will evolve. In order to consider this question, we need the

---

[11] In what follows, we consider the mathematical formulation of thermodynamic laws thus skipping the now historic arguments related to the impossibility of a perpetuum mobile of the first and second kind.

[12] For simplicity, here we are not considering other sources of energy apart from mechanical work and heat (e.g., radiation). If such sources are important for the process under consideration, infinitesimal contributions due to these sources must be added to the right-hand side of (2.99).

notions of reversible and irreversible thermodynamic processes. The postulate that every closed macroscopic system evolves towards the state of a thermo-dynamic equilibrium, which we used as our starting point, in fact postulates irreversibility of all thermodynamic processes since it states that, under given conditions, there is a preferential direction of evolution. The notion of re-versible processes is an idealization which we use to set up a framework of thermodynamic concepts. The process is called reversible if an infinitesimal variation of parameters of the system can be reversed by changing the signs of the (infinitesimal) external influences. It is important that, as follows from this definition, reversibility is associated only with the 'leading orders' of in-finitesimal variations of thermodynamic quantities. This idealized concept is useful on two counts: (a) it allows one to introduce notions convenient for the description of thermodynamic systems (often in the way involving other ideal-ized notions like, for example, that of an infinitesimal fluid particle), and (b) it gives the guidelines for experiments aimed at obtaining relationships based on this concept. In many cases (though not always, see for example Sedov 1997), very slow 'equilibrium processes' in which the system passes through a continuous succession of equilibrium states can in practice be regarded as reversible.

The second law of thermodynamics states that for every thermodynamic system there exist a function S_m of the parameters of state, called the entropy, and a parameter T characterizing the system, called the absolute temperature, such that for $\delta Q^{(e)}$ in reversible processes one has

$$dS_m = \frac{\delta Q^{(e)}}{T}, \tag{2.103}$$

whilst for irreversible processes

$$dS_m > \frac{\delta Q^{(e)}}{T}. \tag{2.104}$$

The difference

$$T dS_m - \delta Q^{(e)} = \delta Q' \geq 0 \tag{2.105}$$

is known as the uncompensated heat.

Both, the first and second laws of thermodynamics are regarded as state-ments accumulating experimental observations. In experimental terms, they relate the internal energy U_m and entropy S_m, which cannot be measured directly, with measurable mechanical characteristics, K_m, $\delta W^{(e)}$, and a mea-surable thermal parameter T. In terms of modelling, equation (2.103) provides an 'integrating factor' for external heat $\delta Q^{(e)}$ and (2.104) gives the direction in which the system has to evolve and hence a constraint which the constitutive relations describing its relaxation have to satisfy.

2.3.2 Application to fluid

Now, we will specify the concepts introduced above for the fluid which in a general case is assumed to be compressible. We already have two conservation laws, that is the conservation of mass, i.e., (2.13)

$$\frac{\partial \rho}{\partial t} + \nabla \cdot (\rho \mathbf{u}) = 0, \tag{2.106}$$

and conservation of momentum (2.31)

$$\rho \left(\frac{\partial \mathbf{u}}{\partial t} + \mathbf{u} \cdot \nabla \mathbf{u} \right) = \rho \mathbf{F} + \nabla \cdot \mathbf{P}, \tag{2.107}$$

where at this stage the stress tensor $\mathbf{P}$ is not specified, and we will have in mind only that it is symmetric. To close this set of equations, one has to consider thermodynamics of the fluid. We will define the total energy of a given mass of fluid confined in a material volume $V(t)$ by

$$\int_{V(t)} \rho \left(\frac{|\mathbf{u}|^2}{2} + U_m \right) dV,$$

where $|\mathbf{u}|^2/2$ and U_m are the kinetic and internal energies per unit mass, respectively. The above definition implies an assumption that U_m, which is yet to be specified, is additive. This assumption will be relaxed once we have the energy balance equation in a differential form applicable to infinitesimal fluid particles. However, at this stage it is convenient to illustrate (2.99) for a finite mass of fluid.

For a finite volume of fluid in the field of external body forces $\mathbf{F}$ with stresses described by the stress tensor $\mathbf{P}$ the first law of thermodynamics (2.99) can be written as

$$d \int_{V(t)} \rho \left(\frac{|\mathbf{u}|^2}{2} + U_m \right) dV = \int_{V(t)} \rho \mathbf{F} \cdot \mathbf{u} \, dt \, dV + \int_{S(t)} \mathbf{n} \cdot \mathbf{P} \cdot \mathbf{u} \, dt \, dS$$

$$- \int_{S(t)} \mathbf{q} \cdot \mathbf{n} \, dt \, dS. \tag{2.108}$$

Here $\mathbf{q}$ is the heat flux per unit area from the surrounding fluid to volume $V(t)$, and $\mathbf{n}$, as before, is the outward normal to surface $S(t)$ confining $V(t)$. The first and second terms on the right-hand side describe the work done on the field of displacement $\mathbf{u} \, dt$ by the body forces $\mathbf{F}$ acting on $V(t)$ and stresses $\mathbf{P}$ acting on $S(t)$ from outside and hence represent $\delta W^{(e)}$. The last term on the right-hand side of (2.108) is, by definition, $\delta Q^{(e)}$.

After the standard procedure of obtaining a differential equation from the integral one, that is after dividing (2.108) by dt, applying (2.8) to the time derivative on the left-hand side, Gauss' theorem to integrals over $S(t)$ and using that $V(t)$ is an arbitrary material volume so that an integral over it is equal to zero if and only if the integrand is identically zero, we arrive at the energy conservation law in the differential form

$$\frac{\partial}{\partial t}\left[\rho\left(\frac{|\mathbf{u}|^2}{2}+U_m\right)\right]+\nabla\cdot\left[\rho\mathbf{u}\left(\frac{|\mathbf{u}|^2}{2}+U_m\right)-\mathbf{P}\cdot\mathbf{u}+\mathbf{q}\right]=\rho\mathbf{F}\cdot\mathbf{u}. \quad (2.109)$$

This equation states that the local variation of energy per unit *volume*, $E_v = \rho|\mathbf{u}|^2/2+\rho U_m$, results from the divergence of the energy flux and the work done by body forces. It should be noted that the flux of energy,

$$\mathbf{J}_E = \rho\mathbf{u}\left(\frac{|\mathbf{u}|^2}{2}+U_m\right)-\mathbf{P}\cdot\mathbf{u}+\mathbf{q},$$

as we defined it includes, in addition to the convective flux, $\rho\mathbf{u}(|\mathbf{u}|^2/2+U_m)$, also the work done by surface forces (i.e., stresses), $-\mathbf{P}\cdot\mathbf{u}$, and the heat flux, $\mathbf{q}$.

Equation (2.109) can also be presented in the form (2.99), that is as a balance of the energy variation, work and heat per unit mass. Indeed, using the continuity equation (2.106) and having in mind that for a function f associated with a moving fluid particle one has[13]

$$\frac{Df}{Dt}\,dt = df, \quad (2.110)$$

we can rewrite (2.109) as

$$d\left(\frac{|\mathbf{u}|^2}{2}+U_m\right) = \frac{1}{\rho}\nabla\cdot(\mathbf{P}\cdot\mathbf{u})\,dt+\mathbf{F}\cdot\mathbf{u}\,dt-\frac{1}{\rho}\nabla\cdot\mathbf{q}\,dt. \quad (2.111)$$

Here

$$\delta W^{(e)} = \frac{1}{\rho}\nabla\cdot(\mathbf{P}\cdot\mathbf{u})\,dt+\mathbf{F}\cdot\mathbf{u}\,dt$$

is the elementary work done by external surface and body forces, and

$$\delta Q^{(e)} = -\frac{1}{\rho}\nabla\cdot\mathbf{q}\,dt$$

is an infinitesimal amount of heat. The latter is defined in such a way that $\nabla\cdot\mathbf{q}<0$ (a sink in the vector field $\mathbf{q}$) corresponds to $\delta Q^{(e)}>0$, that is heat *added* to the fluid.

The kinetic energy balance equation (2.101) can be derived for a moving fluid in exactly the same way as in classical mechanics of a material point,

---

[13]In some books, e.g., Sedov (1997), the notation d/dt is used instead of D/Dt for the substantive derivative thus allowing one to deal with differentials associated with moving fluid particles in a standard way and making (2.110) obvious.

that is one has to multiply the equation of motion (2.107) by an elementary displacement $\mathbf{u}\, dt$ and, using (2.110), will arrive at

$$d\left(\frac{|\mathbf{u}|^2}{2}\right) = \mathbf{F} \cdot \mathbf{u}\, dt + \frac{1}{\rho}\nabla \cdot (\mathbf{P} \cdot \mathbf{u})\, dt - \frac{1}{\rho}\mathbf{P} : (\nabla\mathbf{u})\, dt. \qquad (2.112)$$

The left-hand side of this equation represents the differential of the kinetic energy per unit mass associated with a moving fluid particle whilst the first two terms on the right-hand side describe work done by external body forces and stresses. It should be pointed out that the kinetic energy balance in an *integral* form would in general include also work done by the body forces *internal* to the volume. For example, in the case of a self-gravitating gas one will have to take into account gravitational interactions between material elements comprising the volume. As the size of the volume decreases, so does the contribution from internal body forces so that for an infinitesimal fluid particle, that is when we arrive at a differential equation, all body forces are external.

The last term in (2.112), which describes work by internal stresses $\delta W^{(i)}$, can be written down as follows. Since the stress tensor $\mathbf{P}$ is symmetric, one can show, similarly to what we did at the end of §2.2.3, that its convolution with the antisymmetric part of $\nabla\mathbf{u}$ is identically zero, that is

$$p^{ij}(\nabla_i u_j - \nabla_j u_i) = 0.$$

Now, given this, one can take the stress tensor in a general form as $\mathbf{P} = -p\mathbf{I} + \boldsymbol{\sigma}$ (where $\boldsymbol{\sigma}$ is yet to be specified) and, making use of the continuity equation (2.106), write $\delta W^{(i)}$ down as a sum of the work done by the pressure on compressing the fluid and the work of viscous stresses on deforming it:

$$\delta W^{(i)} \equiv -\frac{1}{\rho}\mathbf{P} : (\nabla\mathbf{u})\, dt = -\frac{1}{2\rho}\mathbf{P} : [\nabla\mathbf{u} + (\nabla\mathbf{u})^*]\, dt$$

$$= \frac{1}{\rho}(p\mathbf{I} - \boldsymbol{\sigma}) : \mathbf{E}\, dt = \frac{p}{\rho}\mathbf{I} : \mathbf{E}\, dt - \frac{p}{\rho}\boldsymbol{\sigma} : \mathbf{E}\, dt$$

$$= \frac{p}{\rho}(\nabla \cdot \mathbf{u})\, dt - \frac{1}{\rho}\boldsymbol{\sigma} : \mathbf{E}\, dt = -\frac{p}{\rho^2}\frac{D\rho}{Dt}\, dt - \frac{1}{\rho}\boldsymbol{\sigma} : \mathbf{E}\, dt$$

$$= -\frac{p}{\rho^2}\, d\rho - \frac{1}{\rho}\boldsymbol{\sigma} : \mathbf{E}\, dt. \qquad (2.113)$$

Here, as before, $\mathbf{E} = \frac{1}{2}[\nabla\mathbf{u} + (\nabla\mathbf{u})^*]$ is the rate-of-strain tensor and $\mathbf{I} : \mathbf{E} \equiv e_i^i = \nabla \cdot \mathbf{u}$ is its first invariant. Equation (2.112) now takes the form

$$d\left(\frac{|\mathbf{u}|^2}{2}\right) = \mathbf{F} \cdot \mathbf{u}\, dt + \frac{1}{\rho}\nabla \cdot (\mathbf{P} \cdot \mathbf{u})\, dt - \frac{p}{\rho^2}\, d\rho - \frac{1}{\rho}\boldsymbol{\sigma} : \mathbf{E}\, dt. \qquad (2.114)$$

Subtracting (2.112) from (2.111) we obtain a particular form of equation (2.102) per unit mass of the fluid

$$dU_m = \delta Q^{(e)} + \frac{p}{\rho^2}\, d\rho + \frac{1}{\rho}\sigma^{ij} e_{ij}\, dt,$$

or, using (2.105) to represent $\delta Q^{(e)}$,

$$dU_m = T\,dS_m + \frac{p}{\rho^2}\,d\rho + \frac{1}{\rho}\sigma^{ij}e_{ij}\,dt - \delta Q'. \qquad (2.115)$$

For reversible equilibrium processes it becomes simply

$$dU_m = T\,dS_m + \frac{p}{\rho^2}\,d\rho. \qquad (2.116)$$

This equation, known as Gibbs' formula, suggests that by specifying U_m as a function of S_m and ρ for a particular fluid, one will have two equations of state

$$T = \left(\frac{\partial U_m}{\partial S_m}\right)_\rho, \qquad p = \rho^2\left(\frac{\partial U_m}{\partial \rho}\right)_{S_m}, \qquad (2.117)$$

which will relate all the parameters we used so far to characterize the fluid. (The subscripts ρ and S_m indicate the parameters kept constant as the derivative of U_m is taken.)

It should be pointed out that the above procedure contains an implicit assumption. For a two-parametric fluid one can obviously choose ρ and S_m as parameters of state and hence, after specifying $U_m(\rho, S_m)$, will have (2.117) for the equilibrium states and reversible equilibrium processes. However, it is an assumption that (2.117) will also hold for *irreversible* processes where, as one possibility, the state, if this concept can at all be introduced, could be characterized by a greater number of parameters of state than in equilibrium. The assumption that the internal energy is the same function in irreversible processes as in equilibrium is known as the *hypothesis of local equilibrium*. Experiments show that it holds for most systems in a wide range of parameters.

Another important point to be noted here is that comparison between (2.115) and (2.116) might lead one to the conclusion that

$$\frac{1}{\rho}\sigma^{ij}e_{ij}\,dt = \delta Q',$$

and hence for the simple thermodynamic system we are considering here irreversibility is due entirely to the work of viscous stresses. This is not the case, however. Indeed, in deriving (2.116) for reversible equilibrium processes, introducing U_m as a function of S_m and ρ, and postulating that this function also represents internal energy in irreversible processes we tacitly assumed that heat is added to the system in a reversible way, or, more specifically, that there is no spatial redistribution of heat. As we will see later, heat conductivity, that is the only nonmechanical source of energy we introduced in (2.108), is irreversible. As a result, even if the fluid is thermally insulated from the environment, the redistribution of heat in the fluid increases its entropy. In order to take this effect into account we need to consider the balance of entropy per unit volume and apply the second law of thermodynamics.

For an incompressible fluid p is determined by the dynamics of the flow and not thermodynamically. Then, U_m becomes a function of one parameter and, given that $dU_m = T\, dS_m$, one can specify U_m as a function of either S_m or T. Gibbs' formula (2.116) for a moving particle obviously has the form

$$\frac{DU_m}{Dt}\, dt = T \frac{DS_m}{Dt}\, dt + \frac{p}{\rho^2} \frac{D\rho}{Dt}\, dt. \qquad (2.118)$$

After multiplying (2.114), where, according to (2.110), the differential on the left-hand side can be written down as a product of the substantive derivative and dt, and (2.118) by ρ/dt, we subtract them both from (2.109) and make use of the continuity equation to take ρ under the sign of differentiation. As a result, one obtains that

$$\frac{\partial(\rho S_m)}{\partial t} + \nabla \cdot (\rho \mathbf{u} S_m) + \frac{1}{T} \nabla \cdot \mathbf{q} = \frac{1}{T} \sigma^{ij} e_{ij}. \qquad (2.119)$$

This equation expresses the balance of entropy per unit volume ($S_v = \rho S_m$) and accounts for the entropy production due to redistribution of heat. We will write this equation down in a slightly different form which makes all sources of entropy appear explicitly. To do this, in addition to the convective entropy flux, $\rho \mathbf{u} S_m$, we have to specify what is assumed to be the flux of entropy associated with heat transfer. We will make this essential assumption by splitting the third term on the left-hand side of (2.119) into two parts:

$$\frac{1}{T} \nabla \cdot \mathbf{q} = \nabla \cdot \frac{\mathbf{q}}{T} + \frac{1}{T^2} \mathbf{q} \cdot \nabla T,$$

where the form of the flux, $\mathbf{q}/T$, is suggested by (2.103). Now, equation (2.119) becomes

$$\frac{\partial(\rho S_m)}{\partial t} + \nabla \cdot \left(\rho \mathbf{u} S_m + \frac{\mathbf{q}}{T} \right) = \frac{1}{T} \sigma^{ij} e_{ij} - \frac{q^i}{T^2} \nabla_i T. \qquad (2.120)$$

The two terms on the left-hand side of this equation correspond to the local variation of entropy per unit volume and the divergence of its flux,

$$\mathbf{J}_S = \rho \mathbf{u} S_m + \frac{\mathbf{q}}{T},$$

whilst the right-hand side represents the rate of entropy production,

$$\dot{S} = \frac{1}{T} \boldsymbol{\sigma} : \mathbf{E} - \frac{1}{T^2} \mathbf{q} \cdot \nabla T. \qquad (2.121)$$

According to the second law of thermodynamics, the rate of entropy production must be nonnegative:

$$\frac{1}{T} \boldsymbol{\sigma} : \mathbf{E} - \frac{1}{T^2} \mathbf{q} \cdot \nabla T \geq 0. \qquad (2.122)$$

This fundamental constraint determines the direction of all thermodynamic processes and has to be satisfied by the constitutive equations specifying $\boldsymbol{\sigma}$ and $\mathbf{q}$. The latter are needed to close the system which at the moment includes the continuity equation (2.106), the equation of motion (2.107), the total energy balance equation (2.109) (or the entropy balance equation (2.120)) and the equations of state (2.117).

Assuming that the two terms in (2.122) correspond to independent physical factors, or more precisely that the possible cross-effects can be neglected, one can look for constitutive equations which would make nonnegative *each* of the two terms in (2.122):

$$\frac{1}{T}\boldsymbol{\sigma} : \mathbf{E} \geq 0, \qquad -\frac{1}{T^2}\mathbf{q} \cdot \nabla T \geq 0. \qquad (2.123)$$

This assumption is often justified when one is dealing with dissipative processes described by tensors of different rank, like in our case where heat transfer is associated with vector fields $\mathbf{q}$ and ∇T while work of viscous stresses is expressed in terms of tensors of the second rank $\boldsymbol{\sigma}$ and $\mathbf{E}$. For dissipative processes described by tensors of the same rank, for example vectors when, say, heat transfer and redistribution of components in a multicomponent system are considered, cross-effects are often important, and it is the *sum* of the dissipative terms in the entropy production which has to be made nonnegative.

Constitutive equations are usually formulated as generalizations of experimental observations of dissipative processes so that positiveness of the entropy production corresponding to these processes becomes an *a posteriori* test. However, an analysis of known constitutive equations suggests that many of them could be obtained in the framework of a general procedure which we will illustrate for the case of a two-parametric fluid considered here.

Each of the additive terms in the expression for the entropy production can be regarded as a product of a 'thermodynamic flux' and a 'thermodynamic force' which drives the 'flux'. If one assumes that the 'flux' is directly proportional to the 'force', then their product becomes a quadratic form which can be made positively determined by the appropriate choice of the coefficients. As a result, one will have a linear constitutive equation which will make the corresponding term in the entropy production nonnegative for all thermodynamic processes.

Applying this procedure to the second term in (2.122), that is regarding $\mathbf{q}$ as a "thermodynamic flux" and ∇T as the corresponding "thermodynamic force", we have

$$q^i = B^{ij}\nabla_j T.$$

If the fluid is isotropic, one can show that the nondiagonal components of tensor B^{ij} are zero while the diagonal ones are equal. Thus, we arrive at a constitutive equation in the form

$$\mathbf{q} = -\kappa\nabla T, \qquad (2.124)$$

where the coefficient of thermal conductivity κ is a material parameter, which in a general case is a function of parameters of state. The second inequality in (2.123) holds for any positive value of κ:

$$-\frac{1}{T^2}\mathbf{q}\cdot\nabla T = \frac{\kappa|\nabla T|^2}{T^2} \geq 0.$$

Equation (2.124), known as the Fourier law, provides an accurate description for the steady process of heat conduction in isotropic media, and in many practically important cases κ can be regarded as a material constant.

If U_m is a function of temperature only, which is the case, for example, for an incompressible fluid, then for the fluid at rest equation (2.109) combined with the Fourier law (2.124) reduces to the classical equation of heat conduction

$$\frac{\partial T}{\partial t} = \bar{\kappa}\Delta T, \tag{2.125}$$

where $\bar{\kappa} = \kappa/(\rho\, dU_m/dT)$.

As has been mentioned earlier, if in addition to heat conduction there are other dissipative processes involving other vector fields (for example, diffusion of components in a multicomponent medium), then cross-effects between them may become important. As a result, thermodynamic fluxes will be proportional not only to their 'own' thermodynamic forces but also to thermodynamic forces associated with other thermodynamic fluxes. The number of different coefficients describing cross-effects can be reduced by considering reciprocal relationships based on the so-called principle of microscopic reversibility (Onsager 1931a, 1931b).

In order to satisfy the first inequality in (2.123), we consider the stress and the rate-of-strain tensors as a thermodynamic force-flux pair and, assuming a linear dependence between them, arrive at equation (2.61) considered earlier:

$$\sigma^{ij} = A^{ij\alpha\beta}e_{\alpha\beta}.$$

As shown in §2.2.6, in the case of an isotropic fluid this equation reduces to (2.65), that is

$$\sigma^{ij} = (\zeta - \tfrac{2}{3}\mu)g^{ij}e_k^k + 2\mu e^{ij}. \tag{2.126}$$

Now, we have to find the constraints on ζ and μ for the first inequality (2.123) to hold.

Using (2.126), one has

$$\boldsymbol{\sigma} : \mathbf{E} \equiv \sigma^{ij}e_{ij} = (\zeta - \tfrac{2}{3}\mu)(e_k^k)^2 + 2\mu e^{ij}e_{ij} = \zeta I_1^2 + 2\mu\left(I_2 - \frac{I_1^2}{3}\right), \tag{2.127}$$

where $I_1 = \mathbf{I} : \mathbf{E} = e_i^i$ and $I_2 = \mathbf{E} : \mathbf{E} = e^{ij}e_{ij}$ are the first and second invariants of the rate-of-strain tensor (see Appendix A). In the principal axes of the rate-of-strain tensor, the invariants are expressed in terms of the principal (diagonal) components e_{ii} $(i = 1, 2, 3)$ as

$$I_1 = e_{11} + e_{22} + e_{33} \quad \text{and} \quad I_1^2 = (e_{11})^2 + (e_{22})^2 + (e_{33})^2,$$

so that

$$I_2 - \frac{I_1^2}{3} = \left(e_{11} - \frac{I_1}{3} \right)^2 + \left(e_{22} - \frac{I_1}{3} \right)^2 + \left(e_{33} - \frac{I_1}{3} \right)^2 \geq 0. \tag{2.128}$$

Given that I_1 and I_2 are invariant with respect to the coordinate system, one has that $I_2 - I_1^2/3 \geq 0$ for all coordinate systems and arbitrary deformations. Thus, it follows from (2.127) and (2.128) that the first inequality (2.123) holds for arbitrary deformations if

$$\zeta \geq 0, \quad \mu \geq 0. \tag{2.129}$$

Indeed, by considering separately (a) deformations which keep the density constant ($\nabla \cdot \mathbf{u} = 0$, that is $e_i^i = 0$) and (b) uniform compression ($e_{ij} = kg_{ij}$) one can make I_1^2 and $I_2 - I_1^2/3$, which both are always nonnegative, equal to zero independently and hence for $\boldsymbol{\sigma} : \mathbf{E} \geq 0$ to hold both coefficients ζ and μ must be nonnegative.

Thus, thermodynamic considerations allow one to arrive at a closed set of equations for a two-parametric fluid (which, physically, corresponds to a wide class of fluids) by adding to the general conservation laws (2.106), (2.107), (2.109) also the equations of state (2.117) and constitutive equations (2.124), (2.126) as well as to obtain constraints $\kappa > 0$ and (2.129), which must be satisfied by the material constants. To apply this set of equations to a particular fluid, one has to specify U_m as a function of S_m and ρ and set the values of material constants characterizing the fluid. The internal energy U_m can also be specified as a function of parameters of state other than S_m and ρ. In this case, however, the internal energy will no longer be a 'thermodynamic potential' generating the equations of state via simple differentiation, like in (2.117), and hence extra equations relating thermodynamic parameters will become necessary to specify the system.

In many situations, it is convenient to formulate equations of state using thermodynamic potentials other than the internal energy. These potentials can be obtained by the Legendre transform of U_m, that is $U_m \mapsto U_m - \sum X_i(\partial U_m/\partial X_i)$, where X_i are the arguments of U_m. For a two-parametric system one can define the free energy

$$F_m = U_m - TS_m,$$

so that

$$dF_m = -S_m \, dT + \frac{p}{\rho^2} \, d\rho,$$

and the equations of state will take the form

$$S_m = -\left(\frac{\partial F_m}{\partial T}\right)_\rho, \qquad p = \rho^2 \left(\frac{\partial F_m}{\partial \rho}\right)_T.$$

These equations use measurable quantities T and ρ as parameters of state which is convenient for experimental determination of thermodynamic properties of particular systems.

The enthalpy H_m is defined by

$$H_m = U_m + \frac{p}{\rho},$$

so that

$$dH_m = T\,dS_m + \frac{dp}{\rho},$$

and hence

$$T = \left(\frac{\partial H_m}{\partial S_m}\right)_p, \qquad \frac{1}{\rho} = \left(\frac{\partial H_m}{\partial p}\right)_{S_m}.$$

The enthalpy appears naturally in aerodynamics, in the analysis of compressible flows with shock waves.

Finally,

$$\Psi_m = U_m - TS_m + \frac{p}{\rho} \tag{2.130}$$

is known as the chemical (or Gibbs') potential for which one has

$$d\Psi_m = -S_m\,dT + \frac{dp}{\rho}$$

and consequently

$$S_m = -\left(\frac{\partial \Psi_m}{\partial T}\right)_p, \qquad \frac{1}{\rho} = \left(\frac{\partial \Psi_m}{\partial p}\right)_T.$$

For the internal energy per unit volume, $U_v = \rho U_m$, one obviously has

$$dU_v = T\,dS_v + \Psi_m\,d\rho.$$

This equation replaces (2.116) and is convenient for multiphase systems with mass exchange across interfaces.

As we have already mentioned, equations (2.70), (2.71) for a viscous incompressible liquid form a closed set sufficient to solve the mechanical part of the problem, that is to find the velocity field and the pressure distribution in the bulk, without knowing thermal characteristics of the flow.[14] On the

---

[14]This is the case provided that the temperature variations are not too large so that the viscosity coefficient μ, which depends on temperature, can be approximately treated as a material constant for the given fluid.

other hand, in order to find the distributions of thermal parameters, one has to solve the mechanical problem first and only then consider the evolution of temperature on the given background flow. This splitting of the two sides of the problem appears to be possible due to the assumption of incompressibility. This assumption also simplifies Gibbs' formula (2.116) which now takes the form $dU_m = T\,dS_m$, and hence shows that the internal energy can be specified as a function of either S_m or T. The term $\boldsymbol{\sigma} : \mathbf{E}$ describing the rate of dissipation of mechanical energy into heat due to viscous friction per unit volume for an incompressible fluid takes the form

$$\boldsymbol{\sigma} : \mathbf{E} = 2\mu e_{ij}e^{ij} = \tfrac{1}{2}\mu(\nabla_i u_j + \nabla_j u_i)(\nabla^i u^j + \nabla^j u^i)$$

$$= \mu[\nabla\mathbf{u} : \nabla\mathbf{u} + \nabla\mathbf{u} : (\nabla\mathbf{u})^*]. \qquad (2.131)$$

The exposition of irreversible thermodynamics in this section can be regarded as formalization of experimental observations with a minimum of transparent theoretical assumptions binding them together. We will use this approach and the notions we have introduced in Chapter 4 to deal with nonequilibrium states of interfaces.

2.4   Classical boundary conditions

2.4.1   Boundaries and boundary conditions

The continuum approximation introduced in §§2.1.2 and 2.1.3 was motivated by the convenience of using well-developed tools, the differential calculus and theory of partial differential equations, to model the fluid. The choice of mathematical tools in its turn determines the nature of concepts we have to use and the way in which the mathematical problems aimed at describing the fluid's behaviour have to be formulated. In the framework of fluid mechanics, the problem formulation includes the following three elements:

(i) A set of differential equations describing the distribution of the fluid's parameters in the bulk;

(ii) A domain in the space of independent variables $\mathbf{r}$ and t where the bulk equations have to be solved;

(iii) Boundary, initial and other conditions necessary to uniquely determine the solution or a class of solutions.

All these elements are intimately connected and, from the modelling viewpoint, interdependent. Here, we briefly discuss some aspects of this interdependence.

The bulk equations incorporate what we call 'the laws of nature' governing the parameters we would like to model. Once these equations are formulated, it is their mathematical structure and properties that determine the ways of formulating mathematical problems. In particular, the type and order of the bulk equations determine the number of boundary (initial/other) conditions required and, depending on the domain, the boundaries where the conditions have to be posed and constraints they have to satisfy. To put it pictorially, the bulk equations ask specific 'questions' about what is not in them and the 'answers' have to be given by the boundary (initial/other) conditions. These conditions must therefore provide only the information the bulk equations ask for.

We have to make this methodological point, which is almost a commonplace in the theory of partial differential equations, since when it comes to the use of this theory in continuum mechanics, there appears a certain degree of arbitrariness. In a number of books (e.g., Milne-Thomson 1968), the boundary conditions are formulated and discussed in their own right from the viewpoint of what physics they represent *before* any bulk equations are even considered. This methodological confusion seems innocent until one finds that from textbooks this approach tends to slip into research papers, where, as one example, one can come across higher-order terms kept artificially in simplified asymptotic equations only out of the need to satisfy some pre-set boundary conditions.

Another variant of a methodologically inconsistent approach to the boundary conditions is to formulate them from the viewpoint of 'what actually happens on the boundary', as opposed to considering (a) what the *bulk equations* see as the boundary and hence what information about the physical processes at the boundary they require, and (b) the concepts through which one has to incorporate these processes. This approach can be disorientating, especially when one is dealing with situations where conventional problem formulations lead to singularities and some 'extra' physics has to be taken into account.

In specifying the domain, we often have to make simplifying assumptions. For example, the actual solid surface is normally rough and chemically inhomogeneous on a macroscopic length scale which at the same time is small compared to the characteristic length scale of the flow to be described. The details of the surface structure are often both unknown and unimportant whereas some average characteristics associated with the deviation from an 'ideal' surface play a role in the flow dynamics. Then, one can try to simplify the problem by replacing the actual, rough and/or chemically inhomogeneous, solid surface with a smooth 'effective' boundary and hence, in a general case, has to reformulate the boundary conditions applied at the actual surface into some 'effective' conditions applied at the artificial 'effective' boundary. In many cases, this problem is far from trivial and can even lead to paradoxical results.

A number of questions arise when initially or as a result of evolution the spatial boundary of the domain turns out to be only piece-wise smooth. How should one describe and interpret singularities (angles/cusps/points) in the shape of the domain? If these singularities are associated with singularities in the distributions of the flow parameters, what is acceptable from the physical point of view? Can/should the angles/cusps/points be artificially (e.g., numerically) smoothed away or will it also throw away the essential physics of the phenomenon to be modelled? These are only a few questions one faces in addressing the problem.

The three elements of the problem formulation we listed earlier as (i)–(iii) imply that our way of modelling splits the physics of a phenomenon under consideration into that described by the bulk equations and the 'remainder' incorporated into the boundary conditions at the external boundaries and internal interfaces. This poses a fundamental question about the intrinsic limitations inherent in this scheme itself. An associated practical question is to what extent these limitations could be accommodated by the appropriate alteration of the boundary conditions and how this problem should be addressed in a regular way. This problem becomes increasingly important for modelling of flows on small length scales.

Finally, there is an emerging problem of conditions which would link solutions in domains where the flows are described in terms of different sets of concepts. This is the case, for example, when the fluid-mechanical description of a liquid has to be linked via an interface with the molecular-dynamic one in the neighbouring gas. Such a situation can take place, for example, if the interface evolves into a cusp where one has a macroscopic length scale in one direction and an infinitesimal one normal to it.

These problems will be discussed later, as they arise in the modelling of particular flows. In this chapter, we will consider the classical boundary conditions at liquid-fluid interfaces and solid boundaries and conditions at the lines along which these boundaries intersect. Experiments show that solutions to mathematical problems formulated using these conditions are very accurate in describing a great number of flows. The result is that these conditions are often treated as already settled, almost set in stone. Here our goal is to highlight the assumptions behind the classical boundary conditions which will help us later to re-examine the standard problem formulations in the situations where there is either no solution to the problem or a solution exhibiting properties unacceptable from a physical point of view.

2.4.2 Kinematic and dynamic conditions on liquid-fluid interfaces

The notion of an 'interface' or a 'surface of discontinuity' arises as a useful simplification when, instead of modelling a thin layer where parameters vary steeply and often some physical factors additional to those accounted for in the bulk become important, we replace this layer with a geometric surface of zero thickness and incorporate the physics associated with this layer into boundary conditions. This general idea is behind interfaces as diverse as

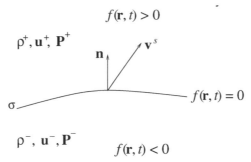

$f(\mathbf{r}, t) > 0$

$\rho^{+}, \mathbf{u}^{+}, \mathbf{P}^{+}$

$\mathbf{n}$ $\quad$ $\mathbf{v}^{s}$

σ $\qquad$ $f(\mathbf{r}, t) = 0$

$\rho^{-}, \mathbf{u}^{-}, \mathbf{P}^{-}$ $\qquad$ $f(\mathbf{r}, t) < 0$

FIGURE 2.9: A definition sketch for a liquid-fluid interface.

shock waves, propagating fronts of chemical reactions, liquid-fluid interfaces and many others. We will consider this idea here in its application to an interface between two fluids in the framework of classical fluid mechanics modelling. In Chapter 4, we return to the problem of boundary conditions at interfaces and examine it in more detail in the light of singularities arising in the mathematical description of some flows.

When two fluids are brought in contact, the molecules in a layer separating them begin to experience nonsymmetric influence from the bulk phases. Although this layer is very thin (a few nanometers for most fluids), the inter-molecular forces acting on it from the bulk phases are so strong that asymmetry in these forces is sufficient for the resultant dynamic properties of the layer to be often important for the overall dynamics of the system. As we have already discussed in §§2.1.2 and 2.1.3, all molecular length scales, including those associated with intermolecular forces, become infinitesimal in the continuum approximation and hence a sharp interface as well as the division of forces into body forces and stresses are the key elements forming the continuum mechanics modelling.

When the interfacial layer is modelled as a geometric surface, i.e., an 'interface' separating two fluids, the boundary conditions for the bulk parameters to be formulated on this surface have to incorporate both the universal conservation laws as well as the specific physics of the processes in the interfacial layer for a particular system. We will consider boundary conditions required for the Navier-Stokes equations (2.70), (2.71). These equations will be assumed to describe (incompressible) liquids on both sides of the interface.

Let $f(\mathbf{r}, t) = 0$ be the equation of a moving smooth interface separating fluid 1 in the region where $f > 0$ and fluid 2 where $f < 0$. We will mark the bulk parameters of fluids 1 and 2 with the superscript $+$ and $-$, respectively; $\mathbf{n} = \nabla f / |\nabla f|$ is a unit normal pointing from fluid 2 to fluid 1 (Fig. 2.9).

If δl is an elementary distance travelled in the normal direction to the interface in time δt by a point with position vector $\mathbf{r}$ located in the interface at moment t, then one has $f(\mathbf{r}, t) = 0$ and $f(\mathbf{r} + \delta l\, \mathbf{n}, t + \delta t) = 0$. Then, dividing a Taylor expansion,

$$0 = f(\mathbf{r} + \delta l\, \mathbf{n}, t + \delta t) = \frac{\partial f}{\partial t}(\mathbf{r}, t)\, \delta t + \delta l\, \mathbf{n} \cdot \nabla f(\mathbf{r}, t) + o(\delta l, \delta t) \quad \text{as } \delta l, \delta t \to 0,$$

by δt, taking the limit $\delta t \to 0$ and introducing the normal projection of the interface velocity,

$$\mathbf{v}^s \cdot \mathbf{n} \equiv v_n^s = \lim_{\delta t \to 0} \frac{\delta l}{\delta t},$$

we obtain

$$\frac{\partial f}{\partial t} + \mathbf{v}^s \cdot \nabla f = 0. \tag{2.132}$$

This equation merely specifies the shape of the interface in terms of its normal velocity and holds for any smooth surface moving in space.

The conservation laws can be translated into the corresponding boundary conditions in several broadly equivalent ways. Here, we will use an approach based on considering fluxes across the interface.

Given that the interfacial layer is thin and the density there is of the same order as in the bulk, one may assume that under normal circumstances the role of an interface as a sink or source of mass can be neglected compared to the mass fluxes across the boundary. As a result, in the general case of a permeable interface we have the continuity of mass flux across it

$$\rho^+(\mathbf{u}^+ - \mathbf{v}^s) \cdot \mathbf{n} = \rho^-(\mathbf{u}^- - \mathbf{v}^s) \cdot \mathbf{n} \qquad \text{at } f(\mathbf{r}, t) = 0. \tag{2.133}$$

However, if a solution to a particular problem obtained using this condition gives that at some point the mass fluxes across the boundary and the velocity of the boundary itself vanish (i.e., when a stagnation point for the velocity field is located on the boundary), then, if there is a source/sink of mass at this point due to mass exchange between the interface and the bulk, one will often need to take it into account.

Condition (2.133) alone is not sufficient, however, since it does not allow one to actually determine the mass flux. For example, equation (2.133) holds for both permeable and impermeable interfaces. Therefore, in addition to (2.133) we have to prescribe the mass flux itself,

$$\rho^+(\mathbf{u}^+ - \mathbf{v}^s) \cdot \mathbf{n} = \chi, \tag{2.134}$$

where χ has to be specified in terms of parameters determining a particular physical mechanism responsible for mass transfer across the interface, such as, for example, chemical reactions, evaporation-condensation, mutual dissolution of fluids, etc. Naturally, χ does not have to be an algebraic function of its arguments, and its mathematical structure can in its turn require more information about the process in terms of appropriate equations involving the bulk and, possibly, surface parameters.

In a particular case of an impermeable interface, one has $\chi = 0$, and (2.134) combined with (2.133) now gives

$$(\mathbf{u}^{\pm} - \mathbf{v}^{s}) \cdot \mathbf{n} = 0. \qquad (2.135)$$

Then, one can eliminate the normal component of $\mathbf{v}^{s}$ from (2.132) and have two kinematic conditions

$$\mathbf{u}^{+} \cdot \mathbf{n} = \mathbf{u}^{-} \cdot \mathbf{n}, \qquad \frac{\partial f}{\partial t} + \mathbf{u}^{+} \cdot \nabla f = 0, \qquad \text{at } f(\mathbf{r}, t) = 0, \qquad (2.136)$$

which, respectively, relate the normal projections of the bulk velocities evaluated at the interface and allow one to determine the evolution of its shape. If the interface is steady in some coordinate system, (2.136) become simply

$$\mathbf{u}^{\pm} \cdot \mathbf{n} = 0. \qquad (2.137)$$

The momentum conservation law translates into the boundary conditions in a similar way. The flux of momentum is determined via the momentum flux tensor $\mathbf{\Pi}$ introduced by (2.73). The momentum fluxes in fluids 1 and 2 across a moving interface are given by $\mathbf{n} \cdot \mathbf{\Pi}^{+}$ and $\mathbf{n} \cdot \mathbf{\Pi}^{-}$, where

$$\mathbf{\Pi}^{\pm} = \rho^{\pm}(\mathbf{u}^{\pm} - \mathbf{v}^{s})(\mathbf{u}^{\pm} - \mathbf{v}^{s}) - \mathbf{P}^{\pm}. \qquad (2.138)$$

If the source/sink of momentum due to the interface can be neglected compared with $\mathbf{n} \cdot \mathbf{\Pi}^{\pm}$, then the momentum flux across an interface is continuous:

$$\mathbf{n} \cdot \mathbf{\Pi}^{+} = \mathbf{n} \cdot \mathbf{\Pi}^{-}. \qquad (2.139)$$

This is almost always the case, for example, when one is dealing with flows associated with very large length scales (tidal waves, lava flows, large-scale free-surface flows in industry, etc.). However, in many situations the dynamics of the interfaces contributes significantly to the overall dynamics of the system, and we will consider this contribution here.

The source/sink of momentum due to the presence of an interface has the following two components. The first one is associated with external forces, $\mathbf{F}^{s}$, acting on the interface. This component can be important when the interface possesses properties which in a dynamic sense would 'compensate' its negligible thickness. For example, if the interface is electrically charged and/or conducts electric current whilst the rest of the fluid is electrically neutral and nonconducting, an external electromagnetic field will be able to exert a force on the interface which can significantly influence the dynamics of the system and hence will have to be accounted for in the boundary conditions.

If the interface has no properties which would magnify the influence of external forces, then, due to the negligible thickness of the interfacial layer, these forces will play a minor role and can be neglected. For example, the effect of gravity is proportional to the mass of liquid contained in the interfacial layer and hence practically never plays any role in the interfacial dynamics.

The second way in which an interface can contribute to the overall dynamics of the system is through its intrinsic dynamic properties the most important being the surface tension. The concept of the 'surface tension' has been introduced by Segner in 1751 (see Bikerman 1977/78), who also tried, though unsuccessfully, to give its mathematical description.

Physically, the surface tension appears as a result of an asymmetric action on the interfacial layer of intermolecular forces from the bulk phases. These forces are singularly strong compared to those considered in fluid mechanics so that their strength compensates the negligible thickness of the interfacial layer making the resultant dynamic effect finite.

From a mechanical point of view, the surface tension is a force exerted by a part of the interface on a line confining it along the inward normal to the line lying in the plane tangential to the interface (Fig. 2.10a). The surface tension can be seen as a two-dimensional analogue of pressure, but unlike pressure which, when positive, always tries to expand the volume of fluid, the surface tension is trying to contract the interface. Quantitatively, the surface tension is characterized by the surface tension coefficient, σ, which is defined as the magnitude of the force acting on line per unit length. The experimentally measured equilibrium values of σ for some liquid-fluid pairs are given in Appendix D. It is also known from experiments that σ depends on temperature and, if surfactants are added to the system, on their concentration. In some systems, the surface tension can depend also on the intensity of an external electromagnetic field.

It should be emphasized here that in the 3-dimensional space the surface tension is a *concentrated* force that produces a finite force while acting on a *line*. On the other hand, the bulk stress is a *distributed* force which generates a finite force only when acting on an *area*. This important distinction reflects two points: (a) the fact that the surface tension appears as a result of singularly strong forces of nonhydrodynamic origin acting on the interfacial layer, and (b) our modelling of the interfacial layer as a two-dimensional interfacial surface of zero thickness resulting from the continuum approximation. The distinction between concentrated and distributed forces will help us to understand how to remedy some singularities in mathematical models we will consider later.

Mathematically, σ is a function defined along the interface and, in the framework of fluid-mechanical modelling, it has to be included in a 2-dimensional *surface* stress tensor $\mathbf{P}^s$. In order to imbed $\mathbf{P}^s$ into the 3-dimensional space, it is convenient to use a tensor $(\mathbf{I} - \mathbf{nn})$, where as before $\mathbf{I}$ is the metric tensor in space, and $\mathbf{n}$ is a unit normal to the interface. This tensor generates a metric on the surface and singles out the tangential components of vectors: if $\mathbf{a} = a_n\mathbf{n} + \mathbf{a}_{\parallel}$, where $\mathbf{a}_{\parallel}$ is tangential to the interface, one obviously has $(\mathbf{I} - \mathbf{nn}) \cdot \mathbf{a} = \mathbf{a}_{\parallel}$.

Taking

$$\mathbf{P}^s = \sigma(\mathbf{I} - \mathbf{nn}), \qquad (2.140)$$

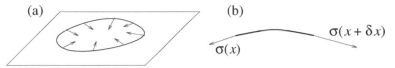

FIGURE 2.10: (a) The surface tension with which an element of the interface acts on its boundary is normal to the boundary and tangential to the interface; it tries to minimize the area of this element. (b) The resultant action of the surface tension on a surface element from a surrounding surface has a normal component if the interface is not flat, and a tangential component if the surface tension varies along the interface.

we have $\mathbf{n} \cdot \mathbf{P}^s = 0$, whilst for any two unit vectors $\mathbf{t}_1$ and $\mathbf{t}_2$ lying in the interface and normal to each other $\mathbf{t}_1 \cdot \mathbf{P}^s \cdot \mathbf{t}_2 = 0$ and $\mathbf{t}_1 \cdot \mathbf{P}^s \cdot \mathbf{t}_1 = \sigma$. Hence for a line lying in the interface and normal to $\mathbf{t}_1$ the stress tensor defined by (2.140) describes a force directed along $\mathbf{t}_1$ with the magnitude σ per unit length of the line. This is exactly how we defined the surface tension. In what follows, it is important to remember also that since σ is defined only on the interface, its derivative in the direction normal to it is zero by definition, that is $\mathbf{n} \cdot \nabla \sigma = 0$. This constraint allows one to use σ formally as a function of all three space coordinates. The same applies to other surface characteristics if the interface possesses other mechanical and/or thermal properties, and in particular one has $\mathbf{n} \cdot \nabla \mathbf{n} = 0$.

Consider the momentum flux across an interface. Given the requirement of momentum conservation and taking into account the contribution from the interface, we now have

$$\mathbf{n} \cdot (\mathbf{\Pi}^+ - \mathbf{\Pi}^-) = \nabla \cdot \mathbf{P}^s + \mathbf{F}^s \tag{2.141}$$

instead of (2.139). Here $\mathbf{F}^s$ is the density of external forces per unit area acting on the interface; the contribution from the surface stress in the form of $\nabla \cdot \mathbf{P}^s$ is analogous to the corresponding contribution of bulk stresses in (2.31).

Substituting expressions (2.138) and (2.140) for $\mathbf{\Pi}^\pm$ and $\mathbf{P}^s$ and making use of the mass flux continuity condition (2.133), we have

$$\rho^+ \mathbf{u}^+ (\mathbf{u}^+ - \mathbf{v}^s) \cdot \mathbf{n} - \rho^- \mathbf{u}^- (\mathbf{u}^- - \mathbf{v}^s) \cdot \mathbf{n} - \mathbf{n} \cdot (\mathbf{P}^+ - \mathbf{P}^-)$$

$$= \nabla \sigma - \sigma \mathbf{n} \nabla \cdot \mathbf{n} + \mathbf{F}^s. \tag{2.142}$$

The first two terms on the left-hand side written down as

$$\rho^+ \mathbf{u}^+ (\mathbf{u}^+ - \mathbf{v}^s) \cdot \mathbf{n} - \rho^- \mathbf{u}^- (\mathbf{u}^- - \mathbf{v}^s) \cdot \mathbf{n} = \chi (\mathbf{u}^+ - \mathbf{u}^-),$$

show that mass transfer across an interface has an impact on the momentum balance when (a) it is significant in itself, and (b) it is associated with a considerable jump in the bulk velocity. This is the case, for example, in shock waves. On the contrary, for liquid-fluid interfaces the effect of mass transfer on the momentum balance is practically always negligible compared to capillary effects and especially to the bulk stress.

The first two terms on the right-hand side of (2.142) give the tangential and normal components of the force acting on a surface element due to surface tension and its gradient (Fig. 2.10b). In writing them down we used that (a) $(\mathbf{I} - \mathbf{nn}) \cdot \nabla\sigma = \nabla\sigma$ since the surface-tension gradient is directed along the interface and (b) $\mathbf{n} \cdot \nabla\mathbf{n} = 0$.

In the simplest case of an impermeable interface ($\chi = 0$) and negligible external surface forces ($\mathbf{F}^s = 0$), the normal projection of (2.142),

$$p^- - p^+ + \mathbf{n} \cdot (\boldsymbol{\sigma}^+ - \boldsymbol{\sigma}^-) \cdot \mathbf{n} = \sigma\nabla \cdot \mathbf{n}, \qquad (2.143)$$

is known as the *capillarity equation*. Here $p^\pm$ are pressures in the two fluids evaluated at the interface and $\boldsymbol{\sigma}^\pm = \mu^\pm[\nabla\mathbf{u}^\pm + (\nabla\mathbf{u}^\pm)^*]$ are viscous stresses; $\sigma\nabla \cdot \mathbf{n}$ is known as the *capillary* (or Laplacian) *pressure*, where $\nabla \cdot \mathbf{n}$ is the mean curvature of the interface. If in the Cartesian coordinates x, y, z one can represent the free surface explicitly as $f(\mathbf{r}, t) = h(x, y, t) - z = 0$, then

$$\nabla \cdot \mathbf{n} = \nabla \cdot \left(\frac{\nabla h}{\sqrt{1 + |\nabla h|^2}} \right). \qquad (2.144)$$

The tangential projection of (2.142) has the form

$$\mathbf{n} \cdot (\boldsymbol{\sigma}^- - \boldsymbol{\sigma}^+) \cdot (\mathbf{I} - \mathbf{nn}) = \nabla\sigma. \qquad (2.145)$$

This condition is the key element in the mathematical description of the Marangoni effect (Marangoni & Stefanelli 1872, 1873), where the flow is driven by the surface-tension gradient resulting from a nonuniform distribution of temperature and/or surfactant concentration along the interface. In order to model this effect, we need to bring in additional equations describing the temperature and/or surfactant concentration distributions in the bulk and on the interface as well as the surface equation of state specifying how σ depends on these factors. If fluid 2 is assumed to be inviscid and hence one has

$$\mathbf{n} \cdot \boldsymbol{\sigma}^- \cdot (\mathbf{I} - \mathbf{nn}) = 0,$$

the Marangoni flow arises from the balance of the shear stress resisting the motion and the surface tension gradient driving it:

$$\mathbf{n} \cdot \boldsymbol{\sigma}^+ \cdot (\mathbf{I} - \mathbf{nn}) + \nabla\sigma = 0. \qquad (2.146)$$

In the absence of the surface-tension gradient, one has simply the continuity of tangential stress across the interface:

$$\mathbf{n} \cdot \boldsymbol{\sigma}^+ \cdot (\mathbf{I} - \mathbf{nn}) = \mathbf{n} \cdot \boldsymbol{\sigma}^- \cdot (\mathbf{I} - \mathbf{nn}). \qquad (2.147)$$

Here we set aside an important question as to whether the surface-tension gradient can arise in a surfactant-free system with no significant thermal effects, i.e., due to the flow itself and, if this is the case, how to model it. This question will be addressed in the subsequent chapters.

Thus, the two conservation laws together with the kinematic equation of motion for the interface allowed us to formulate five scalar boundary conditions which, in the simplest case of an impermeable interface, no external surface forces and constant surface tension, are given by (2.136), (2.143) and (2.147). These conditions are not sufficient, however. Indeed, equations (2.70), (2.71) in both fluids require six scalar conditions on an interface with a prescribed shape. Given that in our case the interface itself is not known, this means that two scalar boundary conditions are missing in our formulation. (In the case of compressible fluids, we would derive an addition condition resulting from the energy balance requirement, but since there would also be more unknowns, the problem would not go away.)

In order to complete the set of boundary conditions, we need to consider how the tangential components of the bulk velocity on the opposite sides of the interface are related. This condition and assumptions behind it are less obvious than those described earlier, and here we will consider in more detail how it is conventionally formulated, what assumptions about the interfacial layer are implied and in what situations it might become inadequate.

The standard way of formulating the missing condition is to assume that the tangential component of the bulk velocity is continuous across the interface:

$$\mathbf{u}_{\parallel}^{+} = \mathbf{u}_{\parallel}^{-}. \tag{2.148}$$

Here for brevity we used the notation $\mathbf{u}_{\parallel}^{\pm} = \mathbf{u}^{\pm} \cdot (\mathbf{I} - \mathbf{nn})$ for the tangential projection of velocities.

The justification for condition (2.148) is based on an assumption that the two fluids separated by an interface are physically similar in a sense that the interface as a material surface is essentially similar to a material surface in the bulk of a viscous fluid. Then,

> Any discontinuity in velocity across a material surface would lead almost immediately (through molecular transport) to a very large stress at the surface of such a direction as to tend to eliminate the relative velocity of the two masses; the condition of continuity of the velocity is thus not an exact law, but a statement of what may be expected to happen approximately, in normal circumstances (Batchelor 1967, p. 149).

Quantification of this argument allowing one to identify possible *abnormal* circumstances under which (2.148) will become inadequate can be found in Barenblatt & Chernyi (1963) where some general implications of discontinuities in dissipative media are examined. The gist of Barenblatt & Chernyi's result is that the tangential velocity variation across a layer is proportional

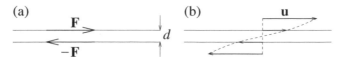

FIGURE 2.11:   A torque formed by forces external to a layer (a) creates a variation in the tangential velocity across it (b). When the layer collapses into a surface, as for an interfacial layer in the continuum approximation, the velocity difference across it is determined by the limit of Fd as $d \to 0$.

to the angular momentum of the torque acting on it from outside (Fig. 2.11), that is to (a) the magnitude of forces trying to create the velocity difference and (b) the thickness of the layer. As the layer's thickness goes to zero, the velocity variation across it due to finite forces will also tend to zero so that in this limit one will arrive at (2.148). However, if in this limit the angular momentum of the torque remains finite, that is the magnitude of the forces tends to infinity inversely proportional to the thickness of the layer, one will end up with the tangential velocity discontinuous across a surface. In our case, the forces acting on the interfacial layer are the tangential components of viscous stresses from the two fluids, i.e., $\mathbf{n} \cdot \boldsymbol{\sigma}^{\pm} \cdot (\mathbf{I} - \mathbf{nn})$. Thus, for finite tangential stresses acting on an interfacial layer, whose thickness in the continuum approximation is equal to zero, the boundary condition (2.148) becomes justified. At the same time, if a problem formulated using (2.148) leads to a solution with a singularity in the tangential stress distribution at the interface, this will be in conflict with the assumptions about the solution inherent in the problem formulation which then will have to be revisited. In §4.3 we will show how condition (2.148) follows from the derivation of boundary conditions at interfaces using methods of irreversible thermodynamics.

The Navier-Stokes equations (2.70), (2.71) in the regions $f < 0$ and $f > 0$ together with boundary conditions (2.136), (2.143), (2.147) and (2.148), i.e.,

$$\frac{\partial f}{\partial t} + \mathbf{u}^{+} \cdot \nabla f = 0, \tag{2.149}$$

$$\mathbf{u}^{+} \cdot \mathbf{n} = \mathbf{u}^{-} \cdot \mathbf{n}, \tag{2.150}$$

$$p^{-} - p^{+} + \mathbf{n} \cdot (\boldsymbol{\sigma}^{+} - \boldsymbol{\sigma}^{-}) \cdot \mathbf{n} = \sigma \nabla \cdot \mathbf{n}, \tag{2.151}$$

$$\mathbf{n} \cdot \boldsymbol{\sigma}^{+} \cdot (\mathbf{I} - \mathbf{nn}) = \mathbf{n} \cdot \boldsymbol{\sigma}^{-} \cdot (\mathbf{I} - \mathbf{nn}), \tag{2.152}$$

$$\mathbf{u}_{\parallel}^{+} = \mathbf{u}_{\parallel}^{-}, \tag{2.153}$$

allow one to find the bulk parameters in both fluids as well as the time-dependent shape of an interface between them in the simplest case of an impermeable interface with constant surface tension and no external surface forces. If these simplifications are relaxed, one will have to bring in additional equations and boundary conditions specifying χ, σ, $\mathbf{F}^{s}$, the distributions of temperature, concentrations of surfactants, etc.

If fluid 2 is inviscid, then in $f < 0$ one will have to use (2.40), (2.41) instead of (2.70), (2.71) and hence need fewer boundary conditions at the interface between the two fluids. The problem formulation will then require the kinematic condition (2.136) together with the dynamic conditions (2.143) and (2.147), which will now take a simpler form:

$$p^- - p^+ + \mathbf{n} \cdot \boldsymbol{\sigma}^+ \cdot \mathbf{n} = \sigma \nabla \cdot \mathbf{n}, \tag{2.154}$$

$$\mathbf{n} \cdot \boldsymbol{\sigma}^+ \cdot (\mathbf{I} - \mathbf{nn}) = 0. \tag{2.155}$$

The tangential velocities on the two sides of the interface are not related and have to be found.

Finally, at an interface between two ideal incompressible fluids one has (2.136) and the capillarity equation

$$p^- - p^+ = \sigma \nabla \cdot \mathbf{n}, \tag{2.156}$$

which now has the same form as in the static situation. One must have in mind, however, that pressures in this condition are dynamic and have to be determined.

2.4.3 Conditions on solid boundaries

If the solid in contact with the fluid is deformable, then to describe the fluid motion one has to consider the continuum mechanics problems for the fluid and the solid simultaneously and formulate the boundary conditions at the fluid-solid interface in conceptually the same way as at the liquid-fluid interface. Given that solid mechanics and fluid mechanics have considerably different mathematical frameworks, formulating a tractable mathematical problem for this type of interactions becomes a nontrivial task.

However, if elastic and plastic deformations of the solid are not important in the processes to be studied, the shape and velocity of the solid boundary are known and can be prescribed for the fluid-mechanical part of the problem. This apparent simplification compared to the liquid-fluid interface, where both of these characteristics have to be found, brings in also some fundamental theoretical difficulties. The actual liquid-solid interface, where the boundary conditions must reflect the essential physics of the liquid-solid interaction, is almost invariably rough and chemically heterogeneous on a macroscopic length scale (Fig. 2.12). In mathematical modelling, we neither can nor wish to incorporate all details of the solid surface structure and have to replace the actual liquid-solid interface with an 'effective' smooth boundary. Hence it becomes necessary to fuse the boundary conditions on the actual surface and the relevant average characteristics of this surface into some 'effective' boundary conditions on the 'effective' smooth boundary. This already nontrivial problem becomes even more challenging after one realizes that quantification of the very term 'roughness' as a characteristic of the surface will depend on the phenomenon to be modelled and differs, for example, for turbulence and dynamic wetting. We will consider some aspects of this problem later (§5.8.1), and here it will be assumed that on the fluid-mechanical length scale the solid surface can be regarded as smooth. This idealization is applicable, for example, to surfaces like the sides of crystals, cleaved mica or the walls of some microscopic capillaries; it can also be applied (though with some caution) to laminar flows over polished surfaces.

If $\mathbf{U}$ is the velocity of the solid boundary S, then the generally accepted boundary condition for the velocity field $\mathbf{u}$ applied on S is given by

$$\mathbf{u} = \mathbf{U}, \qquad \mathbf{r} \in S \tag{2.157}$$

and is known as the *no-slip condition*. Now the interface separates a liquid and a solid, that is two media essentially different from a physical point of view, and the only justification for the no-slip condition one can find in the literature is that the solutions obtained using this condition appear to be accurate in describing many experimental observations:

> The absence of slip at a rigid wall is now amply confirmed by direct observation and by correctness of its many consequences under normal conditions (Batchelor 1967, p. 149).

These 'normal conditions' have been gradually established by considering flows in capillaries (e.g., Whetham 1890), resistance of a sphere moving in a viscous liquid at low Reynolds numbers (e.g., Lamb 1932, pp. 594–616) and other model flows, most of which are now instrumental in viscometry. The 'abnormal conditions' are generally seen to be those where one has a rarefied gas flowing past a solid surface with the Knundsen effect associated with the finiteness of intermolecular distances leading to a certain degree of slip.

It should be emphasized, however, that even leaving aside the case of a rarefied gas and considering a *liquid*-solid interface, we cannot regard the no-slip condition as settled for the following two reasons.

Firstly, there are a number of quite 'normal' flows where the mathematical problem formulated using the no-slip condition has *no solution* so that the very question of how accurate the no-slip condition is for these flows becomes meaningless. We will discuss such flows in detail in Chapters 3 and 5.

Secondly, some experiments show that even for the standard capillary flow there are situations where one can measure significant deviations from the Poiseuille law (e.g., Churaev, Sobolev & Somov 1984). A review of experimental studies regarding slip of Newtonian liquids at solid surfaces can be found in Neto et al. (2005).

It is also important to note that although the ultimate validation for all boundary conditions, as well as the bulk equations themselves, comes from how accurate the solutions obtained on their basis are in describing experimental observations, their justification on purely empirical grounds, such as the one we see in the case of the no-slip condition, provides no conceptual framework for generalizations required to incorporate other physical effects.

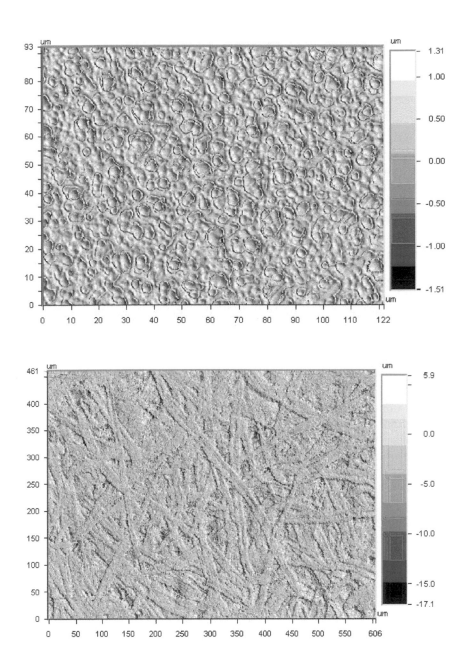

FIGURE 2.12: Magnified views of real solid surfaces. (Courtesy of A. Clarke).

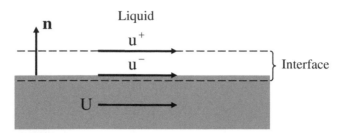

FIGURE 2.13:   A sketch illustrating the structure of a liquid-solid interface. Physically, an interfacial layer, modeled in the continuum approximation
as a zero-thickness 'interface', comprises two sublayers where physical properties of the media differ from those of the bulk phases. If the shape and velocity
of the solid is prescribed and hence the solid is excluded from consideration,
the role of an interface is played by the liquid sublayer.

Consider the actual assumptions behind the no-slip condition (2.157). Physically, a liquid-solid interface, as well as a liquid-fluid one, is a thin layer
(modelled as a surface) where the molecules of the two media experience non-
symmetic influence of intermolecular forces from the bulk phases. This layer
is formed by two sublayers (Fig. 2.13) and, strictly speaking, physical properties of the two materials in each of the sublayers differ from those in the
bulk (see Derjaguin, Churaev & Muller 1987, ch. 7 for a review of experiments). Therefore, in a general case of a deformable solid the problems in the
fluid and in the solid are coupled and have to be solved simultaneously. If,
as we assumed at the very beginning, processes in the solid are not essential
for the fluid-mechanical phenomena to be modelled, the shape and velocity
of the solid substrate can be prescribed and the liquid-solid interface can be
regarded as comprising only the liquid sublayer with the material of the solid
manifesting itself via the equilibrium parameters of this sublayer. The physical origin of the specific 'surface' properties in the liquid sublayer is the same
as the origin of the surface tension in a liquid-fluid interface, and, as we will
see in the next section, when this sublayer is not properly accounted for, one
can arrive at fundamentally incorrect results.

The velocity required as a boundary condition by the bulk equations in the
liquid is that on the *liquid*-facing side of the interface. If we use $\mathbf{u}^+$ and $\mathbf{u}^-$
to denote, respectively, the velocities on the liquid-facing and solid-facing side
of the liquid-solid interface whereas $\mathbf{U}$, as before, denotes the velocity of the
solid, then the no-slip condition (2.157) becomes the result of the following
four assumptions.

(i) The solid substrate is impermeable at a microscopic level and there are
     no chemical reactions on its surface so that

$$(\mathbf{u}^- - \mathbf{U}) \cdot \mathbf{n} = 0. \qquad (2.158)$$

(ii) There is no *actual* slip on the solid surface:

$$\mathbf{u}_\parallel^- = \mathbf{U}_\parallel. \tag{2.159}$$

This seems not to be the case for some liquid-solid pairs, at least potentially, as indicated, for example, by molecular dynamics simulations (Thompson & Robbins 1990).

(iii) There is no net mass exchange between the interface and the bulk as a result, for example, of the interface formation process:

$$(\mathbf{u}^+ - \mathbf{u}^-) \cdot \mathbf{n} = 0. \tag{2.160}$$

In all 'normal' situations considered in classical fluid mechanics the interfaces are invariably assumed to have been already formed.

(iv) External forces acting on the interface do not cause *apparent* slip, that is the difference between the tangential components of velocities on the opposite sides of the interface:

$$\mathbf{u}_\parallel^+ = \mathbf{u}_\parallel^-. \tag{2.161}$$

This condition is similar to (2.148) and results from the same arguments. It also implies that in the situations where the use of the no-slip condition leads to singularities in the tangential-stress distribution at a solid boundary, the solution is in conflict with the assumptions lying at the basis of the problem formulation.

Thus, once the liquid-solid interface is treated in conceptually the same way as the liquid-fluid one, we see that the no-slip condition appears as a result of a number of assumptions about the interfacial layer. Each of these assumptions can become inadequate under certain conditions.

If the liquid in contact with a solid is assumed to be inviscid, the bulk equations (2.40), (2.41) require only a condition on the normal component of velocity. The impermeability conditions usually formulated at a solid surface

$$(\mathbf{u} - \mathbf{U}) \cdot \mathbf{n} = 0$$

implies (2.158) and (2.160) in terms of assumptions about the interfacial layer.

2.4.4   Conditions at three-phase contact lines

If two boundaries intersect, there appears a region where the molecules are subject to the action of intermolecular forces from three bulk phases. In the continuum approximation, all molecular length scales are negligible, and this region becomes a one-dimensional three-phase *contact line*. If at least one of the intersecting boundaries is a free surface and its surface tension is

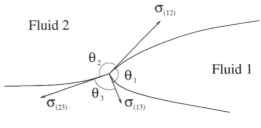

Fluid 3

FIGURE 2.14:   The balance of forces acting on a three-phase contact line in a liquid/liquid/fluid system.

important for the processes to be described, the capillarity equation (2.143) (or its simplifications (2.154), (2.156)) becomes an equation determining the free-surface shape. The differential operator (2.144) expressing the free-surface curvature as a function of the free-surface position is elliptic, and hence to specify the free-surface shape we need one boundary condition at every point of the contact line.

In classical fluid mechanics, conditions at contact lines are considered only at equilibrium, i.e., in the situations where the contact line is not moving with respect to the bulk phases (Batchelor 1967). We will formulate these conditions here, leaving the dynamic situation for the subsequent chapters.

In the case of a liquid-liquid-fluid system, the triple line is free to move in any direction so that in equilibrium the vector sum of all forces acting on it has to be zero. As we have already mentioned in §2.4.2, the bulk stresses, being distributed forces, make no contribution to the balance of forces acting on a line so that the state of equilibrium corresponds to the configuration in which the concentrated forces, that is surface tensions, are in balance. This situation is illustrated in Fig. 2.14 where the low indices are put in brackets to distinguish the surface tensions acting along the boundary between two phases from components of the viscous stress tensor $\boldsymbol{\sigma}$; θ_i is the angle at which the interfaces confining the ith fluid intersect at the contact line. These angles can be defined as the angles between the tangents to the interfaces in the cross-section normal to the triple line as in Fig. 2.14, or, equivalently and more formally, as the angles between *normals* to the interfaces.

After projecting the surface tension balance on two nonparallel directions, one can obtain two scalar equations relating $\sigma_{(ij)}$ $(i, j = 1, 2, 3; i \neq j)$. In particular, projecting this balance on the directions of two of the surface tensions one obtains

$$\sigma_{(12)} + \sigma_{(23)} \cos \theta_2 + \sigma_{(13)} \cos \theta_1 = 0, \qquad (2.162)$$

$$\sigma_{(23)} + \sigma_{(12)} \cos \theta_2 + \sigma_{(13)} \cos \theta_3 = 0. \qquad (2.163)$$

The third equation,

$$\sigma_{(13)} + \sigma_{(12)} \cos \theta_1 + \sigma_{(23)} \cos \theta_3 = 0$$

is obviously a consequence of (2.162) and (2.163). If we neglect the mass of the triple line and the interfaces, then equations (2.162) and (2.163) will hold for the dynamic situation as well, though in the dynamic case the surface tensions in these equations are not necessarily the same as in equilibrium.

The force balance (2.162), (2.163) is only possible if each of the surface tensions is not greater than the sum of the other two. If this is not the case and

$$|\sigma_{(ij)}| > |\sigma_{(jk)}| + |\sigma_{(ik)}| \qquad (i \neq j \neq k \neq i)$$

for some i, j, k, the equilibrium configuration shown in Fig. 2.14 becomes impossible and, physically, the kth fluid will tend to spread out until it forms a layer whose thickness is on a molecular scale. In the continuum approximation, such thickness is infinitesimal so that, from the viewpoint of fluid mechanics, the kth fluid ceases to be a 'bulk' phase and the presence of a microscopic layer of this fluid has to be taken into account via the modified properties of the interface between the ith and the jth fluid. A typical example of such a situation is a film of oil on the surface of water.

If a free surface meets a solid boundary, the situation becomes conceptually more complex since, from a physical point of view, the interface between a liquid and a solid includes sublayers in both media, as has already been schematically shown in Fig. 2.13, so that, strictly speaking, the surface tension in this interface includes contributions from both sublayers. Correspondingly, the contact line is, strictly speaking, the region comprising molecules of all three phases. If processes in the solid are not essential for the fluid-mechanical part of the problem, the liquid-solid interface can be regarded as comprising only liquid sublayer with the solid as a geometric constraint determining its shape and manifesting itself via a reaction force that acts on the contact line to prevent it from moving in the direction normal to the solid surface. The material of the solid, i.e., the intermolecular forces with which it acts on the molecules of the liquid, will determine the equilibrium parameters of the liquid-solid interface. Now, the surface tension of the liquid-solid interface to be considered in fluid-mechanical problems results from the tangential forces acting in the liquid sublayer. Unlike the liquid-fluid interface, the surface tension in the liquid-solid interface as defined above can be positive or negative depending on the intermolecular forces between the solid and the liquid (Gibbs 1928). Correspondingly, the 'contact line' is now a region subject to intermolecular forces from all three bulk phases that comprises only the molecules of the two fluids (Fig. 2.15a).

The contact line can move only parallel to the solid surface so that the equilibrium condition is the balance of the tangential projections of forces acting on the contact line (Fig. 2.15b):

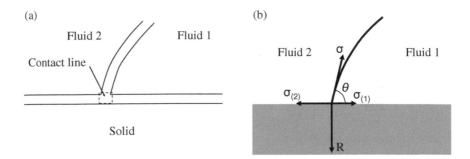

FIGURE 2.15: Sketch of fluid-fluid-solid system and Young's force diagram. If the solid is excluded from consideration so that its shape and velocity are prescribed, the three-phase-interaction zone (the 'contact line') is formed only by the molecules of the fluids that are subject to intermolecular forces from the three bulk phases (a). The solid manifests itself via the reaction force (b) and the equilibrium values of parameters characterizing the liquid-solid interfaces. These interfaces are also understood as thin layers in the *fluids* adjacent to the solid boundary and modelled as mathematical surfaces.

$$\sigma \cos \theta = \sigma_{(2)} - \sigma_{(1)}. \tag{2.164}$$

This fundamental relationship, known as the Young equation (Young 1805), defines the concept of the *contact angle*, θ, between a free surface and a solid boundary in macroscopic fluid mechanics. The fact that the contact angle is *defined* in terms of a force balance is important since it emphasizes that we have already taken the continuum limit in which the interfaces became mathematical surfaces of zero thickness with concentrated forces (surface tensions) acting along them. The Young equation has been originally formulated for hydrostatics and below, when we refer to the static situation, we will use the notation θ_s and call it the 'static' (or 'equilibrium') contact angle.

In equilibrium, the contact angle measured through the liquid in a liquid/gas/solid system defines the property referred to as 'wettability'. If fluid 1 is a liquid and

$$\sigma \le \sigma_{(2)} - \sigma_{(1)}, \tag{2.165}$$

then the liquid tends to spread out to form a microscopic layer. Formally, one can say that in this case the equilibrium contact angle is $0°$. This situation is referred to as the case of 'complete' (or 'perfect') wetting. The equilibrium contact angle between $0°$ and $180°$ corresponds to 'partial' or 'incomplete' wetting, and if it is equal to $180°$ (this case is yet to be discovered), this is 'non-wetting' (Ablett 1923).

If one of the contacting fluids, say fluid 2, is a gas where, as we have mentioned in §2.1.1, the average intermolecular distance is much larger than

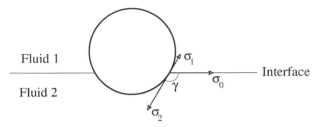

FIGURE 2.16: Young's force diagram for a floating spherical ball in Finn's (2006) counterexample.

the range of intermolecular forces, then, theoretically, both the solid-gas and the liquid-gas interfaces have no sublayer in the gas, and hence $\sigma_{(2)} = 0$. However, in real systems the solid-gas interface is almost invariably imperfect. It contains adsorbed molecules of the gas and/or vapour that modify the interfacial properties of the solid surface so that $\sigma_{(2)}$ can be nonzero, though still very small. An important point to be emphasized here is that $\sigma_{(2)}$ should not be confused with the 'surface tension of the solid' (e.g., Marmur 2006). The latter is difficult to define and, theoretically, it can be large, but, as pointed out by Gibbs (1928, p. 329), it is *not* the quantity involved in the Young equation (2.164).

It is also noteworthy that in static experiments $\sigma_{(1)}$ and $\sigma_{(2)}$ cannot be measured separately so that usually it is the postulated Young equation together with the measurements of σ and θ that allow one to determine the *difference* $\sigma_{(2)} - \sigma_{(1)}$. As we will see in Chapter 5, in the dynamic situation $\sigma_{(1)}$ and $\sigma_{(2)}$ act differently, depending on the direction of the contact-line motion, and this effect can be used for their experimental investigation.

The normal projection of the balance of forces acting on the contact line (Fig. 2.15b) allows one, if necessary, to find the reaction force **R** exerted on the contact line by the solid. The magnitude of this force is given by

$$R = \sigma \sin \theta. \tag{2.166}$$

The reaction force received little attention in the literature, and Maxwell's (1875, p. 566) words that "the surface tensions normal to the surface are balanced by the resistance of the solid" essentially sum up the conventional view on this matter. However, a recent paper by Finn (2006) makes it instructive to revisit this topic to illustrate the concepts we discussed earlier.

2.4.4.1 Reaction force and Finn's (2006) 'counterexample'

Finn (2006) considered a model situation of "a spherical ball floating in equilibrium, in an infinite liquid bath, in zero gravity" (Fig. 2.16). Then, the bulk pressures in both fluids are constant and equal so that the only solution for the equations of hydrostatics is that the free surface is a plane meeting the solid at an angle γ (in this section we use Finn's original notation). According to Young, this angle depends on the materials of the system and hence, in a general case, is not equal to 0, π or $\pi/2$. Commenting on this solution, Finn writes:

> For this solution there is a normal force per unit length on the latitude circle, $\sigma_N = \sigma_0 \sin\gamma$ and hence a vertical force $\sigma_V = \sigma_0 \sin\gamma \cos\gamma$ pulling downward, with no reaction force to compensate.
>
> *We are in trouble; the ball cannot be in equilibrium.* (Finn 2006, 047102-2)

In other words, the tangential components of σ_0, σ_1 and σ_2 balance each other (Fig. 2.16) whereas the normal component of σ_0 remains the only force acting on the contact line. Apart from the horizontal component of this force that gives no net contribution to the force balance, for $\gamma \neq \pi/2$ there is also a vertical component that produces a force pulling downward, and there is nothing to balance it since the equal pressures in the bulk give no net contribution to the force acting on the ball.

The above argument has led Finn to a conclusion that "Young reasoned incorrectly", his original construction in terms of tensions was an "artefact, depending on the notion of solid/liquid interfacial tension that nobody understood and which there was no way to measure", and that the reason why Young's assertion has "survived" so long is that this equation can be reformulated in terms of the surface energies that, according to Finn, are conceptually much clearer (though no easier to measure experimentally).

As we will see below, the above reasoning and the conclusion derived from it are fundamentally incorrect. It is precisely the understanding of what the tension of a liquid-solid interface is that allows one to avoid confusion, resolve Finn's "trouble" in an elementary way and extend Young's analysis to a dynamic situation.

In order to emphasize that the liquid-solid interface has two sides and that, importantly, the interface featuring in the Young equation (2.164) is a microscopic layer of *liquid* adjacent to the solid surface, we will show it graphically in Fig. 2.17a,b as a thin layer. The surface tension of the liquid-solid interface is the tension in this layer. Gibbs (1928, p. 329) called it "the superficial tension of the fluid in contact with the solid" to distinguish it from the surface tension of the interface as a whole, i.e., the layer comprising molecules of both the liquid and the solid. It is important to emphasize that this "superficial tension" acts *everywhere* along the liquid-solid interface and not just at the contact line.

Consider the forces acting on the outer boundary of the liquid-solid interface, i.e., on the ball together with the interface (Fig. 2.17a). These forces are: the surface tension acting from the free surface on the contact line and the hydrostatic pressures from the two fluids. Since in equilibrium the pressures in the two fluids are the same, there is no net force on the liquid-facing side of

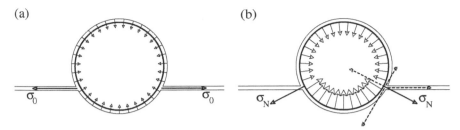

FIGURE 2.17: Sketch illustrating the resolution of Finn's (2006) paradox. (a) Forces on the outer boundary of the liquid-solid interface. (b) Forces acting on the solid sphere itself. In both cases, the total force is zero. The dashed arrows show the Young's diagram of forces acting on the contact line (three phase interaction zone) from the three interfaces and the reaction force exerted on the contact line by the solid. σ_N is the force acting *on* the solid from the contact line; it is equal to the normal projection of the surface tension acting along the free surface, and its net downward action is compensated exactly by the upward force due to the difference in the pressures on the top and bottom segments of the sphere.

the interface due to this factor. The (constant) surface tension σ_0 acting on the closed contact line from a planar free surface also produces no net force so that, for the ball together with the liquid-solid interface, there is no paradox; the system is in perfect equilibrium.

Now, consider the forces acting on the ball itself (Fig. 2.17b). As we cross the contact line (i.e., the three-phase-interaction zone), the tangential component of σ_0 becomes balanced by σ_1 and σ_2, as the Young equation describes, so that one has only the force normal to the ball's surface. It is this projection that features in Finn's argument. The magnitude of the normal force per unit length of the contact line is obviously $\sigma_N = \sigma_0 \sin\gamma$, and it is only the vertical projection of it of magnitude $-\sigma_0 \sin\gamma\cos\gamma$ that enters the balance of forces acting on the ball. The total force, which for the configuration shown in Fig. 2.17b is in the downward direction, is then

$$-2\pi r \sigma_0 \sin^2\gamma\cos\gamma, \tag{2.167}$$

where r is the ball's radius.

The crucial point to realize is that the normal stress per unit area acting on the solid surface is *not* equal to the pressure in the bulk of the fluids. The liquid-solid interface is under tension (which can be positive or negative), and this tension is different for the upper and the lower segment. Then, the capillary pressure proportional to the surface tension and the interface's curvature must be added to the hydrostatic pressure of the fluid. This is completely analogous to the normal-stress difference one has across a curved liquid-fluid interface. Thus, the spherical segment of the solid boundary located above the free surface experiences an additional pressure $2\sigma_1/r$ whereas the lower segment, where for the configuration shown in Fig. 2.17 the surface tension is larger, experiences a greater additional pressure $2\sigma_2/r$. Integrating the vertical projections of these pressures in spherical coordinates we find that an additional force acting downward on the upper segment is given by

$$\int_0^\gamma \int_0^{2\pi} \left(\frac{2\sigma_1}{r}\right) \cos\theta\, r^2 \sin\theta\, d\phi\, d\theta = 2\pi r \sigma_1 \sin^2\gamma, \qquad (2.168)$$

and an additional force acting upward on the lower segment is

$$-\int_\gamma^{2\pi} \int_0^{2\pi} \left(\frac{2\sigma_2}{r}\right) \cos\theta\, r^2 \sin\theta\, d\phi\, d\theta = 2\pi r \sigma_2 \sin^2\gamma. \qquad (2.169)$$

Here ϕ and θ are spherical coordinates.

Then, the total force in the upward direction is

$$2\pi r \sin^2\gamma(\sigma_2 - \sigma_1). \qquad (2.170)$$

Given that in Finn's paper the contact angle is measured through fluid 2, we have Young's equation in the form

$$\sigma_1 = \sigma_2 + \sigma_0 \cos\gamma. \qquad (2.171)$$

Now, after substituting (2.171) in (2.170) we obtain (2.167). Thus, the downward and the upward forces acting on the solid balance each other, and the ball is in perfect equilibrium. Then, there are absolutely no grounds for dismissing Young's equation in terms of tensions and the force diagram as a long-surviving "artefact".

The origin of confusion is an implicit assumption that, although there may be some surface *energy* associated with the liquid-solid interface (and hence Young's equation can be reformulated in terms of the surface energies, as Finn does), there is no surface *tension*. The latter is assumed to appear only at the contact line. This is clearly a misunderstanding: as the very term '*surface tension*' suggests, it is a force associated with 'tension' of a *surface* and not something suddenly appearing only at a contact line. As mentioned earlier, the existence of this tension everywhere along the liquid-solid boundary has already been known to Gibbs (1928).

In order to have a better feel for this concept with respect to the liquid-solid interface, one can imagine a rubber film under tension (which can be positive or negative) covering the solid surface. (The very notion of 'surface tension' was first introduced by Segner using a comparison of an interface to a stretched membrane.) For a planar surface in equilibrium the presence of this 'film' makes no difference from a dynamic viewpoint unless the contact lines are involved. However, if an interface is curved, as Finn's "counterexample" shows, then, even in equilibrium, the surface tension enters the force balance not only at contact lines but also via the capillary pressure.

It is also important to point out that the Young's equation formulated in terms of surface energies and the Young's equation formulated in terms of surface tensions are equivalent only in equilibrium, i.e., when there is no energy dissipation and the former can be derived by minimizing the total energy of the system. However, when the system is out of equilibrium and the contact line is moving with respect to the solid surface, it is the Young's equation in terms of the surface tensions that allows one to proceed with the mathematical modelling of the process. In this situation, both the surface tensions and the surface energies deviate from their equilibrium values, their distributions along the interfaces are nonuniform and, in a general case, coupled with the bulk flow. The (dynamic) contact angle becomes a boundary condition for the equation determining the free-surface shape, and it can be specified locally in terms of the (dynamic) surface tensions evaluated at the contact line in a straightforward way (see Chapter 5) whereas the variational approach required to specify the contact angle in terms of the surface energies will face difficulties of principle when it has to deal with the dissipation of energy in the bulk due to fluid's viscosity and the nonuniform distributions of the surface energies along the interfaces coupled with the bulk flow.

2.5 Physically meaningful solutions and paradoxes of modelling

2.5.1 'Standard model'

The Navier-Stokes equations (2.70), (2.71) for the bulk parameters, boundary conditions (2.136), (2.143), (2.147), (2.148) at free interfaces, the no-slip condition (2.157) at solid boundaries and conditions (2.162), (2.163), (2.164) at different types of contact lines define the physical objects as they are described in classical fluid mechanics for flows of incompressible viscous fluids with impermeable capillary interfaces. For brevity we will refer to these elements as the 'standard model'.[15] In order to describe a particular flow, one has to formulate some conditions which will set the above objects in motion and, where necessary, specify a particular class of solutions to be studied. The mathematical problems resulting from this procedure are nonlinear and notoriously difficult to solve analytically so that at present only a handful of exact analytical solutions dealing with very simple flow configurations are known (see Wang 1989, 1991, Drazin & Riley 2006 for reviews). Asymptotic

---

[15]This model has been in existence for more than a century so that it is by far more standard and tested than the 'standard model' of particle physics.

methods make it possible to find approximate analytic solutions in many more cases though, by the very nature of these methods, in a very narrow region of the corresponding parameter space. In most situations, solutions are beyond the reach of analytic techniques and have to be obtained numerically.

This constructive programme of describing particular flows presumes that the corresponding problems formulated in the framework of the standard model have solutions and, once found, these solutions will indeed describe the flows in question in a physically realistic way. However, this is not always the case, and in many situations one has to revise the way in which interfaces and contact lines are modelled, examine the assumptions behind the corresponding boundary conditions and incorporate additional physical effects. Below we discuss these aspects of modelling in detail, but before we proceed it is necessary to make a digression and mention two issues concerning the Navier-Stokes equations themselves.

2.5.1.1   On the Navier-Stokes equations

As has already been mentioned in §2.4.1, the boundary conditions provide information about the 'outside world', i.e., about physical processes not accounted for in the bulk equations, and when we consider how this additional information specifies the bulk flow and whether or not it is sufficient and, more importantly, self-consistent, it is tacitly assumed that the Navier-Stokes equations on their own, that is in an unbounded domain, remain physically reasonable at all times and adequate in describing the effects they incorporate. Here we are talking not about the limits of applicability of these equations or their accuracy; the question is about their most fundamental properties. One aspect of this question is the problem of existence of a solution for $t > 0$ with an arbitrary solenoid velocity field as the initial condition. Clearly, if there is a finite-time blowup of a solution starting from some smooth physically reasonable initial distribution of velocity, this would mean that one has to revise many issues associated with the Navier-Stokes equations. Although it is a widely accepted belief in the fluid-mechanical community that there is no finite-time blowup scenario inherent in the Navier-Stokes equations and this opinion is invariably supported by numerical computations, from a rigorous mathematical point of view this is still an open question. This problem remains a focus of a substantial research effort aimed at finding a proof of the existence theorem in one form or another with only partial success so far. There are a number of results proving the existence of solutions for different cases (see Ladyzhenskaya 1975, Majda & Bertozzi 2001 and references therein), which include solvability for two-dimensional flows and, under certain conditions, three-dimensional flow over a finite time interval (Ladyzhenskaya 1969). In what follows, we will be dealing with the situations where the question of solvability does not arise.

Another aspect of the general question concerning physical meaningfulness of the Navier-Stokes equations has to do with propagation of disturbances. It can be easily shown that for equations (2.70), (2.71) the speed of disturbances is infinite. One source of this is the condition of incompressibility which makes the speed of sound infinite for both viscous and ideal fluids. If necessary, this can be remedied in a natural way by allowing for the density variations, that is replacing (2.70) with (2.13) and closing the set of equations (2.13), (2.71) in the way described earlier in this chapter.

A more difficult issue is t-parabolicity of the equation of motion (2.71) and the resulting infinite speed of propagation of shear disturbances. This feature is similar to that inherent in the classical heat conduction equation (2.125) where, as is well known, the speed with which thermal disturbances propagate is also infinite.

In the theory of heat conduction, the first attempt to remedy the situation has been made by Cattaneo (1948, 1958) who introduced a short but finite time lag between the temperature gradient and the temperature rise it causes. The result was the so-called 'telegraphic equation' of heat conduction with the second order time derivative and a small parameter in front of it present on the left-hand side of (2.125).

These ideas for thermal processes have been developed in a number of works (e.g., Vernotte 1958, Lykov 1965) and transferred to fluid dynamics where they took different forms, from the telegraphic equation as an ad hoc model for propagation of disturbances (Weymann 1967) and the same equation but resulting from what effectively amounts to a revision of Newton's second law (Kuznetsov 1993) to a more sophisticated but no less radical approach (Ruggeri 1983, 1989) based on the so-called 'extended thermodynamics' (Müller & Ruggeri 1998). The gist of the latter is that it extends the set of independent parameters of state for nonequilibrium states to include viscous stresses σ and thermal flux $\mathbf{q}$ and, motivated by Grad's (1949) momentum method known in the kinetic theory of rarefied gases, postulates two complementary balance laws for some vector and tensor quantities to be added to those of mass, momentum and energy. Constitutive equations will then have to relate, via a set of algebraic equations, the extended set of parameters of state with vector and tensor quantities and fluxes appearing in all balance laws, including the new ones. This scheme makes the speed of propagation of disturbances finite at the expense of introducing a whole set of new thermodynamic laws of supposedly the same fundamental nature as the conservation of mass, momentum and energy. It should be noted, however, that at present the validity and physical meaning of these new laws are far from being clear, especially for liquids as opposed to rarefied gases.

It should be pointed out that although in the Navier-Stokes equations the speed of propagation of disturbances is infinite and from a theoretical viewpoint this is an undeniable flaw of the model which must be understood, at present there are no experiments indicating discrepancies between what the Navier-Stokes equations predict for unsteady flows and what is actually measured. More significantly, experiments do not indicate the presence of additional similarity parameters which would almost invariably result from modifications of the Navier-Stokes equations. Although both arguments are of course empirical and heavily rely on the accuracy of measurements and the range of parameters accessible to them, they can be also considered as statements broadly specifying the range of validity and the degree of accuracy associated with the Navier-Stokes equations. It is also worth mentioning that known modifications of the Navier-Stokes equations do not suggest possible experiments where the new physical effects they incorporate could be detected.

Thus, there are still unresolved issues concerning the unsteady Navier-Stokes equations, and, having these issues in mind, in what follows we will be dealing only with the situations where the Navier-Stokes equations can be relied upon and regarded as physically adequate.

2.5.2 Analysis of results

Now, we can return to our constructive programme. After a problem has been solved, analytically or numerically, the results have to be interpreted, first of all, from the point of view of whether or not the solution is physically meaningful. This can be a nontrivial task since in many situations complexity of the solution, especially if it has been obtained numerically, makes any intuitive assessment based on common sense and previous experience impossible and there are no sufficiently accurate experimental data for quantitative validation of the theory. Therefore, as a starting point, one has to formulate some general criteria a physically meaningful solution must satisfy. Experimental observations, if available at this stage, can be regarded only as qualitative.

Consider the most basic set of the necessary conditions to be applied to a mathematical theory aimed at describing fluid motion. These criteria include mathematical and modelling aspects as well as a 'descriptive element'. First of all, a mathematical theory must ensure that:

(A) The flow is described as a solution of a mathematically well-posed problem. In fluid mechanics, the most important elements of this requirement are the existence of a solution and its stability with respect to initial/boundary conditions; the requirement of uniqueness can be relaxed when one considers a class of flows, for example waves or local solutions, provided that one can select a unique solution from this class by completing the formulation to a well-posed problem.

Historically, the very concept of well-posedness emerged as the most general necessary condition of physical meaningfulness to be satisfied by problem formulations in mathematical physics. In fluid mechanics, this condition helps to identify which aspects of the flow are known and what has to be determined. For example, in considering the equilibrium of a liquid drop on a solid surface one has to decide whether to specify the position of the contact line with the contact angle to be determined by the solution or to prescribe the contact angle distribution along the contact line and allow the latter to find its position. The two problem formulations correspond to two physically different situations and answer different questions; then condition A helps us to avoid overspecification and ill-posedness of the problem.[16] The way we relaxed the condition of uniqueness allows one to study *classes* of flows which is often what is required in fluid mechanics.

Condition A allows for both regular and singular solutions. In order to narrow the set of solutions down to physically meaningful ones, we have to introduce the basic criterion of self-consistency and require that:

(B) The solution remains within the limits of applicability of the model specified via the assumptions made in its formulation. In particular, the flow parameters, that is the components of velocity and pressure, remain finite everywhere in the flow domain and on its boundaries at all time.

A mathematically obvious side of this criterion is the requirement of regularity, that is finiteness of the flow parameters. In nature, the velocities and pressures in all fluid motions are always finite, and in formulating a mathematical model of a certain natural phenomenon we always presume finiteness of the parameters to be modelled. Therefore, a solution obtained in the framework of this model should have no infinities in the distributions of these parameters. Physically, infinite values of velocity and/or pressure would imply infinite energies of the underlying molecular motion, which is clearly not acceptable.

This aspect of criterion B deserves a bit more attention, however, given that singularities, like, for example, point sources of mass or momentum, dipoles, vortices, etc. are routinely used in mathematical models, and we must understand where and in what sense they are acceptable. Looking at the singularities used in classical fluid mechanics, like those mentioned above, one can see that they are a 'price' one pays for simplifications in representing the flow *in the far field*. If we 'zoom in' and consider the near-field flow, the singularities disappear: a source of mass or momentum acquires a finite size and some distributions of finite velocity and pressure across it (with the intensity of the source in the far field being an average of those); a dipole becomes, say, a body of a finite size oscillating with a finite frequency and amplitude, etc. In other words, the singularity is removed by bringing in 'extra' physics, which in these elementary examples is merely the macroscopic features which have been collapsed into a line or a point to simplify the far-field modelling. A singularity in the near field, that is a genuine infinity in the distribution of physical parameters, would have been clearly unacceptable

---

[16]It can be shown that specification of both the position of the contact line and the contact angle on some part of the boundary with no conditions on the other leads to Hadamard's classical example of an ill-posed problem for Laplace's equation.

as contradicting all known experiments, in particular, as implying infinite energies on the molecular level.

In the literature, in connection with the singularities of flow parameter one occasionally comes across arguments suggesting a so-called 'molecular cut-off'. This approach is, strictly speaking, neither mathematical nor physical, and we will return to it later in the context of particular problems.

The physical side of criterion B is more subtle than the mathematically obvious one. Considering the limits of applicability of the model we assess whether or not an obtained solution indicates the potential importance of physical effects neglected in the process of the problem formulation. For example, the initial assumption that the flow is continuous and describable in terms of the standard model can be undermined by the solution where, say, the range over which the parameters vary might indicate the occurrence of cavitation, phase transition or significant variations of what was assumed to be material constants. Clearly, the physical side of criterion B provides only a necessary condition of validity of the model since our assumptions concern the known effects whereas in reality there could be other, a priori unknown, physical factors playing an important, if not dominant, role in the process in question.

Finally, in assessing the results, especially those of numerical computations, it is desirable to have a qualitative criterion specific to the particular phenomenon one is investigating. This criterion can be seen as a rough 'sanity check'. In fluid mechanics, the simplest descriptive criterion is that:

(C) The flow kinematics in the solution is qualitatively the same as that observed in experiments.

Clearly, if the model leads to the flow kinematics that is qualitatively wrong, one can hardly expect it to describe other features of the flow, in particular the underlying dynamics, in a correct way. The qualitative nature of criterion C implies assessment in broad 'topological' terms, such as the number of stagnation points in the flow field, their location with respect to characteristic features of the flow domain, topology of the domain and its variation, etc.

The 'ABC-criterion' we have formulated provides only the *necessary* conditions to be satisfied by a physically meaningful solution. If a solution does satisfy these conditions, one can move on to the next stage involving experimental verification, first, of qualitative predictions the solution makes for the dependence of the characteristics it describes on the similarity parameters (or, if in experiments these parameters cannot be varied independently, on material constants) and finally its full quantitative validation. On the other hand, if a solution *fails* the ABC-criterion, this outcome conclusively invalidates this solution and one will have to examine simplifications that have been made to obtain it and ultimately to revise the problem formulation, i.e., the model itself.

2.5.3   Paradoxes of modelling

In most cases, problems formulated in the framework of the standard model
have solutions and these solutions routinely satisfy the ABC-criterion of §2.5.2.
This situation is at the base of success of fluid mechanics as a mathematical
discipline. From this point, further exposition of fluid mechanics usually takes
the form of a description of particular results accumulated in the course of its
history.

The question one has to ask is:

- Is the standard model *always* adequate, at least in terms of the basic
  ABC-criterion, for the flows that, on the descriptive level, seem to be
  well within its limits of applicability?

Surprisingly, the answer to this questions is that there are entire *classes* of
quite 'ordinary' flows where the standard fluid-mechanical model fails one or
more requirements of the ABC-criterion formulated in §2.5.2. This is a really
paradoxical situation given that the standard model is made of elements which
have been verified experimentally, separately and in some combinations, for
numerous flows.

It should be emphasized that here we are talking about paradoxes inherent
in the model itself as opposed to *apparent* paradoxes, i.e., "plausible arguments which yield conclusions at variance with physical observation" (Birkhoff
1950). The latter result from the simplifying assumptions made in the process
of obtaining the solution to a problem and vanish once these assumptions are
relaxed and the solution is looked for in a wider class (see Goldshtik 1990
for a review). By contrast, the paradox inherent in the model itself indicates
inadequacy of the model and not the means used to find a solution.

Let us begin with the following examples:

(a) *Flows with moving contact lines in liquid/fluid/solid systems.* The key
element of this wide class of flows is the process by which one viscous
fluid displaces another from a solid surface. In gas/liquid/solid systems,
it is also known as 'dynamic wetting' or the spreading of a liquid over a
solid surface. As we will show in the next chapter, in the framework of
the standard model: the corresponding mathematical problem has no
solution.

(b) *Flows involving the 'sliding-plate problem'.* These can be also referred to
as flows with moving liquid/solid/solid contact lines. If two solid walls,
being part of a boundary confining a volume of fluid, meet at an angle
and the contact line formed at their intersection moves with respect to
one of them, then the solution describing the fluid flow in the framework
of the standard model is singular: the pressure at the corner and the
forces acting from the fluid on the solids are unbounded (Batchelor 1967,
p. 224). The sliding-plate problem is an element of a number of flows,
including the so-called driven-cavity flow, where a viscous liquid filling

a rectangular cavity is driven by a plate on top sliding parallel to itself (Shankar & Deshpande 2000).

(c) *Steady convergent flows with singularities of free-surface curvature.* In experiments (e.g., Joseph et al. 1991), a free surface in a sufficiently strong convergent flow folds to form a line where the normal to the free surface is not uniquely defined. One can see immediately that in the standard model there will be a concentrated force due to the surface tension acting on this line and this force cannot be balanced by finite stresses from the fluid. Experiments also show that the formation of this line singularity corresponds to a qualitative change in the flow kinematics: the fluid particles no longer stay on the free surface at all times as assumed by the kinematic boundary condition of the standard model (Jeong & Moffatt 1992).

Two more wide and partially overlapping classes of flows can be broadly described as flows with topological transitions of the flow domain and flows with emerging finite-time singularities of the free-surface curvature. They include:

(d) *Coalescence of liquid volumes.* Once the standard boundary conditions on the free surface are applied for the flow immediately after the moment when the two liquid volumes touch, the resulting solution gives that the onset of coalescence is associated with singularities in the distributions of the flow parameters.

(e) *Capillary breakup of liquid threads.* For this process there is a solution in the framework of the standard model which predicts, however, that both pressure in the vanishing neck of the thread and the axial velocity of the fluid will tend to infinity as the breakup is approached.

(f) *Rupture of liquid films.* For both free liquid sheets and liquid films on a solid substrate there is no solution in the framework of the standard model describing the process of their rupture.

Every paradox of the kind outlined means that some physics, essential to the phenomenon in question, is missing in the standard model. In order to remove the paradox, one has to identify this missing physics and incorporate it into the mathematical model in a way which must be both self-consistent and compatible with the conceptual framework of fluid mechanics. On the face of it, it seems that the classes of flows we have listed are very different so that each of them is associated with its own 'extra' physics. In the subsequent chapters, we consider these paradoxes in a wider physical context trying to identify common physical mechanisms behind them. Once found, these mechanisms have then to be described in terms of macroscopic fluid mechanics and lead to the minimal self-consistent generalization of the standard model. In order to make this point in the most transparent way, in each case we are going to analyze, the roots of the corresponding paradox will be examined independently thus showing how the same physics emerges from different guises.

Chapter 3

Moving contact lines: An overview

In this chapter, we begin by describing the essence of the famous/infamous 'moving contact-line problem', i.e., the problem arising when the process of displacement of one fluid by another immiscible fluid from a solid surface is modeled in the framework of classical fluid mechanics. Then we will review the relevant experiments dealing with various aspects of the liquid-fluid displacement process and analyze the theories proposed in the literature as possible solutions to the problem. The theories will be examined against both the basic criteria to be satisfied by a mathematical theory that we formulated in §2.5.2 and the experimental results. Finally, we will identify the physical mechanism which is at the core of the liquid-fluid displacement phenomenon and discuss its role in other 'paradoxical flows' outlined at the end of the previous chapter.

3.1 Essence of the problem

3.1.1 Spreading of liquids on solids

Consider the simplest though fully representative case of the steady spreading of a viscous liquid over a smooth solid surface where the displaced medium is an inviscid gas (or vacuum). For simplicity, in the subsequent analysis body forces are neglected, which in the general case means that the characteristic length scale L in the vicinity of the contact line satisfies the condition $\rho F L^2/(\mu U) \ll 1$, where F is the characteristic magnitude of body forces. The inclusion of body forces will make no difference to the essence of our analysis.

In the coordinate frame moving with the contact line (Fig. 3.1), the solution to the steady Euler equations for the gas subject to the impermeability boundary condition at interfaces gives that the gas is at rest.

In considering the liquid, we will be looking at a two-dimensional steady flow in the vicinity of the moving contact line on a length scale L such that the Reynolds number $Re = \rho L U/\mu$ based on L and the speed U of the contact line with respect to the solid surface is small. Then, to leading order in Re as $Re \rightarrow 0$ one can neglect the left-hand side of (2.71) ending up with a steady Stokes flow.

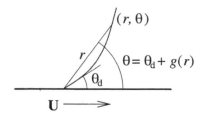

FIGURE 3.1:   A definition sketch for the moving contact-line problem.

Let the liquid occupy a region $0 < r < \infty$, $0 < \theta < \theta_d + g(r)$ in a polar coordinate frame (r, θ) moving with the contact line (Fig. 3.1), where θ_d is the contact angle measured through the liquid and

$$g(r) \to 0 \qquad \text{as } r \to 0. \tag{3.1}$$

Components of the internal unit normal to the free surface will then be given by

$$n_r = \frac{rg'}{(1 + r^2 g'^2)^{1/2}}, \qquad n_\theta = -\frac{1}{(1 + r^2 g'^2)^{1/2}}.$$

Scaling lengths, velocities and pressure (measured from the constant pressure in the gas) with L, U and $\mu U/L$, respectively, and using the notation u and v for the radial and transversal components of velocity, one arrives at the problem of finding a solution to the Stokes equations in the bulk, which in polar coordinates take the form (Appendix B)

$$\frac{1}{r}\frac{\partial(ru)}{\partial r} + \frac{1}{r}\frac{\partial v}{\partial \theta} = 0, \tag{3.2}$$

$$\frac{\partial p}{\partial r} = \Delta u - \frac{u}{r^2} - \frac{2}{r^2}\frac{\partial v}{\partial \theta}, \tag{3.3}$$

$$\frac{1}{r}\frac{\partial p}{\partial \theta} = \Delta v - \frac{v}{r^2} + \frac{2}{r^2}\frac{\partial u}{\partial \theta}, \tag{3.4}$$

where

$$\Delta = \frac{\partial^2}{\partial r^2} + \frac{1}{r}\frac{\partial}{\partial r} + \frac{1}{r^2}\frac{\partial^2}{\partial \theta^2},$$

subject to the no-slip condition (2.157) on the solid boundary, i.e.,

$$u = 1, \quad v = 0 \qquad (\theta = 0,\ r > 0), \tag{3.5}$$

together with the standard kinematic and dynamic conditions at the free surface, which include zero normal velocity (2.137), i.e.,

$$rg'u - v = 0, \qquad (\theta = \theta_d + g(r),\ r > 0), \tag{3.6}$$

zero tangential stress (2.155), i.e.,

$$p_{n\tau} \equiv \tfrac{1}{2}(p_{rr} - p_{\theta\theta})\sin 2\varphi - p_{r\theta}\cos 2\varphi = 0, \tag{3.7}$$

and the balance of normal stress and capillary pressure, i.e., (2.154)

$$Ca\,p_{nn} \equiv Ca\left[\tfrac{1}{2}(p_{rr} + p_{\theta\theta}) + \tfrac{1}{2}(p_{\theta\theta} - p_{rr})\cos 2\varphi - p_{r\theta}\sin 2\varphi\right]$$

$$= \frac{2g' + rg'' + r^2 g'^3}{(1 + r^2 g'^2)^{3/2}} \qquad (\theta = \theta_d + g(r),\ r > 0). \tag{3.8}$$

Here, $Ca = \mu U/\sigma$ is the capillary number, where σ is the surface tension of the liquid-gas interface,

$$p_{rr} = -p + 2\frac{\partial u}{\partial r}, \quad p_{\theta\theta} = -p + 2\left(\frac{1}{r}\frac{\partial v}{\partial\theta} + \frac{u}{r}\right), \quad p_{r\theta} = \frac{1}{r}\frac{\partial u}{\partial\theta} + r\frac{\partial}{\partial r}\left(\frac{v}{r}\right) \tag{3.9}$$

are the dimensionless physical components of the stress tensor in polar coordinates, and

$$\sin 2\varphi = \frac{2rg'}{1 + r^2 g'^2}, \qquad \cos 2\varphi = \frac{1 - r^2 g'^2}{1 + r^2 g'^2}$$

are just convenient notations (geometrically, φ is the angle between the tangent to the free surface at (r, θ) and the radius-vector).

It is important to emphasize that in this formulation the contact angle, θ_d, cannot be determined; it is independent of the flow field and has to be prescribed separately.

Given that θ_d is prescribed so that the leading order term in the asymptotic expansion as $r \to 0$ for the free-surface shape is known *a priori*, and assuming that the solution exists, we can employ the following iterative procedure to find the near-field asymptotics.

Introducing the stream function ψ by

$$u = \frac{1}{r}\frac{\partial\psi}{\partial\theta}, \qquad v = -\frac{\partial\psi}{\partial r}$$

and eliminating p from (3.3), (3.4), we have that as $r \to 0$ the stream function has to satisfy a biharmonic equation in a wedge,

$$\Delta^2\psi = 0 \qquad (0 < \theta < \theta_d, r > 0), \tag{3.10}$$

with the boundary conditions (3.5), (3.6), (3.7) becoming

$$\psi = 0 \qquad \text{at } \theta = 0, \theta_d, \tag{3.11}$$

$$\frac{1}{r}\frac{\partial\psi}{\partial\theta} = 1 \qquad \text{at } \theta = 0, \tag{3.12}$$

$$\frac{\partial^2\psi}{\partial\theta^2} = 0 \qquad \text{at } \theta = \theta_d. \tag{3.13}$$

In other words, the velocity field can be found prior to the pressure distribution and the correction to the free surface shape. After ψ is found, one can determine the leading term of the asymptotics for p as $r \to 0$ from (3.3), (3.4), which now become

$$\frac{\partial p}{\partial r} = \left(\frac{1}{r} \frac{\partial^3}{\partial r^2 \partial \theta} + \frac{1}{r^3} \frac{\partial^3}{\partial \theta^3} + \frac{1}{r^2} \frac{\partial^2}{\partial r \partial \theta} \right) \psi, \tag{3.14}$$

$$\frac{\partial p}{\partial \theta} = -\left(r \frac{\partial^3}{\partial r^3} + \frac{\partial^2}{\partial r^2} + \frac{1}{r} \frac{\partial^3}{\partial r \partial \theta^2} - \frac{1}{r} \frac{\partial}{\partial r} - \frac{2}{r^2} \frac{\partial^2}{\partial \theta^2} \right) \psi. \tag{3.15}$$

Then, using the following expressions for the components of the stress tensor in terms of pressure and the stream function,

$$p_{rr} = -p + 2L(\psi), \quad p_{\theta\theta} = -p - 2L(\psi), \quad p_{r\theta} = M(\psi),$$

where

$$L = \frac{1}{r} \frac{\partial^2}{\partial r \partial \theta} - \frac{1}{r^2} \frac{\partial}{\partial \theta}, \qquad M = \frac{1}{r^2} \frac{\partial^2}{\partial \theta^2} - \frac{\partial^2}{\partial r^2} + \frac{1}{r} \frac{\partial}{\partial r},$$

so that at $\theta = \theta_d + g(r)$ the normal stress is given by

$$p_{nn} = -p - 2L(\psi) \cos 2\varphi - M(\psi) \sin 2\varphi,$$

and one can find the leading term of an asymptotic expansion for $g(r)$ as $r \to 0$ from (3.8). Then, given this new free-surface shape, the procedure can be repeated to find a correction to the stream function, etc.

As is known, the biharmonic equation (3.10) has a series of solutions of the form

$$\psi_\lambda = r^\lambda F_\lambda(\theta) \tag{3.16}$$

for λ real or complex, and the form of the no-slip condition (3.12) suggests $\lambda = 1$. The corresponding solution of (3.10) is given by

$$\psi_1 = r[(A_1 + A_2\theta) \sin \theta + (A_3 + A_4\theta) \cos \theta] \tag{3.17}$$

and after satisfying the boundary conditions (3.11), (3.12), (3.13) one arrives at (Moffatt 1964)

$$\psi_1 = rF_1(\theta) = \frac{r}{\sin \theta_d \cos \theta_d - \theta_d}[(\theta - \theta_d) \sin \theta - \theta \sin(\theta - \theta_d) \cos \theta_d]. \tag{3.18}$$

Then, from (3.14), (3.15) as $r \to 0$ the pressure takes the form

$$p = -\frac{2[\sin \theta - \sin(\theta - \theta_d) \cos \theta_d]}{r(\theta_d - \sin \theta_d \cos \theta_d)} + O(1), \tag{3.19}$$

and, since $L(\psi_1) \equiv 0$, $M(\psi_1) = O(r^{-1})$ and $\sin 2\varphi = o(1)$, the left-hand side of (3.8) becomes

$$Ca\, p_{nn} = -Ca\, p + o(r^{-1})$$

$$= \frac{2\,Ca\sin\theta_d}{r(\theta_d - \sin\theta_d\,\cos\theta_d)} + o(r^{-1}) \qquad \text{as } r \to 0. \qquad (3.20)$$

Given that $g(r)$ is supposed to satisfy (3.1) by definition, to leading order as $r \to 0$ the right-hand side of (3.8) becomes simply $(rg)''$. Now, integrating

$$(rg)'' = \frac{2\,Ca\sin\theta_d}{r(\theta_d - \sin\theta_d\,\cos\theta_d)}$$

we find that

$$g(r) = \frac{2\,Ca\sin\theta_d}{\theta_d - \sin\theta_d\,\cos\theta_d}\,\ln r \to \infty, \qquad \text{as } r \to 0,$$

which contradicts (3.1).

Thus, the problem with the classical formulation is that *the solution does not exist*, at least not in a physically-relevant class of smooth bounded functions. The formulation itself includes a prescribed parameter, θ_d, whose value, as we will see later, is intimately related with the flow.

The failure of the standard model to provide a solution can be illustrated with the following qualitative arguments. The no-slip condition on the solid surface allows the solid substrate to remove the liquid from the vicinity of the contact line, thus reducing pressure there. The capillary pressure (in the normal-stress boundary condition) tries to compensate the difference between a constant pressure in the gas and the reduced pressure in the liquid by bending the free surface towards the gas. However, the no-slip condition is so efficient in removing the liquid and reducing the pressure that the capillary pressure has to bend the free surface so strongly that the latter simply cannot meet the solid surface to form a contact angle.

It is instructive to look also at a simplified formulation where condition (3.8), which we could not satisfy, is dropped and the free-surface shape is prescribed. (As we will see later, this simplification is formalised in some theories where extra attractive forces, singular at $r = 0$, are added to the normal-stress balance (3.8) to force the free surface to meet the solid boundary.) Then, the above deficiency of the classical formulation manifests itself in a different way. In this formulation, we no longer need to satisfy the normal-stress boundary condition so that the solution exists and, to leading order as $r \to 0$, the stream function is again given by (3.18). This is all one needs to calculate the tangential stress acting on the solid from the liquid:

$$p_{r\theta}|_{\theta=0} = \frac{1}{r}\frac{\partial u}{\partial \theta}\bigg|_{\theta=0} = \frac{1}{r^2}\frac{\partial^2 \psi_1}{\partial \theta^2}\bigg|_{\theta=0} = \frac{1}{r}F_1''(0) = -\frac{2\sin^2\theta_d}{r(\theta_d - \sin\theta_d\,\cos\theta_d)}.$$

Since the tangential force acting on a finite area of the solid adjacent to the contact line is the integral of $p_{r\theta}$, this force appears to be infinite. Thus, according to the *simplified* formulation, no finite force can immerse a solid into a liquid which is clearly not what one can observe in experiments.

This outcome is in fact what one should expect given that the conservation of momentum of any control volume is incorporated in the equations of motion and, when applied to the volume embracing the contact line, this basic law ensures that the singularity of the normal stress on the free surface and that of the tangential stress on the solid boundary have the same type. Then, they are either both nonintegrable and hence the normal-stress boundary condition cannot be satisfied (nonexistence of the solution in the general case) and the tangential force acting on the finite area of the solid is infinite (the outcome of the simplified formulation), or they are both integrable and hence the solution exists and the total tangential force acting on the solid surface is finite.

The nonintegrability of the tangential stress can be also interpreted as a consequence of multivaluedness of the velocity field at the contact line (Dussan & Davis 1974). Indeed, both components of velocity in the above 'semi-solution' are functions of θ only and their limit as $r \to 0$ depends on the direction along which the contact line is approached. In particular, the radial component u varies from $u = 1$ at $\theta = 0$ to $u = u_{(12)}(\theta_d, 0)$ at the free surface, where

$$u_{(12)}(\theta_d, 0) = \frac{\sin \theta_d - \theta_d \cos \theta_d}{\sin \theta_d \cos \theta_d - \theta_d}. \tag{3.21}$$

Here, the second argument of $u_{(12)}$ reminds us that the gas-to-liquid viscosity ratio is assumed to be equal to zero.

It is worth pointing out here that it was the *simplified* analysis in terms of integrability of the tangential stress and multivaluedness of velocity that prompted the often used terminology referring to the moving contact-line problem as "the problem of nonintegrable shear-stress singularity". Presenting the problem in this way suggests a direction of research aimed at removing this nonintegrability and leads to a class of models that come to be known collectively as 'slip models'. We will review these models in §3.4.1. Here it is important to emphasize that, as was first pointed out in Shikhmurzaev (2006) and discussed above, in the *general* formulation, the essence of the moving contact-line problem is the *nonexistence* of a solution which suggests a different course of action: one should take a more general view and look for the physics missing in the classical formulation and the way of embedding the liquid-fluid displacement phenomenon in a wider physical context.

3.1.2 Liquid-liquid displacement

If we take into account viscosity of the displaced fluid as well, the problem becomes slightly more cumbersome with its essence remaining unchanged. The same iterative procedure as in §3.1.1 applies and, in the immediate vicinity of the contact line where the Reynolds numbers based on the distance from the contact line are small for both fluids, one has to solve two biharmonic equations for the stream functions in two wedge regions,

$$\Delta^2 \psi^\pm = 0 \qquad (+ \text{ for } 0 < \theta < \theta_d; - \text{ for } \theta_d < \theta < \pi; r > 0), \tag{3.22}$$

subject to the no-slip condition (2.157) at the solid boundary for both fluids

$$\frac{1}{r}\frac{\partial \psi^{\pm}}{\partial \theta} = \pm 1, \qquad \psi^{\pm} = 0 \qquad (\theta = 0, \pi; r > 0) \qquad (3.23)$$

together with conditions at the free surface including the impermeability condition (2.137), zero tangential stress condition (2.147) and the condition of continuity for the tangential velocity (2.148),

$$\psi^{\pm} = 0 \qquad (\theta = \theta_d, \ r > 0), \qquad (3.24)$$

$$\frac{\partial^2 \psi^+}{\partial \theta^2} = k_\mu \frac{\partial^2 \psi^-}{\partial \theta^2} \qquad (\theta = \theta_d, \ r > 0), \qquad (3.25)$$

$$\frac{\partial \psi^+}{\partial \theta} = \frac{\partial \psi^-}{\partial \theta} \qquad (\theta = \theta_d, \ r > 0), \qquad (3.26)$$

where k_μ is the receding-to-advancing fluid viscosity ratio.

The solution of (3.22) satisfying (3.23)–(3.26) is given by (Huh & Scriven 1971)

$$\psi^{\pm} = rF^{\pm}(\theta), \qquad (3.27)$$

$$F^+ = C_1 \theta \sin(\theta - \theta_d) + C_2(\theta - \theta_d) \sin \theta,$$

$$F^- = C_3(\theta - \pi) \sin(\theta - \theta_d) - C_4(\theta - \theta_d) \sin \theta,$$

where constants C_i $(i = 1, \dots, 4)$ are as follows:

$$C_1 = \frac{\cos \theta_1 K(\theta_2) - k_\mu \left[\pi(\sin \theta_1 + \theta_1 \cos \theta_1) + \cos \theta_1 K(\theta_1)\right]}{k_\mu(\theta_2 - \sin \theta_2 \cos \theta_2)K(\theta_1) - (\theta_1 - \sin \theta_1 \cos \theta_1)K(\theta_2)},$$

$$C_2 = \frac{1}{\theta_1}(1 - C_1 \sin \theta_1),$$

$$C_3 = \frac{\theta_2}{K(\theta_2)}\left(\frac{\pi \sin \theta_1}{\theta_1 \theta_2} + C_1 \frac{K(\theta_1)}{\theta_1}\right),$$

$$C_4 = -\frac{1}{\theta_2}(1 + C_3 \sin \theta_2),$$

$$\theta_1 = -\theta_d, \quad \theta_2 = \pi - \theta_d, \quad K(\theta) = \theta^2 - \sin^2 \theta.$$

Then, equations (3.14) and (3.15) allow one to find pressures in the two fluids

$$p^{\pm} = -\frac{1}{r}\left[(F^{\pm})''' + (F^{\pm})'\right] + o(r^{-1}),$$

where the prime denotes differentiation with respect to θ. The normal-stress boundary condition at the interface (2.143) has the form

$$Ca\left[\tfrac{1}{2}(p_{rr}^+ + p_{\theta\theta}^+) + \tfrac{1}{2}(p_{\theta\theta}^+ - p_{rr}^+)\cos 2\varphi - p_{r\theta}^+ \sin 2\varphi\right]$$

$$-k_\mu Ca\left[\tfrac{1}{2}(p_{rr}^- + p_{\theta\theta}^-) + \tfrac{1}{2}(p_{\theta\theta}^- - p_{rr}^-)\cos 2\varphi - p_{r\theta}^- \sin 2\varphi\right]$$

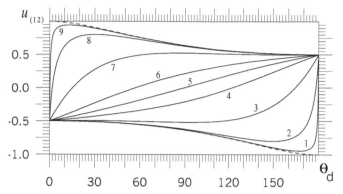

FIGURE 3.2: The free surface velocity versus the dynamic contact angle in a wedge region for different viscosity ratios. Curves 1–9 correspond to $k_\mu = 10^{-3}, 10^{-2}, 10^{-1}, 0.5, 1, 2, 10, 10^2$ and 10^3, respectively. Dashed lines are obtained for $k_\mu = 0, \infty$.

$$= \frac{2g' + rg'' + r^2 g'^3}{(1 + r^2 g'^2)^{3/2}} \qquad (\theta = \theta_d + g(r), \ r > 0), \qquad (3.28)$$

where, as before, Ca is the capillary number based on the viscosity of the advancing fluid. To leading order as $r \to 0$ this equation becomes

$$(rg)'' = \frac{Ca F(\theta)}{r},$$

$$F(\theta) = (F^+)''' + (F^+)' - k_\mu \left[(F^-)''' + (F^-)' \right],$$

so that again

$$g(r) = Ca\, F(\theta) \ln r \to \infty \qquad \text{as } r \to 0$$

contrary to what was assumed in (3.1).

The velocity field is multivalued at the contact line for both fluids with the radial velocity varying from ± 1 on the solid surface to

$$u_{(12)}(\theta_d, k_\mu) = \frac{(\sin\theta_d - \theta_d \cos\theta_d)K(\theta_2) - k_\mu(\sin\theta_2 - \theta_2 \cos\theta_2)K(\theta_d)}{(\sin\theta_d \cos\theta_d - \theta_d)K(\theta_2) + k_\mu(\sin\theta_2 \cos\theta_2 - \theta_2)K(\theta_d)},$$
$$(3.29)$$

at the interface between the fluids. The dependence of $u_{(12)}$ on θ_d for different values of k_μ is shown in Fig. 3.2.

It is noteworthy that, no matter how small the gas-to-liquid viscosity ratio k_μ is, the viscosity of the gas cannot be neglected when $\pi - \theta_d$ becomes sufficiently small. This leads to a nonmonotone dependence of $u_{(12)}$ on θ_d and, as we will show in Chapter 5, plays an important role in the displacement process. At the same time, it is necessary to emphasize that the above expression for $u_{(12)}$ is associated with a planar liquid-fluid interface, i.e., strictly speaking, with zero capillary numbers.

3.1.3 Summary

As we have shown, the mathematical problem formulated in the framework of the standard fluid-mechanical model whose solution is intended to describe the process of displacement of one fluid by another from a solid surface with capillary effects taken into account in the situation where at least one of the fluids is viscous has the following two deficiencies:

- The problem has no solution.

- In the very formulation of the problem, it is assumed that the contact angle formed by the free surface with the solid boundary is not part of the solution and has to be prescribed.

If the formulation is artificially simplified by dropping the normal-stress boundary condition and prescribing the free-surface shape, then a solution exists. However, from a physical point of view, it has a major flaw:

- The bulk pressure evaluated at the contact line and the stress acting on the solid surface are both nonintegrably singular so that the total force between the fluid and the solid is infinite.

All this indicates that some essential physical mechanism is not accounted for in the standard model. For brevity, before it is identified we will refer to this mechanism as 'the specific physics of wetting' to distinguish it from other physical effects, which, under certain conditions, can complement it.

3.2 Experimental observations

3.2.1 Preliminary remarks

The information needed to identify the specific physics of wetting and to test the proposed theories has to be drawn from experimental studies. However, interpretation of experimental results, as well as the unavoidable aspects of experimentation such as reproducibility of measurements, present a number of formidable problems when considered in the context of the moving contact-line problem. We will briefly outline the main issues to have them in mind in what follows.

First of all, it is necessary to note that neither the boundary conditions nor even the contact angle involved in the macroscopic fluid-mechanical modelling are the objects of direct observation. Experimental observations are dealing with averaged quantities where the averaging is associated with the spatial and temporal resolution of the measuring devices. Continuum mechanics, on the other hand, is dealing with a set of idealised concepts (continuous distributions of parameters, sharp interfaces, etc.) resulting from the averaging

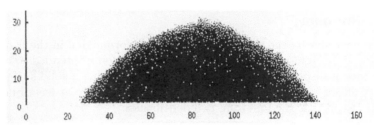

FIGURE 3.3:   Molecular dynamics simulations of a liquid drop in equilibrium on a solid surface illustrate that considerable magnification in virtual or real-life experiments makes it nontrivial to correlate well-defined theoretical concepts of macroscopic fluid mechanics, such as the 'contact angle', with what is actually observed experimentally. (The figure from de Ruijter, Blake & De Coninck 1999; © 1999 American Chemical Society; reproduced with permission.)

implied by the continuum approximation; it also introduces artificial separation of physical effects into those incorporated in the bulk equations and those attributed to the boundary conditions. A key assumption of continuum mechanics is that the two sets of parameters are essentially the same within the accuracy of experiments and the continuum approximation.

In the case of the moving contact-line problem, however, the paradox in the theory is associated with the *limit* $r \to 0$ for the fluid-mechanical parameters and this limit has no experimental equivalent. Indeed, experimental 'zooming in' is an essentially different process, which will, at least in principle, eventually turn 'interfaces' into interfacial layers of a finite thickness and the 'contact line' into a finite three-phase-interaction zone. In other words, it will blow up the very concepts used in macroscopic fluid mechanics rather than provide the direct answer to a theoretical question formulated in terms of these concepts. Further magnification will bring in the full-scale picture of the fluid as a discrete medium (Fig. 3.3). Then one will need some form of averaging to interpret the experiments in continuum-mechanics terms. Thus, in dealing with experimental data and their theoretical description one must have in mind the intrinsic limitations on the accuracy associated with the modelling in the framework of continuum mechanics.

The presentation and interpretation of quantitative data raise another issue. Once the raw data is processed and emerges in the form of relationships between different nondimensional similarity parameters, it is tempting to start analysing these relationships in the same way as one always treats experimental data in classical fluid mechanics. However, there is a catch. Indeed, since some physical mechanism is clearly missing in the standard model, the set of nondimensional similarity parameters based on the dimensional quantities involved in this model is incomplete: once identified, the specific physics of wetting will bring in its own dimensional constants and the corresponding nondimensional similarity parameters.

Experimenters have come to realise this fact empirically (e.g., Gutoff & Kendrick 1982). Thus, if variation of a measurable quantity, say, the contact angle, due to variation of some flow parameter, say, the contact-line speed U, is presented as a relationship between dimensionless variables, say, the contact angle as a function of the capillary number $Ca = \mu U/\sigma$, this means nothing more than a simple scaling of the original 'raw' variables with the appropriate combinations of material constants. If this relationship is interpreted in a broader way, say, as the dependence also on material constants, this would mean that one is making implicit assumptions about the 'hidden' similarity parameters associated with the specific physics of wetting.

In the context of the moving contact-line problem, the unavoidable issue of reproducibility of experimental results acquires a number of specific features. The most serious difficulty concerns specification of the solid substrate. As we have already mentioned in §2.4.3, in reality the solid surface is almost invariably rough and very often chemically inhomogeneous so that, even setting aside the problem of its theoretical modeling with respect to the process of dynamic wetting, its description for the purpose of reproducibility of experiments is a problem in its own right. One macroscopic manifestation of this problem is contact-angle hysteresis. It is also important to note that if the same substrate is used for a succession of experiments, then, in a general case, the state of the solid surface in each run will be different due to the traces of a liquid left after previous runs (Washburn 1921, Bolton & Middleman 1980), adsorption (or absorption) of liquid by the solid substrate (Lam et al. 2001, 2002), possible chemical reactions between the two media (Chaudhury & Whitesides 1992) and other alterations in the state of the solid surface.

Finally, there are ever-present questions of accuracy of experimental measurements and comparability of results obtained for different material systems and possibly using different experimental methods. Given that a theoretical limit $r \to 0$ and experimental 'zooming in' are two essentially different concepts, one will have to deal with finite distances from the contact line and the associated interpretation of experimental data. This brings in the question as to what extent the results obtained using different methods, for example experimental setups with different flow geometries, are directly comparable. The question of comparing experiments for different material systems sends us back to the difficulties of dimensional analysis in the situation where the set of nondimensional similarity parameters is incomplete.

Having in mind the issues outlined above, let us review the main results of experimental studies of flows with moving contact lines. In many cases, experimental papers pursued rather limited objectives dealing either with technical problems of their time or verification of approximate formulae proposed to describe particular aspects of the flow. We are interested in the most fundamental features of liquid-fluid displacement and will be looking for the answers to the following two main questions posed by the problem that emerged from the standard formulation:

- In the search of the physical process to be incorporated in the model to resolve the problem of nonexistence of the solution, we would like to know what specific features of this process are pointed out by experimental observations.

- Since the contact angle plays an important role in specifying the flow domain, we need to know whether the contact angle involved in the macroscopic fluid-mechanical formulation for the moving contact line deviates from the value it has in the static situation. If this is the case, what are the main features characterising its behaviour?

Experimental results together with the basic ABC-criterion of §2.5.2 will help us later to analyse theories proposed to resolve the moving contact-line problem. The two elements we will be particularly interested in are (a) the flow kinematics, which provides the necessary 'descriptive element' for the modelling and is potentially useful in identifying specific features of the physics activated when the contact line begins to move, and (b) the contact angle, which plays the main role in specifying the geometry of the flow domain near the contact line.

3.2.2   Flow kinematics

Most of the early experiments involving moving contact lines considered flows in tubes and dealt with the effects of capillarity as a whole, such as the resistance due to the presence of menisci when an index is pushed through a narrow capillary (West 1911, Yarnold 1938) or the overall motion resulting from the capillary pressure of the meniscus acting as a driving force (Lucas 1918, Washburn 1921, Rideal 1922). These experiments allow one to estimate relative importance of capillarity, viscosity and other factors for the global flow but are of little use as far as the moving contact-line problem is concerned. Gradually, the emphasis in experimental research started to shift towards studying local processes in the immediate vicinity of the moving contact line, in particular the flow kinematics.

3.2.2.1   Advancing contact line

Experiments on flow kinematics are much less straightforward than those on the contact angle and the free-surface shape which are 'observable' features of dynamic wetting, at least at a qualitative level. Early naked-eye observations of the flow kinematics of drops on a tilted plane have been by-products of experiments aimed at measuring the contact angles and, rather confusingly, described the flow as both 'rolling' and 'sliding' (e.g., see Ablett and Sumner discussing experiments by Sumner 1937).

Schwartz, Rader & Huey (1964) were the first to actually visualize the flow inside an index moving in a capillary tube by inserting a 'marker' whose motion could then be traced. They wrote:

The manner in which an individual index moves was followed by using a viscous liquid and inserting a small portion of dye at one end of the index before starting the movement. The index proceeds forward by rolling away from the tube wall and inward towards its center axis at the rear, while rolling outward from its center axis and onto the tube wall at the front. There is no apparent sliding of liquid along the tube wall. < ... > Similar observations showed that partly dyed droplets of high contact angle liquids on tilting plates of Teflon and polyethylene roll rather than slide... (Schwartz, Rader & Huey 1964, p. 253).

These observations confirmed early predictions by West (1911) whose sketch is reproduced in Fig. 3.4a, and Yarnold (1938) that the flow near the moving contact line can be qualitatively described as 'rolling'. More specifically, this flow pattern is illustrated in Fig. 3.4b, where one can see photographs taken by Dussan & Davis (1974) of a drop of honey moving down a tilted Plexiglas plate. A black 'marker' placed on the free surface moves towards the contact line and — this is the defining feature of the rolling motion — reaches it in a finite time, thus suggesting that the 'fluid particles' forming the liquid-gas interface do the same. As the marker (fluid particle) reaches the contact line and the drop rolls forward, the marker (fluid particle) becomes part of the liquid-solid interface. This kinematics is schematically shown in Fig. 3.5. This flow pattern is qualitatively consistent with that described by the stream function (3.18), which is what one should expect, given that no matter what specific physics of wetting is activated in the vicinity of the contact line, away from it the standard model should take over.[1]

It must be noted, however, that the spatial resolution of experiments utilizing macroscopic markers of various kinds (dye, latex particles, microbubbles, etc.) is limited by the marker's size and, if the marker is completely immersed in the liquid, its distance from the interface so that their results should be seen as suggestive rather than conclusive.

A different type of confirmation of the rolling motion comes from a study by Blodgett (1935) of what is now known as the Langmuir-Blodgett deposition of monolayers on solid surfaces. In her experiments on so-called X-layers, Blodgett used a solid plate moving into a pool of liquid where a layer of an insoluble surfactant was present on the liquid-air interface. It has been noted that the plate drags the surfactant molecules with it down into the pool. If one considers this experiment in a fluid-mechanical context and uses a low concentration of surfactant (which is the opposite of what chemical engineers need),

---

[1] The flow pattern observed far away, where the classical boundary conditions apply, depends on the Reynolds number based on the distance from the contact line and, when the Reynolds number is large, it can qualitatively differ from that described by (3.18) (Savelski et al. 1995). A claim that such experiments can invalidate the rolling pattern near the contact line where the Reynolds number tends to zero would lack any justification.

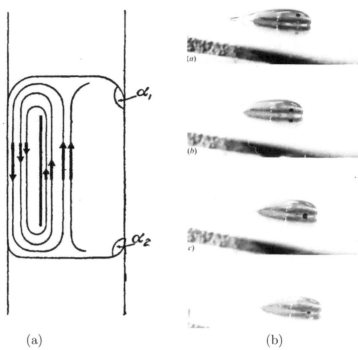

(a)   (b)

FIGURE 3.4: (a) West's (1911) prediction of the flow pattern inside a mercury index moving through a capillary has been qualitatively confirmed by Schwartz, Rader & Huey (1964), who visualized the flow by using a drop of transparent liquid and injecting a little dye into it. (b) Dussan & Davis' (1974) visualization of a drop of honey moving on a Plexiglas surface. One can see how a 'marker' placed on the free surface moves towards the contact line as the drop rolls over the solid (© 1974 Cambridge University Press; reproduced with permission).

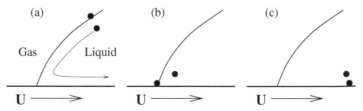

FIGURE 3.5: Sketch illustrating the 'rolling' motion near the moving contact line in the liquid-gas displacement. Two solid particles as 'markers' placed on the free surface and in the bulk move as if the liquid 'rolls over' the solid substrate.

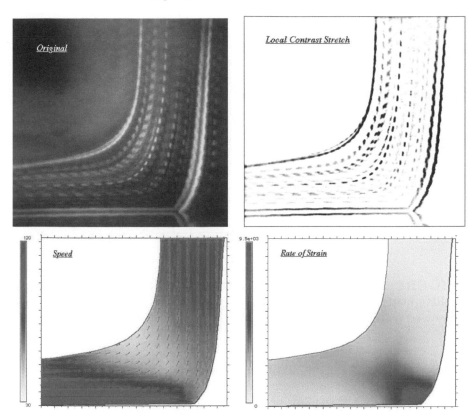

FIGURE 3.6: Visualization of the flow field in curtain coating and the trends in the flow field highlighted by the image analysis (Clarke 1995; © 1995 Elsevier; reproduced with permission). One can see that (a) the flow field gives no indication that there might be a stagnation zone near the contact line and (b) the rate of strain increases as the contact line is approached.

then the molecules of the surfactant can be regarded as 'markers' labelling 'fluid particles' belonging to the free surface. Then, the whole experiment becomes that on flow kinematics and as such has a molecular-level spatial resolution. The problem here is that as the surfactant concentration goes to zero, it becomes more and more difficult to detect these 'markers', whereas at higher concentrations the surfactant can influence the fluid flow via the Marangoni effect.

Some qualitative features of the rolling motion emerge as one considers the *trends* in the flow field visualized using the marker method. Such analysis makes it possible to a large extent to overcome the limitations of the method associated with a relatively low spatial resolution due to the macroscopic size of the markers whilst keeping its advantages as a method of direct observation.

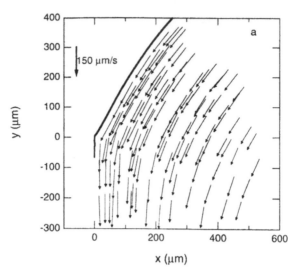

FIGURE 3.7: Vector plot of the measured velocity field clearly shows the rolling motion of the advancing liquid (Chen, Ramé & Garoff 1996; © 1996 Elsevier; reproduced with permission).

Clarke (1995) applied this technique to curtain coating, where the flow has been visualized using microscopic hydrogen bubbles generated by electrolysis and a pulse light source allowing one to see successive positions of the bubbles in the field of view (Fig. 3.6). By analyzing the images Clarke has demonstrated that, in a reference frame where the contact line is at rest, the plug flow in the falling curtain slightly decelerates as it changes its direction but the local velocity distribution does not suggest any low-velocity zone near the contact line (Fig. 3.6). The local rate of strain in the liquid increases as the contact line is approached. In this reference frame, the immediate vicinity of the contact line is associated with more intensive flow than that farther afield and, alongside the rest of the solid surface, this region can be seen as a generator of motion, not an obstacle to it. This region is by no means a stagnation zone.

The specific physics of wetting comes into play in a region which comprises the contact line and is associated with some characteristic length scale. The implicit presence of this length scale in the dimensional analysis has been pointed out by Gutoff & Kendrick (1982).

At distances from the contact line large compared with this length scale the influence of the physics of wetting dies out and the standard model takes over. If the problem of nonexistence of the solution in the standard model is resolved, even artificially, by some alteration of the boundary conditions near the moving contact line, one will see this alteration reflected in the free-surface profile close to the contact line, whereas far away from the contact

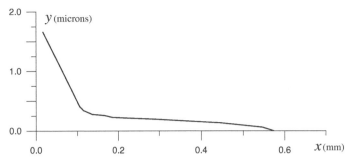

FIGURE 3.8: The profile of a precursor film for the spreading of a 0.43–mg drop of siloxane oil on glass at $t = 180$ min (Beaglehole 1989).

line the profile will tend to that resulting from the standard model. Thus, by comparing the observed free-surface shape with that produced by the standard model (or, more precisely, the modified standard model where the solution exists), one can see when the region of the specific physics of wetting enters or, more precisely, influences the field of view. This study has been carried out by Chen, Ramé & Garoff (1996) who visualized the rolling motion of the fluid (Fig. 3.7, see also Chen, Ramé & Garoff 1997) and, by analysing the free-surface shape, found that the characteristic length scale of the region associated with the specific physics of wetting *increases* with the contact-line speed. This is an important result showing that, as expected, in a general case, the physics of wetting operates on a hydrodynamic length scale and is sensitive to hydrodynamic factors.

3.2.2.2 Precursor film

In the case of perfect wetting, i.e., when $\theta_s = 0$, liquid spreading has some specific features. Experiments show that at sufficiently low contact-line speeds the free-surface profile can be qualitatively described as having two distinct regions: (a) the clearly visible main body of fluid which, as one approaches the solid along the free surface, turns into (b) a very thin (submicron scale) film (Fig. 3.8) whose leading edge can be at a substantial (mm scale) distance from the main body (Hardy 1919, Bascom, Cottington & Singleterry 1964, Williams 1977, Marmur & Lelah 1980, Beaglehole 1989). The latter is known in the literature as the 'precursor' or 'primary' film whilst the motion of the main body of fluid is usually referred to as the 'secondary spreading' (Hardy 1919) to distinguish it from the 'primary' pre-wetting of the solid by the precursor film.

The range of contact-line speeds for which the free-surface profile has the above structure and the precursor film peculiarities were studied experimentally in a number of works by considering so-called 'spontaneous spreading', i.e., the evolution towards equilibrium of a volume of fluid placed in contact with a solid substrate in the situation where gravity and other external factors which might force the contact line to move have negligible effect.

It has been found that the velocity at which the precursor film advances depends on the materials of the gas/liquid/solid system and can vary in a wide range. For example, by analysing the data reported by Bascom, Cottington & Singleterry (1964), who combined interference microscopy and ellipsometry to study the spreading of squalane (kinematic viscosity $\nu = 26.8$ cSt, surface tension $\sigma = 27.6$ dyn cm^{-1}) on stainless steel, one can estimate the speed of propagation of the precursor film for this system to be about 10^{-4} cm s^{-1}, whereas indirect measurements for water on glass (Marmur & Lelah 1980) suggest that there the precursor film velocity is of the order of 10 cm s^{-1}.

The typical thickness of the precursor film moving in front of a macroscopic body of fluid is in the range 500–3000 Å (Bascom, Cottington & Singleterry 1964, Williams 1977, Beaglehole 1989). If in experiments the main body of fluid itself has dimensions in the above range or smaller, for example when a drop of several hundred angstroms in diameter is deposited on a solid surface, its evolution also involves a precursor film whose thickness can then be on or slightly above the molecular scale (Heslot, Cazabat & Levinson 1989, Heslot, Fraysse & Cazabat 1989, Iwamoto & Tanaka 2002). At present it is not clear whether in a general case there is in fact a combination of two distinct precursor films, the 'macroscopic' one next to the main (macroscopic) body of fluid and a 'microscopic' precursor film with the thickness on a molecular scale at the leading edge, or there is just one precursor film with the thickness varying gradually from the macroscopic down to the molecular scale. The two scenarios have different implications for the modelling.

One can demonstrate experimentally that evaporation-condensation plays only a complementary role in the formation of precursor films, irrespective of their thickness, by considering the spreading of low-volatility liquids (Bascom, Cottington & Singleterry 1964, Williams 1977, Heslot, Cazabat & Levinson 1989, Heslot, Fraysse & Cazabat 1989). At the same time, the behaviour of precursor films of considerably different thickness has qualitatively different features. For example, the 'macroscopic' precursor film can develop instabilities along the contact line (Williams 1977) similar to those observed in macroscopic fluid motion whereas the 'microscopic' one exhibits smectic-like ordering on the molecular level (Heslot, Fraysse & Cazabat 1989), thus suggesting a different balance of randomizing and ordering factors than one has in macroscopic volumes of fluid. In terms of fluid-mechanical modelling, this suggests that 'microscopic' precursor films should be accounted for in the boundary conditions, which incorporate all nonhydrodynamic effects, rather than in the bulk equations.

In the situations where the spreading of liquids involves a precursor film, there is at present little information about the flow kinematics so that the modelling of this process, including rich behaviour of the free-surface profile and its dependence on macroscopic factors (Beaglehole 1989), is a nontrivial problem. If the liquid advances at a speed higher than the speed of spontaneous spreading of the precursor film (e.g., when driven by gravity or other external factors), the free-surface profile near the contact line exhibits no qualitative difference from that of a partially wetting fluid.

3.2.2.3 Receding contact line and liquid-liquid displacement

Kinematics of the receding contact-line motion where a liquid is displaced by a gas is less clear-cut than that of dynamic wetting. Although the overall fluid motion is rolling (Schwartz, Rader & Huey 1964, see Fig. 3.4a), details of the flow near the contact line point to a more complicated picture. The main question one has to answer is whether or not the liquid is *completely* displaced by the gas.

Early attempts to address this question go back to Vaillant (1913), who noted that a visible film is sometimes left behind a receding liquid, and Templeton (1956), who reported the presence of an invisible residual film. These qualitative observations, however, do not answer the question whether there is a microscopic film left behind the contact line or simply the main body of liquid is pushed too fast so that the actual contact line is at the trailing edge of a macroscopic film left on the wall.

More recent experiments (e.g., Lam et al. 2001, 2002) have provided substantial clarification of this point. Measurements of the time-evolution of the contact angles in a slow receding motion of two homologous series of fluids (*n*-alkanes and 1-alcohols) on silicon wafer surfaces and their analysis point to "liquid penetration and surface swelling, or at least liquid retention, even on this very hydrophobic surface" (Lam et al. 2001). From the point of view of fluid mechanics, this suggests that the liquid displacement by the gas is not complete: the receding volume of liquid is losing mass and this might have to be accounted for. Given that this process is on the molecular scale, one will then have to modify the boundary conditions at the contact line and/or the liquid-solid interface.

Kinematics of the liquid-liquid displacement is more complex than in the cases of the advancing and receding contact-line motion. Dussan (1977) studied this process by considering a model system where glycerine displaces mineral oil in a narrow (0.635 cm in inner diameter) Plexiglas circular tube. The flow was visualized by injecting a dye mark into the upper (displaced) fluid (Fig. 3.9a). The results are schematically represented in Fig. 3.9b in the reference frame at rest with respect to the interface. One can see that the lower (more viscous) fluid undergoes the already familiar rolling motion as it displaces the upper fluid, which exhibits the so-called 'split-ejection' pattern, where the fluid driven to the contact line by the moving wall (*d* in Fig. 3.9b) and the interface (*e*) is ejected back into the bulk (*f*). This flow pattern near the contact line is qualitatively the same as that described by the stream functions (3.27), i.e., the 'semi-solution' obtained in the framework of the standard model. It must be noted, however, that the spatial resolution of Dussan's (1977) experiment is rather low and some of her conclusions rely on the previous theoretical analysis of the possible flow kinematics in the liquid-

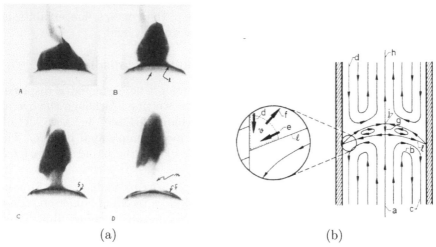

(a)   (b)

FIGURE 3.9:   Visualization of flow with dye mark (a) and its schematic representation (b) in the reference frame at rest with respect to the interface (Dussan 1977; © 1977 AIChE; reproduced with permission). The lower (more viscous) fluid undergoes the rolling motion whilst the upper one exhibits 'split-ejection' as the fluid dragged into the corner (**v**) by the wall and along the interface has to go back into the bulk.

liquid displacement (Dussan & Davis 1974) rather on what has actually been observed.

A more detailed study covering a wider range of parameters has been reported by Brown, Jones & Neustadter (1980), who visualized the flow by injecting polysterene latex particles of very low concentration in the bulk of the two fluids and at the interface. Their results broadly confirm the possibility of the two flow patterns suggested by Huh & Scriven (1971) and Dussan & Davis (1974) and qualitatively described by the stream functions (3.27). The first is the rolling motion of the advancing fluid combined with the split-ejection flow in the receding one (Fig. 3.10b), as in Dussan (1977); the second is the rolling motion of the receding fluid whereas the advancing one exhibits the split-injection flow (Fig. 3.10a). (In the latter, one has to complete the sketch in Fig. 3.10a by drawing streamlines compatible with the given ones in the region near the interface where the authors do not describe the flow.)

An intriguing feature reported by Brown et al. is that in the situation where the rolling motion of the displaced liquid was observed the interface was almost nonflowing. At the same time, the flowing interface (coupled with the rolling motion of the advancing liquid) was always observed together with a residual film of the displaced fluid on the wall. The thickness of this film has not been reported. The authors came to the conclusion that "the very presence of a well-defined three-phase contact line was a dominant factor in determination of the flow regime".

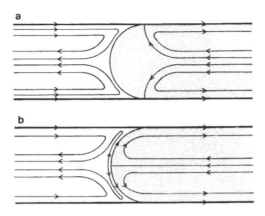

FIGURE 3.10: Schematic representations of the flow patterns observed during immiscible displacement in cylindrical capillaries (Brown, Jones & Neustadter 1980; © 1980 Elsevier; reproduced with permission).

3.2.3 Dynamic contact angle

3.2.3.1 Contact angle measurements

As already mentioned in §2.4.4, the notion of a contact angle between a free surface and a solid boundary has been introduced into macroscopic fluid mechanics by Young (1805) through what has become known as the Young equation (2.164). Since the times of Young, qualitative observations have lead researchers to the view that the dynamic contact angle deviates from the static one. This view has become common wisdom to the extent that, for example, West (1911) mentioned the variation of the contact angle as something so obvious (Fig. 3.4) that he felt no need to justify it experimentally, unlike the no-slip condition which was a controversial issue at the time and needed justification by referring to the appropriate experiments.[2]

Once the emphasis of research started to shift from the effects of capillarity as a whole towards understanding dynamic wetting as a local phenomenon, the experiments on dynamic contact angles have become more quantitative. Early research employed setups with easy access to the contact line (Ablett 1923) and gradually experimenters moved on to other flow configurations. Schematically, the main geometries used in experiments over the years are given in Fig. 3.11. From a physical point of view, not all of them are equivalent and the setup with a rotating cylinder (Fig. 3.11a), originally used by

---

[2] As West pointed out later (see Yarnold 1938), experiments with liquid indices involved measurements of the force needed to move the index and hence, using the Young equation and associating the measured force entirely with the capillary effects, one could calculate the difference between the cosines of the advancing and receding contact angle. Thus, to some extent, the velocity-dependence of the dynamic contact angle has been checked.

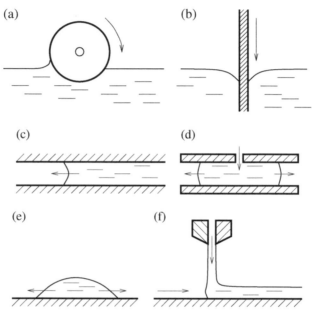

FIGURE 3.11: Some of the geometries used to study dynamic wetting. (a) The rotation of a horizontal cylinder partially submerged into a bath of liquid. (b) The 'dip-tank' experiment: a plate, tape of fiber is steadily immersed into a pool of liquid. (c) The steady flow through a capillary or in a channel formed by two parallel plates. (d) The axially-symmetric spreading of a liquid between two parallel plates where the liquid is introduced through a hole in the top plate. (e) The unsteady spreading of a drop deposited onto a solid surface. (f) Curtain coating in which a liquid sheet impinges on a moving solid substrate.

Ablett (1923) and most notably by Wilkinson (1975) and Bolton & Middle-man (1980), stands out as it qualitatively differs from the other types since in this flow the contact line advances on a solid surface that has already been exposed to the liquid. Then the solid surface becomes 'pre-wet', at least at a microscopic level and possibly even macroscopically (noted first by Vaillant 1913 and later by Templeton 1956), and the presence of a thin layer of liquid on the cylinder is mentioned by the experimenters who used this setup. Experiments (e.g., Sedev et al. 1993) show, however, that the subsequent dynamic wetting can significantly differ from that in the situation where the contact line advances over a dry solid surface so that the results in these two cases should not be directly compared.

The flow geometries shown in Fig. 3.11b–d allow one to study dynamic wetting in its proper meaning, that is the spreading of a liquid over an initially dry solid substrate. The dip-tank experiment (Fig. 3.11b) is perhaps the one used most frequently given that it provides easy access to the contact line and makes it possible to study both advancing and receding contact-line motion in steady and controlled unsteady regimes. Its disadvantage is that the possibilities of varying the flow geometry are very limited (one can only replace a tape by wires of different radii, which for small radii leads to technical complications).

The most flexible flow geometry used in experiments so far is that of curtain coating (Fig. 3.11f), where, in addition to variations of the contact-line speed, one can independently vary the curtain's height, the flow rate and the angle of inclination of the solid substrate with respect to the falling curtain.

As has already been illustrated in Fig. 3.3, in a general case the value of the contact angle obtained in experiments depends on the degree of magnification, that is on how the contact angle has actually been defined and measured. On a macroscopic length scale, the molecular structure of fluid becomes invisible but going from the molecular scale up obviously does not give a specific spatial resolution which would correspond to the theoretical 'contact angle'. The latter remains, as is always the case in continuum mechanics, an approximation to what is measured and has to be interpreted when applied to experimental results.

The experimental ways of measuring the contact angle can be conventionally divided into 'direct' and 'indirect'. The direct methods are based on taking the free-surface profile as it is seen in experiment. Then, using the definition of the contact angle, one can measure the angle between the tangents to this profile and the solid boundary at the contact line. This procedure has been applied by Chen (1988) and Blake, Bracke & Shikhmurzaev (1999) to mention but two examples. A variant of the direct method is to use the free surface as a mirror and interpret how a beam of light is reflected from it (e.g., Ablett 1923, Allain, Ausserré & Rondelez 1985). Alternatively, the edge of the liquid can be analyzed by considering a shadowgraphic image formed by a laser beam after it refracts into the liquid, then reflects from the solid surface and finally refracts out of the liquid (Zhang & Chao 2002). In the direct method, no theoretical assumptions are made about the shapes of the interfaces, and the result, of course, depends on the spatial resolution associated with visualization of the free surface and/or the optical device.

The indirect methods of obtaining the contact angle are based on approximating the free-surface profile using some curve chosen from a certain family on the basis of some theoretical assumptions. After the family is chosen, the parameters specifying the suitable curve are obtained by measuring only the location of a few characteristic points of the free surface. Once the curve is chosen, the contact angle it makes with the solid boundary can be calculated, often analytically. For example, the shape of a meniscus for the flow in a thin capillary (Fig. 3.11c) or that of a spreading drop (Fig. 3.11e) can, under certain conditions, be approximated by a spherical cap. Then, in experiments, it becomes sufficient to find the position of the contact line and the apex of the meniscus or the drop to calculate the contact angle. Clearly, the procedure implies — and heavily relies on the validity of — some theory that is supposed to describe the flow and whose outcome, as far as the free-surface shape is concerned, is the assumed family of profiles. Usually, the assumed profile is that of a static interface where the shape is determined either by capillarity alone (the spherical-cap or cylindrical approximation) or by the combined action of capillarity and gravity. The influence of dynamics on the free-surface shape is neglected and the contact-angle value is a free parameter.

This method with the spherical-cap approximation has been used by Hansen & Toong (1971) and Hoffman (1975) for the flow in a thin capillary, Chen (1988), Dodge (1988) and Foister (1990) for the spreading of drops, by Ngan & Dussan (1982) with a cylindrical approximation for the flow between parallel plates, by Lam et al. (2001, 2002) with automated axisymmetric drop shape analysis, which fits the static profile resulting from the combined action of capillarity and gravity, for the study of sessile drops to mention but a few examples. In the cases where the results obtained by direct and indirect methods have been compared, they appear to be different though rather close (Chen 1988).

In flow geometries such as those shown in Fig. 3.11a,b,f, one has to use the direct method since there is no plausible assumption about the free surface shape based on some simplified theory which one can justify as providing higher accuracy of the obtained values of the contact angle than the direct method. A review of some methods for measuring contact angles can be found in Neumann & Good (1979).

3.2.3.2   Dynamic advancing contact angle versus contact-line speed

It is convenient to begin with the most important case of the contact angle in the process of dynamic wetting, i.e., the situation where a liquid displaces a gas from a solid surface. As in §3.1.1, we will use the term 'dynamic contact angle' referring to the angle between a free surface and a solid boundary measured through the liquid when the contact line is moving. As for the 'static' contact angle, in dealing with experiments one finds out that this concept is not as clear-cut as that of the dynamic contact angle. The reason is that once the contact line stops moving, a number of physical and chemical processes associated with time scales large compared with the hydrodynamic one can take place so that the eventual equilibrium state and the corresponding contact angle will have little relevance to what characterizes the contacting materials as far as fluid mechanics is concerned. Contact angle hysteresis and associated dependence of the static state on the history of liquid-solid interaction is another factor which complicates the situation. We will return to these issues later.

Here we are interested in hydrodynamic processes and consequently a hydrodynamic time scale, and it is convenient to use as the working definition the term 'static contact angle' as the angle formed by the free surface and the solid boundary as the contact line ceases to move. In the literature, the above-defined static contact angle is sometimes referred to as the 'static advancing' contact angle to distinguish it from the 'static receding' contact angle and hence acknowledge the existence of the contact-angle hysteresis.

The main emphasis of experimental research on dynamic contact angles has always been and to a large extent remains focussed on the dependence of the dynamic contact angle on the contact-line speed. These studies involved different materials, conditions of the solid surface, flow geometry, velocity range as well as different methods of determing the contact angle and spatial resolution of the measurements. All these factors make comparative analysis of experimental results nontrivial, sometimes even at a qualitative level. Typical sets of data for the dynamic advancing contact angle are shown in Fig. 3.12 for the case of perfect wetting and in Fig. 3.13 for partial wetting.

For all gas/liquid/solid systems studied, the dynamic contact angle measured in experiments invariably increases with the contact-line speed (Ablett 1923, Yarnold & Mason 1949, Rose & Heins 1962, Elliott & Riddiford 1967, Inverarity 1969, Schwartz & Tejada 1970, 1972, Hansen & Toong 1971, Hoffman 1975, Burley & Kennedy 1976a,b, Zheleznyi & Korneva 1979, Rillaerts & Joos 1980, Gutoff & Kendrick 1982, Ngan & Dussan 1982, Cain et al. 1983, Burley & Jolly 1984, Chen 1988, Bracke, De Voeght & Joos 1989, Ström et al. 1990, Fermigier & Jenffer 1991, Petrov & Petrov 1992a, Blake 1993, Hayes & Ralston 1993, Lee & Chiao 1996, Shen & Ruth 1998, Schneemilch et al. 1998, Blake, Bracke & Shikhmurzaev 1999, Blake & Shikhmurzaev 2002, Petrov et al. 2003a,b, Blake, Dobson & Ruschak 2004, Ramé, Garoff & Willson 2004, Rio et al. 2005, Šikalo et al. 2005, Clarke & Stattersfield 2006, Bayer & Megaridis 2006).

In this respect, there is no qualitative difference between the situations of perfect and partial wetting once the wetting velocity exceeds that of the precursor film spreading (Schwartz & Tejada 1970, 1972). Roughness of the solid substrate, once below a certain level, has no appreciable effect on the contact angle (Schwartz & Tejada 1970, 1972, Fermigier & Jenffer 1991, see Fig. 3.12). As one should expect, the velocity-dependences of the dynamic

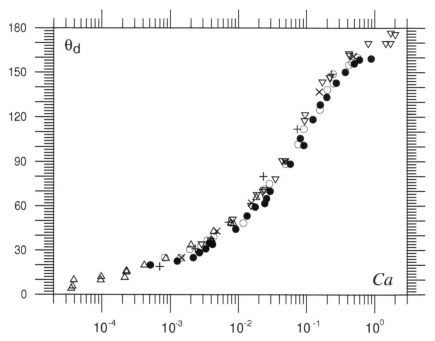

FIGURE 3.12: Dynamic advancing contact angle for some liquid/air/solid systems (perfect wetting, $\theta_s = 0°$). Silicone oils ($\triangle$: SF-96, $\mu = 9.58$ P; ∇: Brookfield Std Viscosity fluid, $\mu = 988$ P) in a glass capillary (Hoffman 1975) and in the dip-tank experiment ($+$: $\mu = 4.80$ P; $\times$: $\mu = 9.95$ P) with a polystyrene tape (Ström et al. 1990). Silicone oil 47V5000 ($\mu = 50$ P) in a standard glass tube ($\bullet$) and in a precision bore tube ($\circ$) (Fermigier & Jenffer 1991).

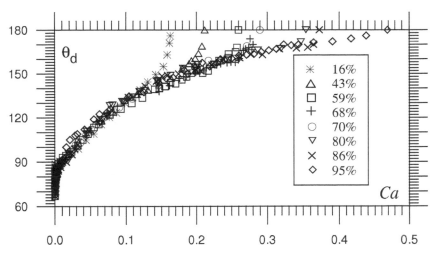

FIGURE 3.13: Dependence of the dynamic advancing contact angle on the capillary number for eight water-glycerol solutions (Blake & Shikhmurzaev 2002). Viscosities: $\mu = 1.5$ cP (16% glycerol concentration), 4.2 (43%), 10 (59%), 19 (68%), 23 (70%), 58 (80%), 104 (86%), 672 (95%).

contact angle for different material systems do not collapse into one curve in dimensionless similarity variables composed of the dimensional parameters involved in the standard model (see e.g., Ramé, Garoff & Willson 2004), though a number of attempts have been made to divine such a 'master curve' (Hoffman 1975, Jiang, Oh & Slattery 1979, Bracke, De Voeght & Joos 1989, Foister 1990, Remoortere & Joos 1993). This point will be illustrated even more explicitly with the effect of 'hydrodynamic assist of dynamic wetting' we will discuss later in this chapter.

3.2.3.2.1 Air entrainment. The maximum speed, $U_{\max}$, with which a liquid can displace a gas from a solid surface is found to be finite. This speed is usually referred to as the maximum speed of wetting or, in industrial applications, the maximum coating speed. As the speed of wetting reaches $U_{\max}$ and the contact angle reaches its maximum value, which is close to 180° for very low gas-to-liquid viscosity ratios and significantly lower than 180° for less viscous liquids, the gas starts to entrain the advancing liquid (Wilkinson 1975, Burley & Kennedy 1976a,b, Blake & Ruschak 1979, Bolton & Middleman 1980, Burley & Jolly 1984, Petrov & Sedev 1985, Buonopane, Gutoff & Rimore 1986, Sedev & Petrov 1988, Bracke, De Voeght & Joos 1989, Blake, Clarke & Ruschak 1994). Given that in experiments the displaced gas is almost invariably air, this effect is known in the literature as 'air entrainment'. The onset of air entrainment is not progressive, but occurs abruptly when the contact-line speed reaches $U_{\max}$. The value of $U_{\max}$ is different for different material systems (e.g., Burley & Jolly 1984), and the form air entrainment

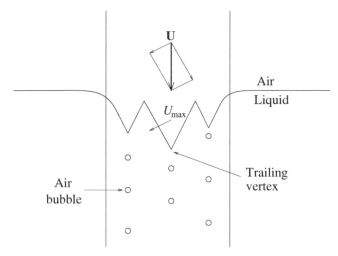

FIGURE 3.14: Schematic representation of the sawtooth contact line obtained when a tape plunges vertically into a pool of liquid at a speed, U, greater than the maximum speed of wetting, $U_{\max}$. The contact line shows both leading and trailing vertices, with air bubbles entrained at the latter (Blake & Ruschak 1979).

takes depends on how much the speed at which the liquid is forced to spread exceeds $U_{\max}$.

The onset of air entraiment has been investigated systematically for the first time by Blake & Ruschak (1979) who found that for the speed of spreading greater than a certain $U_{\max}$ the contact line takes a 'sawtooth' shape (Fig. 3.14) so that the normal component of the contact-line velocity remains equal to $U_{\max}$. Air entrainment occurs in the form of bubbles, usually of a fairly narrow size distribution (Blake 1993) that appear from the trailing vertices of the sawteeth (Fig. 3.14). The maximum speed of wetting imposes limitations on the performance of industrial coating devices (Kistler & Schweizer 1997).

It should be noted here that there is no evidence that the maximum speed of wetting is an intrinsic property of materials of the liquid and the solid substrate it spreads on. The gas (air) plays an essential role and as its influence is reduced, for example by placing the coating device in a vacuum chamber, $U_{\max}$ increases (Benkreira, Khan & Patel 2004).

Another factor related to the onset of air entrainment is roughness of the solid substrate. As the contact angle increases towards $180°$ (i.e., provided the onset of air entrainment is not triggered earlier by the displaced air), the free surface becomes almost parallel to the solid and the roughness elements of the solid substrate can 'puncture' the free surface in front of the advancing contact line thus creating 'nucleation sites' promoting wetting. As a result, the steady advancement of the liquid can break down even in the absence of air in front of the liquid.

3.2.3.2.2 Low-velocity region. At very low contact-line speeds the contact angle behaviour is more complex (Elliott & Riddiford 1967, Inverarity 1969, Johnson, Dettre & Rrandreth 1977, Cain et al. 1983, Blake 1993). In a number of cases, experimenters report that the dynamic contact angle is independent of the contact-line speed in a certain range of (very low) capillary numbers. This range differs not only for different materials, as should be expected, but one can come across essentially different results reported for seemingly the same liquid/gas/solid system. An example of such discrepancy is given in Fig. 3.15, where we reproduce the data reported by Elliott & Riddiford (1967), Johnson, Dettre & Brandreth (1977) and Cain et al. (1983) for water spreading on silicone glass.

An apparent reason for this is that the region of very low contact-line speeds, besides being notoriously difficult for experimental investigation, can involve, in addition to the specific physics of wetting, a number of other physical processes, ranging from evaporation-condensation to some reactions between the fluid and the solid, whose characteristic rate becomes comparable with that of dynamic wetting itself. Indeed, for example, a drop of liquid at rest on a solid surface, that is in the situation where the velocity of wetting is zero, can 'spread' by the evaporation-condensation mechanism alone. This indicates that there must be a range of contact-line speeds where dynamic wetting and evaporation-condensation have comparable effects. This has been demonstrated by Sobolev et al. (2001) who considered the motion of a meniscus formed by a water-nitrogen interface in ultrathin glass capillaries (90–540 nm in diameter). They have found that the (indirectly measured) dynamic contact angle strongly depends on the meniscus velocity at capillary numbers as low as 10^{-8} (the contact-line speed in the μm per second range). This dependence disappears at higher capillary numbers when the meniscus advances over a dry solid surface.

Similar arguments can be considered for other 'additional' physical processes.

The physical mechanisms complementing the specific physics of wetting at very low contact-line speeds are obviously material-dependent and one can alter their role by the suitable choice of the material system, whereas the specific physics of wetting remains an ever present factor. These complementary mechanisms, however, as well as the poorly defined 'state of the solid surface', can considerably influence experimental results and have to be carefully controlled. An early example of controversy resulting from uncontrolled experimental factors goes back to Ablett (1923) and Yarnold & Mason (1949) who studied dynamic contact angles formed by water on paraffin wax. Yarnold & Mason have come to the conclusion that "Ablett's measurements are vitiated by the fact that the time of immersion is disregarded" and hence pointed out

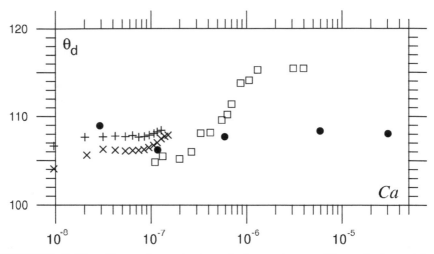

FIGURE 3.15: Dynamic contact angle for water on siliconed glass at low contact-line speeds. □: Elliott & Riddiford (1967); •: Johnson, Dettre & Brandreth (1977); (+) and (×): different runs in Cain et al. (1983).

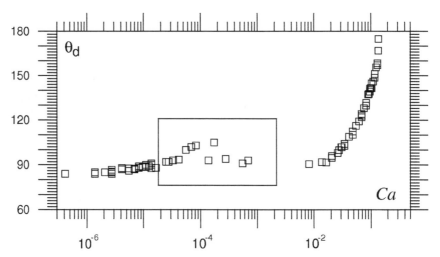

FIGURE 3.16: The velocity-dependence of the dynamic contact angle for water spreading on polyethyleneterephthalate (PET). The rectangle shows the region of unsteady dynamic wetting where the contact angle alternates between two different values (Blake 1993).

that for the system they studied at low contact-line speeds there can be an additional factor to be taken into account in experiments. Subsequently, the role of different forms of retention of a liquid by a solid substrate has been studied in a number of works by Neumann and co-workers (see Lam et al. 2002 and references therein).

Another feature of low-velocity dynamic wetting is that for some systems there appears a range of contact-line speeds where dynamic wetting is intrinsically unsteady. Blake (1993) has reported that in his observations of water spreading on polyethyleneterephthalate (PET) dynamic wetting was steady for the contact-line speed below approximately 1 mm s^{-1} and the contact angle rose steadily on a steep curve, whereas above this speed the contact angle "began to alternate between values on the steep curve and lower values on a much shallower curve. $< \dots >$ The unsteadiness died out at about 10 cm s^{-1}, and the data then rose smoothly on the shallower curve up to the maximum wetting velocity". Blake's results are given in Fig. 3.16, where the region of unsteadiness is inside the rectangle. Pictorially, the process can be described as if one or two 'zippers' move along the contact line switching the contact angle from one position to the other (Blake 1995).

As we have seen, the dynamic contact angle behaviour at very low contact-line speeds is very rich due to a number of additional physical effects coming into play. In our search for the specific physics of wetting, however, this is a disadvantage since these experiments reflect a combined effect of several physical phenomena making identification of the physics we are looking for more difficult than in the case of medium and high contact-line speeds.

3.2.3.3   Nonlocal hydrodynamic effects

An important question with fundamental implications for the mathematical modelling is whether for a given liquid/gas/solid system the dynamic contact angle is a *function* of the contact-line speed or it depends on the flow field/geometry near the moving contact line and hence should be regarded as a *functional* of the flow field.

It is more elucidating to reformulate this question as follows. If considered in the reference frame at rest with respect to the contact line (Fig. 3.1), the velocity of the solid boundary is the main but not the only factor determining the flow near the contact line. For the same velocity of the contact line with respect to the solid surface, the flow field near the contact line can be influenced, for example, by the boundaries confining the flow domain if they are located sufficiently close to the contact line and by conditions on these boundaries. Only when all boundaries are far away from the contact line, the contact-line speed becomes the only parameter determining the flow field near the contact line. Given that the dynamic contact angle is but one of the features resulting from the specific physics of wetting we are looking for and that the size of the region associated with this physics is on the scale of macroscopic fluid mechanics (Chen, Ramé & Garoff 1996), one would

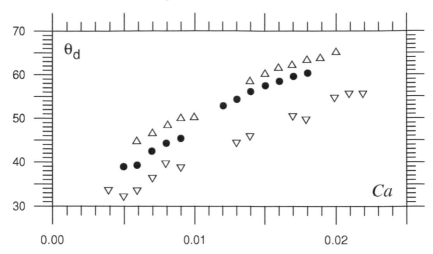

FIGURE 3.17:   Variation of the indirectly measured dynamic contact angle with capillary number during the displacement of air by silicone oil between plane-parallel glass surfaces of nominal separation 0.01 cm ($\bigtriangledown$), 0.07 cm ($\bullet$) and 0.12 cm ($\triangle$) reported by Ngan & Dussan (1982). The oil has viscosity $\mu = 970$ cP; the surface tension of the oil-air interface $\sigma = 19.7$ dyn cm^{-1}.

expect that the contact angle should depend on the flow field/geometry near the moving contact line. Then, the velocity-dependence of the contact angle becomes just a particular manifestation of this general dependence.

This possibility has been examined in several experimental works. Ngan and Dussan (1982) have reported experiments on the motion of a meniscus formed by the silicone oil-air interface between plane-parallel glass plates. This flow configuration is shown in Fig. 3.11c. The objective was to investigate how the dynamic contact angle depends on the distance between the plates that is on the flow geometry. The contact angle was determined indirectly by using the cylindrical approximation of the free-surface shape and then calculating the angle from the measured positions of the contact line and the apex of the meniscus. The results are shown in Fig. 3.17. One can see immediately that the dynamic contact angle does indeed depend on the gap between the plates. Although, due to the parameter range involved and the indirect method of measuring the contact angle, these results are more indicative than conclusive, they certainly suggested that the problem deserves further investigation. Ironically, the findings have been dismissed on the following grounds:

> Since the actual contact angle is a material property of the system, it cannot depend on the geometry of the apparatus in which it is measured. However, it is fairly clear that the angle reported in figure 4 [Fig. 3.17 here — Y.S.] does depend on the width of the slot; hence it cannot be the actual contact angle (Ngan & Dussan

1982, p. 36).

In other words, first, it is presumed as an axiom that the actual *dynamic* contact angle is independent of the flow geometry, that is the very effect the authors are investigating is ruled out a priori as impossible. Then, since the experimental results appear to be in conflict with this 'axiom', they have to be reinterpreted to comply with it. The subsequent arguments in Ngan & Dussan (1982) and in a number of works that followed in its wake were intended to support their conclusion on the basis of the then known theoretical models which, in their turn, also used an assumption that the dynamic contact angle is independent of the flow field and interpret all features of the measured contact angle as 'apparent'.

Obviously, this scheme of reasoning is not flawless, not the least because the basic question as to why the *dynamic* contact angle must be a "material property of the system" remains unaddressed. As we know, the contact angle between a free surface and a solid boundary is introduced into macroscopic fluid mechanics via the Young equation (2.164), which expresses the balance of forces tangential to the solid boundary acting on the contact line. In the static situation, these forces, described as 'surface tensions', are determined by the contacting materials and hence the *static* (equilibrium) contact angle is indeed "a material property of the system". However, there is no reason to presume that when the contact line is moving, the surface tensions in the immediate vicinity of the contact line remain the same as in equilibrium. Furthermore, the velocity-dependence of the contact angle reported in numerous experiments and their analysis in the framework of various theories (to be discussed later) suggest that this is not the case.

Blake, Clarke & Ruschak (1994) considered the situation where the dynamic contact angle approaches 180° and air entrainment takes place. They used the curtain coating setup (Fig. 3.11f) and found that the onset of air entrainment depends strongly on the flow geometry. This has important industrial consequences and a special term 'hydrodynamic assist of dynamic wetting' was coined to emphasize the fact that the flow field can be manipulated to promote wetting and postpone air entrainment in coating processes. It must be noted that air entrainment is by no means an 'apparent' phenomenon and the results of Blake, Clarke & Ruschak (1994) suggest that the dynamic contact angle should also be dependent on the overall flow.

The problem has been studied systematically by Blake, Bracke & Shikhmurzaev (1999), who have also used a curtain coating apparatus and the direct way of measuring the contact angle. In their experiments, the liquid exits from four narrow slots within the die and combines at its lower lip to form a uniform curtain (Fig. 3.18). The curtain falls between two optical windows and impinges vertically upon the moving tape, which is horizontal. Fine, platinum wire cathodes are positioned so as to just touch the front and back surfaces of the curtain, and a copper anode is placed a few centimeters downstream of the impingement zone. A voltage applied between the electrodes

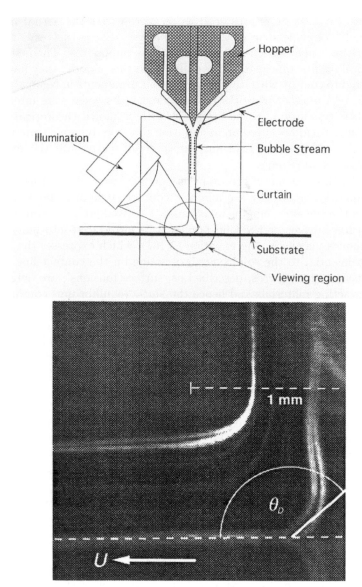

FIGURE 3.18: Top: Apparatus used to visualize the dynamic contact angle during curtain coating. Bottom: Typical image of the impingement zone at the foot of the curtain showing measurement of the dynamic contact angle. The bright lines of the bubble streams are doubled by reflection in the adjacent surfaces.

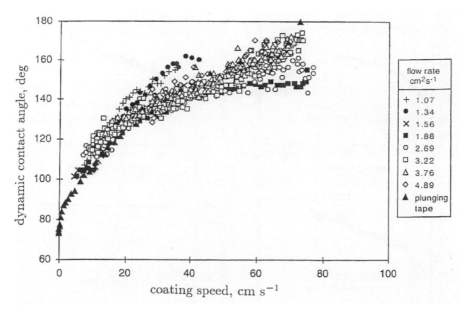

FIGURE 3.19: Dynamic contact angle measured as a function of coating speed at different flow rates for a 3 cm high curtain of 25 cP aqueous glycerol solution on PET tape (Blake, Bracke & Shikhmurzaev 1999). Also shown are data obtained in a separate plunging tape experiment. One can see that for the same coating speed the contact angle varies with the flow rate, that is with the flow geometry.

generates a stream of hydrogen bubbles by electrolysis from the tip of each cathode. A collimated beam of light optically sections the curtain in the plane of the bubbles. The bubbles scatter the light allowing the postion of liquid surfaces to be defined to within about 20 μm, which determines the spatial resolution of the measurements. The contact angle was measured directly as the angle between the tangent to the visualized free surface and the solid substrate (Fig. 3.18).

The raw data obtained with 25 cP aqueous glycerol for a curtain height of 3 cm for different flow rates determining the thickness of the curtain are shown in Fig. 3.19 together with the data for the same liquid obtained in the plunging tape experiment (Fig. 3.11b). The curtain thickness was typically in the range 0.1–1.0 mm. One can see immediately that for one and the same liquid the velocity-dependence of the dynamic contact angle varies with variations of the flow geometry. Interestingly, the data for the plunging tape experiment lie mostly in the centre together with those obtained at the highest flow rates of 3.76 and 4.89 cm^2 s^{-1}. Data for the lowest flow rate(1.07 cm^2 s^{-1}) lie well to the left, i.e., they show the steepest dependence of the contact angle on the contact-line speed. Data for successively higher flow rates then progress to

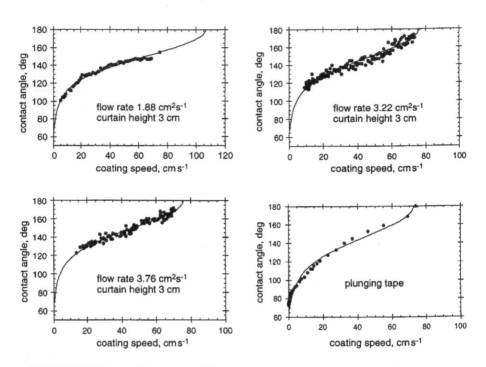

FIGURE 3.20: Data from Fig. 3.19 fitted to equation (3.30).

the right until, at a flow rate of 1.88 cm^2 s^{-1}, shows the weakest dependence on U, i.e., the lowest contact angle for a given contact-line speed. The results then track back to the centre as the flow rate is increased still further.

In order to make the effect more apparent, it is convenient to fit the data for each flow rate to a smooth curve and then to use this curve to generate a regular, 'synthetic' data set. The function

$$\theta_d = \cos^{-1}[\cos\theta_s - m_1\sinh^{-1}(m_2 U)] \qquad (3.30)$$

with two adjustable parameters, m_1 and m_2, provides an accurate approximation for the data as shown in Fig. 3.20). This function originates from the molecular-kinetic theory of wetting (Blake & Haynes 1969), which will be discussed later, and here (3.30) is used simply as a convenient approximating formula.

The synthetic data are shown in Fig. 3.21. The trends noted in Fig. 3.19 are now evident. By plotting the synthetic data as the contact angle vs the flow rate, as in Fig. 3.22, we can illustrate the influence of the flow field/geometry. For high flow rates, the back free surface of the curtain is far away from the contact line (the 'overflooded' regime) and there is no dependence of the contact angle on the flow rate. The dynamic contact angle is equal to that in the plunging tape experiment for the same contact-line speed. As the flow

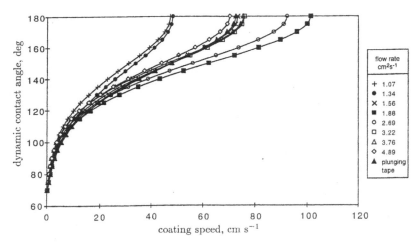

FIGURE 3.21: Dynamic contact angle as a function of coating speed for different flow rates. Synthetic data calculated by fitting equation (3.30) to the data of Fig. 3.19.

rate reduces and consequently the curtain thickness diminishes, the contact angle becomes sensitive to the flow rate, i.e., to the position of the back free surface of the curtain. At a certain flow rate, the contact angle reaches its minimum and then increases again until air entrainment takes place.

The correlation between the dependence of the contact angle on the flow field and the 'hydrodynamic assist of dynamic wetting' reported by Blake, Clarke & Ruschak (1994) in connection with air entrainment is illustrated in Fig. 3.23, where the synthetic data were used to generate a series of contours, for fixed dynamic contact angles, at $10°$ intervals from $90°$ to $180°$. Also shown are the speeds at which the onset of air entrainment (AE) was observed (solid diamonds). Below a critical flow rate (about 1 cm^2 s^{-1} in Fig. 3.23), the curtain becomes unstable and ruptures. One can note that all the contact angle contours and the air-entrainment boundary have broadly the same shape. Given that air entrainment is by no means an 'apparent' phenomenon, one is led to the conclusion that the contact angles measured in experiments originate from the same hydrodynamics as does air entrainment.

Blake, Bracke & Shikhmurzaev (1999) have also shown that other ways of varying the flow geometry one can use in curtain coating, for example variations of the curtain height, also result in a corresponding variation of the dynamic contact angle for the same contact-line speed. Hence it has been demonstrated that the dynamic contact angle directly measured in experiments is not a mere function of the contact-line speed; in mathematical terms it has to be treated as a functional of the flow field. In other words, the contact angle is not a local quantity associated with the contact line and the speed at which it advances across the solid substrate; it is generated as a

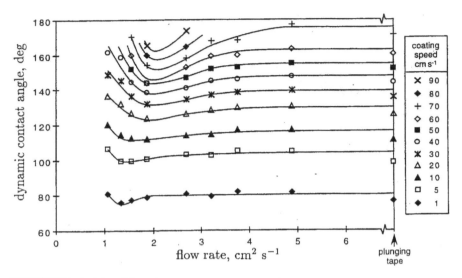

FIGURE 3.22: The dynamic contact angle versus flow rate at different coating speeds obtained using the synthetic data calculated by fitting equation (3.30) to the data of Fig. 3.19.

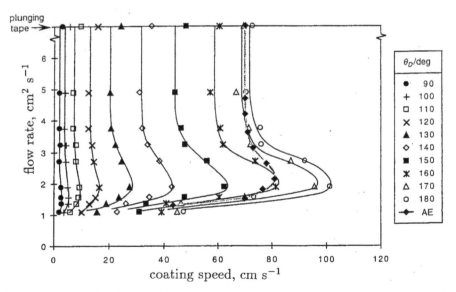

FIGURE 3.23: Map of flow rate versus coating speed showing dynamic contact angle contours plotted using the synthetic data obtained from Fig. 3.19 using equation (3.30). Coating speeds for the onset of air entrainment (AE) are also shown.

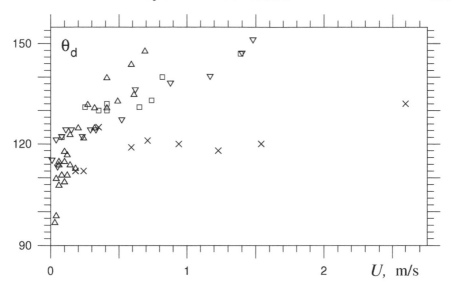

FIGURE 3.24: Dynamic contact angle variation with contact-line speed for water droplets 2 mm in diameter impacting on a partially wettable solid surface at different velocities (Bayer & Megaridis 2006; courtesy of I. S. Bayer). $\square$: $We = 0.16$; $\triangle$: $We = 0.3$; $\triangledown$: $We = 0.9$; $\times$: $We = 1.8$. ($We = \rho U_*^2 D/\sigma$, where U_* is the velocity of impact and D is the nominal diameter of the drop.)

result of interaction of the flow field with the specific physics of wetting which is activated as the contact line starts to move.

The results of Blake et al. (1999) have been corroborated by Clarke & Stattersfield (2006) who repeated the experiments with the direct way of varying the flow field instead of using the 'synthetic' data obtained from different contact angle versus wetting speed curves.

Bayer & Megaridis (2006) who conducted experiments on the drop impact on solid surfaces of different wettabilities for a wide range of parameters have reported similar findings. By eliminating the time from the dynamic contact angle and the wetting speed evolutions during the impact, the authors produced plots of the dynamic contact angle as a function of the wetting speed for different Weber numbers, i.e., different velocities of impact (Fig. 3.24). It was found that this function is not unique, i.e., the velocity dependence of the dynamic contact angle is influenced by the flow field in the drop. Qualitatively, this is exactly what Blake et al. (1999) and Clarke & Stattersfield (2006) have found. It should be noted, however, that the question of accuracy and the spatial resolution of an experiment dealing with an unsteady flow is nontrivial and further experiments along the lines of Bayer & Megaridis' (2006) work are required.

All the above experiments make one reconsider conclusions often drawn measurements with the flow parameters varying in a very narrow range that the relationship between the dynamic contact angle "is local and does not depend much on the bulk" (Rio et al. 2005). On the contrary, it is important that, at a given capillary number, the contact angle *can* be influenced by the bulk flow, and this effect has profound implications for the understanding of dynamic wetting and its mathematical modelling.

3.2.3.4 Contact-angle hysteresis

A contact-line velocity equal to zero marks the transition from the advancing to the receding fluid motion. This transition corresponds to a discontinuity in the velocity-dependence of the measured contact angle. The difference between the contact angle value at which the contact line stops to advance and that at which it starts to recede is known as *contact-angle hysteresis*.

This general term embraces influences coming from two essentially different origins, which can be referred to as 'physico-chemical' and 'macroscopic'. The first one is associated with the liquid/fluid/solid system itself, i.e., the properties of and processes in the contacting materials. The Young equation (2.164) introduces the unique time-independent static contact angle at every point of the contact line as a result of the action of intermolecular forces from the three contacting phases which in macroscopic terms manifest themselves through the surface tensions. Its key assumption (for the case of partial wetting) is that the three contacting interfaces do not change their properties with time due to penetration of liquid into the solid, chemical reactions, surface diffusion of molecular-scale films, adsorption resulting from the evaporation-condensation process, etc. If any of these factors come into play, the 'static' contact angle will evolve with time on the time scale of the corresponding process. As a result the 'static advancing' contact angle measured when the contact line has just stopped moving will differ from the 'static receding' contact angle measured after the solid has been in contact with the liquid for some time so that the above-mentioned processes, which alter the state of the interfaces, have taken place. The receding motion of the liquid also can leave traces of liquid behind the contact line which would alter the state of the solid-gas interface and hence contribute to the resulting value of the 'static receding' contact angle.

The second ('macroscopic') origin of the contact-angle hysteresis observed in experiments comes from the macroscopic nature of measurements and hence the factors, such as roughness and chemical inhomogeneity of the solid surface, which can interfere with them. For example, if the solid substrate is rough on a scale above the molecular one and below the spatial resolution of the measurements, the contact angle formed by the free surface with the *actual* solid boundary will be the same along the contact line, i.e., the one determined by the Young equation. However, the angle between the free surface and an 'effective' smooth solid boundary seen by the optical device on the scale of its spatial resolution will be different (Fig. 3.25a). The latter will vary along

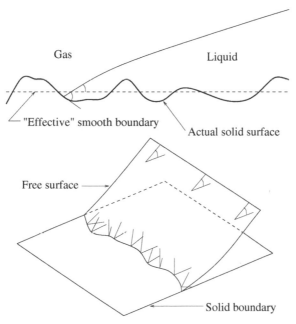

FIGURE 3.25: Top: In the case of a rough solid substrate, the angle formed by the liquid-gas interface with the actual solid surface differs from the one it makes with an 'effective' smooth solid boundary. Bottom: The latter angle varies along the contact line so that the resulting macroscopically measured contact angle is the outcome of averaging on the scale large compared with that of the irregularities of the actual solid surface.

the contact line and produce the measured contact angle as the result of averaging the influences from all roughness elements below the scale of the spatial resolution of the observer (3.25b). This was first noted by Sumner (1937, p. 45) and illustrated in a number of works via experiments with model rough surfaces (e.g., Oliver & Mason 1977, Ramos et al. 2003). Experiments show that as the number of asperities per unit area increases, their isolated influence on the contact line position and the free-surface shape give way to collective effects (Ramos et al. 2003). This resembles the situation one has with dynamic properties in the mechanics of disperse media.

The same argument applies to chemically inhomogeneous surfaces where the actual (Young's) contact angle is determined locally and varies along the contact line as it goes across chemical inhomogeneities of the solid substrate. The macroscopically measured contact angle is again the result of spatial averaging of the contributions from different parts of the contact line.

If the solid substrate is rough and/or chemically inhomogeneous,[3] the situation becomes more complex since the static contact line can assume various positions across irregularities (remaining 'straight' on the macroscopic scale). As a result, the macroscopic static contact angle, which can be called *effective* contact angle, will vary in a certain range. The 'physico-chemical' component of the contact-angle hysteresis also has its range of the (actual) 'static advancing' and 'static receding' contact angles. The combined action of the two sets of factors produces the range of contact-angle hysteresis observed in experiments.

If the contact line is advancing so that the hydrodynamic time and length scale become characteristic scales of the process, under normal circumstances the 'physico-chemical' mechanisms become irrelevant but the macroscopic averaging obviously remains in place and is complemented by time-averaging as the contact line moves across the solid surface irregularities.

In the case of receding contact-line motion, the 'physico-chemical' mechanism is almost always present due to retention of liquid by the solid surface which can take place even for a strongly hydrophobic substrate (Lam et al. 2001). Therefore the extrapolated value of the receding contact angle will depend at least on the outcome of the liquid retention process at very low contact-line velocities and hence will not characterize the static configuration. Roughness and chemical inhomogeneity of the solid surface promote liquid retention as well.

3.2.3.5   Dynamic contact angle in the process of de-wetting

Experiments show that the dynamic contact angle measured through the liquid in the receding contact-line motion decreases with increase in the contact-

---

[3]In using these terms we imply that the scale of roughness/inhomogeneity is above the molecular one. On a molecular scale every solid surface is 'rough' whilst the term 'chemical inhomogeneity' loses its meaning.

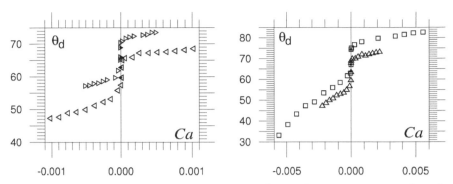

FIGURE 3.26: Dynamic advancing and receding contact angles for methanol (▷), ethanol (◁), 1-butanol (△) and 1-octanol (□) on the amorphous fluoropolymer AF 1600 surface with roughness 0.3–0.4 nm/μm^2 (Petrov et al. 2003b). Black dots mark 'static advancing' and 'static receding' contact angles.

line speed (Elliott & Riddiford 1967, Petrov & Radoev 1982, Kiss & Gölander 1991, Petrov & Petrov 1992, Sedev & Petrov 1992a, Hayes & Ralston 1993, Lee & Chiao 1996, Schneemilch et al. 1998, Petrov et al. 2003a,b, Rio et al. 2005). One has to control experimental conditions very carefully to minimize the impact of the factors responsible for the contact-angle hysteresis.

In Fig. 3.26 we illustrate the velocity-dependence of the dynamic contact angle using the data reported by Petrov et al. (2003b), who investigated a series of well-defined systems in both advancing and receding contact-line motion with a particular emphasis on the transition region. As one can see in this figure, the contact-angle hysteresis is relatively small, varying from 2.8° for methanol to 7.3° for 1-octanol. The data for the advancing and receding flow are nonsymmetric and the low-speed de-wetting is associated with sharp decrease in the contact angle. A noticeable feature is an inflection point in the velocity dependence of the receding contact angle suggesting an interplay of physical factors acting in the opposite directions.

At some contact-line speed, which is usually referred to as the 'maximum de-wetting speed', the measured contact angle becomes equal to zero (Rose & Heins 1962, Rillaerts & Joos 1980, Sedev & Petrov 1991) and further increase in speed leads to a macroscopic film or drops left behind the receding liquid (Blake & Ruschak 1979).

Possible nonlocal hydrodynamic influence on the dynamic receding contact angle has received little attention from experimenters. An attempt to discover such dependence has been made by Sedev & Petrov (1992a), who measured the angles formed on siliconized glass cylinders of different radii as they are withdrawn from a pool of liquid (aqueous glycerol solution). It is reported that the radius of the cylinder has a negligible effect on the measured contact angle, which is presented more generally as the contact angles being "independent of the geometry of the solid surface". It should be noted, however, that, unlike the study of Blake, Bracke & Shikhmurzaev (1999), who have demonstrated the effect of the flow geometry on the dynamic advancing contact angle by varying the flow geometry in the plane of flow, variation of the cylinder radius used by Sedev and Petrov has a relatively weak effect on the flow. It can produce a measurable impact only if, on the length scale associated with the specific physics of wetting, one has a transition from an effectively two-dimensional plane flow to an axisymmetric flow. Given that the radii in Sedev & Petrov's experiment varied in the range 0.013–0.845 cm, it is not surprising that no effect has been discovered. One has to conclude therefore that, as far as experimental evidence is concerned, the question of whether or not the flow field/geometry can have a nonlocal influence on the dynamic receding contact angle remains open.

3.2.3.6 Contact angles in liquid-liquid displacement

For all liquid/liquid/solid systems studied the dynamic contact angle measured through the advancing liquid increases with the contact-line speed (Elliott & Riddiford 1967, Blake & Haynes 1969, Gutoff & Kendrick 1982, Mumley, Radke & Williams 1986a, Foister 1990, Fermigier & Jenffer 1991). Representative trends are shown in Figs 3.27 and 3.28, where we reproduce the data from Foister (1990) and Fermigier & Jenffer (1991) obtained using different experimental techniques.

Foister (1990) considered the spreading of drops of liquid deposited on a solid surface (Fig. 3.11e) as they displace other fluids immiscible with them. The difference with liquid-gas displacement is that for liquid-liquid systems a drop initially rests on a film of the displaced liquid and the spreading starts as the film ruptures. This method, where the contact-line motion is monitored rather than controlled, allowed Foister to study systems with the receding-to-advancing viscosity ratios ranging from 5.9×10^{-3} to 3.18×10^3. It should be noted, however, that although in all experiments the displaced fluids were of the same type (various viscosity grades of polydimethylsiloxane oils), to cover the wide range of viscosity ratios it became necessary to use two different, though "chemically similar", displacing fluids (diglycidyl ether of bisphenol A — EPON 825 and 2-phenoxyoxiran — "phenyl glycidyl ether" or PGE). The contact angle was determined indirectly by using the spherical-cap approximation.

The general trends shown in Fig. 3.27 are broadly consistent with those reported by other workers. At the same time, in analysing Foister's data one should have in mind that the presence of a microscopic residual film is unavoidable in his experiments and that the spherical-cap approximation can be less than accurate for interfaces with low surface tensions (4.76–11.2 dyn cm^{-1} in Foister's experiment) where, for example, the Bond number can be as high as 0.49 (Foister 1990, p. 272).

Fig. 3.28 shows dependence of the dynamic contact angle on the contact-line

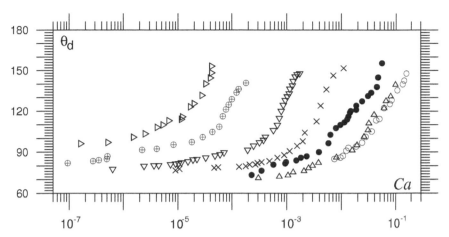

FIGURE 3.27: Dynamic contact angles versus the contact-line speed for different liquid-liquid systems (Foister 1990). The displacing fluids are EPON 825 and PGE; the displaced fluids are various viscosity grades of polydimethylsiloxane oils. $\circ$ — 825/D20, $k_\mu = 5.9 \times 10^{-3}$; $\triangle$ — 825/D100, $k_\mu = 2.39 \times 10^{-2}$; $\bullet$ — 825/D1000, $k_\mu = 2.18 \times 10^{-1}$; $\times$ — PGE/D20, $k_\mu = 5.09$; $\triangledown$ — PGE/D100, $k_\mu = 20.9$; $\oplus$ — PGE/D1000, $k_\mu = 2.09 \times 10^2$; $\triangleright$ — PGE/D12500, $k_\mu = 3.18 \times 10^3$.

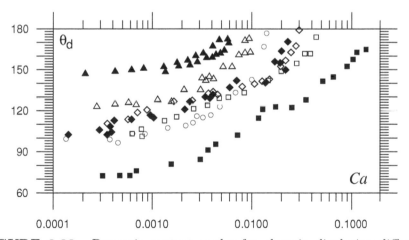

FIGURE 3.28: Dynamic contact angles for glycerin displacing different silicone oils in a precision bore (open symbols) and standard (full symbols) glass capillary tube (Fermigier & Jenffer 1991). Squares: silicone oil 47V2, oil-to-glycerin viscosity ratio $k_\mu = 1.35 \times 10^{-3}$; diamonds: 47V10, $k_\mu = 7.6 \times 10^{-3}$; triangles: 47V100, $k_\mu = 0.1$; circles: 47V1000, $k_\mu = 0.9$.

speed for glycerin displacing various 47V silicone oils in a glass capillary (Fermigier & Jenffer 1991). The displaced oils all had the same chemical structure and differed only by the chain length so that the surface tension of the oil-glycerin interface varied in a very narrow range (25–28 dyn cm^{-1}) whereas the oil-to-glycerin viscosity ratio varied by almost three orders of magnitude $(1.35 \times 10^{-3}\text{–}9.0 \times 10^{-1})$. The contact angle was measured indirectly by using the spherical-cap approximation for the meniscus shape.

As one can see in Fig. 3.28, the contact angle increases with the capillary number, reaching 180° at some critical value of Ca after which the meniscus becomes unstable and "a macroscopic film of the displaced fluid is left on the solid surface". The authors also point out that

> The contact angle hysteresis[4] is of approximately the same order of magnitude in both precision and standard tubes, which have different surface roughness. Hence, the large hysteresis should be attributed to a nonhomogeneous glass surface composition, or to the adsorption of a microscopic liquid film on the solid substrate (Fermigier & Jenffer 1991, p. 233).

It is noteworthy that for liquid-air displacement the same tubes give prcatically identical results (Fig. 3.12).

3.2.4   Summary of experiments

The main experimental findings relevant to the moving contact-line problem can be summarized as follows.

For the process of dynamic wetting, i.e., the displacement of a gas by a liquid from a solid surface, one has:

- The flow kinematics in the vicinity of the moving contact line is 'rolling', so that fluid particles initially located in the liquid-gas interface arrive at the contact line and then move with the solid surface away from it in a finite time.

- Qualitatively, the region near the moving contact line appears in experiments as that of more intensive flow than further afield.

- The contact angle measured through the liquid increases with the contact-line speed and, for a given contact-line speed, depends on the flow field/geometry in the vicinity of the contact line.

- For a given flow geometry, there is a finite contact-line speed (the maximum speed of wetting) at which the steady motion of a straight contact line breaks down and entrainment of the displaced gas into the moving

---

[4]It ranges from 31° to 66° in the precision bore tube and from 41° to 54° in the standard tube with no apparent pattern.

liquid begins. As the influence of the gas is reduced by placing the system in the vacuum chamber, the maximum speed of wetting increases.

- At low contact-line speed a number of physical effects can come into play which can lead, for example, to oscillations of the contact angle between different values.

For receding contact-line motion, i.e., the displacement of a liquid by a gas, one has:

- A microscopic film of the displaced liquid is usually left on the solid substrate behind the moving contact line.

- The contact angle measured though the liquid decreases with the contact-line speed. The low-speed regime is usually associated with a sharp decrease in the contact angle.

- At some finite contact-line speed (the maximum speed of de-wetting) the contact angle becomes equal to zero. This is followed by a macroscopic film left on the solid substrate behind the main body of receding liquid.

For liquid-liquid displacement, the main experimental observations are that:

- The process is normally associated with at least a microscopic film of the displaced fluid left on the solid surface.

- The contact angle measured through the advancing fluid increases with the contact-line speed.

- For a given capillary number (and the same or at least close values of the static contact angle), the dynamic contact angle increases with the receding-to-advancing fluid viscosity ratio.

As one moves away from the contact line, the flow field becomes broadly consistent with that described by the stream functions (3.18) and (3.27) resulting from the standard model.

3.3 Molecular dynamics simulations

Molecular dynamics simulations provide a bridge between physical experiments and macroscopic (hydrodynamic) theories. Although by no means a substitute for real-life experiments and itself requiring experimental validation, molecular-dynamics modeling is becoming a powerful investigative tool allowing one to study independently the role of different physical factors that is difficult if not impossible to do in physical experiments.

The two main components of molecular-dynamics modeling are: (a) the way in which the fluid as a system of interacting molecules is simulated computationally and (b) the way in which the results are interpreted in macroscopic terms. The first of these components depends primarily on the available computer power. At present, computers allow one to study what can be described as 'quasi-macroscopic' systems that in physical terms correspond to the length scales of the order of tens of nanometers. However, even for these very small systems a number of additional simplifications are needed to avoid prohibitively long computations. For example, the Lennard-Jones potential used in most studies,

$$V_{LJ}(r) = 4\epsilon \left[\left(\frac{a}{r} \right)^{12} - \left(\frac{a}{r} \right)^6 \right], \tag{3.31}$$

where ϵ and a are characteristic energy and length scale, has to be truncated, typically at $r = 2.5a$, to reduce the amount of computations, with a suitable potential added to (3.31) to make the force vanish at the cut-off distance. Obviously, this truncation leaves out collective interactions associated with long-range forces that play an important role in dense systems (see §7.7.2). Additional care has to be taken to maintain constant temperature of the system and to model mutual influence of the molecules of the fluid and the solid walls. The assumptions and simplifications that are at present used in molecular-dynamics modeling, as well as the nanometer size of the flow domain, make it more of a numerical experiment rather than a computer equivalent of a real physical system. Nevertheless, the obtained information gives an insight into the picture of dynamic wetting on the molecular level.

The first detailed studies of the liquid-fluid displacement using MDS have been published by Koplik, Banavar & Willemsen (1988, 1989) who considered a steady fluid-fluid displacement in the Poiseuille and Couette flows. As a validation test, the authors reproduced the Poiseuille and Coette flows for a single fluid. It should be noted, however, that this validation has to be taken with caution since the distance between the opposite walls in the computations was less than $15a$ and the total thickness of the fluid-solid interfacial layers, i.e., the layers where the density distribution experienced strong influence from the solid, was actually larger than the 'bulk'.

The immiscibility of the fluids was modelled by adding to (3.31) a potential of the form $(c_i - c_j)^2 4\epsilon(r/a)^{-6}$, where c_i is a "pseudocharge" associated with species i.

It has been shown that the model reproduced the overall kinematics of the fluid-fluid displacement observed in experiment (see Fig. 3.9). The authors have also found that the contact angle measured on the molecular length scale deviates from the static one when the contact line is moving; the advancing contact angle increases and the receding contact angle decreases with the contact-line speed, as is indeed observed in experiments. These results have been confirmed by a number of more recent large-scale MD similations of spreading liquid drops and cylinders (de Ruijter et al. 1997, de Ruijter, Blake

& De Coninck 1999, de Ruijter et al. 1999, de Ruijter, De Coninck & Oshanin 1999, He & Hadjiconstantinou 2003, Heine, Grest & Webb 2003, 2004) and capillary imbibition (Martic et al. 2002, 2004).

It should be noted here that the dynamic contact angle involved in the macroscopic (hydrodynamic) modelling implies a limit $\ell/L \to 0$ for the ratio of molecular-to-macroscopic length scale, and therefore the corresponding angle in the simulations should result from some averaging with the length scale of averaging exceeding the molecular scale by at least an order of magnitude.

Another aspect of MD simulations is the possible insight they might give into the boundary conditions one should use in the macroscopic modelling to resolve the moving contact-line problem. In this respect, the studies reported to date seem to show a certain degree of confusion since they tend to examine the conditions on the *actual* solid surface, i.e., on the *solid*-facing side of the fluid-solid interface rather than on its *fluid*-facing side, which is the boundary for the domain where the bulk equations of fluid mechanics operate. As we have already seen in the last section of §2.4.4, ignoring the fluid-solid interface can lead to unphysical paradoxes.

Koplik, Banavar & Willemsen (1988, 1989) reported slippage of the fluid on the (actual) solid surface near the contact line, and Thompson & Robbins (1989) have been able to quantify it. They found that "the usual no-slip boundary condition broke down within ~ 2 atomic spacings from the contact line". This actually means that, from the point of view of fluid mechanics, the no-slip boundary condition holds *everywhere* on the actual solid surface since in the continuum approximation one has $\ell/L \to 0$ and hence "2 atomic spacings" turn into a point. This, however, does not answer the question about the boundary conditions for the Navier-Stokes equations since it is the interaction of the fluid-solid interfacial layer with the bulk flow and the conditions on its *fluid*-facing side that have to be considered.

A step towards the methodology of fluid mechanics has been made by Thompson, Brinckerhoff & Robbins (1993) who considered the fluid-fluid and the solid-fluid interfaces and verified that the surface tensions in their simulations satisfy Laplace's capillarity equation and the Young equation (2.164) for the static contact angle. However, this analysis has not been extended to the dynamic situation and the boundary conditions on the fluid-facing side of the interfaces, perhaps, as a consequence of insufficient computer resources available at the time.

3.4 Review of theories

Now, we will return to the moving contact-line problem of §§3.1.1, 3.1.2 and briefly analyze the theories proposed to address it. These theories range from the models dealing with the moving contact-line problem as it stands, i.e., in the framework of fluid mechanics, to those, not necessarily fluid-mechanical, intended to describe or interpret particular aspects of the liquid-fluid displacement phenomenon and hence, from the viewpoint of fluid mechanics, can be seen as auxiliary.

3.4.1 Slip models

We will begin with a large group of what is collectively known as 'slip models'. These models operate in the framework of continuum mechanics and are intended to address both aspects of the moving contact-line problem, i.e., the nonexistence of a solution and the contact-angle behaviour. As we remember, the contact angle is the boundary condition for the equation determining the free-surface shape and in the standard model described in §3.1.1 it is not specified and can be prescribed arbitrarily.

3.4.1.1 Removal of singularity

As shown in §3.1.1, the analysis of a simplified formulation, where the normal-stress boundary condition is replaced by prescribing the free-surface shape, presents the moving contact-line problem in a different way, namely as that of the nonintegrable singularity in the distribution of the tangential stress acting from the liquid on the liquid-solid interface. As pointed out in Dussan & Davis (1974), the shear-stress singularity can also be seen as a consequence of the multivaluedness of the velocity field at the contact line, where the limits of the velocity components as $r \to 0$ depend on the direction along which the contact line is approached (see §3.1.1). Therefore, a natural way to remove the multivaluedness of velocity and hence the shear-stress singularity following from the analysis of the *simplified* formulation is to relax the no-slip condition by allowing $u = 0$ at $r = 0$ and assuming that $u \to 1$ as $r \to \infty$, $\theta = 0$. This means that the fluid will slip at the contact line and its vicinity with the no-slip condition restoring either asymptotically as $r \to \infty$ or at some finite distance from the contact-line. This is the key idea behind all slip models, which differ only by particular forms of the slip condition they introduce to ensure this property and their physical motivation.

Consider slip boundary conditions proposed in the literature. Slip models known in the literature broadly fall into two main groups. (Below, we will be using the polar coordinates in the plane of flow introduced in §3.1.1 and, where necessary, an asterisk will mark dimensional variables.) The first one includes models where the radial component of the bulk velocity on the solid surface is prescribed explicitly in the form

$$u^* = f(r^*, k_1, k_2, \dots)$$

where $f(0, \dots) = 0$ and $f(r^*, \dots) \to U$ as $r^* \to \infty$; k_i are constants characterizing physical properties of the contacting media. The proposed expressions for f include algebraic relaxation of u to the velocity of the solid (Dussan 1976, Somalinga & Bose 2000)

$$u^*(r^*, 0) = U \frac{(r^*/s)^k}{1 + (r^*/s)^k}, \qquad k = \tfrac{1}{2}, 1, 2, \tag{3.32}$$

as well as an exponential one (Zhou & Sheng 1990, Sheng & Zhou 1992, Finlow, Kota & Bose 1996, Somalinga & Bose 2000)

$$u^*(r^*, 0) = U[1 - \exp(-r^*/s)], \tag{3.33}$$

where s is a 'slip length'. These boundary conditions have no physical motivation and are used as a simple and convenient way of mimicking the (unidentified) specific physics of wetting and/or to assess its potential importance for the overall flow.

The second group of models follow Huh & Scriven (1971), who revived the boundary condition originally proposed by Navier (1823) as a general one to be satisfied by solutions of the Navier-Stokes equations on solid boundaries. Navier's condition assumed that slip, i.e., the difference between the velocity of the fluid and that of the solid, is proportional to the tangential stress acting from the fluid on the fluid-solid interface. Using the notation of §2.4.3, this condition can be written down in a tensor form as

$$\mathbf{n} \cdot \mathbf{P}^+ \cdot (\mathbf{I} - \mathbf{nn}) = \beta(\mathbf{u}^+ - \mathbf{U}). \tag{3.34}$$

The phenomenological coefficient β is known as the coefficient of sliding friction (Lamb 1932, p. 586). Often it is also referred to as the 'slip coefficient'. The normal projection of (3.34) gives $\mathbf{u}^+ \cdot \mathbf{n} = \mathbf{U} \cdot \mathbf{n}$, i.e., a combination of (2.158) and (2.160). In the literature, the term 'Navier condition' is often used as a reference to the tangential projection of (3.34) only.

The Navier condition seems a natural continuation of the bulk model and it was widely debated in the 19th century. Gradually it was abandoned in favour of no-slip, which became accepted on the grounds that its use provides an accurate description of experiments involving a variety of flows.[5] An additional argument used against the Navier condition in the late 19th and early 20th century is that the implicit presence of a parameter like β has not been detected by the then known experiments. However, the Navier condition is resurrected every time the no-slip one produces physically unacceptable results, and in the past 40 years its main application was the moving contact-line problem.

In the polar coordinates we are using here, (3.34) is given by

$$\frac{\mu}{r^*} \frac{\partial u^*}{\partial \theta} = \beta(u^* - U), \tag{3.35}$$

---

[5] A brief review of hypotheses and beliefs that have eventually led to the no-slip condition on the solid surface can be found in Goldstein (1965, pp. 676–680).

The slip models based on Navier's idea one can find in the literature differ by the expressions they use for β and the physical arguments employed to support them. The most frequently used assumption is that

$$\beta = \text{const} \tag{3.36}$$

(Hocking 1976, 1977, 1981, 1990, 1992, Huh & Mason 1977a, Davis 1980, Lowndes 1980, Levine et al. 1980, Hocking & Rivers 1982, Kröner 1987, Tilton 1988, Zhou & Sheng 1990, Ehrhard & Davis 1991, Haley & Miksis 1991, Smith 1995, Baer et al. 2000).

The proposed physical justification for (3.35), (3.36) is that roughness of the solid substrate leads to trapping of the displaced fluid in pockets of the solid surface topography thus providing a lubricant for the liquid advancing over an 'effective' smooth boundary (Hocking 1976). This argument, however, raises a number of questions such as

(i) How does the contact-line move across the *actual* solid substrate where, as pointed out by Huh & Scriven (1971), one still needs to address the moving contact-line problem?

(ii) Does every topography and not just some model configurations trap the receding fluid?

(iii) How does this argument justify the same boundary condition for the *receding* fluid, that is *ahead* of the moving contact line, as used, for example in Hocking (1977)?

However, regardless of justification, the boundary condition (3.35), (3.36), as well as any other slip condition where $u = 0$ at $r = 0$, removes the singularity at the contact line and makes the problem well-posed.

A variant of the Navier condition used in some works (Greenspan 1978, Greenspan & McCay 1981, Haley & Miksis 1991, Braun et al. 1995, Lopez, Bankoff & Miksis 1996, Eggers 2004) is to assume that β is a function of the distance h between a point on the solid surface and the free surface above it,

$$\beta = \beta(h). \tag{3.37}$$

This assumption implies that the dynamic contact angle is less than 90° and is usually applied together with the lubrication approximation for the Navier-Stokes equations. In specifying (3.37) it is usually assumed that $\beta \propto h^n$, where $n = 1$ (Greenspan 1978, Greenspan & McCay 1982, Haley & Miksis 1991, Lopez, Bankoff & Miksis 1996) and $n = 2$ (Haley & Miksis 1991).

Huh & Mason (1977a) followed an idea advanced by Hansen & Miotto (1957) and assumed that "as the liquid element is brought to the solid surface at the contact line, there will be a relaxation time during which the bonding of liquid molecules to the solid is incomplete" (Huh & Mason 1977a, p. 402). The authors describe the physics in terms of the relaxation time τ required for the fluid's molecules, which "become disoriented" as they arrive the contact line, "to be reoriented at the liquid/solid interface" (*ibid.*). The former state is assumed to be associated with zero tangential stress between the fluid and the solid, whereas the latter corresponds to the no-slip condition. Then, mathematically, one has $\mathbf{n} \cdot \mathbf{P} \cdot (\mathbf{I} - \mathbf{nn}) = 0$ for $x^* < U\tau$ and $\mathbf{u} = \mathbf{U}$ for $x^* \geq U\tau$, where x^* is the distance from the contact line along the solid surface (Fig. 3.1). In terms of the Navier condition (3.35), this 'slip-stick' behaviour means that

$$\beta = \begin{cases} 0, & r^* < U\tau \\ \infty, & r^* \geq U\tau \end{cases}. \tag{3.38}$$

What this condition does mathematically is that it separates the point at which the streamline going along the boundary of the flow domain has a singularity of curvature (at the contact line) from the point where there is a singularity due to the boundary conditions changing their type (from zero tangential stress to no-slip). In the standard model, both these points coincide at the contact line leading to the nonintegrable singularities in the physical properties of the flow field. Once these points are separated, the singularities at each of them become integrable. The flow in the corner near the contact line appears to be driven from outside, i.e., by the no-slip condition which generates the flow at some distance from the contact line.

The physical arguments behind (3.38) suggest that the fluid motion is rolling and the length scale associated with the specific physics of wetting, which in this model is the delayed bonding of the liquid molecules to the solid, increases with the contact-line speed. These are indeed the features observed in experiments (see §3.2.2). Ironically, as was pointed out by Dussan (1979), although the physical motivation of (3.38) implies the rolling motion, the *actual* motion produced by the use of this boundary condition is not rolling and, according to it, the fluid particles belonging to the free surface *never* arrive at the contact line. We will discuss this feature, which is common to many of the slip models, later.

Durbin (1988, 1989) suggested removing the shear-stress singularity by using a 'yield-stress' condition, which limits the maximum shear stress the liquid can impose on the liquid-solid interface. According to his assumption, the no-slip condition holds if the tangential stress experienced by the liquid-solid interface is below a certain prescribed yield stress S_{yield} and otherwise has to be replaced by

$$\frac{\mu}{r^*} \frac{\partial u^*}{\partial \theta} = -S_{\text{yield}}.$$

Mathematically, the yield-stress condition can be presented as a particular form of the Navier condition (3.35), where in this case one has to use

$$\beta = \begin{cases} -\dfrac{S_{\text{yield}}}{u^* - U}, & r^* < s \\ \infty, & r^* \geq s \end{cases}. \tag{3.39}$$

The value of s is determined by the requirement that the tangential stress is continuous along the liquid-solid interface. A variant of Durbin's approach based on an assumption of the fluid becoming shear-thinning near the moving contact line has been proposed by Carre & Woehl (2002).

Finally, to complete the picture of slip conditions, we have to mention 'numerical slip' often used, especially in the volume-of-fluid codes, to relieve the stress singularity (e.g., Renardy, Renardy & Li 2001, van Mourik, Veldman & Dreyer 2005). In this approach, the slip length is linked to the size of the mesh elements, so that the resulting numerical solution becomes mesh-dependent.

3.4.1.2   The contact angle

The modelling of the contact-angle behaviour is the other side of the moving contact-line problem. The contact angle provides a boundary condition for equation (3.8), or more generally (2.143), which determines the free-surface shape, and hence has a far-field effect on the whole flow.

Slip models treat this side of the moving contact-line problem completely separately from the way the shear-stress singularity is addressed. A number of works use a simple assumption that

$$\theta_d \equiv \theta_s, \tag{3.40}$$

that is the dynamic contact angle is always equal to the static one although the contact line is moving (Dussan 1976, Hocking 1977, 1981, 1992, Huh & Mason 1977a, Davis 1980, Hocking & Rivers 1982, Cox 1986, Durbin 1988, 1989, Zhou & Sheng 1990, Eggers 2004). Then, the measured contact angle, which, as we have seen in §3.2.3, is far from being a constant, has to be interpreted as 'apparent'. We will discuss this idea in the next subsection.

Condition (3.40) is, of course, a strong physical assumption. As an alternative to it, in some works the question of the dynamic contact angle value is left open, and the contact angle is simply prescribed as a "specific boundary condition" (Tilton 1988).

Other slip models use a more flexible approach and assume that θ_d is a function of the contact-line speed,

$$\theta_d = f(Ca, \theta_s, \chi_1, \chi_2, \dots). \tag{3.41}$$

Here χ_i are nondimensional similarity parameters formed by the material constants of the system. Clearly, (3.41) includes (3.40) as a particular case. The functional form (3.41) has no physical justification so that in the literature the preference is given to the simplest polynomial dependence

$$Ca = k(\theta_d - \theta_s)^m, \tag{3.42}$$

where k and m are assumed to be some material-related constants to be determined empirically. One can come across $m = 1$ (Greenspan 1978, Davis

1980, Greenspan & McCay 1981, Hocking 1987, 1990, Goodwin & Homsy 1991, Haley & Miksis 1991, Smith 1995, Baer et al. 2000), $m = 2$ (Haley & Miksis 1991), $m = 3$ (Ehrhard & Davis 1991, Haley & Miksis 1991, Lopez, Bankoff & Miksis 1996).

Sometimes the slip models simply prescribe θ_d without specifying its dependence on the contact-line speed and other parameters (Huh & Mason 1977a, Lowndes 1980, Levine et al. 1980, Tilton 1988, Finlow, Kota & Bose 1996, Somalinga & Bose 2000), so that it becomes an empirical adjustable constant.

Given that in all slip models the shear-stress singularity and the contact-angle behaviour are treated separately, any combination or modification of the above assumptions will produce a new theory. For example, Hocking (1977, 1981, 1992) combined (3.35), (3.36) with (3.40); Baer et al. (2000)[6] used (3.35), (3.36) with (3.42) with $m = 1$; Greenspan (1978) applied (3.35) with $\beta \propto h$ together with (3.42) for $m = 1$; Haley & Miksis (1991) extended this approach and combined (3.35), $\beta \propto h^n$ ($n = 0, 1, 2$) with (3.42), $m = 1, 2, 3$; Zhou & Sheng (1990) tried (3.40) coupled with (3.33) or (3.35), (3.36); Huh & Mason (1977a) used $\theta_d = 90°$ combined with (3.35) where β is given either by (3.36) or by (3.38), etc.

There is still room for some combinations and hence new "theories" but the arbitrariness inherent in the very approach, which was recognized as early as in Dussan (1979), suggests examining its basis, first of all, from the viewpoint of the ABC-criterion of physical meaningfulness we have formulated in §2.5.2.

3.4.1.3 Slip models and the ABC-criterion of §2.5.2

Any slip condition removes multivaluedness of the velocity field at the moving contact line and makes the solution exist, which we will demonstrate by considering its asymptotic behaviour as $r \to 0$. Thus, the most basic requirement A of the ABC-criterion of §2.5.2 is met. What we need to examine is whether these models, or any of their generalisations that uses the same approach, can satisfy simultaneously requirements B and C, i.e., can ensure finiteness of velocity *and* pressure as $r \to 0$ (requirement B) and reproduce the flow kinematics in a qualitatively correct way (requirement C). For simplicity, we will be considering the fully representative case where the displaced medium is an inviscid gas (physically, vacuum) so that the correct kinematics means the rolling motion of the advancing fluid observed in experiments. The analysis can be generalised in a straightforward way for the case where both fluids are viscous.

The slip models alter the standard problem formulation by replacing the no-slip condition (3.5) with a slip condition with the impermeability condition, $v = 0$, remaining unchanged. The contact angle is specified separately either by prescribing it, as in the standard model, or by using a particular form of

---

[6]These authors implemented the model with full understanding of its deficiency as a step in building a 3D numerical algorithm capable of handling wetting flows.

(3.41). We will make the slip condition dimensionless using the same scales as in §3.1.1 and express it in terms of the stream function. Then, to leading order as $r \to 0$, all slip conditions with a prescribed distribution of the radial velocity can be represented by

$$\left. \frac{\partial \psi}{\partial \theta} \right|_{\theta=0} = a_\lambda r^\lambda, \tag{3.43}$$

where λ and a_λ are specified in terms of parameters of a particular model. For example, condition (3.32) gives $\lambda = 1 + k$ ($k = 0.5, 1, 2$), $a_\lambda = (L/s)^k$ and (3.33) results in $\lambda = 2$, $a_2 = (L/s)$. If $\lambda = 1$, $a_\lambda = 1$, one arrives at the no-slip condition (3.12).

The boundary condition embracing as $r \to 0$ all models based on the Navier slip condition (3.35) can be represented by

$$\left. \frac{\partial^2 \psi}{\partial \theta^2} \right|_{\theta=0} = -b_\lambda r^\lambda, \tag{3.44}$$

where λ and b_λ again depend on the particular model. For example, condition (3.36) gives $\lambda = 2$, $b_2 = \beta L/\mu$; condition (3.39) results in $\lambda = 2$, $b_2 = S_{\text{yield}} L/(\mu U)$, and for $\beta \propto h^n$. $\lambda = r^{2+n}$ with the corresponding b_λ.

At leading order as $r \to 0$, condition (3.43) or (3.44) replaces (3.12) whilst (3.11) and (3.13) remain unchanged. Given that θ_d is independent of the velocity field, we can use the same iterative procedure as in §3.1.1 and, to leading order as $r \to 0$, the stream function is a solution of the biharmonic equation

$$\Delta^2 \psi = 0 \qquad (0 < \theta < \theta_d, r > 0), \tag{3.45}$$

subject to (3.11), (3.13), that is

$$\psi = 0 \qquad \text{at } \theta = 0, \theta_d, \qquad . \tag{3.46}$$

$$\frac{\partial^2 \psi}{\partial \theta^2} = 0 \qquad \text{at } \theta = \theta_d, \tag{3.47}$$

together with (3.43) or (3.44). The form of this solution is suggested by (3.43) or (3.44) and in both cases is given by

$$\psi_\lambda = r^\lambda F_\lambda(\theta). \tag{3.48}$$

To ensure $u = v = 0$ at $r = 0$ and hence remove multivaluedness of velocity, we must require

$$\lambda > 1. \tag{3.49}$$

Then, in accordance with the ideology of slip models, the tangential stress on the solid boundary and the normal stress on the free surface both become integrable, the normal stress boundary condition (3.8) can be satisfied (and will give a leading-order correction to the free-surface shape), a solution exists, thus satisfying requirement A of the ABC-criterion.

Substituting (3.48) into (3.45), (3.46), (3.47), (3.43) and (3.45), (3.46), (3.47), (3.44) one arrives at

$$F_2(\theta) = B_1 + B_2\theta + B_3 \sin 2\theta + B_4 \cos 2\theta \tag{3.50}$$

and

$$F_\lambda(\theta) = C_1 \sin\theta + C_2 \cos\theta + C_3 \sin[(\lambda - 2)\theta] + C_4 \cos[(\lambda - 2)\theta] \quad \text{for } \lambda \neq 2.$$

For the constants of integration it follows from (3.46) and (3.47) that, if $\lambda = 2$,

$$B_2 = -\frac{1}{\theta_d}B_1, \quad B_3 = B_1 \cot(2\theta_d), \quad B_4 = -B_1, \tag{3.51}$$

where

$$B_1 = a_2 \frac{\theta_d \sin 2\theta_d}{2\theta_d \cos 2\theta_d - \sin\theta_d}$$

for (3.43), and

$$B_1 = -\frac{b_2}{4} \tag{3.52}$$

for (3.44).

For $1 < \lambda \neq 2$, one has

$$C_2 = -C_4 = -C_1 \tan\lambda\theta_d, \quad C_3 = -C_1 \tan\lambda\theta_d \cot[(\lambda - 2)\theta_d],$$

where

$$C_1 = \frac{a_\lambda}{\lambda - (\lambda - 2)\tan\lambda\theta_d \cot[(\lambda - 2)\theta_d]}$$

for (3.43) and

$$C_1 = -\frac{b_\lambda}{4(\lambda - 1)} \cot\lambda\theta_d$$

for (3.44). The above expressions for coefficients indicate that there are a number of special cases, like for example $\theta_d = \pi/\lambda$, where some denominators are equal to zero. In these cases, one needs a slightly different asymptotic solution. Here we are interested in the general case which highlights the properties of the model.

Given (3.49), the velocity components are obviously finite, so that one part of requirement B of the ABC-criterion is satisfied. Let us find conditions required for the pressure to be finite as well. Substituting $\psi_\lambda = r^\lambda F_\lambda(\theta)$ into (3.14), we get

$$\frac{\partial p}{\partial r} = r^{\lambda-3}[F_\lambda'''(\theta) + \lambda^2 F_\lambda'(\theta)]$$

and, for F_λ obtained above, the expression in the square brackets is not identically zero. Then, to satisfy requirement B of the 'ABC-criterion' and make p nonsingular, we need

$$\lambda > 2. \tag{3.53}$$

Now, we have to consider requirement C and find conditions needed for the fluid's motion to be 'rolling'. As $r \to 0$, to leading order, the motion of a fluid particle belonging to the free surface is described by

$$\frac{dr}{dt} = u(r, \theta_d) = r^{\lambda-1} F'_\lambda(\theta_d).$$

Then, for the fluid particle to be able to reach the contact line in a finite time we need

$$\lambda < 2. \tag{3.54}$$

If this condition is not satisfied, the fluid's motion will be 'sliding': the fluid particles belonging to the free surface will never reach the contact line, which then will have to consist of the same fluid particles at all time.

Thus, since (3.53) and (3.54) are clearly incompatible, none of the existing (and possible) slip models is able to satisfy the ABC-criterion of §2.5.2. In particular, for the most popular Navier condition (3.35) with β given by (3.36) we have $\lambda = 2$ and hence it fails to satisfy both (3.53) and (3.54). The flow kinematics will then be qualitatively wrong and the pressure logarithmically singular at the contact line: from (3.14), (3.15) and (3.50), (3.51), (3.52) it follows that

$$p = \frac{b_2}{\theta_d} \ln r + p_0 + o(1) \qquad \text{as } r \to 0, \tag{3.55}$$

where p_0 is a constant of integration. This function is obviously integrable at $r = 0$, and the corresponding correction to the free-surface shape is given by

$$g(r) = -\frac{b_2 Ca}{2\theta_d} r \ln r + r \frac{Ca}{2} \left(b_2 \frac{\theta_d + \sin 2\theta_d}{\theta_d \sin 2\theta_d} - p_0 \right) + o(r). \tag{3.56}$$

It should be noted, however, that (3.56) leads to the free-surface curvature being singular at $r = 0$, whereas the modelling of the liquid-gas interfacial layer as a geometric surface of zero thickness ('interface') implies that the thickness of this layer is negligibly small compared with the radius of curvature.

One also has that $p \to -\infty$ as $r \to 0$, so that p will be below, say, the cavitation pressure in a finite vicinity of the contact line, which is incompatible with the assumption made in the formulation of the model that there is no cavitation.

A general point to be made about the slip models is that even when a slip model does satisfy condition (3.54) required for the rolling motion, the resultant rolling is 'defective' in a sense that it has some qualitative features not observed in experiments. For example, when the contact angle is sufficiently large and one might expect, and indeed observes in experiments, an almost unidirectional flow near the contact line, a slip model will still make a fluid particle belonging to the free surface decelerate to zero velocity at the contact line and then accelerate again to reach the velocity of the solid substrate far away downstream. Then, the region near the contact line will be that of low velocity, and the global flow will feel the contact line as an 'obstacle'

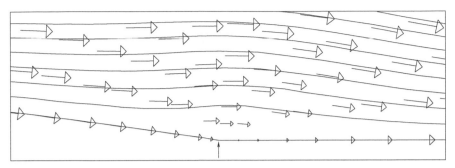

FIGURE 3.29: Flow field near the contact angle computed using the Navier slip condition (3.35) for $\theta_d = 170°$ (courtesy of M.C.T. Wilson). All slip models require that the contact line is a stagnation point in the plane of flow for the fluid motion. As a result, near the contact line one always has a spurious region of low-velocity which is felt as an 'obstacle' by the outer flow: fluid particles 'overcome' this region so that their velocities are directed away from the free surface upstream from the contact line and towards the solid boundary downstream. An arrow pointing upwards indicates the position of the contact line.

(Fig. 3.29). In experiments, as we have seen in §3.2.2, one has exactly the opposite with the contact-line region, alongside the rest of the solid surface, being a source of motion rather than an obstacle to it, and the fluid particles near the contact line move faster than further afield.

Thus, the very approach of slip models appears to be fundamentally flawed as it does not allow one to describe fluid motion as a singularity-free solution with the correct flow kinematics.

Qualitatively, it is easy to understand why this happens. In the slip models, one always has an angle in the streamline going along the interfaces. In order to minimise the impact of this instantaneous change in direction of the flow velocity and hence reduce and ultimately suppress the pressure singularity, whose nonintegrability in the standard model is at the root of the problem, one has to pose a boundary condition which would create a low-velocity region, almost a stagnation zone, near the contact line. Then, the pressure singularity can be suppressed at the expense of imposing a completely wrong kinematics of the flow. The borderline case is $\lambda = 2$, where the pressure singularity is not suppressed yet whereas the flow kinematics is qualitatively wrong already.

3.4.1.4 Slip models and the contact angle description

We have shown that, as far as the near-field dynamics is concerned, the question is not that of choosing the 'right' slip model: in this limit all of them provide an unsatisfactory description of the flow. It is interesting to consider whether or not some of them might be acceptable in some less restrictive sense, namely as a simple and adequate tool for modelling the overall effect

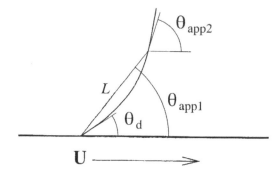

FIGURE 3.30: A definition sketch illustrating the concept of an 'apparent' contact angle. Given the spatial resolution of measurements L, one can define such an angle (a) as the angle θ_{app1} between the the solid and the chord connecting the contact line with a point on the free surface at a distance L from the contact line, and (b) as an angle θ_{app2} between the tangent to the free surface at this point and the solid surface.

produced by the moving contact line on the flow in the far field.

The two key elements of all slip models are (i) the slip condition, which alters the flow field on the length scale comparable with the slip length, and (ii) the way these models describe the dynamic contact angle and through it influence the shape of the flow domain. This influence is felt on a length scale associated with capillary effects, which is usually much greater than the assumed slip length.

Dussan (1976) examined the effect three different slip conditions (3.32) make on the flow far away from the contact line under the assumption that the contact angle is the same. She has come to the conclusion that, as one should have anticipated, for the low-Reynolds-number regime, the choice of a slip condition affects the flow field on the length scale comparable with the slip length s whereas its leading-order influence on the far-field dynamics is associated only with the *value* of s. In the far field, the slip condition produces a minor correction to the overall flow determined by the standard no-slip condition and the shape of the flow domain.

The shape of the flow domain, however, is determined by the dynamic contact angle, which acts as a boundary condition for (3.8). Obviously, the assumption that the dynamic contact angle is always equal to the static one does not describe what is actually observed in experiments, not to mention MD simulations. For this reason, the slip models that rely on this assumption have to introduce an ad hoc scheme of linking theory with experimental measurements. This scheme that ultimately has been adopted by all slip models is based on a concept of 'apparent' contact angle. The basic idea behind this concept is as follows.

All experimental measurements of the contact angle (as well as of other hydrodynamic parameters) have a finite spatial resolution that, in the simplest case, can be characterized by a length scale L. Then, the measured value appears as a result of some spatial averaging associated with this length scale. This length scale is not part of a theory and can be brought into the theory as an artificial adjustable parameter that allows one to introduce an 'apparent' contact angle whose sole purpose is to help to interpret experimental measurements.

If there is one adjustable length scale L, there are two ways of introducing an apparent contact angle (Fig. 3.30). The first one is to define it as an angle θ_{app1} between the solid surface and the segment of length L connecting the contact line and a point on the free surface. A qualitative justification for this definition is that the finiteness of the spatial resolution of measurements actually means that a smooth curve representing the free-surface profile is seen in experiments as a piecewise-linear one with the segments commensurable with the spatial resolution.

The second way of introducing an apparent contact angle is motivated primarily by the convenience of its mathematical analysis and defines it as an angle θ_{app2} between the tangent to the free surface at a point at a distance L from the contact line and the solid surface. Physically, one can see this way of introducing the apparent angle as accounting for the impossibility to zoom in to the contact line itself.

In both cases, L acts as an adjustable parameter reflecting finiteness of the spatial resolution of measurements and, as the accuracy of measurements increases, one has $L \to 0$ and consequently both θ_{app1} and θ_{app2} tend to θ_d, which, in the theory, is given either by (3.40) or, more generally, by (3.41).[7]

Thus, if the 'apparent' angle at some preset distance L from the contact line behaves in the same way as the experimentally measured one, then, given that for $s/L \to 0$ the influence of the particular slip condition at a distance L from the contact line tends to zero, it would be possible to by-pass the moving contact-line problem, as far as the contact-line influence on the overall flow is concerned, and use any of the slip models.

A substantial research effort has gone into obtaining an analytical expression relating θ_{app2} and θ_d which could help experimenters to test slip models, with the added concept of an 'apparent' contact angle, against their data. The idea here is to account for the bending of the free surface due to the stresses generated by the flow and incorporate the leading-order effect of slip which is what makes the solution exist. The technical difficulties associated with this nonlinear problem are too formidable to allow one to obtain a general result so that the main attention has been focussed on the case of zero Reynolds and small capillary numbers.

---

[7]It is worth mentioning that reservations about the notion of 'apparent' contact angle have been expressed at the very beginning of its use: "The apparent contact angle is a rather doubtful concept and rests on the inability to observe the meniscus closely enouth to see the true contact form" (Ludviksson & Lightfoot 1968).

After a number of partial successes (e.g., Dussan V. 1976, Voinov 1976, 1978, Huh & Mason 1977a, Hocking 1977, 1981, Hocking & Rivers 1982), the required expression has been finally derived by Cox (1986). He has shown that for zero-Reynolds-number flow in both advancing and receding fluids, to leading order as $Ca \to 0$ and $s/L \to 0$, the values of θ_{app2} and θ_d are related by

$$g(\theta_{\mathrm{app2}}, k_\mu) = g(\theta_d, k_\mu) + Ca\ln(L/s), \tag{3.57}$$

where

$$g(\theta, k_\mu) = \int\limits_0^\theta \frac{d\theta}{f(\theta, k_\mu)},$$

$$f(\theta, k_\mu) \equiv \frac{2\sin\theta \left\{ k_\mu^2 K(\theta) + 2k_\mu[\theta(\pi - \theta) + \sin^2\theta] + [(\pi - \theta)^2 - \sin^2\theta] \right\}}{k_\mu K(\theta)[(\pi - \theta) + \sin\theta\cos\theta] + [(\pi - \theta)^2 - \sin^2\theta](\theta - \sin\theta\cos\theta)}.$$

Zhou & Sheng (1990) confirmed this result by comparing it with their numerical calculations. Addressing the behaviour of θ_d, Zhou & Sheng have shown that, contrary to assumption (3.40) used by Cox (1986), in order to describe experiments reported by Fermigier & Jenffer (1988) one has to assume that θ_d depends at least on the contact-line speed. Later, Fermigier & Jenffer (1991) have also shown that (3.57) combined with (3.40) is in poor agreement with their data reproduced here in Figs 3.12 and 3.28. These results, however, do not answer the question whether or not the very idea of an 'apparent' contact angle combined with a suitably chosen function in (3.41) is sufficient in principle to describe experimental observations in a general case.

An investigation aimed at clarifying this issue has been carried out by Wilson et al. (2006), who attempted to describe nonlocal hydrodynamic effects, reported earlier by Blake, Bracke & Shikhmurzaev (1999) and discussed in §3.2.3, using slip models (3.33) and (3.35), (3.36). The study employed a finite-element numerical code and made no simplifying assumptions concerning the values of the capillary and Reynolds numbers. The idea was to interpret everything *in favour* of the slip models, by using θ_d together with s (or β) as adjustable parameters with L being specified by the spatial resolution of the measurements ($< 20~\mu$m). Then, if θ_{app1} or θ_{app2} is made equal to the measured value of the contact angle for high flow rate, this gives one constraint on θ_d and s (or β) and for the chosen apparent contact angle as a function of the flow rate we have a one parametric family of curves.

It has been found, firstly, that $\theta_d \not\equiv \theta_s$ and, secondly, that the behaviour of the experimentally measured contact angle shown in Fig. 3.22 cannot be described via the concept of an apparent contact angle irrespective of the way this angle is defined (see Figs 3.31 and 3.32). It is simply that the spatial resolution of the measurements is too high (and hence L is too small) for the apparent contact angle to deviate from θ_d to such an extent that this

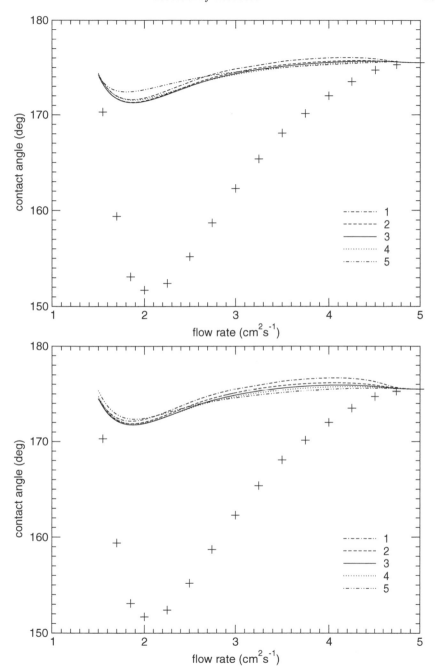

FIGURE 3.31: Variation of θ_{app1} with flow rate for two slip models (Wilson et al. 2006). Top: A prescribed (exponential) slip-velocity distribution (3.33) with various slip lengths; curves 1, 2, 3, 4 and 5 correspond to $s = 0, 01, 0.1$, 1, 10 and 100 μm, respectively. Bottom: The Navier condition (3.35) with various coefficients of sliding friction; curves 1, 2, 3, 4 and 5 correspond to $\beta = 1000, 100, 10, 1$ and 0.1 kg/(cm^2 s). The experimental data $(+++)$ are taken from Fig. 3.22, $U = 70$ cm s^{-1}.

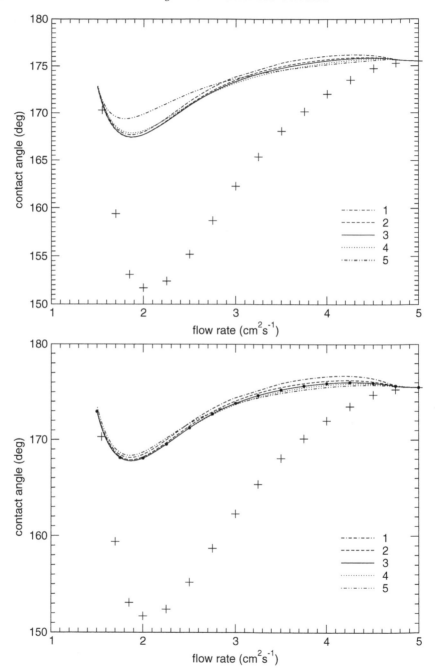

FIGURE 3.32:   Variation of θ_{app2} with flow rate for two slip models (Wilson et al. 2006). Top: A prescribed (exponential) slip-velocity distribution (3.33) with various slip lengths; curves 1, 2, 3, 4 and 5 correspond to $s = 0, 01, 0.1$, 1, 10 and 100 μm, respectively. Bottom: The Navier condition (3.35) with various coefficients of sliding friction; curves 1, 2, 3, 4 and 5 correspond to $\beta = 1000, 100, 10, 1$ and 0.1 kg/(cm^2 s). The experimental data $(+++)$ are taken from Fig. 3.22, $U = 70$ cm s^{-1}.

would account for the experimentally observed effect of 'hydrodynamic assist of dynamic wetting'.

It should be emphasized that the use of θ_d as an adjustable parameter making the apparent contact angle equal to the experimentally measured one for high flow rates (and a fixed contact-line speed) in the curtain coating accounts for *all* possible relationships between θ_d and the contact-line speed given by (3.41). The discrepancy between theory and experiments on the dependence of the contact angle on the flow rate is obvious (Figs 3.31–3.32), and there are no tools left in the slip models, even conceptually, to allow one to improve the fit.

The analysis of Wilson et al. (2006) confirmed the original conjecture by Blake, Bracke & Shikhmurzaev (1999) based on the order-of-magnitude estimates that it is θ_d which, for a given contact-line speed, depends on the flow field/geometry in the vicinity of the moving contact line.

Thus, one has to conclude that, as far as the general case is concerned, experiments on 'hydrodynamic assist of dynamic wetting' invalidate 'slip models' even as a way of mimicking the overall influence of the moving contact line on the flow in the far field.

3.4.2 Models developed for dynamic wetting by thin films

3.4.2.1 Models with a macroscopically pre-wet solid surface: "Tanner's law"

A simple way of by-passing the moving contact-line problem used in many works (e.g., Ludviksson & Lightfoot 1968, Tanner 1979, Starov 1983, Ishimi, Hikita & Esmail 1986, Mumley, Radke & Williams 1986b, Troian et al. 1989, Diez et al. 1994, Wang, Peng & Lee 2003) is to assume that the surface is already covered with a macroscopic film of the same fluid. Physically, this film can be either a precursor film propagating ahead of the rest of the fluid in the case of complete wetting or a residual film left as a result of the previous wetting/de-wetting cycle, as for example in Fig. 3.11a. In both cases, it is essential that the film is assumed to be 'macroscopic' in a sense that it can be described by the same equations as the main body of fluid. Thus, the actual moving contact-line problem associated with how the film first appeared is not considered, and the study is focussed on the subsequent evolution of the flow and the free-surface shape. Sometimes, physically there is no actual precursor film (the case of partial wetting) and it is introduced artificially as a mathematically convenient way of removing the moving contact-lime problem (e.g., Troian et al. 1989).

This 'precursor-film approach' to the moving contact-line problem is usually implemented in the framework of the lubrication approximation, and, to illustrate the main result obtained with its use, it is convenient to introduce this approximation here in the simplest case. The history of this approximation goes back to Reynolds (1886); reviews related to different aspects of its

Moving contact lines: An overview

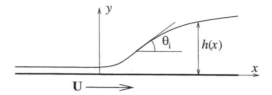

FIGURE 3.33: A definition sketch for a thin-film flow with a macroscopic film in front of the main body of fluid. The apparent contact angle measured at the inflection point of the free-surface profile is the one described by the so-called "Tanner's law" (Tanner 1979).

application can be found elsewhere (e.g., Cameron 1966, Dowson 1979, Oron, Davis & Bankoff 1997).

Consider a steady two-dimensional spreading of a film of fluid preceded by a thinner film of the same fluid in a reference frame where the free surface is at rest and hence the solid is moving with some speed U (Fig. 3.33). If L_y is a characteristic thickness of the film and L_x is a characteristic length scale for the motion in question in the x-direction, then one can introduce their ratio $\epsilon = L_y/L_x$ and by considering an asymptotic limit $\epsilon \to 0$ specify a class of flows, which can be broadly referred to as flows of the boundary-layer type. Particular approximations within this class depend on the assumptions made about other dimensionless parameters that appear in the problem.

The distribution of the flow parameters are described by a solution of the steady Navier-Stokes equations subject to standard boundary conditions (2.137), (2.154) at the free surface and the no-slip condition (2.157) on the solid boundary. Let u, v be the components of the fluid's velocity in a Cartesian coordinate system Oxy in the plane of flow (Fig. 3.33); p is the pressure measured with respect to a constant pressure in the surrounding inviscid gas; $y = h(x)$ is the free-surface profile to be determined. After introducing nondimensional variables

$$\bar{x} = \frac{\epsilon x}{L_y}, \ \bar{y} = \frac{y}{L_y}, \ \bar{u} = \frac{u}{U}, \ \bar{v} = \frac{v}{\epsilon U}, \ \bar{h} = \frac{h}{L_y}, \ \bar{p} = \frac{\epsilon L_y p}{\mu U}, \tag{3.58}$$

we arrive at the following problem. In the bulk, i.e., for $-\infty < \bar{x} < \infty$, $0 < \bar{y} < \bar{h}(\bar{x})$, one has

$$\frac{\partial \bar{u}}{\partial \bar{x}} + \frac{\partial \bar{v}}{\partial \bar{y}} = 0, \tag{3.59}$$

$$\epsilon Re \left(\bar{u} \frac{\partial \bar{u}}{\partial \bar{x}} + \bar{v} \frac{\partial \bar{u}}{\partial \bar{y}} \right) = -\frac{\partial \bar{p}}{\partial \bar{x}} + \epsilon^2 \frac{\partial^2 \bar{u}}{\partial \bar{x}^2} + \frac{\partial^2 \bar{u}}{\partial \bar{y}^2}, \tag{3.60}$$

$$\epsilon^3 Re \left(\bar{u} \frac{\partial \bar{v}}{\partial \bar{x}} + \bar{v} \frac{\partial \bar{v}}{\partial \bar{y}} \right) = -\frac{\partial \bar{p}}{\partial \bar{y}} + \epsilon^2 \left(\epsilon^2 \frac{\partial^2 \bar{v}}{\partial \bar{x}^2} + \frac{\partial^2 \bar{v}}{\partial \bar{y}^2} \right). \tag{3.61}$$

At $\bar{y} = \bar{h}(\bar{x})$ the conditions of impermeability, zero tangential stress and equality of the normal stress and the capillary pressure take the form

$$\bar{u}\bar{h}' - \bar{v} = 0, \tag{3.62}$$

$$(1 - \epsilon^2 \bar{h}'^2) \left(\frac{\partial \bar{u}}{\partial \bar{y}} + \epsilon^2 \frac{\partial \bar{v}}{\partial \bar{x}} \right) - 4\epsilon^2 \bar{h}' \frac{\partial \bar{u}}{\partial \bar{x}} = 0, \tag{3.63}$$

$$-\bar{p} - \frac{2\epsilon^2}{1 + \epsilon^2 \bar{h}'^2} \left[(1 - \epsilon^2 \bar{h}'^2) \frac{\partial \bar{u}}{\partial \bar{x}} + \bar{h}' \left(\frac{\partial \bar{u}}{\partial \bar{y}} + \epsilon^2 \frac{\partial \bar{v}}{\partial \bar{x}} \right) \right] = \frac{\epsilon^3 \bar{h}''}{Ca \left(1 + \epsilon^2 \bar{h}'^2 \right)^{3/2}}. \tag{3.64}$$

On the solid surface, the no-slip and impermeability conditions are given by

$$\bar{u} = 1, \quad \bar{v} = 0 \quad (\bar{y} = 0). \tag{3.65}$$

Here $Re = L_y U / \nu$ is the Reynolds number ($\nu = \mu/\rho$); $Ca = \mu U / \sigma$ is, as before, the capillary number; the prime denotes a derivative with respect to $\bar{x}$. In writing down (3.63) and (3.64) we have used (3.59) to replace $\partial \bar{v} / \partial \bar{y}$ with $\partial \bar{u} / \partial \bar{x}$. Conditions (3.59)–(3.65) must be supplemented with some conditions in the far field, i.e., as $x \to \pm\infty$, to specify a particular solution.

Now the asymptotic limit $\epsilon \to 0$ allows one to consider a number of approximations. Their key feature is that, at leading order in ϵ as $\epsilon \to 0$, the viscous terms in (3.61) vanish, and the Laplacian on the right-hand side of (3.60) becomes simply a second derivative in the y-direction. The lubrication approximation is based on an assumption that

$$Re = O(1) \quad \text{as } \epsilon \to 0 \tag{3.66}$$

so that, at leading order, one can neglect the inertial terms in both (3.60) and (3.61). The remaining assumption has to be made about the asymptotic limit for Ca. In the film spreading problems the capillary effects are often important and to keep the capillary pressure term in (3.64) one has to assume that

$$Ca \epsilon^{-3} \equiv \bar{Ca} = O(1) \quad \text{as } \epsilon \to 0. \tag{3.67}$$

The parameter $\bar{Ca}$ is essentially the normalised capillary number, and condition (3.67) poses limitations on the actual capillary number requiring it to be $O(\epsilon^3)$, i.e., asymptotically very small. Assumptions (3.66), (3.67) together with those following from the scaling (3.58) determine the limits of applicability of the approximation.

After integrating (3.59) across the film, from $\bar{y} = 0$ to $\bar{y} = \bar{h}$, and making use of an identity

$$\frac{d}{dx} \int\limits_{0}^{f(x)} F(x, y) \, dy = \int\limits_{0}^{f(x)} \frac{\partial F}{\partial x} \, dy + F f',$$

the kinematic boundary condition (3.62) and the second (impermeability) condition of (3.65), one arrives at

$$\frac{d}{d\bar{x}} \int_0^{\bar{h}} \bar{u}(\bar{x}, \bar{y}) \, d\bar{y} = 0, \tag{3.68}$$

which is an obvious result stating that for a steady flow the flux across every cross-section of the film is the same.

Using the asymptotic expansions

$$\bar{f} = \sum_{n=0}^{\infty} \epsilon^n f_n \qquad \text{as } \epsilon \to 0 \text{ for } \bar{f} = \bar{u}, \bar{v}, \bar{p}, \bar{h}$$

and considering the problem following from (3.59)–(3.65) for the leading-order terms, we obtain from (3.61) and (3.64) that, at 0th order, pressure in the film is constant across the film and, being equal to the capillary pressure, varies only with $\bar{x}$,

$$p_0 = -\bar{C}a^{-1} h_0''. \tag{3.69}$$

This expression together with (3.60) give that u_0 is a quadratic function of $\bar{y}$ with the boundary condition (3.63) and the first (no-slip) condition of (3.65) specifying the two unknown coefficients in this function:

$$u_0 = 1 - \frac{\bar{y} h_0'''}{\bar{C}a} \left(\frac{\bar{y}}{2} - h_0 \right).$$

Given this expression, equation (3.68) expanded to leading order in ϵ reduces the problem to an ordinary differential equation for h_0:

$$h_0' + \frac{(h_0^3 h_0''')'}{3\bar{C}a} = 0. \tag{3.70}$$

This is the simplest form of the lubrication-approximation equation for the free-surface profile. The only parameter in this equation can be scaled into the independent variable by using $H(\tilde{x}) = h_0(\bar{x})$, where $\tilde{x} = \bar{C}a^{1/3}\bar{x}$, thus giving

$$H' + \tfrac{1}{3}(H^3 H''')' = 0.$$

Solutions of this equation are determined by the boundary conditions in the far field.

Given that there is no actual contact angle in the problem, one can define an apparent contact angle at any point on the free surface. Tanner (1979) considered the angle θ_i at the inflection point in the free-surface profile (Fig. 3.33). If the inflection point is at $\tilde{x} = \tilde{x}_i$ where $H(\tilde{x}_i) = k$, then, after going back to the dimensional variables, this gives

$$U = \frac{\sigma}{\mu k} (\tan \theta_i)^3.$$

Since the lubrication approximation is valid for small slopes, $\tan \theta_i \approx \theta_i$, and one arrives at what has become known as "Tanner's law":

$$U = A\theta_i^3. \tag{3.71}$$

It should be emphasized that this "law" is a *result* that follows from the scaling of the lubrication approximation and it has been obtained in the framework of a particular problem with no actual contact angle; it is not a boundary condition, like particular forms of (3.41) in the slip models, which is supposed to reflect assumptions made about the physics of dynamic wetting. As is always the case with boundary conditions, this physics must be *additional* to the physics already incorporated in the bulk equations, whereas (3.71) follows from the latter. Therefore, the use of "Tanner's law" as a boundary condition in a problem where the contact line is actually present would be a fundamental mistake on the conceptual level.

3.4.2.2 Continuum models with intermolecular forces

In studying the spreading of thin films, one faces a serious methodological difficulty related to the distinction between 'macroscopic' and 'microscopic' films as they are modelled in the framework of continuum mechanics. The essence of the problem is as follows.

For a film, the characteristic length scales in the directions tangential to the solid are, by definition, macroscopic, whereas the film's thickness is much smaller. The latter can vary along the film and become comparable with the range of intermolecular forces so that the opposite interfaces 'feel' each other's presence via these forces. As we remember (§§2.1.2, 2.1.3), the continuum approximation means the 0th-order approximation in the ratio of molecular-to-macroscopic length scales so that, if a film's thickness is on the molecular scale, then this film is 'invisible' from the point of view of continuum mechanics. Such a film becomes 'microscopic' in a sense that, in the framework of continuum mechanics, it has to be modelled as a two-dimensional continuum and characterised by its surface parameters per unit area rather than a three-dimensional 'macroscopic' phase like the bulk of the fluid. The question whether, physically, for a film of spatially varying thickness the transition from it being 'macroscopic' to 'microscopic' takes place gradually, i.e., on a macroscopic length scale, or, as some experiments seem to suggest, this transition is a steep one remains open and requires further investigation.

In the literature, the mathematical modelling of the spreading of thin films is approached in the following way. The film is treated as three-dimensional so that the equations of fluid mechanics apply. Then, the two sides of the moving contact-line problem described in §3.1.1 have to be addressed, and this is done by using one of the slip models we described earlier. In order to account for the microscopic thickness of the film, a potential body force is added to the right-hand side of the equations of motion formulated in the framework of continuum mechanics to describe the effect of long-range intermolecular forces (Lopez, Miller & Ruckenstein 1976, Ruckenstein & Dunn 1977, Neogi & Miller 1982, de Gennes 1985, Ruckenstein 1992, Hocking 1993, Treviño, Ferro-Fontán & Méndez 1998, Pismen, Rubinstein & Bazhlekov 2000, Chebbi 2000, Pismen & Rubinstein 2001, Pismen 2002). The resulting equations are then simplified using the lubrication approximation[8] and studied for different particular cases.

In terms of the problem formulation we considered earlier, this means the replacement

$$\bar{p} \mapsto \bar{p} - \bar{\phi} \tag{3.72}$$

in (3.60), (3.61) and (3.64), where $\bar{\phi} = \epsilon L_y \phi / (\mu U)$ is the dimensionless potential of the body force. If there is a moving contact line in the problem, one also prescribes the velocity dependence of the contact angle and replaces (3.65) with an appropriate slip condition.

The body force added to the continuum description in these models is obtained by integrating the force exerted on a molecule in the fluid by molecules of the solid substrate. The latter force is assumed to be the van der Waals force due to fluctuations of electromagnetic fields of interacting molecules (Dzyaloshinskii, Lifshitz & Pitaevskii 1960; see §7.7.2). It is calculated either explicitly (Dzyaloshinskii et al. 1960) by considering the physical mechanisms involved, or in a simplified way via the London potential (Miller & Ruckenstein 1974). The most common expression used for the potential is given by

$$\phi = \frac{A}{6\pi y^3}, \tag{3.73}$$

where, as before, y is the distance from the solid surface and A is a material constant. One can also come across modifications of (3.73) aimed at accounting for a finite angle between the free surface and the solid substrate (e.g., Wu & Wong 2004). In some works (e.g., Glasner 2003), the potential in (3.72) is assumed to have the form

$$\phi = \frac{A}{y^m} - \frac{B}{y^n} \tag{3.74}$$

to account for both attractive and repulsive forces.

The forces corresponding to (3.73) and (3.74) are obviously singular at $y = 0$ so that their use requires an artificial "cut-off" at a molecular distance from the solid surface (e.g., Pismen & Rubinstein 2001) to prevent the model from displaying clearly unphysical features.

Thus, the approach to the modelling of the spreading of thin films described here has the same shortcomings as the slip models it relies upon to address

---

[8]If the lubrication approximation is applied without addressing the shear-stress singularity at the contact line, as, for example, in Lopez, Miller & Ruckenstein (1976), where the no-slip condition is not relaxed, then the lubrication theory will be approximating a solution with a nonintegrable shear-stress singularity.

the moving contact-line problem. In addition, it explicitly introduces inter-molecular forces and a molecular "cut-off" length scale into the continuum mechanics modelling, which is fundamentally incompatible with the contin-uum approximation described in §§2.1.2 and 2.1.3.

It should be also noted that the procedure of "cut-off" required by this approach lies outside mathematics. From a mathematical viewpoint, if one "cuts off" the contact line, that is excludes a certain region around the contact line from consideration, then the domain where the bulk equations operate will have the border of this region as part of its boundary. On this boundary one will then have to pose some boundary conditions since otherwise the mathematical problem in the bulk would simply be not formulated. Given that the cut-off boundary is completely artificial and hence has no precise location, it is not associated with any specific physics additional to the physics described by the bulk equations, and it is this additional physics that always goes into the boundary conditions and specifies the solution. Therefore, apart from other arguments, the fact that a "cut-off" is needed, makes the models using it, at best, semi-qualitative.

3.4.3 Diffuse-interface models

Another direction of research into the spreading of liquids on solids is to consider the flow on a length scale comparable with the thickness of interfacial layers. This approach has been proposed by Seppecher (1996) who assumed that the thickness of an interfacial layer between the advancing liquid and the receding gas is on a macroscopic, not molecular, length scale and hence this layer itself can be described in the framework of continuum mechanics. As a result, one has a situation where there is no 'boundary' between the displacing and the displaced fluids and the same bulk equations describe both of them as well as the transitional zone (interfacial layer) between them. In deriving these equations, Seppecher postulated that the free energy per unit volume, $F_v = \rho F_m$, is given by the expression used in Cahn (1977) to model the equilibrium state of a liquid-gas interface

$$F_v = W(\rho) + \frac{\kappa}{2}(\nabla\rho)^2, \tag{3.75}$$

where, for a given temperature, W is a prescribed positive function with two minima corresponding to the densities of the bulk phases. The term $(\kappa/2)(\nabla\rho)^2$ is what tends to reduce the transition zone between the phases; in equilibrium, this term determines the thickness of an interface.

The subsequent derivation of the model in Seppecher (1996) is carried out in the framework of irreversible thermodynamics with a number of simplifying assumptions aimed at recovering the Navier-Stokes equations away from the interfacial layer. The resulting set of equations for a steady flow takes the form

$$\nabla \cdot (\rho\mathbf{u}) = 0, \tag{3.76}$$

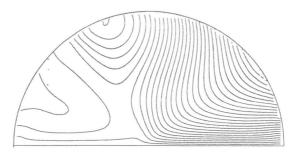

FIGURE 3.34: The streamlines produced by a diffuse-interface model (Seppecher 1996; © 1996 Elsevier; reproduced with permission). The liquid-gas interface appears to be permeable allowing a flux from the receding into the advancing fluid near the solid boundary.

$$\mathbf{u} \cdot \nabla \mathbf{u} = -\nabla \left(\frac{dW}{d\rho} - \kappa \Delta \rho \right) + \frac{\lambda + \mu}{\rho} \nabla (\nabla \cdot \mathbf{u}) + \nu \Delta \mathbf{u}, \qquad (3.77)$$

where λ and μ are the viscosity coefficients, and $\nu = \mu/\rho$. These equations are then solved numerically with the no-slip boundary condition for $\mathbf{u}$ everywhere on the solid surface and some additional conditions ensuring matching with the known solution in the far field.

The flow field produced by the computations is shown in Fig. 3.34. One can see that near the solid there is flow from the receding gas into the advancing liquid so that, kinematically, there is no actual contact line. Alternatively, if the interfacial layer is defined as that formed by the isodensity curves (which do reach the solid boundary), one has that the model describes a *permeable* interface with a flux of mass across it.

The contact angle is treated by Seppecher as the 'apparent' one formed between the solid surface and the tangent to a (collapsing) family of isodensity curves far away from it. This angle is found to increase almost linearly with the capillary number and the slope of this dependence is independent of the value of the static contact angle.

Comparing these results with experiments reviewed in §§3.2.2, 3.2.3, one can conclude that both, the flow kinematics and the dynamic contact angle behaviour produced by the model, do not correspond to what is observed experimentally.

More importantly, there is a fundamental flaw inherent in the very approach of diffuse-interface models when they are applied to immiscible fluids. These models operate in the framework of continuum mechanics and, by considering an interface as a finite-thickness transition zone between two bulk phases, imply that the thickness of an interfacial layer is on a macroscopic length scale. It is this implicit assumption that allows the interfacial layer's thickness remain finite after the continuum approximation is taken. In other words, the interfacial layer's thickness is assumed to be in no way related to the range of intermolecular forces and the length scale associated with the molecular motion. Physically, however, it is the range of intermolecular forces that determines the thickness of an interfacial layer and hence becomes negligible in the continuum approximation (see §2.1.3). Thus, the diffuse-interface approach separates the interfacial layers with the free surface energy postulated to describe them (eq. (3.75)) from the actual physics that determines their existence and lies behind (3.75).

The description of an equilibrium state of interfacial layers in the framework of thermodynamics accounting for their finite thickness was first developed by van der Waals (1893) and later rederived by Cahn & Hilliard (1958). These works consider minimization of the free energy of a liquid-gas interfacial layer assuming that the local density ρ is a continuous function across this layer. Cahn (1977) has also applied this approach to the 'critical point wetting', i.e., again an equilibrium state of a two-phase system in contact with a solid near the critical point for the two phases. An important point to note about these and related studies is that, although they consider a continuous density field on a scale of the interfacial layer thickness, which physically is determined by the range of intermolecular forces, the thermodynamic framework they use does not involve the macroscopic concept of stresses. As pointed out in §2.1.3, once we introduce stresses, this corresponds to the isotropic 0th-order approximation in the ratio of molecular-to-macroscopic scales and has as its consequence the introduction of the notions of bulk phases separated by 'interfaces' modelled as geometric surfaces. The details of the interfacial layer structure determined by the action of intermolecular forces from the bulk phases and molecular motion inside them, which are accounted for in the thermodynamics of van der Waals and Cahn & Hilliard, collapse into macroscopic (average) properties of the two-dimensional interfaces. In other words, the thickness of interfaces is 'invisible' from the viewpoint of continuum mechanics, which operates on a macroscopic length scale. We will discuss these issues further in Chapter 4.

It should also be pointed out that in the continuum approximation the three-phase-interaction region becomes a one-dimensional 'contact line'. As shown by Merchant & Keller (1992), the classical Young equation (2.164) relating the contact angle with the surface tensions acting on the contact line holds on the scale large compared with the range of intermolecular forces, whereas within this range the 'contact line' becomes a three-dimensional region where the three contacting phases interact and the (microscopic) 'contact angle', if one can define it, differs from the macroscopic one involved in the Young equation.

Jacqmin (2000) realized that the issues related to length scales make the diffuse-interface approach, in his words, "mildly inconsistent" in dealing with the moving contact-line problem and considered the fluid-fluid displacement in the case where the fluids are two-component and partially miscible. The thickness of a transition zone between them is then determined by the mutual diffusivity of the components and can be on a macroscopic scale. Then, one could consider the flow in the framework of multicomponent fluid dynamics combining sharp interfaces with extended diffusion layers in both fluids due to their mutual solvability. Instead, the analysis in Jacqmin (2000) has been carried out using the Navier-Stokes equations amended with diffusion terms resulting from the same Cahn-Hilliard formalism as in Seppecher (1996). A similar study has been reported almost simultaneously by Chen, Jasnow & Viñals (2000). An assumption that the interface near the solid boundary remains in equilibrium in the process of the fluid-fluid displacement resulted in the condition $\theta_d = \theta_s$ for the flow on the macroscopic length scale, which coincides with (3.40) of the slip models. This implies that experimental observations have to be interpreted in terms of the 'apparent' contact angle in the way discussed in §3.4.1. A possible modification of this condition discussed in Jacqmin (2000) leads to condition (3.41) also used by the slip models. The flow field obtained in computations shows the same qualitative pattern as in Seppecher (1996) with the interface being permeable near the solid wall.

Pismen & Pomeau (2000) and Pismen (2001) applied a model similar to that of Seppecher (1996) in the lubrication approximation to the spreading of thin films and arrived at qualitatively similar results with the flux from the receding into the advancing fluid through the isodensity curves in the vicinity of the solid boundary.

Villanueva & Amberg (2006) and later Ding & Spelt (2007) found that in the diffuse-interface model the contact-line speed decreases with the thickness of the diffuse interface. The 'effective' slip that allows the contact line to move appears to be proportional to the diffuse interface thickness (Ding & Spelt 2007). In other words, one has that either, physically, the interfacial layer's thickness is on a macroscopic scale or, in the model, the contact line stops moving in the limit of a sharp interface, i.e., in the continuum approximation.

A full-scale numerical study of the diffuse-interface model has been reported by Khatavkar, Anderson & Meijer (2007) who considered an unsteady problem of the drop spreading on a horizontal solid surface. Interestingly, in carrying out a partial comparison of their computations with the experimental data of Zosel (1993), the authors found that the final radius of the drop in their computations depends not only on the static contact angle, as the classical hydrostatic description would give, but also on the Péclet number, which is the convective-to-diffusive rate ratio defined in terms of parameters of the model and the drop. Furthermore, the faster the drop spreads (i.e., the lower the Péclet number is), the smaller the final radius becomes. None of these qualitative features has been reported in experiments, and it would be curious to know what elements of the model are responsible for their appearance.

3.4.4 Other models

3.4.4.1 Molecular-kinetic theories

Cherry & Holmes (1969) and Blake & Haynes (1969) considered molecular kinetics of the liquid-fluid displacement in the framework of a simple variant of Eyring's theory of absolute reaction rates (Glasstone, Laidler & Eyring 1941). This approach can be described as follows.

At equilibrium, the contact line is not moving, but at the molecular level the three-phase-interaction zone, which is macroscopically seen as the 'contact line', is in the state of vigorous molecular motion with molecules of both fluids fighting for the same adsorption sites on the solid surface. Following Eyring, one can associate the frequency of molecular displacement in one direction κ_+ and in the reverse direction κ_- with some activation energies $E_\pm$:

$$\kappa_\pm = K_\pm \exp\left(-\frac{E_\pm}{NkT}\right). \tag{3.78}$$

Here N is the number of adsorption sites involved in the process per unit area, T is the absolute temperature, k is the Boltzmann constant. If, macroscopically, the contact line is moving so that there is a net displacement of one fluid by the other, at the molecular level this can be interpreted as preferential adsorption of molecules of one of the fluids due to changes in the activation energies. The resulting takeover of the adsorption sites by molecules of the advancing fluid produces the corresponding macroscopic speed of the contact line. Blake & Haynes assumed that, when the contact line is moving, the activation energy for the adsorption of the advancing fluid is reduced and that for the receding fluid is increased by the same amount and equated this amount to the work done by the (macroscopic) force

$$F = \sigma(\cos\theta_s - \cos\theta_d) \tag{3.79}$$

per unit length of the contact line. This force is due to the "out-of-balance interfacial tension forces acting on the wetting line" (Blake 1993, p. 268) and one arrives at (3.79) by comparing the Young equation (2.164) applied to both the static and dynamic situations and assuming that the fluid-fluid surface tension remains unchanged when the contact line is moving. The resulting expression for the velocity-dependence of the dynamic contact angle has the form:

$$\theta_d = \cos^{-1}[\cos\theta_s - m_1 \sinh^{-1}(m_2 U)], \tag{3.80}$$

where the constants m_1 and m_2 can be expressed in terms of temperature and material properties of the system. We have already used (3.80) as an approximating formula for the experimental data obtained in the curtain coating and plunging tape experiments (Fig. 3.20). Once m_1 and m_2 are related to the material constants, this provides an insight into the qualitative role played by different molecular factors in determining the dynamic contact angle.

In the fluid-mechanical context, the velocity-dependence of the dynamic contact angle given by (3.80) becomes a particular form of the function f in equation (3.41) employed in different forms by the slip models. For example, Petrov & Petrov (1992) combined (3.80) with Cox's formula (3.57) for one of the 'apparent' contact angles; van Mourik, Veldman & Dreyer (2005) used (3.80) in a volume-of-fluid code that essentially implements 'numerical slip'; etc.

An attempt to use molecular-kinetics arguments to formulate a slip boundary condition has been made by Ruckenstein & Dunn (1977) and Ruckenstein (1992). As in the later developed diffuse-interface models, it was assumed that slippage originates due to the surface diffusion, which is driven by the gradient of the chemical potential in the bulk phase. The resulting expression for the slip velocity relates it to the gradient of the sum of the bulk pressure and the potential of intermolecular forces introduced in the same way as in the models developed for thin-film flows.

3.4.4.2   Hybrid atomistic-continuum models

In the literature, one can also come across "hybrid atomistic-continuum" models for the contact-line motion (e.g., Hadjiconstantinou 1999a,b; Flekkoy, Wagner & Feder 2000, Nie et al. 2004). Their approach is based on the use of molecular dynamics simulations in the near field, i.e., near the three-phase-interaction region (which becomes a 'contact line' only on a macroscopic length scale), and the continuum mechanics modelling of the flow in the far field. The two descriptions are then "coupled" at some intermediate distance or over an "overlap region".

In this approach, the description of the flow near the three-phase-interaction region is simply a numerical experiment, similar to full-scale molecular dynamics simulations of the whole flow (see §3.3), and it has its own assumptions as well as limitations. However, the idea of 'coupling' the molecular dynamics description with the continuum mechanics in the form of a 'hybrid' model encounters a problem of principle.

Firstly, one has to remember that molecular dynamics simulations operate on a scale comparable with the molecular one whereas the latter is zero in the continuum approximation. There is simply no overlapping between the two scales where both approaches would be valid and 'coupling' could be performed.

Secondly, the two approaches operate with fundamentally different *sets of concepts*. In continuum mechanics, they result from the continuum approximation/thermodynamic limit whereas molecular dynamics simulations operate with the primitive variables that are not associated with any limit. Consequently, there is no justifiable way of matching these concepts. One specific difficulty on this way has been pointed out by Schofield & Henderson (1982). As a result of the above problems of principle the 'hybrid' models generate more questions about, to use a recent neologism, 'patching' of the two con-

stituent methods than they provide answers about the flows they are applied
to.

3.4.5 Summary

As follows from our review of theories, only the slip models address the moving contact-line problem as it stands and remain strictly within the methodology and conceptual framework of continuum mechanics. These models can
be summarily characterized as follows:

- Each of them leads to a well-posed mathematical problem by removing
  the multivaluedness of the velocity field at the moving contact line by
  replacing the no-slip boundary condition with a particular form of slip
  along the solid boundary in the vicinity of the contact line.

- None of the slip models can simultaneously remove the pressure singularity at the contact line and reproduce, at least qualitatively, the
  kinematics of the flow that is actually observed in experiments ('rolling'
  motion).

- These models fail to describe the dependence of the dynamic contact
  angle on the flow field/geometry near the moving contact line observed
  experimentally ('hydrodynamic assist of dynamic wetting').

The reviewed theories address the modelling of the spreading of liquids on
solid surfaces on an ad hoc basis, as a special 'one-off' problem, and hence
do not provide a conceptual framework that could be used for other 'paradoxical flows'. As a result, these theories cannot be tested independently by
considering a different type of flow.

3.5 The key to the moving contact-line problem

The fact that there is no solution to the moving contact-line problem in
the framework of the standard model indicates that some physical mechanism
is not accounted for in the problem formulation. An important question to
answer is whether this mechanism is specific to the process of the liquid-fluid
displacement from a solid surface, thus making this process an exotic special
case in fluid mechanics, or whether the spreading of a liquid over a solid substrate is a particular case of a more general physical phenomenon and can
be described as a by-product of a theory of this phenomenon. Embedding
the moving contact-line problem into a wider physical context would make
it possible to replace semiempirical ad hoc treatment of the problem with a

physically justified theory and to test this theory, qualitatively and quantitatively, by applying it to different processes within this general class of flows. This would also provide a conceptual framework for incorporating additional physical effects which may be important in particular applications. Let us try to indentify this class of flows.

First of all, we can notice immediately that the very term 'dynamic wetting', often used to describe the spreading of a liquid on a solid substrate, implies the formation of a new, 'wetted', solid surface. In other words, it refers to the formation of what is in fluid-mechanical terms a new, freshly-formed liquid-solid interface. This suggests the general class of physical phenomena we are looking for, namely flows with *formation* and/or disappearance of interfaces. As one can find in every textbook on fluid mechanics, the classical boundary conditions described in §§2.4.2–2.4.4 and solutions of all particular problems obtained on their basis imply interfaces which are already formed. If capillarity is considered, these interfaces are characterized by their surface tensions regarded as material constants determined by the pair of contacting media. Therefore, it becomes clear that if the interface formation or disappearance is a key element in a flow to be studied, the standard model is bound to fail simply because it does not take this process into account. In particular, as we have pointed out in §2.4.3, one of the assumptions behind the no-slip boundary condition on the solid surface is that there is no interface formation process, which would have been associated with a mass exchange between the interface and the bulk.

Let us consider how the above idea correlates with the qualitative features of the liquid-fluid displacement observed experimentally (§§3.2.2, 3.2.3).

Experiments show that the contact angle deviates from its static value when the contact line is in motion. The notion of the contact angle is introduced in macroscopic fluid mechanics via the Young equation (2.164), i.e.,

$$\sigma \cos \theta = \sigma_{(2)} - \sigma_{(1)}, \tag{3.81}$$

which, from a mechanical point of view, expresses the balance of forces acting on the contact line in the direction normal to the contact line and tangential to the solid surface. Therefore, one has to conclude that, for the contact angle to vary when the contact line is moving, at least one of the surface tensions σ, $\sigma_{(1)}$, $\sigma_{(2)}$ involved in (3.81) must have a different value than in the static situation.[9] At the same time, the liquid spreading as an interface formation process suggests that the yet forming liquid-solid interface is out of equilibrium at the contact line, and its interfacial parameters, first of all the surface tension, have some dynamic (i.e., nonequilibrium) values there.

---

[9] Strictly speaking, for the moving contact line one has to include, in addition to the surface tensions, also the momentum fluxes due to the motion of interfaces. As we will see in Chapter 4, this contribution in all practical situations is negligible and the Young equation still has the form (3.81) when the contact line is moving with the surface tension being, of course, the dynamic ones.

This dynamic value of the liquid-solid surface tension inevitably leads to the deviation of the dynamic contact angle from the static (equilibrium) one.

The fluid for 'building' the fresh liquid-solid interface has to be brought to the contact line by the flow field, so that the flow field will influence the surface tension acting on the contact line from the liquid-solid interface and hence the dynamic contact angle. This is the essence of nonlocal hydrodynamic effect on the contact angle ('hydrodynamic assist of dynamic wetting') we described in §3.2.3. The leading role in the influence of the flow field belongs, naturally, to the contact-line speed, as was indeed shown by experiments.

The 'rolling' motion of the fluid observed in experiments (§3.2.2) suggests that one has simultaneously the disappearance of the liquid-gas interface and the formation of the liquid-solid one as the fluid particles belonging to the free surface arrive at the contact line to become material points of the liquid-solid interface and gradually relax to their new equilibrium state. It should be noted, however, that the liquid-gas interface is just one source of 'material' for building the fresh liquid-solid interface, and the rest of it is provided by the bulk, which makes the interface formation dependent on the bulk flow.

Thus, the qualitative features observed in experiments on dynamic wetting map naturally onto what one would expect from the description of the flow as an interface formation/disappearance process. The problem now is to develop a mathematical theory of flows with forming/disappearing interfaces in the framework of continuum mechanics.

It is interesting to note that attempts to bring in the physics of interface formation in dealing with the moving contact-line problem can be traced back to several of the theories we considered earlier. Indeed, Hansen & Miotto (1957) described this idea in words in terms of reorientation of the molecules belonging to an element of the liquid-gas interface as it arrives at the contact line and becomes an element of the liquid-solid interface. Twenty years later, Huh & Mason (1977a) tried to incorporate this idea into a fluid-mechanical model by introducing a relaxation time required for the no-slip boundary condition to be switched on. However, as pointed out first by Dussan (1979) and restated in a more general form in §3.4.1, the actual kinematics of the flow produced by Huh & Mason's model contradicts the physical assumption of Hansen & Miotto it is based on. The essence of Cherry & Holmes' (1969) and Blake & Haynes' (1969) works is to model in a simple way the kinetics of interface formation, though decoupling it from hydrodynamic factors and hence not addressing the moving contact-line problem as it stands in continuum mechanics.

If we turn to the free-surface flows outlined in §2.5.3 whose description by means of the standard model leads to unphysical paradoxes, it seems that all of them are in fact particular fluid motions with forming or disappearing interfaces. As we will show later, this is indeed the case. Hence a mathematical theory incorporating the interface formation/disappearance phenomenon into the framework of continuum mechanics should allow one to tackle all these paradoxes in a unified self-consistent way. It should be possible also to test the theory by considering independent experiments and determine the values of material constants involved from different independent measurements.

In the next chapter, we will describe the simplest mathematical theory incorporating the process of interface formation into the boundary conditions for the Navier-Stokes equations.

Chapter 4

Boundary conditions on forming interfaces

In this chapter, we introduce a conceptual framework for incorporating the processes of formation and disappearance of interfaces during flow into the boundary conditions for the equations of macroscopic fluid mechanics. The simplest mathematical theory developed within this framework is considered with a particular attention to the assumptions involved in its derivation.

4.1   Modelling of interfaces

As we discussed in §§2.1 and 2.4, in the continuum approximation, which is at the core of fluid mechanics, the 'interfaces' are always modelled as mathematical surfaces of zero thickness separating the 'bulk' phases. In a general case, these interfaces possess their own physical properties, such as the surface tension, and hence have to be regarded as two-dimensional continua, in other words as 'surface phases'. In reality, of course, an 'interface' is a layer of a finite thickness between the two 'bulk' media, and its surface properties are integrals across this layer of the corresponding distributions of microscopic parameters. Physically, the interfacial properties arise due to asymmetry in the intermolecular forces acting on the molecules of the interfacial layer from the bulk phases, and hence the thickness of this layer is determined by the range of these forces. This range and hence the interfacial layer's thickness are on a molecular scale and, being much smaller than a characteristic length scale of the bulk flow, become negligible in the continuum approximation. However, the intermolecular forces that give rise to the surface properties are much stronger than those considered in macroscopic fluid mechanics of the bulk phases so that the 'product' of these two factors, which in the mathematical language one would describe as an infinitesimal thickness of the interface and singularly-large forces acting inside it, is finite. As a result, an interfacial layer can play a significant role in the overall dynamics of the fluid via capillary effects. In the literature, the surface properties that are comparable with the corresponding bulk properties are often, for brevity, referred to as 'singular' to emphasize that, in a particular respect, the influence of the interfacial layer is disproportional to its thickness.

The general problem of how to model interfaces mathematically is a difficult one since the interfaces are physical systems with strong anisotropy involving both macroscopic and molecular scales as well as forces of differing nature. There are two main approaches to this general problem. They pursue different goals and can be broadly labeled as 'structural' and 'integral' (or 'macroscopic'). The first one is aimed at describing the internal structure of the interface treated as a three-dimensional anisotropic system. Its goal (and the principal difficulty) is to model explicitly how the factors that give rise to this structure, namely the asymmetry of intermolecular forces from the bulk phases and the intermolecular interactions within the interfacial layer, are related with the measurable physical parameters. As all macroscopic characteristics of the interfaces, the latter are obtained via a suitable averaging across the interfacial layer of the appropriate microscopic quantities.

The integral approach considers the interface as a two-dimensional 'surface phase' endowed with macroscopic characteristics, such as the surface tension, and the ultimate objective is to formulate boundary conditions for the equations describing the bulk phases that would allow one to investigate various process in systems involving interfaces. In broad terms, the integral approach pursues the goal of incorporating interfaces as elements in the general framework of continuum mechanics whereas the structural approach is that of reductionism along the lines of statistical physics and molecular dynamics simulations. An essential constraint on the integral approach is that, as is always the case with boundary conditions, it has to provide both constitutive equations for the interfaces and to model the interactions between the surface and bulk phases in terms of the parameters featuring in the bulk model(s), thus ensuring consistency of the overall description.

The origin of both approaches can be traced back to the works of Laplace (1806, 1807). As far as the structural approach is concerned, Laplace was the first to calculate the equilibrium surface tension in terms of additive rapidly decaying intermolecular forces (in his words, forces that are "sensible only at insensible distances") assuming that, microscopically, an interface corresponds to a step-change in density. From the integral viewpoint, he also found how the hydrostatic pressure inside a spherical drop is related with the drop's radius and the surface tension on its free surface. Later he generalized this result for the case of an interface with different principal radii of curvature, thus essentially arriving at what is now known as the capillary pressure and the normal-stress boundary condition.

The first thermodynamic theory of capillarity was developed by Gibbs in 1875–77 (see Gibbs 1928) who considered the equilibrium of two homogeneous bulk phases separated by what he for brevity called "a surface of discontinuity". He emphasizes, however, that this is not a mathematically defined geometric surface where the bulk properties are discontinuous, as in Laplace's treatment, but a "non-homogeneous film which separates homogeneous or nearly homogeneous masses" (p. 219). In the terminology we use here, Gibbs'

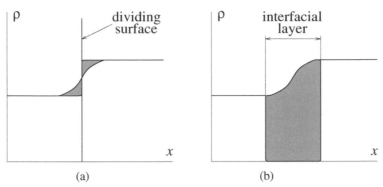

FIGURE 4.1: A sketch of the density distribution across an interfacial layer illustrating different ways of introducing surface variables. In Gibbs' theory, the distributions of the bulk parameters are extended into the interfacial layer up to the so-called 'dividing surface' (a) and the 'excess' quantities (the shadow areas) are attributed to this dividing surface as surface parameters. In the continuum mechanics approximation, the interfacial layer's thickness is negligible compared with the characteristic length scale the model operates on, so that the layer as a whole 'collapses' into a mathematical surface. Then, the surface properties appear as integrals across the layer of the corresponding bulk distributions (the shadow area in (b)). In (a), the contributions of the two bulk phases to the 'excess' interfacial properties specified geometrically by the position of the dividing surface. In (b), this has to be specified by additional physical assumptions about the properties of the interfacial layer, so that the interface becomes *defined* by its physical properties.

"surface of discontinuity" is in fact an interfacial layer. In his analysis which is essentially macroscopic in nature, Gibbs considers the (equilibrium) structure of this layer in descriptive terms, without detailing the physics responsible for its formation, and the role of structural arguments is to help identify the parameters that determine the state of the interface and illustrate their physical meaning. The analysis is based on the thermodynamic principle of equilibrium as a state with the maximum entropy for a given internal energy and considers the variations of the internal energy in a volume embracing the interfacial layer and adjacent parts of the bulk phases. The main result in Gibbs' study is the extension of (equilibrium) thermodynamics to systems with interfaces, thus creating the conceptual framework used in different variants in all subsequent works in this area.

In order to reduce the 3-dimensional interfacial layer to a 2-dimensional 'interface', Gibbs introduces an artificial construction he calls the "dividing surface" (Fig. 4.1a). The dividing surface is a mathematical surface located inside the interfacial layer, and it formally divides the system into two bulk phases. If the homogeneous distributions of the bulk parameters on both sides of the interfacial layer are extrapolated up to the dividing surface, there appear the so-called "excess" quantities resulting from the interfacial layer having a structure (Fig. 4.1a. Gibbs assigns these "excess" quantities to the dividing surface, thus making it a 'surface phase' endowed with certain physical properties.

The key feature of this scheme is that it is *exact* in a sense that it introduces an interface and interfacial parameters without using any approximation, such as the continuum limit. This is often regarded as an advantage of the scheme. However, the dividing surface technique is ultimately a purely geometric way of defining what are essentially physical quantities. In this connection, a question one has to ask is whether *different* interfacial properties can be introduced in a meaningful way using *one* dividing surface. For example, from considering adsorption one has to choose the position of the dividing surface in such a way that the excess masses per unit area realistically represent the actual composition of the interfacial layer, whereas the arguments based on the distribution of stresses across the layer require the dividing surface to be the surface of tension. The two surfaces do not necessarily coincide, and the difference between them becomes negligible only in the continuum approximation when the interfacial layer's thickness itself vanishes as this layer becomes a mathematical surface separating two bulk phases.

Another important point is that the excess quantities themselves depend on, and are very sensitive to, the choice one makes about the location of the dividing surface. In this connection Defay, Prigogine & Bellemans (1966) write:

> It is immediately apparent that the actual values of these quantities will in general be sensitive to the precise choice of the position of the dividing surface, and this property endows the Gibbs treatment with an apparent abstract character. <...> In discussing the application of theoretical equations to the interpretation of experimental results, Gibbs took great care not to define the surface of division in terms of any geometrically defined position, but rather by a convention based on the adsorptions themselves. <...> The extreme sensitivity of the adsorption Γ_i to the position attributed to the dividing surface, together with the impossibility of defining any surface experimentally with the required precision, has led to the conclusion that to describe surface phenomena in terms experimental quantities it is necessary to choose quantities whose values are invariant with respect to the position of the dividing surface (pp. 24–26).

The above difficulties also vanish in the continuum approximation where the interface as a thermodynamic system becomes *defined* by the assumptions one makes about its thermodynamic properties. This is essentially similar to how thermodynamics is applied to 'fluid particles' in continuous media (see §2.3).

Since our ultimate goal here is to incorporate interfaces into the general framework of fluid mechanics and consider processes where the system is out of equilibrium, both dynamically and thermodynamically, the use of concepts introduced via the method of dividing surface, not to mention that of several dividing surfaces (Rusanov 2005), presents obvious difficulties. For example, if one considers mass convection along the surface phase, then, physically, this will involve all the matter in the interfacial layer and not just the "excess" bits associated with one dividing surface or another.

A number of authors, notably Bakker (1928), Guggenheim (1940) and Defay, Prigogine & Bellemans (1966), describe an approach intended to overcome the shortcomings inherent in the dividing surface method by considering the interface as a layer of a finite thickness between two bulk phases (Fig. 4.1b). However, like in Gibbs' treatment, they also aim at an exact description and hence face the question about the exact location of the interfacial layer's boundaries. This question is essentially similar to the one these authors pose about the location of Gibbs' dividing surface, though now it is less acute since the layer's boundaries are placed where the distributions of parameters are almost uniform, at least in equilibrium.

More important, however, is that thermodynamic properties of a finite layer become dependent on the values of the intensive bulk parameters (p and T) inside the layer, thus making it, in the words of Defay et al. (1966), a "non-autonomous" phase. In Gibbs' more formal treatment this problem does not arise. Tellingly, in order to overcome the ambiguity associated with the location of the interfacial layer's boundaries, Guggenheim (1940) has to look for thermodynamic relationships that remain invariant when the boundaries are shifted, thus essentially *defining* the interface by its thermodynamic properties. As we will see later, this is how the interfaces are introduced in the genuinely integral approach. Even more telling is that the dependence of the (macroscopic) interfacial properties on p and the fact that p varies across a curved interface make Guggenheim resort to approximate "practical" arguments which are essentially equivalent to the continuum approximation. In the consistently applied continuum limit the above difficulties do not arise in principle but at a price that the treatment is no longer exact. In other words, the 'interface' as a well-defined concept and the associated integral (macroscopic) characteristics of such interface appear only in the leading-order approximation in the ratio of molecular-to-macroscopic length scales, ℓ/L, as $\ell/L \to 0$. The reason is that if the interfacial layer's thickness, which is on the scale of ℓ, is not negligible compared to the length scale of interest, as is the case in the above models as well as in Gibbs' treatment, then it becomes, explicitly or implicitly, an essential parameter, a part of what characterizes the interfacial layer's structure. However, once the structural arguments are brought into the frame, it becomes necessary to consider not only the thickness of the interfacial layer but also the actual distributions of parameters inside this layer, i.e., to use the full-scale structural approach in a consistent way.

The first structural thermodynamic theory that considers the actual physics in the interfacial layer was developed by van der Waals (1893) who based his analysis on an assumption that the density varies continuously across the interfacial layer separating the two phases.[1] Van der Waals justifies the condition of thermodynamic equilibrium as the state with the minimum of free energy at a given temperature, which is equivalent to the starting point in Gibbs' theory. The structure of the interfacial layer is investigated by considering the internal energy in the interfacial layer, where the density varies continuously from a (uniform) density in one bulk phase to that in the other, as a function of the local density and the densities in the neighbouring phases. This function is calculated by treating the internal energy as a potential energy associated with the force field acting on a unit mass at every point in the interfacial layer from elementary masses of the surrounding medium, similar to how one would consider, say, a self-gravitating gas. The force between two elementary masses is assumed to depend, in unspecified way, on the distance between them and is not influenced by the distribution of other masses.[2] Using a number of simplifying assumptions, van der Waals shows that the local free energy inside the interfacial layer, to leading order, is the sum of two terms, one dependent on the local density and the other proportional to its second derivative in the direction normal to the interfacial layer (see equation (3.75) in Chapter 3). A comparison with Gibbs (1928) allows him to correlate microscopic distributions with macroscopic quantities in Gibbs' theory.

More than 60 years after the publication of van der Waals' original paper, his results were rediscovered by Cahn & Hilliard (1958). Since then van der Waals' approach, which is essentially that of statistical physics, has been developed (e.g., Triezenberg & Zwanzig 1972, Lovett et al. 1973, Bongiorno & Davis 1975) and applied by many authors to various equilibrium systems and critical phenomena (e.g., Cahn 1977, Widom 1978, Teletzke, Scriven & Davis 1982; see also Rowlinson & Widom 1982 and references therein). This approach is also used in numerical codes based on the Lattice-Boltzmann method.

A structural analysis significantly differing from that of van der Waals was presented by Brenner (1979) who considered the equilibrium of a multicomponent system with an interface. The goal of his study was to relate microscopic parameters characterising the interfacial layer with macroscopic quantities describing the 'interface' by using the matched asymptotic expansions in the ratio of the characteristic thickness of the layer to the macroscopic dimension

---

[1] In the first section of his work, van der Waals writes: "According to Gibbs' theory, capillary phenomena are present only if there is a discontinuity between the portions of fluid that are face-to-face. The chapter of his paper that discusses capillarity carries, in fact, the title, 'Influence of surfaces of discontinuity, etc'." However, as noted earlier, Gibbs used the term "surface of discontinuity" for brevity and his analysis does not rely on the fluid properties being actually discontinuous on the microscopic level.

[2] Strictly speaking, in what is now referred to as van der Waals interactions collective effects can be neglected only for a rarefied medium and have to be taken into account if the medium is dense; see §7.7.2.

in the bulk phases as a small parameter. The author begins by introducing potential "external" forces which act on molecules of each component inside the interfacial layer and disappear as one moves into the bulk phases. Physically, therefore these are forces due to the asymmetry of the influence on the molecules inside the interfacial layer of the bulk phases. Then, following the idea used by Einstein (1956) in his analysis of Brownian motion, Brenner considers the dynamic equilibrium and equates the diffusion flux arising from the concentration gradient and a convective flux due to the external forces. However, in the subsequent analysis of the microstructure of the interfacial layer the author (a) brings in the equation of hydrostatics, which involves the pressure gradient and is essentially macroscopic, and (b) uses Gibbs' thermodynamic equalities which again are macroscopic in their nature. By contrast, van der Waals' analysis is based on first principles and does not involve equations of macroscopic dynamics and thermodynamics.

From the viewpoint of fluid mechanics, the results of the equilibrium structural analysis are of little interest since not only in the equilibrium (static) situations but also in most flows the interfaces are completely characterized by their surface tensions being material constants. The structural analysis simply relates these constants to the underlying physical factors, whereas for fluid mechanics purposes one can use the values of surface tensions measured experimentally. Similarly, in dealing with the bulk fluid statistical physics is aimed at calculating transport coefficients whereas fluid mechanics, in most cases, treats them as material constants. There are situations, however, where these 'material constants' become variables, and one has to extend the model to account for their dependence on other characteristics of the flow.

In order to describe situations where the interface is driven out thermodynamic equilibrium, one has to model the mass, momentum and energy between the interfacial layer and the bulk phases as well as fluxes of mass, momentum and energy along the interfacial layer. The results presented in a macroscopic form will then become boundary conditions for the continuum mechanics equations for the bulk phases and describe the coupling between the bulk and the interfacial parameters that will determine the distributions of the latter along the interface. In the structural description of interfaces a step from the equilibrium to a nonequilibrium situation encounters fundamental difficulties. To make this step it becomes necessary to consider the distributions of fluxes inside the interfacial layer on a molecular level, i.e., one has not only to overcome yet unresolved difficulties of principle inherent in the kinetic theory of liquids associated with multi-particle interactions but also extend the analysis to an anisotropic situation, where in one direction the parameters vary on a molecular scale. At present, this problem is very far from its resolution.

The problem becomes much less formidable in the 'macroscopic' approach to the modelling of interfaces, where one introduces the notion of an 'interface' by applying the continuum limit so that from the start the interface is considered as a two-dimensional continuum ('surface phase') and is characterized by its macroscopic 'surface' properties, i.e., properties defined per unit area. Physically, these properties are integrals across the interfacial layer of the corresponding microscopic distributions (Fig. 4.1b) but this is just an illustration of their meaning since, thermodynamically, the interface is *defined* by the postulated relationships between its thermodynamic parameters and constitutive equations for its transport properties. As is always the case in continuum mechanics, all molecular effects are incorporated in phenomenological material constants appearing as transport coefficients and parameters in the constitutive equations. Then, the interface driven out of equilibrium presents no conceptual difficulties, and one can employ the methods of irreversible thermodynamics, similarly to how Gibbs applied equilibrium thermodynamics in his, descriptively structural but essentially macroscopic, analysis of the equilibrium state of interfaces.

Waldmann (1967) was the first to apply methods of irreversible thermodynamics to the derivation of the boundary conditions between two immiscible fluids. He considers the possibility of singular momentum and heat fluxes along the interface (here the term 'singular' is used with the meaning explained at the beginning of this section) and, after deriving an expression for the entropy production and examining it in the way described in §2.3 for the bulk phases, arrives at a set of linear constitutive equations for the fluxes along and across the interface in terms of the velocity and temperature fields on its two sides. Although Waldmann allows for singular fluxes along the interface and hence rheological properties of interfaces described earlier in a purely hydrodynamic way by Boussinesq (1913) and Scriven (1960), he does not consider the possibility of singular density of any quantity so that capillary effects, associated with singular energies and stresses in the surface phase, become excluded from his analysis. These effects were brought into the frame by Bedeaux, Albano & Mazur (1976) who assumed that a fluid-fluid interface can possess singular stress (i.e., the surface tension), energy and entropy as well as conduct singular heat fluxes along it. This extension allows the authors to obtain the equation of capillarity, the surface analogue of the Fourier law of heat conduction as well as a number of linear relationships between thermodynamic fluxes and forces expressed in terms of the temperature and velocity distributions.

It is important to note that in their analysis Bedeaux et al. neglected the mass density at the interface from the very beginning ("as is customary in the surface thermodynamics for such systems", p. 439) and hence excluded from consideration the mass exchange between the interface and the bulk phases which is the key element of the interface formation process we are interested in here. Obviously, the mass per unit area of the interfacial layer (i.e., the 'surface density') is always very small compared with the mass per unit volume in the bulk phase since the interfacial layer's thickness is on a molecular rather than a macroscopic scale. Hence practically always the inertial properties of interfaces may be neglected. In most cases (but, as we will see later, not always!), one may also neglect the mass exchange between the interface and the bulk in the boundary condition for the *bulk* phase. Then, for an impermeable interface one has an equality between the normal components of the bulk and surface velocities (condition (2.135) in §2.4). However, this exchange, if it takes place, is always important in the equations for the *surface* phase, where the surface density together with temperature can be used as the parameters of state. As a result, although the inertial properties of the interface are practically always negligible compared with those of the bulk phases, the surface density and its evolution determine the state of the interface and hence influence the bulk flow via the surface tension, which is a 'singular' interfacial property.

The work of Bedeaux et al. paved the way for a series of papers with formal generalisations of their results to multicomponent systems with singular excess densities of all components, singular multipole moments of excess densities, etc. (e.g., Kovac 1977, Napolitano 1978, 1979, Ronis, Bedeaux & Oppenheim 1978, Albano, Bedeaux & Vlieger 1979, Ronis & Oppenheim 1983). These models examine the formal structure of equations resulting from the extension of the set of governing parameters defining the state and transport mechanisms in the interfaces and in this sense have a certain methodological value. On the other hand, they do not fully specify the interfaces via the appropriate equations of state and hence do not provide a full set of boundary conditions needed to solve fluid mechanical problems involving the effects they account for. For this reason none of these models in their full complexity has ever been applied in fluid mechanics and verified experimentally.

Shikhmurzaev (1993) developed the approach proposed by Bedeaux et al. in a different way. His goal was to formulate the *simplest* (i.e., irreducible) self-consistent theory specifically addressing the process of interface formation, use it to describe 'paradoxical' flows in the framework of macroscopic fluid mechanics and verify the results experimentally. Below, we will follow this theory and highlight its main assumptions as points where, if experiments indicate such a necessity, the simplest model could be generalised.

4.2   Conservation laws

Consider the boundary conditions accounting for the process of interface formation in the framework of the macroscopic approach. The derivation will require the use of irreversible thermodynamics outlined in §2.3 and generally follows and expands the analysis of Shikhmurzaev (1993). We will begin with the simplest cases of liquid-gas and liquid-solid interfaces and then consider a more complex situation of an interface between two immiscible liquids.

As in §2.4.2, let $f(\mathbf{r}, t) = 0$ be the equation of a fluid-fluid interface with a unit normal $\mathbf{n} = \nabla f / |\nabla f|$ pointing from fluid 2 occupying the region $f < 0$

to fluid 1 corresponding to $f > 0$ (Fig. 4.2). The superscripts $+$ and $-$ refer to the bulk parameters in fluid 1 and 2, respectively. As in the classical case, the boundary conditions at the interface will have to incorporate both the general conservation laws for mass, momentum and energy and the physical mechanisms specific to the type of interfaces and processes we are considering.

The conditions following from the conservation laws can be derived in two ways. The first one is to apply the integral form of equations expressing the conservation laws to a small control volume embracing the interface and then take a limit as the size of the volume in the direction normal to the interface goes to zero. The integral equations make it possible to consider functions discontinuous across the interface and point out the assumptions that have to be made about the integrals over the vanishing lateral sides of the control volume and over the volume itself as one introduces surface properties of the interface.

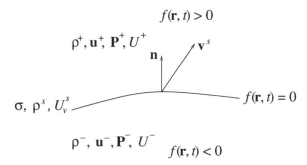

FIGURE 4.2: A definition sketch for a fluid-fluid interface.

An equivalent and mathematically more convenient way of describing a system with an interface across which the bulk parameters can be discontinuous and which itself can possess some surface properties and hence be regarded as a surface phase is by using the technique of generalized functions. This technique, first applied to the study of interfaces by Bedeaux et al. (1976), allows one to consider the conservation laws in the form of partial differential equations and arrive at the corresponding equations for the bulk and surface phases in a more straightforward way. Below, we will take this second route.

The required mathematical background is as follows. First, we will introduce two Heaviside functions

$$\theta^+(f) = \begin{cases} 1, f > 0 \\ 0, f \leq 0 \end{cases}, \qquad \theta^-(f) = \begin{cases} 0, f \geq 0 \\ 1, f < 0 \end{cases},$$

and then, given that $d\theta^\pm/df = \pm\delta(f)$, where δ is the delta-function, define a 'surface' delta-function

$$\delta^s(\mathbf{r}, t) = \delta(f)\,|\nabla f|$$

so that

$$\nabla\theta^\pm = \pm\delta(f)\nabla f - \perp\delta^s(\mathbf{r}, t)\,\mathbf{n}. \tag{4.1}$$

Since f always satisfies (2.132), i.e.,

$$\frac{\partial f}{\partial t} + \mathbf{v}^s \cdot \nabla f = 0, \tag{4.2}$$

which simply introduces the normal component of the interface velocity $\mathbf{v}^s$, and hence for the time derivative of f one has

$$\frac{\partial f}{\partial t} = -|\nabla f|\,\mathbf{v}^s \cdot \mathbf{n}.$$

The time derivative of $\theta^\pm$ can be expressed as

$$\frac{\partial\theta^\pm}{\partial t} = \pm\delta(f)\frac{\partial f}{\partial t} = \mp\delta^s(\mathbf{r}, t)\,\mathbf{v}^s \cdot \mathbf{n}. \tag{4.3}$$

The 'surface' parameters which characterize properties of the interface are defined only along it; if considered as functions of all three spatial coordinates they will then have to satisfy the constraint that their derivatives in the direction normal to the interface are equal to zero. In particular, this means that

$$\mathbf{n} \cdot \nabla\mathbf{v}^s = \mathbf{n} \cdot \nabla\mathbf{n} = 0. \tag{4.4}$$

After introducing the notation

$$\frac{D^\pm}{Dt} = \frac{\partial}{\partial t} + \mathbf{u}^\pm \cdot \nabla, \qquad \frac{D^s}{Dt} = \frac{\partial}{\partial t} + \mathbf{v}^s \cdot \nabla$$

for the substantive derivatives associated with the bulk and surface velocities and using (4.2) and (4.4), we have that $D^s\theta^\pm/Dt = 0$ and $D^s\delta^s/Dt = 0$. The latter gives an expression for the time derivative of δ^s:

$$\frac{\partial\delta^s}{\partial t} = -\mathbf{v}^s \cdot \nabla\delta^s. \tag{4.5}$$

An important mathematical result proven by Bedeaux, Albano & Mazur (1976) is that if for two surface quantities $\mathbf{A}^s$ and $\mathbf{B}^s$ one has

$$\mathbf{A}^s\delta^s + \mathbf{B}^s \cdot \nabla\delta^s = 0, \tag{4.6}$$

it then follows that

$$\mathbf{A}^s = \mathbf{B}^s \cdot \mathbf{n} = 0. \tag{4.7}$$

This result allows one to obtain separate equations for the surface tensors of different rank involved in equations of the form of (4.6).

Now, we can begin with the derivation of equations for the surface phase. Introduce the densities of mass and momentum defined at every point in space:

$$\rho(\mathbf{r}, t) = \rho^+(\mathbf{r}, t)\theta^+(f) + \rho^s(\mathbf{r}, t)\delta^s(\mathbf{r}, t) + \rho^-(\mathbf{r}, t)\theta^-(f), \tag{4.8}$$

$$\rho\mathbf{u} = \rho^+\mathbf{u}^+\theta^+ + \rho^s\mathbf{v}^s\delta^s + \rho^-\mathbf{u}^-\theta^-, \qquad \mathbf{u} = \begin{cases} \mathbf{u}^+, \ f > 0 \\ \mathbf{v}^s, \ f = 0 \\ \mathbf{u}^-, \ f < 0 \end{cases}. \qquad (4.9)$$

Here ρ is defined as mass per unit volume and hence ρ^s, which we will call the surface density, is the mass of the interfacial layer (modelled here as a geometric interfacial surface) contained in a unit volume.

Substituting expressions (4.8) and (4.9) into the continuity equation (2.106), using (4.1), (4.3) and (4.5) to express the derivatives of generalised functions and making the coefficients in front of θ^+, θ^- and δ^s separately zero, we obtain the continuity equations for the bulk and surface phases:

$$\frac{\partial\rho^\pm}{\partial t} + \nabla \cdot (\rho^\pm\mathbf{u}^\pm) = 0,$$

$$\frac{\partial\rho^s}{\partial t} + \nabla \cdot (\rho^s\mathbf{v}^s) = -\rho^+(\mathbf{u}^+ - \mathbf{v}^s) \cdot \mathbf{n} + \rho^-(\mathbf{u}^- - \mathbf{v}^s) \cdot \mathbf{n}. \qquad (4.10)$$

The last equation is a generalization of (2.133) accounting for variation in the surface density and the divergence of the surface mass flux along the interface. The mass fluxes between the surface phase and the bulk, $\rho^\pm(\mathbf{u}^\pm - \mathbf{v}^s) \cdot \mathbf{n}$, have yet to be specified.

It is necessary to emphasize that the surface density introduced by (4.8) is the mass per unit area of the whole of the interfacial layer (Fig. 4.1) and not the excess density. For example, the surface velocity $\mathbf{v}^s$ is obviously the average velocity with which the whole mass in the interfacial layer is carried and not the "excess" bits. Thus, when we make a step from the equilibrium thermodynamics of interfaces to their dynamics, it becomes necessary to use the self-consistent approach of continuum mechanics and not the artificial construction of the "dividing surface".

The stress tensor $\mathbf{P}$ and the field of body forces $\rho\mathbf{F}$ can be introduced in the same way as the densities of mass and momentum:

$$\mathbf{P} = \mathbf{P}^+\theta^+ + \mathbf{P}^s\delta^s + \mathbf{P}^-\theta^-, \qquad \rho\mathbf{F} = \rho^+\mathbf{F}^+\theta^+ + \rho^s\mathbf{F}^s\delta^s + \rho^-\mathbf{F}^-\theta^-. \quad (4.11)$$

After substituting these expressions together with (4.8), (4.9) into (2.107) and making the coefficients in front of θ^+, θ^-, δ^s and, according to (4.6) and (4.7), the normal component of the coefficient in front of $\nabla\delta^s$ separately zero, we arrive at

$$\rho^\pm\frac{D^\pm\mathbf{u}^\pm}{Dt} = \nabla \cdot \mathbf{P}^\pm + \rho^\pm\mathbf{F}^\pm$$

in the bulk together with the following equations for the surface phase:

$$\rho^s\frac{D^s\mathbf{v}^s}{Dt} + \rho^+(\mathbf{u}^+ - \mathbf{v}^s)(\mathbf{u}^+ - \mathbf{v}^s) \cdot \mathbf{n} - \rho^-(\mathbf{u}^- - \mathbf{v}^s)(\mathbf{u}^- - \mathbf{v}^s) \cdot \mathbf{n}$$

$$= \nabla \cdot \mathbf{P}^s + \mathbf{n} \cdot (\mathbf{P}^+ - \mathbf{P}^-) + \rho^s\mathbf{F}^s, \qquad (4.12)$$

$$\mathbf{n} \cdot \mathbf{P}^s = 0. \qquad (4.13)$$

Using (4.10), one can write the first of these equations also in the form

$$\frac{\partial(\rho^s \mathbf{v}^s)}{\partial t} + \nabla \cdot (\rho^s \mathbf{v}^s \mathbf{v}^s) + \rho^+ \mathbf{u}^+ (\mathbf{u}^+ - \mathbf{v}^s) \cdot \mathbf{n} - \rho^- \mathbf{u}^- (\mathbf{u}^- - \mathbf{v}^s) \cdot \mathbf{n}$$

$$= \nabla \cdot \mathbf{P}^s + \mathbf{n} \cdot (\mathbf{P}^+ - \mathbf{P}^-) + \rho^s \mathbf{F}^s, \qquad (4.14)$$

which is a generalization of (2.142) we considered earlier. Equation (4.13) establishes that $\mathbf{P}^s$ indeed describes the surface stress acting at every point in the plane tangential to the interface. The arguments similar to those used in §2.2.3 give that $\mathbf{P}^s$ is symmetric.

The internal energy per unit volume, U_v, and the heat flux, $\mathbf{q}$, can be introduced by

$$U_v = U_v^+ \theta^+ + U_v^s \delta^s + U_v^- \theta_-, \qquad \mathbf{q} = \mathbf{q}^+ \theta^+ + \mathbf{q}^s \delta^s + \mathbf{q}^- \theta^-. \qquad (4.15)$$

Repeating the procedure used for the mass and momentum conservation laws, that is, substituting expressions (4.15) together with (4.8), (4.9), (4.11) into the energy balance equation (2.109), using (4.1), (4.3), (4.5) for the derivatives of generalised functions as well as (4.13) for $\mathbf{P}^s$ and making the coefficients in front of θ^+, θ^-, δ^s and $\mathbf{n} \cdot \nabla \delta^s$ separately zero, we arrive at

$$\frac{\partial}{\partial t}\left(\frac{\rho^\pm (\mathbf{u}^\pm)^2}{2} + U_v^\pm\right) + \nabla \cdot \left[\mathbf{u}^\pm \left(\frac{\rho^\pm (\mathbf{u}^\pm)^2}{2} + U_v^\pm\right) - \mathbf{P}^\pm \cdot \mathbf{u}^\pm + \mathbf{q}^\pm\right]$$

$$= \rho^\pm \mathbf{F}^\pm \cdot \mathbf{u}^\pm \qquad (4.16)$$

for the bulk phases together with

$$\frac{\partial}{\partial t}\left(\frac{\rho^s (\mathbf{v}^s)^2}{2} + U_v^s\right) + \nabla \cdot \left[\mathbf{v}^s \left(\frac{\rho^s (\mathbf{v}^s)^2}{2} + U_v^s\right) - \mathbf{P}^s \cdot \mathbf{v}^s + \mathbf{q}^s\right]$$

$$+ \left(\frac{\rho^+ (\mathbf{u}^+)^2}{2} + U_v^+\right)(\mathbf{u}^+ - \mathbf{v}^s) \cdot \mathbf{n} - \left(\frac{\rho^- (\mathbf{u}^-)^2}{2} + U_v^-\right)(\mathbf{u}^- - \mathbf{v}^s) \cdot \mathbf{n}$$

$$- \mathbf{n} \cdot (\mathbf{P}^+ \cdot \mathbf{u}^+ - \mathbf{P}^- \cdot \mathbf{u}^-) + (\mathbf{q}^+ - \mathbf{q}^-) \cdot \mathbf{n} = \rho^s \mathbf{F}^s \cdot \mathbf{v}^s, \qquad (4.17)$$

$$\mathbf{q}^s \cdot \mathbf{n} = 0. \qquad (4.18)$$

for the surface phase. Equation (4.17) expresses the surface energy balance and, in addition to the local variation of the total surface energy and the divergence of its surface flux, also takes into account the energy fluxes from/into the bulk phase, work of bulk stresses, heat fluxes from the bulk and work done by surface forces. Equation (4.18) states that, as expected, $\mathbf{q}^s$ is a surface heat flux vector lying in the plane tangential to the interface. The derivative of $\mathbf{q}^s$ in the direction normal to the interface as well as the corresponding derivatives of all other surface parameters are zero by definition.

A sum of (4.10) multiplied by $(\mathbf{v}^s)^2/2$ and (4.12) multiplied by $\mathbf{v}^s$ yields the kinetic energy balance equation

$$\frac{\partial}{\partial t}\left(\frac{\rho^s(\mathbf{v}^s)^2}{2}\right) + \nabla\cdot\left(\mathbf{v}^s\frac{\rho^s(\mathbf{v}^s)^2}{2}\right) + \rho^+(\mathbf{u}^+ - \tfrac{1}{2}\mathbf{v}^s)\cdot\mathbf{v}^s(\mathbf{u}^+ - \mathbf{v}^s)\cdot\mathbf{n}$$

$$-\rho^-(\mathbf{u}^- - \tfrac{1}{2}\mathbf{v}^s)\cdot\mathbf{v}^s(\mathbf{u}^- - \mathbf{v}^s)\cdot\mathbf{n} - \mathbf{n}\cdot(\mathbf{P}^+ - \mathbf{P}^-)\cdot\mathbf{v}^s + \mathbf{P}^s : \nabla\mathbf{v}^s$$

$$= \nabla\cdot(\mathbf{P}^s\cdot\mathbf{v}^s) + \rho^s\mathbf{F}^s\cdot\mathbf{v}^s, \tag{4.19}$$

which is analogous to (2.112) we had in the bulk. Subtracting this equation from (4.17), we arrive at

$$\frac{\partial U_v^s}{\partial t} + \nabla\cdot(\mathbf{v}^s U_v^s) + U_v^+(\mathbf{u}^+ - \mathbf{v}^s)\cdot\mathbf{n} - U_v^-(\mathbf{u}^- - \mathbf{v}^s)\cdot\mathbf{n}$$

$$+\tfrac{1}{2}\rho^+(\mathbf{u}^+ - \mathbf{v}^s)^2(\mathbf{u}^+ - \mathbf{v}^s)\cdot\mathbf{n} - \tfrac{1}{2}\rho^-(\mathbf{u}^- - \mathbf{v}^s)^2(\mathbf{u}^- - \mathbf{v}^s)\cdot\mathbf{n}$$

$$-\mathbf{P}^s : \nabla\mathbf{v}^s - \mathbf{n}\cdot\mathbf{P}^+\cdot(\mathbf{u}^+ - \mathbf{v}^s) + \mathbf{n}\cdot\mathbf{P}^-\cdot(\mathbf{u}^- - \mathbf{v}^s)$$

$$+(\mathbf{q}^+ - \mathbf{q}^-)\cdot\mathbf{n} + \nabla\cdot\mathbf{q}^s = 0, \tag{4.20}$$

which describes the balance of the surface internal energy.

So far, we made no assumptions about properties of the bulk and surface phases nor have we used any simplifications resulting from the order-of-magnitude analysis of the parameters involved. The above equations basically split the conservation laws into the bulk and surface parts and formally introduce the surface parameters and fluxes of mass, momentum and energy between the three phases.

4.3 Liquid-gas and liquid-solid interfaces

Now, we have to consider the specific physics of liquid-gas and liquid-solid interfaces to determine the fluxes between the surface and bulk phases and formulate the corresponding constitutive equations. We will also make simplifications of the resulting equations based on an analysis of the characteristic values of the parameters involved to arrive at the simplest model that captures the physics we are interested in.

Physically, by a 'liquid-solid interface' we will understand a thin layer of *liquid* adjacent to the solid surface which experiences nonsymmetric action from the bulk phases via intermolecular forces and hence displays 'surface' properties.[3] Similarly, a 'liquid-gas interface' is the corresponding layer of liquid adjacent to the space occupied by the gas. Thus, the common feature of the two interfaces essential for our modelling is that both are formed by molecules of the liquid only. For definiteness, we will consider fluid 1 to be the liquid and fluid 2 will later become a solid or a gas. The boundary condition on the liquid-gas and liquid-solid interfaces will of course be different.

Thus, since there is no flux of matter between fluid 2 into the surface phase, we have

$$(\mathbf{u}^- - \mathbf{v}^s) \cdot \mathbf{n} = 0. \tag{4.21}$$

The key assumption about the thermodynamics of the surface phase is as follows. We will assume the surface phase to be a two-parametric thermodynamic system with Gibbs's formula given by

$$dU_v^s = T^s \, dS_v^s + \Psi^s \, d\rho^s, \tag{4.22}$$

where T^s, S_v^s and Ψ^s are the temperature, entropy and chemical potential of the surface phase. For this equation to specify thermodynamic properties of the surface phase one has to actually prescribe U_v^s as a function of S_v^s and ρ^s. If U_v^s is prescribed, say, as a function of T^s and ρ^s, it will no longer be a thermodynamic potential and, to fully specify thermodynamics of the system, additional constitutive equations will be needed. A similar situation one has, for example, with a thermodynamically perfect gas, where the internal energy (per unit mass) is assumed to be a function of temperature only. It should be reminded here that the subscript v in (4.22) indicates that we are considering the surface quantities of a part of the interface contained in a unit volume.

It is important to emphasize that equation (4.22) incorporates several assumptions. Firstly, in general, the state of an interface could be determined by a larger number of parameters due, for example, to the anisotropy of its structure. The corresponding generalization is rather straightforward, and we will not consider it here. Secondly, once the interface is out of equilibrium, to fully specify its state one might need not only ρ^s and T but also a number of additional 'dynamic' parameters. However, the same argument applies also to the 'fluid particle' in the bulk where we use the same 'hypothesis of local equilibrium' (§2.3.2). Therefore, one can expect that (4.22) holds under the same conditions as the equations of state used in the bulk phases. One can say that (4.22) as well as the corresponding expressions in the bulk *define* the media with an interface as a thermodynamic system. As always, the accuracy of

---

[3]It should be noted that, as described in §§2.4.3–2.4.4, the solid also experiences action of intermolecular forces from the liquid and hence, physically, an interfacial layer between the two media comprises two sublayers adjacent to the liquid-solid border and formed by the molecules of the liquid and the solid, respectively. Here we assume the solid to be undeformable thus neglecting the effect of the liquid on it. If deformations in the solid phase are important (e.g., Extrand & Kumagai 1996), the part of the interface lying in the solid as well as the elasticity for the bulk of the solid will have to be brought into consideration.

the resulting model will be determined by how well it describes experimental data.

By definition, we have

$$\rho^s \Psi^s = U_v^s - T^s S_v^s - \sigma. \tag{4.23}$$

Thermodynamically, the surface tension, σ, is therefore defined as the 'surface pressure' taken with the negative sign, and hence (4.23) is similar to equation (2.130) relating thermodynamic potentials and parameters of state in the bulk. It should be noted that now ρ^s is the surface density of the molecules of fluid 1 and there can be mass exchange between the two phases.

We will assume that in the bulk fluid 1 is also a two-parametric thermodynamic system where

$$dU_v^+ = T^+ dS_v^+ + \Psi_m^+ d\rho^+ \tag{4.24}$$

is Gibbs's formula and, by definition,

$$U_v^+ = \rho^+ \Psi_m^+ + T^+ S_v^+ - p^+. \tag{4.25}$$

4.3.1 Dissipative mechanisms

Now, we can start deriving the entropy balance equation. After using (4.21), (4.22), (4.10), (4.23), (4.25) and a simple identity

$$\mathbf{n} \cdot \mathbf{P}^+ \cdot (\mathbf{u}^+ - \mathbf{v}^s) - \mathbf{n} \cdot \mathbf{P}^- \cdot (\mathbf{u}^- - \mathbf{v}^s) = \tfrac{1}{2}\mathbf{n} \cdot (\mathbf{P}^+ + \mathbf{P}^-) \cdot (\mathbf{u}^+ - \mathbf{u}^-)$$

$$-\mathbf{n} \cdot (\mathbf{P}^+ - \mathbf{P}^-) \cdot \left[\mathbf{v}^s - \tfrac{1}{2}(\mathbf{u}^+ + \mathbf{u}^-)\right], \tag{4.26}$$

we can write down equation (4.20) in the form

$$T^s \frac{\partial S_v^s}{\partial t} + T^s \nabla \cdot (\mathbf{v}^s S_v^s) + T^+ S_v^+ (\mathbf{u}^+ - \mathbf{v}^s) \cdot \mathbf{n} + \tfrac{1}{2}\rho^+(\mathbf{u}^+ - \mathbf{v}^s)^2(\mathbf{u}^+ - \mathbf{v}^s) \cdot \mathbf{n}$$

$$+\sigma \nabla \cdot \mathbf{v}^s - \mathbf{P}^s : \nabla \mathbf{v}^s + (\Psi_m^+ - \Psi^s)\rho^+(\mathbf{u}^+ - \mathbf{v}^s) \cdot \mathbf{n} - p^+(\mathbf{u}^+ - \mathbf{v}^s) \cdot \mathbf{n}$$

$$+\mathbf{n} \cdot (\mathbf{P}^+ - \mathbf{P}^-) \cdot \left[\mathbf{v}^s - \tfrac{1}{2}(\mathbf{u}^+ + \mathbf{u}^-)\right] - \tfrac{1}{2}\mathbf{n} \cdot (\mathbf{P}^+ + \mathbf{P}^-) \cdot (\mathbf{u}^+ - \mathbf{u}^-)$$

$$+(\mathbf{q}^+ - \mathbf{q}^-) \cdot \mathbf{n} + \nabla \cdot \mathbf{q}^s = 0. \tag{4.27}$$

This equation already has some terms which we can attribute to the reversible entropy flux and the source of entropy. However, many terms still have to be dealt with to clarify their physical meaning. Before doing this, let us first consider some auxiliary relationships.

First, given (4.4) and the symmetry of $\mathbf{P}^s$, one has

$$\mathbf{P}^s : \nabla \mathbf{v}^s - \sigma \nabla \cdot \mathbf{v}^s = \mathbf{P}^s : \nabla \mathbf{v}^s - \sigma(\mathbf{I} - \mathbf{nn}) : \nabla \mathbf{v}^s$$

$$= \tfrac{1}{2}\left[\mathbf{P}^s - \sigma(\mathbf{I} - \mathbf{nn})\right] : \left[\nabla \mathbf{v}^s + (\nabla \mathbf{v}^s)^*\right]$$

$$\equiv \left[\mathbf{P}^s - \sigma(\mathbf{I} - \mathbf{nn})\right] : \mathbf{E}^s. \tag{4.28}$$

The last expression can be interpreted as work per unit time done by surface viscous stresses represented by the tensor $[\mathbf{P}^s - \sigma(\mathbf{I} - \mathbf{nn})]$ on the velocity field described by the surface rate-of-strain tensor $\mathbf{E}^s = \frac{1}{2}[\nabla \mathbf{v}^s + (\nabla \mathbf{v}^s)^*]$. This is analogous to $\boldsymbol{\sigma} : \mathbf{E}$ for the work of viscous stresses in the bulk we had in (2.122).

Using (4.21), we also have

$$\frac{1}{2}\mathbf{n} \cdot (\mathbf{P}^+ + \mathbf{P}^-) \cdot (\mathbf{u}^+ - \mathbf{u}^-) - \mathbf{n} \cdot (\mathbf{P}^+ - \mathbf{P}^-) \cdot \left[\mathbf{v}^s - \frac{1}{2}(\mathbf{u}^+ + \mathbf{u}^-)\right]$$

$$= \frac{1}{2}\mathbf{n} \cdot (\mathbf{P}^+ + \mathbf{P}^-) \cdot (\mathbf{I} - \mathbf{nn}) \cdot (\mathbf{u}_\parallel^+ - \mathbf{u}_\parallel^-) - \mathbf{n} \cdot (\mathbf{P}^+ - \mathbf{P}^-) \cdot (\mathbf{I} - \mathbf{nn}) \cdot \left[\mathbf{v}_\parallel^s - \frac{1}{2}(\mathbf{u}_\parallel^+ + \mathbf{u}_\parallel^-)\right]$$

$$+ \mathbf{n} \cdot \mathbf{P}^+ \cdot \mathbf{n}(\mathbf{u}^+ - \mathbf{v}^s) \cdot \mathbf{n}, \qquad (4.29)$$

where, as in §2.4.2, for brevity we use the notation $\mathbf{u}_\parallel^\pm = \mathbf{u}^\pm \cdot (\mathbf{I} - \mathbf{nn})$ and $\mathbf{v}_\parallel^s = \mathbf{v}^s \cdot (\mathbf{I} - \mathbf{nn})$ for the components of velocities parallel to the interface.

Finally,

$$\mathbf{n} \cdot \mathbf{P}^+ \cdot \mathbf{n}(\mathbf{u}^+ - \mathbf{v}^s) \cdot \mathbf{n} + p^+(\mathbf{u}^+ - \mathbf{v}^s) \cdot \mathbf{n} = \mathbf{n} \cdot \boldsymbol{\sigma}^+ \cdot \mathbf{n}(\mathbf{u}^+ - \mathbf{v}^s) \cdot \mathbf{n}, \quad (4.30)$$

where, as before, $\boldsymbol{\sigma}^+ = \mathbf{P}^+ + p^+ \mathbf{I}$ is the viscous stress in fluid 1, and

$$\nabla \cdot \mathbf{q}^s = T^s \nabla \cdot \left(\frac{\mathbf{q}^s}{T^s}\right) + \frac{1}{T^s}\mathbf{q}^s \cdot \nabla T^s. \qquad (4.31)$$

Prior to proceeding with the derivation we will make a simplifying assumption concerning the temperature distribution. So far, for generality we allowed for the surface temperature T^s being different from the temperatures T^+ and T^- of the bulk phases evaluated at the interface. In other words, we assumed that, although the thickness of the interfacial layer is negligible compared with the characteristic length scales of the bulk phases, the temperature variation across it can be finite. Physically, this is possible if the heat flux across an interface is singularly high or the thermal conductivity of the interfacial layer is singularly low. Then, these singular features would 'compensate' the infinitesimal thickness of the interfacial layer and result in a finite temperature difference across it. In what follows we will exclude this case from our consideration and assume that at the interface

$$T^s = T^+ = T^- = T. \qquad (4.32)$$

This imposes the corresponding limitations on the use of the resulting model: if a solution obtained in its framework has a singularity at the interface in the heat flux distribution, this will indicate that the model is being used outside its limits of applicability and the assumption (4.32) will have to be revisited.

Now, using (4.28)–(4.32), we can write (4.27) down as

$$\frac{\partial S_v^s}{\partial t} + \nabla \cdot \left(\mathbf{v}^s S_v^s + \frac{\mathbf{q}^s}{T}\right) + S_v^+(\mathbf{u}^+ - \mathbf{v}^s) \cdot \mathbf{n}$$

$$+ \left(\frac{\mathbf{q}^+}{T} - \frac{\mathbf{q}^-}{T} \right) \cdot \mathbf{n} + \frac{\rho^+ (\mathbf{u}^+ - \mathbf{v}^s)^2}{2T} (\mathbf{u}^+ - \mathbf{v}^s) \cdot \mathbf{n} = \dot{S}^s \qquad (4.33)$$

where

$$\begin{aligned}
\dot{S}^s = {}& \frac{1}{2T} [\mathbf{P}^s - \sigma(\mathbf{I} - \mathbf{nn})] : [\nabla \mathbf{v}^s + (\nabla \mathbf{v}^s)^*] \\
& + \frac{1}{T} \rho^+ (\mathbf{u}^+ - \mathbf{v}^s) \cdot \mathbf{n} \left(\Psi^s - \Psi_m^+ + \frac{1}{\rho^+} \mathbf{n} \cdot \boldsymbol{\sigma}^+ \cdot \mathbf{n} \right) \\
& + \frac{1}{2T} \mathbf{n} \cdot (\mathbf{P}^+ + \mathbf{P}^-) \cdot (\mathbf{I} - \mathbf{nn}) \cdot (\mathbf{u}_\parallel^+ - \mathbf{u}_\parallel^-) \\
& - \frac{1}{T} \mathbf{n} \cdot (\mathbf{P}^+ - \mathbf{P}^-) \cdot (\mathbf{I} - \mathbf{nn}) \cdot [\mathbf{v}_\parallel^s - \tfrac{1}{2}(\mathbf{u}_\parallel^+ + \mathbf{u}_\parallel^-)] \\
& - \frac{1}{T^2} \mathbf{q}^s \cdot \nabla T, \qquad (4.34)
\end{aligned}$$

is the entropy production in the surface phase. As in §2.3.2, we will make $\dot{S}^s$ nonnegative by presenting the terms on the right-hand side of (4.34) as products of thermodynamic forces and fluxes and assuming that the fluxes are proportional to the forces to make $\dot{S}^s$ a positively determined quadratic form. Being interested in the principal effects, we will make a further simplification and neglect all cross-coefficients thus making the thermodynamic fluxes proportional only to their 'own' thermodynamic forces.

The first term on the right-hand side of (4.34) is completely analogous to the corresponding term in (2.121) and, given the two-dimensional isotropy of the surface phase, leads to a two-dimensional analogue of equation (2.64):

$$\mathbf{P}^s = \sigma(\mathbf{I} - \mathbf{nn}) + \lambda^s (\mathbf{I} - \mathbf{nn}) \nabla \cdot \mathbf{v}^s + \mu^s [\nabla \mathbf{v}^s + (\nabla \mathbf{v}^s)^*]. \qquad (4.35)$$

Here, λ^s and μ^s are the surface viscosity coefficients analogous to λ and μ in the bulk phase. Equation (4.35) recovers the rheological properties of the 'Newtonian interface' considered by Scriven (1960).

The next three terms yield:

$$\rho^+ (\mathbf{u}^+ - \mathbf{v}^s) \cdot \mathbf{n} = k_\rho \left(\Psi^s - \Psi_m^+ + \frac{1}{\rho^+} \mathbf{n} \cdot \boldsymbol{\sigma}^+ \cdot \mathbf{n} \right), \qquad (4.36)$$

$$\tfrac{1}{2} \mathbf{n} \cdot (\mathbf{P}^+ + \mathbf{P}^-) \cdot (\mathbf{I} - \mathbf{nn}) = \beta (\mathbf{u}_\parallel^+ - \mathbf{u}_\parallel^-), \qquad (4.37)$$

$$\mathbf{n} \cdot (\mathbf{P}^+ - \mathbf{P}^-) \cdot (\mathbf{I} - \mathbf{nn}) = -\alpha^{-1} [\mathbf{v}_\parallel^s - \tfrac{1}{2}(\mathbf{u}_\parallel^+ + \mathbf{u}_\parallel^-)]. \qquad (4.38)$$

Here k_ρ, β and α are positive coefficients depending, in a general case, on the parameters determining the state of the interface.

Finally, the last term in (4.34) gives the surface analogue of the Fourier law (2.124)

$$\mathbf{q}^s = -\kappa^s \nabla T \cdot (\mathbf{I} - \mathbf{nn}), \qquad (4.39)$$

where κ^s is the thermal conductivity of the interface. Here we have taken into account that $\mathbf{q}^s$ is a surface vector, whereas T is defined and continuous everywhere in space.

Now, to close the system of equations for the surface phase it is necessary only to specify one of the thermodynamic potentials, for example, $U_v^s(\rho^s, T)$, which for the general case is a formidable problem in its own right. It should also be noted that the formally derived model we considered so far, after it is completed with a suitable thermodynamic potential, is unworkable. Although we made several simplifications on the way, the model still leaves us with a formulation which, if applied to a real physical situation, will include terms differing by several orders of magnitude in the same equations and will be experimentally unverifiable. In order to arrive at an irreducible self-consistent model we will consider further simplifications based on the order-of-magnitude analysis of the physical effects that the formal model derived above takes into account. As a working test we will use a qualitative criterion that the physical meaning of the resulting equations should be clear even without looking back on their derivation.

4.3.2 Estimates and further simplifications

In order to make simplifications, we need to make assumptions about the interfacial layer. In most simple fluids, the thickness of this layer ℓ can be estimated at $\ell \sim 1\text{–}3$ nm, and we will assume that microscopically all parameters there, apart from the stress that, macroscopically, results in the surface tension, are of the same order of magnitude as the corresponding quantities in the bulk. As is always the case in the fluid mechanics modelling, the simplifications we will make below as well as those made above are subject to the a posteriori assessment of validity.

Let U be the characteristic velocity and L be the characteristic length scale in the bulk and along the interface. In equation (4.35) the surface viscosity coefficients can be estimated as $\lambda^s, \mu^s \sim \ell\mu$ (μ is the dynamic viscosity of fluid 1) so that the ratio of the last two terms on the right-hand side to the first one is of order $(\ell/L)\,Ca$, where $Ca = \mu U/\sigma$ is the capillary number. For capillary numbers of $O(1)$ and $L \sim 1$ mm one has $(\ell/L)\,Ca < 10^{-5}$ so that the two-dimensional surface phase can be regarded as hydrodynamically ideal, i.e.,

$$\mathbf{P}^s = \sigma(\mathbf{I} - \mathbf{nn}), \tag{4.40}$$

as in classical fluid mechanics (see equation (2.140) in §2.4.2). The physical origin of this simplification is quite obvious: the surface tension results from the strong forces of nonhydrodynamic origin that act on the interfacial layer from the bulk phases and magnify its dynamic role whereas the terms associated with the surface viscosity are of hydrodynamic nature and hence proportional to the interfacial layer's thickness, which is negligible in the continuum limit.

In equation (4.36) we can compare the last two terms in brackets on the right-hand side. Using the ratio of characteristic pressure to characteristic density p/ρ as an estimate for Ψ_m^+, the ratio of the last two terms has the order of $\mu U/(Lp)$. For $\mu \sim 1$–10^3 mPa s, $U \sim 10$ cm s^{-1}, $L \sim 1$ mm and the atmospheric pressure $p \sim 10^5$ Pa, one has $\mu U/(Lp) \leq 10^{-4}$ and hence may neglect the hydrodynamic contribution compared to that from the chemical potential. This simplification is bringing in a certain limit of applicability of the resulting model: if in using the model one arrives at a solution where the neglected hydrodynamic factors are singular, these factors will have to be brought back into the picture. At this stage, however, the simplifying assumption we are making is justified since it is a well known fact in physical chemistry that in the circumstances studied to date adsorption-desorption processes are driven by the difference in the chemical potentials and not by hydrodynamic stresses.

In equilibrium, we have

$$\Psi^s(\rho_e^s, T) = \Psi_m^+(T, \rho^+),$$

where ρ_e^s is the equilibrium surface density. In a general case, it depends on the thermodynamic parameters specifying the equilibrium state, in our case T and ρ^+ as well as on the materials of the bulk phases. If, as we will assume, the fluid in the bulk is incompressible, then the bulk phase is a one-parametric system so that ρ_e^s, besides depending on the materials of the system, becomes a function of temperature only. For many processes where the variations of temperature are sufficiently small ρ_e^s can be regarded as a material constant for a given liquid-gas (or liquid-solid) interface.

Now, assuming that Ψ^s can be expanded in a Taylor series about ρ_e^s, we have that to leading order the right-hand side of (4.36) can be written down as

$$\rho^+(\mathbf{u}^+ - \mathbf{v}^s) \cdot \mathbf{n} = (\rho^s - \rho_e^s)\tau^{-1}, \tag{4.41}$$

where

$$\tau = \frac{1}{k_\rho \left(\dfrac{\partial \Psi^s}{\partial \rho^s}\right)_T} \tag{4.42}$$

is a relaxation time. In obtaining (4.41) we assumed that the deviation of ρ^s from ρ_e^s is in some sense small and the derivative in (4.42) is taken at the point (ρ_e^s, T) in the parameter space. However, we can relax these constraints by postulating the functional form of (4.41) and treating τ as a phenomenological parameter (instead of k_ρ), whose value and possible dependence on the parameters of state must be determined experimentally.[4] As we will see later, in many cases τ can be treated as a material constant, which makes the functional form (4.41) convenient for practical applications.

---

[4]This phenomenological extension of the obtained formula is similar to the extension of the Newtonian rheology in the bulk, derived assuming that the rate of strain is in some sense small, to finite rates of strain.

Using (4.41) as well as (4.21), we can now write down the surface mass balance equation (4.10) in the form

$$\frac{\partial \rho^s}{\partial t} + \nabla \cdot (\rho^s \mathbf{v}^s) = -\frac{\rho^s - \rho^s_e}{\tau}. \tag{4.43}$$

This equation shows that, in the absence of motion, the surface density relaxes exponentially to its equilibrium value. It is important to emphasize once again that the value of ρ^s_e (and, in a general case, of τ) will be different for the liquid-gas and liquid-solid interface. Below, where necessary, we will use the subscripts 1 to mark the parameters of the liquid-solid interface.

Equation (4.37) states that the tangential velocity difference across an interfacial layer is proportional to an external torque acting on it. This equation illustrates the arguments we considered in §2.4.2 and is behind the boundary conditions (2.148) and (2.161) used in classical fluid mechanics. Indeed, for the coefficient of sliding friction β one has an estimate $\beta \sim \mu/\ell$ since (4.37) is analogous to what one has for the Couetta flow in a channel of width ℓ of a fluid with viscosity μ. Then, the ratio of the terms on the left-hand side and on the right-hand side of (4.37) is of order ℓ/L, which is negligibly small for a macroscopic L, and hence we end up with $\mathbf{u}^+_\parallel \approx \mathbf{u}^-_\parallel$. However, by using this condition one arrives at a singularity in the tangential stress (which can be interpreted as $L \to 0$ in the above estimate), the condition of continuity of the tangential velocity becomes unjustified and it becomes necessary to use (4.37) as it stands.

Now, we can turn to the surface momentum balance equation (4.12), where, given (4.21), we drop the last term on the left-hand side. (For simplicity we will also drop $\rho^s \mathbf{F}^s$; if necessary, this term can be restored.) If ρ is the characteristic density in the bulk, then for the surface density one can use an estimate $\rho^s \sim \rho\ell$. Then, the ratio of the convective term $\rho^s D^s \mathbf{v}^s/Dt$ in (4.12) to $\nabla \cdot \mathbf{P}^s$, which, using (4.40) can be written as

$$\nabla \cdot \mathbf{P}^s = \nabla \sigma - \sigma \mathbf{n} \nabla \cdot \mathbf{n}, \tag{4.44}$$

has the order of $\ell \rho U^2/\sigma$. For the typical conditions of capillary flows ($U \sim 1$ m s^{-1} or less), $\rho \sim 1$ g cm^{-3} and $\sigma \sim 10^2$ mN m^{-1} one has $\ell \rho U^2/\sigma \leq 10^{-4}$. Thus, the surface convective momentum can be neglected compared with capillary effects.

In order to estimate the relative importance of the second term in (4.12), we need to consider the normal and tangential components of this equation and compare

$$\rho^+[(\mathbf{u}^+ - \mathbf{v}^s) \cdot \mathbf{n}]^2 \quad \text{with} \quad \sigma \nabla \cdot \mathbf{n} \tag{4.45}$$

and

$$\rho^+(\mathbf{u}^+_\parallel - \mathbf{v}^s_\parallel)(\mathbf{u}^+ - \mathbf{v}^s) \cdot \mathbf{n} \quad \text{with} \quad \nabla \sigma. \tag{4.46}$$

Here we used (4.44) to decompose $\nabla \cdot \mathbf{P}^s$ into the normal and tangential components. Now, we must use as characteristic the length scale associated with the relaxation of the surface properties to their equilibrium values $l = U\tau$, where τ is the relaxation time introduced earlier. Using (4.41) for the normal flux and (4.37) for the tangential velocity difference, we have that the ratios of the first terms in (4.45) and (4.46) to the second ones are in both cases of order $\rho U \ell^2 / (\tau\sigma)$. Experiments show (Blake & Shikhmurzaev 2002; see §§5.3.3, 5.4.1) that τ is proportional to viscosity in the bulk. This is actually what one would expect, given that τ arises from a mechanism which is essentially of diffusive nature. Even taking the lowest estimate for τ ($\sim 10^{-9}$ s, Blake & Shikhmurzaev 2002), one arrives at $\rho U \ell^2 / (\tau\sigma) \leq 10^{-4}$ and hence may neglect the contribution of the second term in (4.12) to the momentum balance of the surface phase.

Thus, given the above estimates, the momentum balance equation (4.12) takes the standard form

$$\nabla\sigma - \sigma \mathbf{n}\nabla \cdot \mathbf{n} + \mathbf{n} \cdot (\mathbf{P}^+ - \mathbf{P}^-) = 0. \tag{4.47}$$

Using this equation we can now write (4.38) down as

$$\mathbf{v}_\parallel^s = \tfrac{1}{2}(\mathbf{u}_\parallel^+ + \mathbf{u}_\parallel^-) + \alpha\nabla\sigma. \tag{4.48}$$

One can also use equation (4.47) to eliminate $\mathbf{P}^-$ from (4.37) to arrive at a generalization of condition (3.34) of §3.4.1:

$$\mathbf{n} \cdot \mathbf{P}^+ \cdot (\mathbf{I} - \mathbf{nn}) + \tfrac{1}{2}\nabla\sigma = \beta(\mathbf{u}_\parallel^+ - \mathbf{u}_\parallel^-). \tag{4.49}$$

This is known as a *generalized Navier condition*.

Another useful formula that we will need later can be obtained by eliminating $\mathbf{P}^+$ and $\mathbf{u}^-$ from (4.37), (4.47) and (4.48) to give

$$(1 + 4\alpha\beta)\nabla\sigma - 2\mathbf{n} \cdot \mathbf{P}^- \cdot (\mathbf{I} - \mathbf{nn}) = 4\beta(\mathbf{v}_\parallel^s - \mathbf{u}_\parallel^+). \tag{4.50}$$

This equation relates the bulk velocity on one side of an interface with the tangential stress on the other.

4.3.3 Surface equation of state

In order to close the model, one has to specify the surface equations of state, say, via one of the thermodynamic potentials. Such specification together with the transport and dissipative mechanisms considered above *defines* the surface phase as a thermodynamic system similarly to the way in which the equations of state for the 'fluid particle' and the mechanisms of dissipation define thermodynamics of the bulk phase. The model as a whole has then to be validated against experimental data.

If thermodynamic properties of the surface phase are specified by prescribing, say, the free energy $F_v^s = F_v^s(\rho^s, T)$, then one can calculate all other parameters using the known thermodynamic identities and definitions. In particular, the surface tension can be expressed as $\sigma(\rho^s, T) = F_v^s - \rho^s (\partial F_v^s / \partial\rho^s)_T$, and it is a function of the same parameters that are assumed to determine the state of the interface. For a general process involving the surface phase one has also to specify thermal properties of the bulk phases that determine $q^{\pm}$ in the entropy balance equation (4.33).

Ultimately, the equations of state for the surface phase as a two-parametric thermodynamic system have to be determined experimentally. Elements of the structural analysis of the interfacial layer can help to narrow down the variety of functional forms these equations may take though one would face many of the issues inherent in the structural approach to the modelling of interfaces we discussed in §4.1. A possible way of resolving the problem is to employ molecular dynamics simulations and use the macroscopic dissipative mechanisms considered in the previous section as a test ensuring compatibility of the two approaches. However, so far there are no conclusive results obtained on this route.

The problem of closure is considerably simplified if, instead of aiming at a model intended to describe all kinds of thermodynamic processes involving the surface phase, we consider a particular one. Then, in the plane of the surface parameters of state, say, (ρ^s, T), this process will corresponds to a curve, which can be parametrized using ρ^s as the parameter. As a result, instead of the unknown general dependence $\sigma = \sigma(\rho^s, T)$ suitable for all processes, one will have $\sigma = \sigma(\rho^s)$ along this curve. In other words, we will assume that the process of the interface formation we are interested in can be modelled as a 'surface barotropic' process similar to the barotropic processes in gas dynamics (§2.2.4).

Qualitatively, the form of the function $\sigma = \sigma(\rho^s)$ can be constructed by considering the interface in a few characteristic situations. At some surface density ρ_e^s corresponding to the equilibrium state of the interfacial layer between a liquid and a gas one has a certain equilibrium value of the liquid-gas surface tension $\sigma_e \equiv \sigma(\rho_e^s) > 0$. The value of ρ_e^s is lower than the surface density one would have for a layer of the same thickness in the bulk since, as is known from both experiments and molecular dynamic simulations (Fig. 3.3), in contact with a gas the liquid becomes rarefied. The surface density in the interfacial layer between the liquid and a solid is higher than ρ_e^s, and the matter in this layer is often compressed: this is the case when the liquid wets the surface and the (static) contact angle formed by the free surface with the solid boundary is much less than 90°. For the compressed interfacial layer one has a negative surface tension, effectively becoming a 'surface pressure'.[5] Thus, for the equilibrium surface density in the liquid-solid interface $\rho_{1e}^s > \rho_e^s$ one has $\sigma_{1e} \equiv \sigma(\rho_{1e}^s) < \sigma(\rho_e^s)$. Obviously, for $\rho^s = 0$ one has $\sigma(0) = 0$. The

---

[5]It is necessary to remember that here, as before, by the 'liquid-solid interface' we understand the layer of the liquid adjacent to the solid boundary, not the whole interface which comprises also a layer in the solid adjacent to the boundary (see Fig. 2.13 and 2.15).

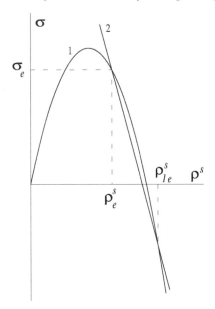

FIGURE 4.3:   Qualitative dependence of the surface tension on the surface density for barotropic processes.

simplest dependence satisfying conditions

$$\sigma(0) = 0, \quad \sigma(\rho_e^s) > 0, \quad \sigma(\rho_{1e}^s) < \sigma(\rho_e^s) \quad \text{for } \rho_{1e}^s > \rho_e^s$$

is a parabola shown in Fig. 4.3, curve 1, i.e.,

$$\sigma = a\rho^s - b(\rho^s)^2, \tag{4.51}$$

where

$$a = \frac{(\rho_{1e}^s)^2 \sigma_e - (\rho_e^s)^2 \sigma_{1e}}{\rho_e^s \rho_{1e}^s (\rho_{1e}^s - \rho_e^s)}, \qquad b = \frac{\rho_{1e}^s \sigma_e - \rho_e^s \sigma_{1e}}{\rho_e^s \rho_{1e}^s (\rho_{1e}^s - \rho_e^s)}.$$

The equation of state (4.51) is a simple way of accounting for the decrease in the surface tension from its equilibrium value for a liquid-gas interface as a result of compression or extreme rarefaction of the surface phase.

For the processes where the surface density varies between ρ_e^s and ρ_{1e}^s it is convenient to use a linear equation of state (Fig. 4.3)

$$\sigma = \gamma(\rho_{(0)}^s - \rho^s), \tag{4.52}$$

which is the simplest way of reflecting the general tendency that, in this region, compression of the surface phase corresponds to a decrease in the surface tension. The parameter γ is a phenomenological constant describing compressibility of the liquid. It should be noted that both (4.51) and (4.52) are intended to describe only the main effect, i.e., the qualitative correlation between σ and ρ^s in the processes of interface formation, so that it is impossible to assess their accuracy a priori. As we will see later, the model incorporating these equations provides a surprisingly accurate description of the main features observed in such processes experimentally and at the moment there is no experimental indication of how these simple equations of state could be refined.

The equations of state (4.51) or (4.52) close the set of surface equations, which are also the boundary conditions for the equations describing the bulk phases, and allow one to find the parameters without considering the internal energy balance (4.20) and the entropy balance (4.33). The barotropic closure also implies that no energy and entropy balance conditions are required at the contact lines (see §4.3.5).

In the next section, we will bring the results together and formulate the simplest model that takes into account the processes of interface formation.

4.3.4   The simplest (irreducible) model

In this section, for the Navier-Stokes equations,

$$\nabla \cdot \mathbf{u} = 0, \quad \rho \left(\frac{\partial \mathbf{u}}{\partial t} + \mathbf{u} \cdot \nabla \mathbf{u} \right) = -\nabla p + \mu \nabla^2 \mathbf{u}, \qquad (4.53)$$

describing the bulk flow we will formulate the boundary conditions at liquid-gas and liquid-solid interfaces in the simplest case of an inviscid gas at constant pressure and a smooth chemically homogeneous solid surface. The simplicity of these two cases comes from the fact that the problem has to be solved only for the equations describing the liquid; on the gas-facing side of the liquid-gas interface one has a prescribed stress (constant in the normal direction and zero in the tangential) and the velocity, if necessary, can be found separately, whereas on the solid-facing side of the liquid-solid interface the velocity is prescribed and the stress, if necessary, can be found.

Using the following expressions for the stress tensors in the gas and the liquid, $\mathbf{P}^- = -p_g \mathbf{I}$ (here p_g is the pressure in the gas) and $\mathbf{P}^+ = -p\mathbf{I} + \mu[\nabla \mathbf{u} + (\nabla \mathbf{u})^*]$, we can write down equations (4.2), (4.47), (4.41), (4.43), (4.50), (4.51) as boundary conditions on the liquid-gas interface:

$$\frac{\partial f}{\partial t} + \mathbf{v}^s \cdot \nabla f = 0, \qquad (4.54)$$

$$p_g - p + \mu \mathbf{n} \cdot [\nabla \mathbf{u} + (\nabla \mathbf{u})^*] \cdot \mathbf{n} = \sigma \nabla \cdot \mathbf{n}, \qquad (4.55)$$

$$\mu \mathbf{n} \cdot [\nabla \mathbf{u} + (\nabla \mathbf{u})^*] \cdot (\mathbf{I} - \mathbf{nn}) + \nabla \sigma = 0, \qquad (4.56)$$

$$\rho(\mathbf{u} - \mathbf{v}^s) \cdot \mathbf{n} = \frac{\rho^s - \rho_e^s}{\tau}, \qquad (4.57)$$

$$\frac{\partial \rho^s}{\partial t} + \nabla \cdot (\rho^s \mathbf{v}^s) = -\frac{\rho^s - \rho_e^s}{\tau}, \qquad (4.58)$$

$$(1 + 4\alpha\beta)\nabla\sigma = 4\beta(\mathbf{v}_\parallel^s - \mathbf{u}_\parallel), \tag{4.59}$$

$$\sigma = a\rho^s - b(\rho^s)^2. \tag{4.60}$$

The last equation can be replaced with (4.52), i.e.,

$$\sigma = \gamma(\rho_{(0)}^s - \rho^s), \tag{4.61}$$

if the process to be modelled does not involve extreme rarefaction (i.e., very rapid expansion) of the surface phase.

Equations (4.54)–(4.56) are the standard ones (see (2.132), (2.154), (2.146)); they determine the evolution of the free surface and express the balance the normal and tangential stress acting on it. It should be noted, however, that, unlike the standard model, where the liquid-gas interface is a material surface consisting of the same 'fluid particles' at all time and hence has the normal velocity equal to the normal velocity of the fluid evaluated at the interface (see (2.136)), the situation here is different. Indeed, equation (4.57) takes into account mass exchange between the bulk and the surface phase as a boundary condition for the normal component of the bulk velocity, whereas (4.58) accounts for this mass exchange in the mass balance for the surface density. Equation (4.59) shows that the difference between the tangential components of $\mathbf{v}^s$ and $\mathbf{u}$ is proportional to the surface tension gradient, which, according to (4.56), together with the tangential stress forms a torque that creates the velocity difference across the interfacial layer.

Now, consider a smooth solid surface moving with a given velocity $\mathbf{U}$. To obtain the boundary conditions for (4.53) on (the liquid-facing side of) the liquid-solid interface we need to put $\mathbf{u}^- = \mathbf{U}$, where required, in (4.21), (4.49), (4.41), (4.43), (4.48) and (4.51) ending up with:

$$(\mathbf{v}^s - \mathbf{U}) \cdot \mathbf{n} = 0, \tag{4.62}$$

$$\mu\mathbf{n} \cdot [\nabla\mathbf{u} + (\nabla\mathbf{u})^*] \cdot (\mathbf{I} - \mathbf{n}\mathbf{n}) + \tfrac{1}{2}\nabla\sigma = \beta_1(\mathbf{u}_\parallel - \mathbf{U}_\parallel), \tag{4.63}$$

$$\rho(\mathbf{u} - \mathbf{v}^s) \cdot \mathbf{n} = \frac{\rho^s - \rho_{1e}^s}{\tau_1}, \tag{4.64}$$

$$\frac{\partial\rho^s}{\partial t} + \nabla \cdot (\rho^s\mathbf{v}^s) = -\frac{\rho^s - \rho_{1e}^s}{\tau_1}, \tag{4.65}$$

$$\mathbf{v}_\parallel^s = \tfrac{1}{2}(\mathbf{u}_\parallel + \mathbf{U}_\parallel) + \alpha_1\nabla\sigma, \tag{4.66}$$

$$\sigma = a\rho^s - b(\rho^s)^2. \tag{4.67}$$

In most cases, the last equation can be replaced with the linear one:

$$\sigma = \gamma(\rho_{(0)}^s - \rho^s). \tag{4.68}$$

As shown by its derivation at the end of §4.3.2, the generalized Navier condition (4.63) results from two premises:

(i) The velocity difference across an interfacial layer being proportional to the external torque acting on it, as described by (4.37);

(ii) The sum of all forces acting on the interface being equal to zero, as described by (4.47).

Then, after eliminating the drag force acting on the interfacial layer from the solid, one arrives at (4.63). This condition shows that there can be *apparent* slip on a solid substrate, i.e., a difference between the tangential component of the bulk velocity **u** on the *liquid*-facing side of the liquid-solid interface and the corresponding component of the velocity of the solid surface. At the same time, the derivation of (4.63) used an assumption that there is no *actual* slip on the solid surface so that the velocity on the *solid*-facing side of the liquid-solid interface is equal to that of the solid, $\mathbf{u}^- = \mathbf{U}$. This is complete agreement with the results of molecular dynamics simulations reviewed in §3.3.

One should have in mind, however, that, potentially, there can be situations where the assumption of no-slip on the solid surface might have to be revised to incorporate actual slip ($\mathbf{u}_\parallel^- \neq \mathbf{U}_\parallel$) and/or chemical reactions between the liquid and the solid ($\mathbf{u}^- \cdot \mathbf{n} \neq \mathbf{U} \cdot \mathbf{n}$).

Importantly, the generalized Navier condition (4.63), as well as all other boundary conditions we formulated on the solid boundary and the free surface, is (a) *local*, i.e., involving only the values of parameters and their gradients at the points where this condition is applied, and (b) *universal*, i.e., not specific to any particular problem it can be applied to. In this connection, it is worth mentioning another generalization of the Navier condition announced recently by Qian, Wang & Sheng (2003, 2006) which, by contrast, is neither local nor universal. This new condition assumes that "the slip velocity is proportional to the total tangential stress — the sum of the viscous stress and the uncompensated Young stress; the latter arises from the deviation of the fluid-fluid interface from its static configuration" (Qian et al. 2006, p. 335). This means that the proposed condition is (a) *ad hoc* as it is intended for a particular problem (moving contact-line problem), and (b) *nonlocal* as it involves the state of the contact line ("uncompensated Young stress") whereas the contact line can be at a considerable distance from the point at which this generalization of the Navier condition is applied. How the contact line is able to instantly 'communicate' about its presence and changes in its state to all points on the solid surface is not clear.[6] Furthermore, unlike the generalized Navier condition (4.63), the condition proposed by Qian et al. deals with *actual* slip on the solid surface whereas, according to MD simulations (§3.3), there is no slip on the solid-facing side of the liquid-solid interface apart from

---

[6]Mathematically, this obvious unphysicality arises from the application of a variational approach, which is inherent in the study of optimization and equilibria, to a dynamic process.

a couple of molecular diameters from the contact line — the distance which is invisible in the continuum modeling.

Conditions (4.64) and (4.65) account for the mass exchange between the bulk and the liquid-solid interface in the process of its formation. Although the solid substrate itself is impermeable for the liquid, for the bulk flow described by (4.53) the boundary is permeable since there can be a normal flux between the bulk and the surface phase. In most situations, however, this flux is negligibly small; it can become important when otherwise a stagnation point in the flow field would be located on the solid surface (see §5.1).

The physical meaning of equation (4.66) is easy to understand by comparing the flow inside an interfacial layer with the Couette-Poiseuille flow between parallel walls moving at different velocities with a superimposed pressure gradient. The velocity of the mean flow in the channel will be given by an identical equation with the role of the surface tension gradient played by the pressure gradient taken with a negative sign. This analogy suggests that $\alpha \propto \ell/\mu \propto \beta^{-1}$.

Naturally, the boundary condition (4.54)–(4.60) and (4.62)–(4.67) have the standard ones as their limiting case. Indeed, as $\tau U/L \to 0$, equation (4.58) gives $\rho^s \equiv \rho^s_e$ and hence, according to (4.60) or (4.61), one has $\sigma \equiv \text{const}$. Then, conditions (4.57) and (4.59) give $\mathbf{u} = \mathbf{v}^s$, which makes (4.54)–(4.56) identical to (2.136), (2.154), (2.155). Similarly, the limit $\tau_1 U/L \to 0$ reduces (4.62)–(4.67) to the Navier condition (3.34), which, given the estimates we had for the coefficient of sliding friction, in the limit $\beta^{-1} \to 0$ reduces to the no-slip condition (2.157).

It is necessary to emphasize that estimates for the phenomenological coefficients involving the ratio ℓ/L in our model do not contradict the continuum approximation: the stresses that otherwise would have been infinite act as a magnifying factor (essentially meaning $L \to 0$); the phenomenological nature of the coefficients also implies that, ultimately, their values have to be determined experimentally.

In (4.54)–(4.60) and (4.62)–(4.67) we allowed for different values of the phenomenological constants α, β and τ for the liquid-gas and liquid-solid interface. The difference can be important in the situations where, for example, the microscopic roughness or inhomogeneity of the solid surface plays a significant role in the dissipative mechanisms associated with these constants. In particular, the formation of the liquid-solid interface depends on the adsorption sites on the solid surface and hence τ_1 can be considerably larger than τ. However, it seems reasonable to start with the simplest possible model and assume that

$$\alpha_1 = \alpha, \qquad \beta_1 = \beta, \qquad \tau_1 = \tau. \tag{4.69}$$

If experiments indicate that (4.69) leads to poor accuracy or incompatibility of results obtained from independent experiments, this assumption will be the first to be revised.

As follows from the derivation, the coefficients characterizing properties of the interfaces depend on the surface parameters of state, and in some situations, especially for rapidly stretched interfaces, their treatment as material properties becomes inappropriate.

It is important to note that the model (4.53)–(4.59), (4.61))–(4.66), (4.69) is the *simplest* one that follows from the macroscopic irreversible thermodynamics of a system with interfaces, and its further simplification would be at the expense of self-consistency. This irreducible model can be seen as a starting point for generalisations which have to be directed by experimental data.

The boundary conditions (4.54)–(4.60) and (4.62)–(4.67) are themselves partial differential equations along the interfaces for the surface variables and to solve them (naturally, together with the bulk equations (4.53)) we need initial and boundary conditions. Apart from the situations where the boundary conditions are formulated 'at infinity' to study local effects, one has to formulate boundary conditions at the lines where the interfaces intersect. As always, these conditions bring in the physics *additional* to that incorporated in the equations operating along the interfaces, and we will consider the main principles of formulating such conditions in the next section.

4.3.4.1 The case of a viscous gas

If the viscosity in the gas phase is to be taken into account, then boundary conditions (4.55), (4.56) and (4.59) have to be replaced with the two projections of (4.47) and equation (4.48), respectively. Then, one needs to bring in the equations of motion for the gas phase which, for small Mach numbers typical for capillary flows, reduce to the same Navier-Stokes equations as in the liquid (4.53) but, naturally, with different values for ρ and μ. It is important to note that, once the gas viscosity is brought into consideration, one will face in the gas phase the same moving contact-line problem associated with the no-slip boundary condition on the solid surface as we had in the liquid phase (see §3.1). However, for the gas this problem has to be addressed differently than for the liquid. The boundary conditions we derived for the equations describing the flow in the liquid, together with conditions one has to formulate at lines of contact of various types (see §4.3.5), allow one to remove the singularities inherent in the standard model. However, one cannot derive similar boundary conditions for the gas phase by considering the dynamics of the solid-gas interface as a layer of gas adjacent to the solid surface: physically, the intermolecular distance in the gas is much larger than the range of intermolecular forces that are responsible for the formation of an interfacial layer and there is no interface as such associated with the gas sublayer adjacent to the solid substrate. To remedy the situation, it becomes necessary to consider the Knudsen effect, i.e., slip due to the finiteness of the ratio of the intermolecular distance in the gas phase to the macroscopic dimension (such as the distance to the moving contact line). As an approximation of this effect

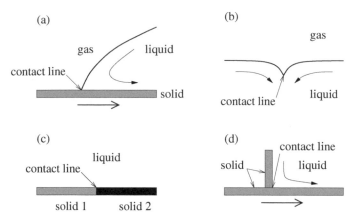

FIGURE 4.4: The main types of contact lines involving liquid-gas and liquid-solid interfaces.

one can use the Navier condtion (3.34).

4.3.5 Conditions at contact lines

From a mathematical viewpoint, 'contact lines' can be defined as boundaries confining regions of interfaces where the distributions of surface parameters are described by a certain set of differential equations. This general definition corresponds to a number of different physical situations. The main types of contact lines involving liquid-gas and liquid-solid interfaces are as follows:

(a) The contact line formed at the intersection of a free surface with a solid boundary (Fig. 4.4a). This situation is central to the wetting processes considered in Chapter 3, and this type of contact line is usually referred to as *the* contact line.

(b) Cusps and corners formed at the free surface (Fig. 4.4b). This situation is typical, for example, to the initial stage of the coalescence processes.

(c) The line formed at the contact of two solids with different physical properties, in particular, with different wettabilities (Fig. 4.4c).

(d) The line formed at the intersection of two solid boundaries (Fig. 4.4d).

Mathematically, conditions at contact lines are boundary conditions for (4.54)–(4.60) and (4.62)–(4.67), i.e., the partial differential equations describing the distributions of the surface parameters along interfaces. In deriving these equations, we narrowed the class of surface phenomena under consideration down to barotropic (in particular, isothermal) processes by using the barotropic surface equation of state (4.51) or (4.52). As a result of this closure, the boundary conditions at contact lines have to operate only with fluxes of mass and momentum (i.e., the balance of forces); no 'extra' condition associated with the energy fluxes can be imposed. This is similar to the situation one has in aerodynamics for the boundary conditions on a shock wave, where the boundary condition on the energy flux would become redundant if the two-parametric equations of state for the gas on both sides of the shock wave are replaced with a barotropic closure making the equations of state one-parametric.

Another point to be made here is that the contact lines sketched in Figs 4.4c and d are, strictly speaking, geometric idealizations. If one zooms in to these contact lines whilst still remaining within the range of length scales associated with continuum mechanics, the picture will change: an unavoidable in practice gap between the two solids in Fig. 4.4d will reduce this contact line to the one shown in Fig. 4.4a, whereas the contact line in Fig. 4.4c will become a finite-size transition zone where the solid surface wettability varies between the values corresponding to the two solids (Sprittles & Shikhmurzaev 2007). Therefore, if one simplifies the problems by introducing the contact lines of the types sketched in Figs 4.4c,d, it is reasonable to expect that a physically adequate model will exhibit some singularities in the physical properties. The key requirement here is that these singularities should be integrable.

We will illustrate the main aspects of formulating boundary conditions at contact lines by using the gas-liquid-solid contact line (Fig. 4.4a) as a representative example.

4.3.5.1 Balance of mass fluxes

The model with a barotropic closure requires boundary conditions formulated in terms of mass and momentum fluxes into and out of the contact line. The form of these conditions will depend on the assumptions one makes about the macroscopic outcome of the physical processes taking place inside the three-phase-interaction region, which in the continuum approximation is modelled as the 'contact line'.

Neglecting the 'line density' of mass in the contact line, we can state that the total of all mass fluxes coming into and going out of the contact line is zero. These fluxes are associated, first of all, with the liquid-gas and liquid-solid interfaces, which in the contact-line motion literally go into or come out of the contact line. Secondly, there can be mass exchange between the contact line and the bulk phase. Finally, there can also be fluxes into and out of the contact line along the gas-solid interface due to microscopic precursor or residual films (see below).

Let $\mathbf{e}_1$ and $\mathbf{e}_2$ be unit vectors normal to the contact line and tangential to the liquid-gas and liquid-solid interfaces, respectively (Fig. 4.5a). Then, in the coordinate frame moving with the contact line the simplest mass balance boundary condition at the contact line is given by

$$(\rho^s \mathbf{v}^s)|_G \cdot \mathbf{e}_1 + (\rho^s \mathbf{v}^s)|_S \cdot \mathbf{e}_2 = 0. \tag{4.70}$$

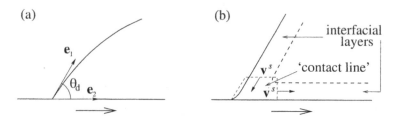

FIGURE 4.5: A sketch of the moving contact line (a) and (b) an illustration of the mass balance boundary condition (4.70), where the 'interfaces' are shown as interfacial layers of a finite thickness and the 'contact line' as a three-phase-interaction zone.

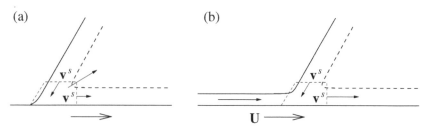

FIGURE 4.6: The moving contact line with a mass exchange between the three-phase-interaction zone and the bulk (a) and with a microscopic film in front of (or, in the motion is reversed, behind) the main body of fluid (b).

Here the subscripts G and S mark the limiting values as the contact line is approached along the free surface and the liquid-solid interface, respectively. Condition (4.70) is schematically illustrated in Fig. 4.5b, where the interfaces are shown as interfacial layers of a finite thickness.

In a more general case, one has to take into account mass exchange between the three-phase-interaction region and the bulk of the fluid (Fig. 4.6a):

$$(\rho^s \mathbf{v}^s)|_G \cdot \mathbf{e}_1 + (\rho^s \mathbf{v}^s)|_S \cdot \mathbf{e}_2 + Q = 0. \tag{4.71}$$

Here Q has to be specified in terms the surface and bulk parameters assumed to be responsible for the mass exchange.

An important special case is the situation where there is a microscopic film between the solid and the gas phase (Fig. 4.6b). Here the term 'microscopic' means that its thickness is on a scale comparable with the thickness of the interfacial layers so that this film is not describable by the bulk equations and has to be modelled as a two-dimensional surface phase. This film could be of different origin: it could be a residual film left behind a receding contact line, a film in front of the advancing fluid left after the previous wetting-dewetting cycle, a precursor film if the main body of fluid moves with respect to the solid surface with velocity different from that of the film, a film resulting from condensation, etc. The presence of such a film modifies properties of the gas-solid interface and manifests itself, first of all, in the mass balance condition at the contact line. If the surface density of the film ρ^s_{film} is known or determined from some microscopic theory of its formation and the film is at rest with respect to the solid surface, then, instead of the simplest mass balance condition (4.70), one now has

$$(\rho^s \mathbf{v}^s)|_G \cdot \mathbf{e}_1 + (\rho^s \mathbf{v}^s)|_S \cdot \mathbf{e}_2 - \rho^s_{\text{film}} \mathbf{U} \cdot \mathbf{e}_2 = 0. \qquad (4.72)$$

4.3.5.2   Balance of forces

Given that inertial properties of the contact line as well as convective momentum fluxes along the interfaces are negligible compared with the surface tensions, the balance of the tangential projections of forces acting on the contact line takes the form of the Young equation

$$\sigma(\rho^s)|_G \cos\theta_d = \sigma_{(2)} - \sigma(\rho^s)|_S, \qquad (4.73)$$

where, as before, θ_d is the dynamic contact angle ($\cos\theta_d = \mathbf{e}_1 \cdot \mathbf{e}_2$); $\sigma(\rho^s)|_G$ and $\sigma(\rho^s)|_S$ are the surface tensions of the liquid-gas and the liquid-solid interfaces, respectively, evaluated at the contact line. The difference between this equation and (2.164) of §2.4.4 is that in (4.73) the surface tensions are *dynamic* so that they depend on the distribution of the surface parameters along the interfaces, which in their turn are coupled with the bulk flow.

The first term on the right-hand side of (4.73) has already been discussed in connection with the standard Young equation (2.164). For a solid-gas interface with no adsorbed molecules of the gas or vapour, one has $\sigma_{(2)} = 0$. As we will see in Chapter 5, a nonzero value of $\sigma_{(2)}$ will manifest itself only when the contact angle is close to 180°.

It is important to emphasize that the bulk stresses make no contribution to the balance of forces acting on the contact line. Indeed, the concept of 'stress' introduced in §2.1.3 and quantified in §2.2.2 describes a *distributed* force, i.e., a force acting on an *area*. Unlike the surface tensions, which are concentrated forces acting on *lines*, the force due to the bulk stress acting on a line is always zero.

4.3.5.3   On the 'line phase', 'line tension' and 'additional' boundary conditions

In formulating the boundary conditions at the contact lines, we need to consider a general question as to whether or not one should treat the contact line as a thermodynamic system in its own right (a 'line phase') whose thermodynamic and dynamic properties could be comparable with those of the surface phases and hence have to be taken into account. In particular, do we need the concepts of 'line energy' and consequently 'line tension', analogous to the surface energy and tension, to model liquid-solid interactions? If we do, then the logical way forward would be to develop thermodynamics of the 'line phase', similar to thermodynamics for the surface phases we considered in this chapter, and derive equations describing the distributions of the 'line properties' along the contact line. Then, the question of boundary conditions will arise at the four-phase-interaction *points*, such as, for example, the points formed when two drops of immiscible liquids sitting on a solid surface are pressed against each other.

The notion of 'line tension' has been used in the literature (e.g., Amirfazli et al. 1998) to interpret some experimental observations. One can also come across theoretical works formally developing the line thermodynamics (e.g., Bedeaux 2004, Rusanov 2005 and references therein), micromechanics of the 'line tension' (e.g., Vignes-Adler & Brenner 1985) and its applications (e.g., Marmur 1998) so that this route is being explored. However, the very basis of this approach suffers from a fundamental inconsistency.

As we discussed in §§2.4.2 and 4.1, the reason why the properties of interfaces can play a role to be taken into account in the macroscopic dynamics of the bulk phases is that, as one goes one dimension down, from the three-dimensional bulk phases to a two-dimensional 'interface', there appear forces (asymmetric intermolecular forces from the bulk phases acting on molecules in the interfacial layer) that are singularly strong compared with those operating in the bulk hydrodynamics. These forces act as a magnifying factor making the dynamic role of the interfacial layer comparable with that of the bulk phases despite the interfacial layer being negligibly thin compared with characteristic dimensions of the bulk phase.

Once we go one dimension down, from the 'surface phase' to the 'line phase', we again need a 'magnifying' physical factor to make the dynamic role of the line comparable with that of the interface. However, when we go from an interface to a line, there appear no new forces that need to be singularly strong compared with the intermolecular forces acting on the interfacial layers from the bulk phases that are already taken into account in the properties of the interfaces. Thus, in the continuum approximation of fluid mechanics the "line tension" and other singular line quantities, such as "line energy" (Bedeaux 2004) or even "line viscosity" (Gouin 2001), are artificial concepts introducing quantities which are fundamentally below the accuracy of the macroscopic model. Importantly, the inclusion of artificial mathematical constructions resulting from considering singular 'line phases' as 'additional' boundary conditions at contact lines would distort the mathematical structure of the resulting problems and lead to completely erroneous results.

Experimental issues involved in measuring the 'line tension' have been examined by de Gennes, Brochard-Wyart & Quéré (2003) who, in a section of their book emphatically entitled "The Myth of the Line Tension", point out fundamental inconsistencies in the experimental works reporting abnormally high values of the 'line tension'. Recent experiments on nanodrops reported by Checco & Guenoun (2003) confirm this conclusion and show that, in their setting, the effects often attributed to the 'line tension' arise due to the influence of the substrate heterogeneity.

However, regardless of all the above arguments, research into the 'line tension' and even the 'point tension' is, in the words of Rusanov (2005), "in full play" in theoretical and experimental literature.

4.4 Liquid-liquid interfaces

Unlike the cases considered so far, an interface between immiscible liquids is more difficult to model since it is composed of molecules of both fluids and both fluids are involved in the exchange of mass, momentum and energy between the bulk and the surface phases. One way of addressing the problem is to make an assumption about the structure of the interface and model it as a two-phase system consisting of two spatially separated surface 'sub-phases' (Shikhmurzaev 1997a). An alternative and more fruitful approach which we will take here is to treat the surface phase as a multicomponent system. The main advantage of this route is that it requires no assumptions about the interfacial layer's structure. This makes it more general from a thermodynamic viewpoint and at the same time can be made simpler mathematically.

Let subscripts 1 and 2 mark surface parameters corresponding to fluid 1 and 2, respectively. The key simplifying assumption we will use for the interface is that the difference between the surface velocities of the two components is negligible, so that $\mathbf{v}_1^s = \mathbf{v}_2^s = \mathbf{v}^s$. Then, given that, by definition, $\rho^s = \rho_1^s + \rho_2^s$, we can split the surface continuity equation (4.10) into two:

$$\frac{\partial \rho_i^s}{\partial t} + \nabla \cdot (\rho_i^s \mathbf{v}^s) = \mp \rho^\pm (\mathbf{u}^\pm - \mathbf{v}^s) \cdot \mathbf{n}, \qquad (i = 1, 2). \qquad (4.74)$$

Here the upper and lower signs correspond to $i = 1$ and $i = 2$, respectively. As before, the surface momentum balance equation is given by (4.12), and below we will simplify it in the same way as in the case of a liquid-gas interface.

Instead of (4.22), we now postulate Gibbs's formula in the form

$$dU_v^s = T\,dS_v^s + \Psi_1^s\,d\rho_1^s + \Psi_2^s\,d\rho_2^s. \qquad (4.75)$$

Hereafter, for simplicity, we assume $T^+ = T^- = T_1^s = T_2^s = T$. The chemical potentials Ψ_1^s and Ψ_2^s are now functions of both ρ_1^s and ρ_2^s (as well as of S_v^s or T, depending whether one uses U_v^s or the free energy $F_v^s = U_v^s - TS_v^s$ as a thermodynamic potential). We will also use the definition

$$\sigma = U_v^s - TS_v^s - \Psi_1^s\rho_1^s - \Psi_2^s\rho_2^s, \qquad (4.76)$$

which is a straightforward generalization of (4.23). As before, the subscript v refers to quantities per unit volume.

Finally, we will assume that both liquids are two-parametric thermodynamic systems and, in addition to (4.24), also postulate

$$dU_v^- = T\,dS_v^- + \Psi_m^-\,d\rho^-, \qquad (4.77)$$

and, by definition,

$$U_v^- = \rho^- \Psi_m^- + T S_v^- - p^-. \tag{4.78}$$

Now, after simple algebra one can write equation (4.20) down as:

$$\frac{\partial S_v^s}{\partial t} + \nabla \cdot \left(\mathbf{v}^s S_v^s + \frac{\mathbf{q}^s}{T} \right) + S_v^+ (\mathbf{u}^+ - \mathbf{v}^s) \cdot \mathbf{n} - S_v^s (\mathbf{u}^- - \mathbf{v}^s) \cdot \mathbf{n}$$

$$+ \frac{\rho^+ (\mathbf{u}^+ - \mathbf{v}^s)^2}{2T} (\mathbf{u}^+ - \mathbf{v}^s) \cdot \mathbf{n} - \frac{\rho^- (\mathbf{u}^- - \mathbf{v}^s)^2}{2T} (\mathbf{u}^- - \mathbf{v}^s) \cdot \mathbf{n}$$

$$+ \frac{1}{T} (\mathbf{q}^+ - \mathbf{q}^-) \cdot \mathbf{n} = \dot{S}^s, \tag{4.79}$$

where

$$\dot{S}^s = \frac{1}{T} \rho^+ (\mathbf{u}^+ - \mathbf{v}^s) \cdot \mathbf{n} \left(\Psi_1^s - \Psi_m^+ + \frac{1}{\rho^+} \mathbf{n} \cdot \boldsymbol{\sigma}^+ \cdot \mathbf{n} \right)$$

$$- \frac{1}{T} \rho^- (\mathbf{u}^- - \mathbf{v}^s) \cdot \mathbf{n} \left(\Psi_2^s - \Psi_m^- + \frac{1}{\rho^-} \mathbf{n} \cdot \boldsymbol{\sigma}^- \cdot \mathbf{n} \right)$$

$$+ \frac{1}{2T} [\mathbf{P}^s - \sigma(\mathbf{I} - \mathbf{nn})] : [\nabla \mathbf{v}^s + (\nabla \mathbf{v}^s)^*]$$

$$+ \frac{1}{2T} \mathbf{n} \cdot (\mathbf{P}^+ + \mathbf{P}^-) \cdot (\mathbf{I} - \mathbf{nn}) \cdot (\mathbf{u}_\parallel^+ - \mathbf{u}_\parallel^-)$$

$$- \frac{1}{T} \mathbf{n} \cdot (\mathbf{P}^+ - \mathbf{P}^-) \cdot (\mathbf{I} - \mathbf{nn}) \cdot [\mathbf{v}_\parallel^s - \tfrac{1}{2}(\mathbf{u}_\parallel^+ + \mathbf{u}_\parallel^-)]$$

$$- \frac{1}{T^2} \mathbf{q}^s \cdot \nabla T, \tag{4.80}$$

and, as before, $\boldsymbol{\sigma}^\pm = \mu^\pm [\nabla \mathbf{u}^\pm + (\nabla \mathbf{u}^\pm)^*]$. Equation (4.79) is a natural generalization of (4.33) and, as is clear from its form, is the simplest one. We will again obtain constitutive equations by making the right-hand side of (4.80) a positively determined quadratic form and, looking for the simplest model, neglect all cross-effects and make the same simplifications based on estimates as we used earlier.

The third term on the right-hand side of (4.80) leads to the constitutive equation of a 'Newtonian interface' (2.64) and, after neglecting surface viscosities, results in (4.40), i.e.,

$$\mathbf{P}^s = \sigma(\mathbf{I} - \mathbf{nn}). \tag{4.81}$$

Then, the momentum balance equation (4.12), where, as before, we neglect the convective momentum fluxes, takes the standard form (4.47), i.e.,

$$\nabla \sigma - \sigma \mathbf{n} \nabla \cdot \mathbf{n} + \mathbf{n} \cdot (\mathbf{P}^+ - \mathbf{P}^-) = 0. \tag{4.82}$$

The last three terms in (4.80), which coincide exactly with the corresponding terms in (4.34), yield (4.37), (4.48) and (4.39), i.e.,

$$\tfrac{1}{2} \mathbf{n} \cdot (\mathbf{P}^+ + \mathbf{P}^-) \cdot (\mathbf{I} - \mathbf{nn}) = \beta(\mathbf{u}_\parallel^+ - \mathbf{u}_\parallel^-), \tag{4.83}$$

$$\mathbf{v}_{\parallel}^s = \tfrac{1}{2}(\mathbf{u}_{\parallel}^+ + \mathbf{u}_{\parallel}^-) + \alpha\nabla\sigma, \qquad (4.84)$$

$$\mathbf{q}^s = -\kappa^s\nabla T \cdot (\mathbf{I} - \mathbf{nn}). \qquad (4.85)$$

The first two terms in (4.80) require a bit of caution. On the one hand, they are similar to the corresponding term in (4.34) so that, again assuming the proportionality of thermodynamic fluxes and forces, we obtain

$$\rho^\pm(\mathbf{u}^\pm - \mathbf{v}^s) \cdot \mathbf{n} = \pm k_i \left(\Psi_i^s - \Psi_m^\pm + \frac{1}{\rho^\pm}\mathbf{n} \cdot \boldsymbol{\sigma}^\pm \cdot \mathbf{n} \right), \qquad (i = 1, 2), \quad (4.86)$$

where, as before, the upper and lower signs correspond to $i = 1$ and $i = 2$, respectively. However, there is also an important difference from the cases considered earlier: according to (4.75), both Ψ_1^s and Ψ_2^s depend on both ρ_1^s and ρ_2^s. Then, using as surface parameters of state ρ_1^s, ρ_2^s and T (which is equivalent to using, instead of U_v^s the free energy $F_v^s = U_v^s - TS_v^s$ as a thermodynamic potential) and neglecting as before the hydrodynamic contribution to the mass exchange between the interface and the bulk, we can approximately write (4.86) down as

$$\rho^+(\mathbf{u}^+ - \mathbf{v}^s) \cdot \mathbf{n} = k_1 \left[\left(\frac{\partial\Psi_1^s}{\partial\rho_1^s}\right)_{\rho_2^s,T} (\rho_1^s - \rho_{1e}^s) + \left(\frac{\partial\Psi_1^s}{\partial\rho_2^s}\right)_{\rho_1^s,T} (\rho_2^s - \rho_{2e}^s) \right]$$

$$= \frac{\rho_1^s - \rho_{1e}^s}{\tau_{11}} + \frac{\rho_2^s - \rho_{2e}^s}{\tau_{12}},$$

$$\rho^-(\mathbf{u}^- - \mathbf{v}^s) \cdot \mathbf{n} = -k_2 \left[\left(\frac{\partial\Psi_2^s}{\partial\rho_1^s}\right)_{\rho_2^s,T} (\rho_1^s - \rho_{1e}^s) + \left(\frac{\partial\Psi_2^s}{\partial\rho_2^s}\right)_{\rho_1^s,T} (\rho_2^s - \rho_{2e}^s) \right]$$

$$= -\frac{\rho_1^s - \rho_{1e}^s}{\tau_{21}} - \frac{\rho_2^s - \rho_{2e}^s}{\tau_{22}},$$

so that equations (4.74) take the form

$$\frac{\partial\rho_i^s}{\partial t} + \nabla \cdot (\rho_i^s\mathbf{v}^s) = -\frac{\rho_1^s - \rho_{1e}^s}{\tau_{i1}} - \frac{\rho_2^s - \rho_{2e}^s}{\tau_{i2}}, \qquad (i = 1, 2). \qquad (4.87)$$

Here ρ_{1e}^s and ρ_{2e}^s are equilibrium surface densities of the two components in the surface phase and the matrix of coefficients τ_{ij} $(i, j = 1, 2)$ is a generalization of the surface-tension-relaxation time τ. It is important to emphasize that the above dependence of the mass exchange of both components with the interface on the deviations of both surface densities from their equilibrium values is the principal effect as it was obtained after we neglected all cross-effects between thermodynamic fluxes and forces in (4.80). If the cross-effects associated with the first two terms of (4.80) are included, this will lead only to a slight variation of the values of τ_{ij}.

A reasonable assumption about the values of the relaxation times is that τ_{ij} are considerably larger than τ_{ii} so that adsorption/desorption of a component is driven primarily by its own deficiency/excess in the surface phase and not by the deficiency/excess of the other component.

For the normal components of the bulk velocities at the interface one obviously has the boundary conditions

$$\rho^{\pm}(\mathbf{u}^{\pm} - \mathbf{v}^s) \cdot \mathbf{n} = \pm \frac{\rho_1^s - \rho_{ie}^s}{\tau_{i1}} \pm \frac{\rho_2^s - \rho_{2e}^s}{\tau_{i2}}, \qquad (i = 1, 2). \qquad (4.88)$$

Finally, we can use the simplest generalization of the linear surface equation of state (4.52),

$$\sigma = \gamma_1(\rho_{1(0)}^s - \rho_1^s) + \gamma_2(\rho_{2(0)}^s - \rho_2^s), \qquad (4.89)$$

that closes the system. An important constraint on the parameters is that the equilibrium surface tension is positive, i.e.,

$$\sigma_e = \gamma_1(\rho_{1(0)}^s - \rho_{1e}^s) + \gamma_2(\rho_{2(0)}^s - \rho_{2e}^s) > 0.$$

Equations (4.2), (4.82), (4.83), (4.84), (4.87), (4.88) and (4.89) form the simplest complete set of boundary conditions to be satisfied by the bulk and surface parameters on a forming interface between immiscible liquids. The equilibrium state is determined by the values of ρ_{1e}^s and ρ_{2e}^s which are regarded as material constants. In particular, if the bulk and surface parameters of the two fluids are the same and $\rho_{ie}^s = \rho_{i(0)}^s$ $(i = 1, 2)$, the above equations describe the process of disappearance of an interface. This happens, for example, when two free surfaces becomes trapped between two volumes of the same fluid and the fluid particles that used to form these free surfaces have to turn into 'ordinary' particles of the bulk fluid. Equations (4.2), (4.82), (4.83), (4.84), (4.87)–(4.89) can then be interpreted as describing two sub-layers (former free surfaces) glued together. We will consider this situation in Chapter 6.

4.4.1   Conditions at contact lines

As in the case of liquid-gas and liquid-solid interfaces, for the barotropic closure (4.89) the boundary conditions at contact lines have to operate only with fluxes of mass and the balance of forces. Compared with the cases involving one-component systems, an important difference now is that the conditions on mass fluxes at a contact lines must be compatible with the physical assumptions made about the interfaces that meet at the contact line. To illustrate the issues arising in formulating these conditions, consider two main situations involving contact lines confining liquid-liquid interfaces. These situations are schematically shown in Fig. 4.7.

4.4.1.1   Contact lines formed by three liquid-fluid interfaces

At a contact line formed by three liquid-liquid interfaces Fig. 4.7a one needs three mass balance conditions and two conditions expressing the balance of

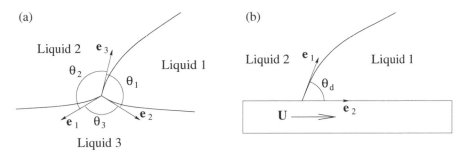

FIGURE 4.7: Contact lines involving liquid-liquid interfaces.

forces acting on the contact line projected on two non-parallel directions. The latter have the form, for example, of (2.162) and (2.163), i.e.,

$$\sigma_{(12)} + \sigma_{(23)} \cos\theta_2 + \sigma_{(13)} \cos\theta_1 = 0, \tag{4.90}$$

$$\sigma_{(23)} + \sigma_{(12)} \cos\theta_2 + \sigma_{(13)} \cos\theta_3 = 0. \tag{4.91}$$

Here the subscripts (ij) $(i, j = 1, 2, 3)$ indicate the limit of a quantity as the contact line is approached along the interface between the ith and the jth fluid.

The mass balance conditions depend on the assumptions one makes about the contact line, in particular, about the separation of the two components as the surface mass flux reaches the contact line. One extreme case is the situation where the contact line offers no resistance to the separation of components and there is no 'leak' from the contact line into the bulk. Then, one has

$$\mathbf{e}_2 \cdot (\rho_1^s \mathbf{v}^s)|_{(13)} + \mathbf{e}_3 \cdot (\rho_1^s \mathbf{v}^s)|_{(12)} = 0, \tag{4.92}$$

$$\mathbf{e}_1 \cdot (\rho_2^s \mathbf{v}^s)|_{(23)} + \mathbf{e}_3 \cdot (\rho_2^s \mathbf{v}^s)|_{(12)} = 0, \tag{4.93}$$

$$\mathbf{e}_2 \cdot (\rho_3^s \mathbf{v}^s)|_{(13)} + \mathbf{e}_1 \cdot (\rho_3^s \mathbf{v}^s)|_{(23)} = 0. \tag{4.94}$$

It should be noted that conditions (4.92)–(4.94) only *allow* for the mass fluxes across the contact line; whether or not there will actually be any fluxes will be determined by the solution of a particular problem involving these conditions.

The opposite extreme case corresponds to no fluxes through the contact line and, instead of (4.92)–(4.94), we then have

$$\mathbf{e}_1 \cdot \mathbf{v}^s_{(23)} = \mathbf{e}_2 \cdot \mathbf{v}^s_{(13)} = \mathbf{e}_3 \cdot \mathbf{v}^s_{(12)} = 0. \tag{4.95}$$

Between these two extreme cases one can have a variety of intermediate situations, where additional physical mechanisms have to be specified.

4.4.1.2   Moving contact line

In the case of the moving contact line (Fig. 4.7b), the force balance condition is again given by the Young equation

$$\sigma_{(12)} \cos \theta_d = \sigma_{(2)} - \sigma_{(1)}. \tag{4.96}$$

In addition to (4.96) we need two conditions on mass fluxes of the two components. If (a) the liquid-solid interfaces that meet at the contact line are one-component systems, (b) there is no mass exchange between the contact line and the bulk, and (c) there is no resistance to the separation of components at the contact line, then one has the conditions expressing continuity of the surface mass fluxes across the contact line,

$$\mathbf{e}_1 \cdot (\rho_i^s \mathbf{v}^s)|_{(12)} + (-1)^{i+1} \mathbf{e}_2 \cdot (\rho^s \mathbf{v}^s)|_{(i)} = 0, \qquad (i = 1, 2), \tag{4.97}$$

which are analogous to (4.92)–(4.94).

As in §4.3.5, these conditions can be generalized to account for a possible flux into/out of the bulk which requires additional physical assumptions. One situation where such a generalization becomes necessary is the case of what experimenters qualitatively describe as incomplete displacement of one fluid by another. Then, physically, there can be different scenaria. The simplest one is when the receding liquid leaves behind a macroscopic film. Then, from the modelling viewpoint, we have no difficulties since there is simply no moving contact line as the advancing fluid flows on top of a macroscopic layer of anther fluid.

The problem becomes nontrivial when traces of the receding fluid are microscopic. Then, formally, for the receding liquid we have

$$\mathbf{e}_1 \cdot (\rho_2^s \mathbf{v}^s)|_{(12)} - \mathbf{e}_2 \cdot (\rho_2^s \mathbf{v}^s)|_{S2} + \rho_{\text{res}}^s \mathbf{U} \cdot \mathbf{e}_2 = 0, \tag{4.98}$$

where the subscript $S2$ means the limit as the contact line is approached along the receding liquid-solid interface and ρ_{res}^s is the surface density of 2nd component that goes into the advancing liquid-solid interface (the subscript *res* stands for 'residual'). For the 1st component one has the same condition for the surface flux as before, i.e., (4.97) for $i = 1$.

Condition (4.98) has important implications. Since according to it the advancing liquid is no longer a one-component system, one has to specify the role played by the second component, adding, if necessary, additional surface and bulk equations. Physically, there are two main situations which we will outline with the emphasis only on the dominant effects.

In the first case, the term 'incomplete displacement' means that the receding liquid leaves behind only an adsorbed microscopic film (typically, a monolayer) on the solid surface. As is known to experimenters (e.g., Fox, Hare & Zisman 1955), the presence of such a film can change the wetting properties of the solid, which essentially means that now the advancing liquid is spreading over a modified solid surface. In the simplest model of the liquid-solid interface summarized in §4.3.4, the solid surface wettability is accounted for in the value of the equilibrium surface density which now will have to be modified. The equilibrium contact angle on such a surface will also be different than on the 'pure' solid substrate.

Another situation is where there is no distinct structural separation between the two components in the advancing liquid-solid interface so that the interface becomes a two-component system. Then, the molecules of fluid 2 essentially plays the role of a surfactant. This 'surfactant' will be insoluble or soluble depending on the relative properties of the three media. If it is insoluble, the simplest way of modelling its presence will be to add its surface density to that of fluid 1 in the surface equation of state (4.68) and to add the surface continuity equation (4.87) for $i = 2$ with $\tau_{21} = \tau_{22} = \infty$ to (4.62)–(4.66). An obvious generalization of this equation is to replace this convection equation with convection-diffusion allowing molecules of fluid 2 to diffuse in the fluid 1/solid surface phase.

If molecules of fluid 2 entrained in fluid 1 behave like a soluble surfactant, then, in addition to modifying as described the surface equation of state (4.68), one must (a) specify the desorption/adsorption rate of this component and use it instead of the current two terms on the right-hand side of (4.87) for $i = 2$ and (b) formulate a convective diffusion equation describing the distribution of component 2 in the bulk of fluid 1. For the latter, the above desorption/adsorption rate will be a boundary condition. Strictly speaking, the presence of molecules of fluid 2 in the bulk brings another complication: now there can be also adsorption of these molecules from the bulk of fluid 1 on the liquid-liquid interface so that the boundary conditions on this interface will have to be modified.

The above two cases show that, in general, boundary conditions at interfaces and at contact lines should not be assembled in a formal way to produce a problem formulation; for the outcome to be physically meaningful, the problem formulation has to be checked from the viewpoint of physical self-consistency and compatibility of assumptions that are incorporated in the boundary conditions as well as in the equations describing the bulk properties.

4.5 Summary

In the previous sections, we introduced a conceptual framework allowing one to develop mathematical models for fluid motion with forming or/and disappearing interfaces of various types. The simplest models formulated within this framework are applicable to flows involving liquid-gas, liquid-solid and liquid-liquid interfaces in different configurations. The model comprises three components:

(i) The Navier-Stokes equations (4.53) describing the bulk flow.

(ii) Boundary conditions (4.54)–(4.60) (or (4.54)–(4.59), (4.61)) on the liquid-gas interface, (4.62)–(4.67) (or (4.62)–(4.66), (4.68)) on the liquid-solid interface, and (4.2), (4.82), (4.83), (4.84), (4.87), (4.88) and (4.89) on a liquid-liquid interface. These boundary conditions are themselves partial differential equations along the interfaces coupled with the bulk flow and require boundary conditions at the contact lines confining the interfaces.

(iii) Conditions at lines of contact of various types linking the distributions of the surface parameters along different interfaces. Examples of such conditions are given in §§4.3.5 and 4.4.1.

The model is intended to remove unphysical singularities inherent in the mathematical description of a number of capillary flows in the framework of the standard approach and allow one to find all bulk and surface parameters, including the shape of the free surface and the angles it forms with solid boundaries, as part of the solution. As emphasized at the end of the previous section, the problem formulation involving forming/disappearing interfaces and contact lines has to be checked from the viewpoint of compatibility of the assumptions underlying its elements.

4.6   Open questions and possible generalizations

The models described in this chapter are the *simplest* ones that can be formulated self-consistently in the framework of a phenomenological approach to account for the process of interface formation. Nevertheless, as we will see later, these models are surprisingly accurate at describing experiments, which justifies their practical use. Although there is no experimental evidence yet that might suggest how the simplest models should be generalized, a number of theoretical points require further research. Leaving aside relatively straightforward extensions of the models via phenomenological coefficients, cross-effects in the constitutive equations, the interface rheology and, in the case of a two-component interface, surface diffusion, consider some points of principle.

The principal simplification that holds the key to possible generalization of the models as far as additional physical and chemical effects are concerned is the surface equation of state. In the simplest case, the interface as a thermo-dynamic system has not been fully specified and we closed the set of equations by assuming that, in the processes in question, variations of the surface tension are related with variations in the surface densities, which, in the present context, determine the degree to which the interface is formed.

The second most important assumption has been made when we considered the process of interface formation and neglected the hydrodynamic factors compared with chemical potentials, thus implying that in the solutions for particular flows obtained using the resulting equations these factors are not singular. This assumption allowed us to reduce the chemical potential difference to the deviation of the surface densities from their equilibrium values and the relaxation times. In other words, this is an assumption that the process of relaxation of the interface to its equilibrium state is exponential and depends on the bulk flow only implicitly as it affects distributions of the surface parameters.

Both main simplifications require further research, firstly, to find the actual functional form of the surface equation of state and the adsorption fluxes in isothermal processes and, secondly, to incorporate additional physical and chemical factors. The two most important directions of research are the study of thermal effects, including phase transitions, and the role played by surfactants. Thermal processes require full specification of thermodynamic properties of the interface and, at present, the most promising 'semi-experimental' way of investigating them is via molecular dynamics simulations. For a two-parametric model of a one-component surface phase this is a relatively simple task conceptually but the level of technical difficulties, especially those associated with numerical noise, is such that, to date, there is very little information one could use to develop a theoretical model.

It should be noted that for a two-parametric equation of state (for a one-component surface phase), as well as in the general case for a liquid-liquid interface, one will need additional boundary condition(s) at the contact line for the energy flux(es). This condition(s) must have the structure of the energy balance equation(s) similar to what we have for the surface mass flux(es). As pointed out in §4.3.5.3, an attempt to obtain the energy balance condition by bringing in a 'line phase' with the energy and dynamic characteristics comparable with those of the surface phases would lead to incorrect results.

The modelling of the interface formation phenomena in the presence of surfactants seems sufficiently simple, at least for low surfactant concentrations. A complication that one should have in mind though is that relatively large molecules of surfactants can make the interface rheology nontrivial so that, in particular, coefficients α and β in the simplest models could no longer be treated as material constants.

Chapter 5

Moving contact lines: Mathematical description

As we concluded in Chapter 3, the moving contact-line problem arises as a consequence of the fact that the standard model does not account for the process of formation of interfaces and hence leads to unphysical results in the situations where this process is important. In Chapter 4, we considered the thermodynamic approach to the modelling of the interface formation process and derived the simplest models that can be formulated within its framework. In this chapter, we will examine some general properties of these models and give illustrative examples describing how to apply the models to particular problems.

5.1 Flow in the immediate vicinity of a moving contact line

Before we start applying the interface formation theory developed in the previous chapter to particular problems involving moving contact lines, it is necessary to verify that it satisfies the basic 'ABC-criterion' formulated in §2.5.2. The criterion requires that (a) mathematical problems formulated in the framework of this theory are well-posed, (b) the solution to these problems remain within the limits of applicability of the theory, in particular, predict finite values of velocities and pressures in the flow domain and on its boundaries, and (c) the kinematic picture of the flow is qualitatively correct. A way of testing these requirements is to consider the asymptotic behaviour of the solution for the liquid spreading on a flat solid surface as the distance to the contact line tends to zero. This is how we have already shown inadequacy of the standard formulation (§3.1.1) and the 'slip models' (§3.4.1).

5.1.1 Problem formulation

Consider a two-dimensional steady flow in the vicinity of the moving contact line on a length scale L such that the Reynolds number $Re = \rho L U / \mu$ based on L and the speed of the contact line with respect to the solid surface, U, is small. Then, as in §3.1.1, to leading order in Re as $Re \to 0$ one may neglect the convective terms in (4.53) and consider the Stokes equations, which in a dimensionless form are given by

$$\frac{1}{r}\frac{\partial(ru)}{\partial r} + \frac{1}{r}\frac{\partial v}{\partial \theta} = 0, \tag{5.1}$$

$$\frac{\partial p}{\partial r} = \Delta u - \frac{u}{r^2} - \frac{2}{r^2}\frac{\partial v}{\partial \theta}, \tag{5.2}$$

$$\frac{1}{r}\frac{\partial p}{\partial \theta} = \Delta v - \frac{v}{r^2} + \frac{2}{r^2}\frac{\partial u}{\partial \theta}, \tag{5.3}$$

where

$$\Delta = \frac{\partial^2}{\partial r^2} + \frac{1}{r}\frac{\partial}{\partial r} + \frac{1}{r^2}\frac{\partial^2}{\partial \theta^2}.$$

Here, as in §3.1.1, we use a polar coordinate system shown in Fig. 3.1 and the notation u and v for the radial and transversal components of velocity. The lengths, velocities and pressure (measured from the constant pressure in the gas) are scaled with L, U and $\mu U/L$, respectively.

We will consider the flow in a coordinate frame moving with the contact line where the liquid occupies the region $0 < r < \infty$, $0 < \theta < \theta_d + g(r)$, $g(r) \to 0$ as $r \to 0$ (Fig. 3.1). For a steady free surface the surface velocity $\mathbf{v}^s$ has only the tangential component v^s and hence the kinematic boundary condition (4.54) is satisfied automatically. We will consider the surface equation of state in the simplest form (4.61) connecting the equilibrium states of the liquid-gas and liquid-solid interface (Fig. 4.3). After scaling the surface densities and the surface tension with $\rho_{(0)}^s$ and σ_e, respectively, this equation is given by

$$\sigma = \lambda(1 - \rho^s), \tag{5.4}$$

where $\lambda = \gamma \rho_{(0)}^s/\sigma_e$.

The boundary conditions (4.55)–(4.59) on the free surface ($\theta = \theta_d + g(r)$, $r > 0$) take the following form:

$$Ca\, p_{nn} = \frac{\sigma(2g' + rg'' + r^2 g'^3)}{(1 + r^2 g'^2)^{3/2}}, \tag{5.5}$$

$$Ca\, p_{n\tau} + \frac{1}{(1 + r^2 g'^2)^{1/2}}\frac{d\sigma}{dr} = 0, \tag{5.6}$$

$$\frac{rg'u - v}{(1 + r^2 g'^2)^{1/2}} = Q(\rho^s - \rho_G^s), \tag{5.7}$$

$$\frac{\epsilon}{(1 + r^2 g'^2)^{1/2}}\frac{d(\rho^s v^s)}{dr} = -(\rho^s - \rho_G^s), \tag{5.8}$$

$$\frac{\epsilon}{(1 + r^2 g'^2)^{1/2}}\frac{d\sigma}{dr} = 4\lambda V^2 \left[v^s - \frac{u + rg'v}{(1 + r^2 g'^2)^{1/2}} \right]. \tag{5.9}$$

Here

$$Ca = \frac{\mu U}{\sigma_e}, \quad Q = \frac{\rho^s_{(0)}}{\tau \rho U}, \quad \epsilon = \frac{\tau U}{L}, \quad V^2 = \frac{\tau \beta U^2}{\gamma \rho^s_{(0)}(1 + 4\alpha\beta)}, \quad \rho^s_G = \frac{\rho^s_e}{\rho^s_{(0)}},$$

and the expressions for the tangential and normal stresses $p_{n\tau}$ and p_{nn} are given by (3.7) and (3.8).

Using for simplicity the equalities (4.69) for the phenomenological constants, one can write down the boundary conditions (4.63)–(4.66) on the liquid-solid interface ($\theta = 0$, $r > 0$) as:

$$\epsilon Ca\, p_{r\theta} + \frac{\epsilon}{2} \frac{d\sigma}{dr} = \lambda(1 + 4A)V^2(u - 1), \tag{5.10}$$

$$v = Q(\rho^s - \rho^s_S), \tag{5.11}$$

$$\epsilon \frac{d(\rho^s v^s)}{dr} = -(\rho^s - \rho^s_S), \tag{5.12}$$

$$v^s = \tfrac{1}{2}(u + 1) + \frac{\epsilon A}{\lambda(1 + 4A)V^2} \frac{d\sigma}{dr}. \tag{5.13}$$

Here $\rho^s_S = \rho^s_{1e}/\rho^s_{(0)}$, $A = \alpha\beta$, and $p_{r\theta}$ is given by (3.9).

We will consider the simplest boundary conditions (4.70), (4.73) linking the distributions of the surface parameters at the contact line:

$$(\rho^s v^s)|_{r\to 0,\, \theta=\theta_d} + (\rho^s v^s)|_{r\to 0,\, \theta=0} = 0, \tag{5.14}$$

$$\sigma|_{r\to 0,\, \theta=\theta_d} \cos\theta_d = \sigma_{sg} - \sigma|_{r\to 0,\, \theta=0}, \tag{5.15}$$

where $\sigma_{sg} = \sigma_{(2)}/\sigma_e$.

Equations (5.4)–(5.15) together with boundary conditions in the far field, which we will not need for the near-field asymptotic analysis, provide a complete set of boundary conditions for (5.1)–(5.3) allowing one to obtain both the flow and the dynamic contact angle. What we need to check first is that this problem has an asymptotic solution in the near field, i.e., as $r \to 0$. If this is the case, the next step is to verify that u, v and p remain finite as $r \to 0$. As far as the interfacial dynamics is concerned, it is already clear from condition (5.14) that the flow is 'rolling' (or, more precisely, allowed to be 'rolling') and the fluid particles belonging to the free surface can cross the contact line to become elements of the liquid-solid interface. What we need to know in addition to this is whether or not the bulk flow is rolling as well, i.e., whether or not the contact line is a stagnation point for the bulk flow and hence whether or not there is a spurious low-velocity region near the contact line, like the one that is always produced by slip models. The absence of the stagnation point at the contact line and the low-velocity region would mean that kinematics predicted by the model is qualitatively the same as that observed in experiments described in §3.2.2.

5.1.2 Asymptotics as $r \to 0$

We will be looking for an asymptotic solution as $r \to 0$ in the following form. For the surface parameters and the correction to the free-surface shape we will use expansions

$$\sigma(r) = \sum_{n=0}^{\infty} r^n \sigma_{in}, \quad \rho^s(r) = \sum_{n=0}^{\infty} r^n \rho_{in}^s, \quad v^s(r) = \sum_{n=0}^{\infty} r^n v_{in}^s, \qquad (5.16)$$

$$g(r) = \sum_{n=1}^{\infty} r^n g_n, \qquad (5.17)$$

where $i = 1$ for the liquid-gas and $i = 2$ for the liquid-solid interface.

Coefficients in expansions (5.16) for σ and ρ^s are related via the surface equation of state (5.4):

$$\sigma_{in} = \lambda(1 - \rho_{in}^s), \ (i = 1, 2; n = 0); \quad \sigma_{in} = -\lambda \rho_{in}^s, \ (i = 1, 2; n \geq 1). \quad (5.18)$$

(If a more general form of the surface equation of state (4.60) is used, one will have to expand σ in Taylor series about ρ_{10}^s and ρ_{20}^s and, after substituting (5.16) into these expansions, end up with cumbersome algebraic equations relating σ_{in} and ρ_{in}^s. This would only be a technical alteration in the present context.)

As in §3.1.1, a solution to (5.1)–(5.3) as $r \to 0$ can be found by solving a biharmonic equation for the stream function ψ in a wedge region

$$\Delta^2 \psi = 0 \qquad (0 < \theta < \theta_d, r > 0), \qquad (5.19)$$

and then obtaining the pressure from

$$\frac{\partial p}{\partial r} = \left(\frac{1}{r} \frac{\partial^3}{\partial r^2 \partial \theta} + \frac{1}{r^3} \frac{\partial^3}{\partial \theta^3} + \frac{1}{r^2} \frac{\partial^2}{\partial r \partial \theta} \right) \psi, \qquad (5.20)$$

$$\frac{\partial p}{\partial \theta} = -\left(r \frac{\partial^3}{\partial r^3} + \frac{\partial^2}{\partial r^2} + \frac{1}{r} \frac{\partial^3}{\partial r \partial \theta^2} - \frac{1}{r} \frac{\partial}{\partial r} - \frac{2}{r^2} \frac{\partial^2}{\partial \theta^2} \right) \psi. \qquad (5.21)$$

For the stream function and the bulk pressure we will be looking for a solution in the form

$$\psi(r, \theta) = \sum_{n=1}^{\infty} \psi_n(r, \theta) = \sum_{n=1}^{\infty} r^n F_n(\theta), \qquad p(r, \theta) = \sum_{n=0}^{\infty} r^n p_n(\theta). \qquad (5.22)$$

In order to demonstrate that our model satisfies criteria A–C of §2.5.2, it is sufficient to show that (i) ψ_1 is nontrivial and hence the flow in the immediate vicinity of the contact line is genuinely 'rolling', with no spurious low-velocity region near the contact line, and that (ii) ψ_1 and ψ_2 are compatible with a regular pressure represented by (5.22), where one should expect p_0 to be a constant.

An inspection of the boundary conditions (5.7)–(5.15) shows that an asymptotics for ψ_1 and ψ_2 will also include

$$p_0, \ \theta_d, \ g_1, \ \sigma_{in}, \ \rho_{in}^s, \ v_{in}^s \quad (i = 1, 2; \ n = 0, 1). \tag{5.23}$$

It is important to note that, whilst in the case of slip models one could preset θ_d arbitrarily and then consequently determine the stream function in a unique way, find the pressure distribution and calculate $g(r)$, the situation with the present model is different. Indeed, since the contact angle is part of the (global) solution, we are going to obtain a set of algebraic equations which will include θ_d alongside with other coefficients of the asymptotic expansions. (θ_d is in fact the first term in the expansion for the free-surface shape where g_1 is the second term.) Then, matching the asymptotic expansions with the global flow will provide information required to determine the coefficients of the expansions, including the contact angle. This makes θ_d dependent not only on the contact-line speed but also on the flow field/geometry, which is the key feature of the experimentally-observed nonlocal dependence of the dynamic contact angle on the flow field ('hydrodynamic assist of dynamic wetting') described in §3.2.3.

Consider the degree of arbitrariness one must have in an asymptotic solution as $r \to 0$. In the slip models, there is one undetermined constant in the near-field asymptotics. For example, in the case of the Navier condition (3.35), (3.36), the asymptotic solution as $r \to 0$ is given by $\psi_2 = r^2 F_2(\theta)$, (3.50), (3.51), (3.52), (3.55), (3.56), where θ_d is prescribed, as is always the case in these models, and p_0 is arbitrary. The latter has to be determined by matching with the outer flow. Thus, the degree of arbitrariness for the slip model based on the Navier condition is 1. If, as in our case, there are surface parameters whose variations along the interfaces are governed by ordinary differential equations, then the degree of arbitrariness in the near-field asymptotics will increase by the order of the set of the ordinary differential equations for these parameters.[1] Given that in our case σ and ρ^s are related by an algebraic equation of state, for the surface variables ρ^s, σ and v^s the order of the set of ordinary differential equations is equal to 2 and hence we should obtain 15 algebraic equations for the 12 coefficients listed in (5.23). This will show that θ_d is indeed a functional of the flow field and that we have the correct number of conditions at the contact line.

A general solution for ψ_1 is given by (3.17), that is,

$$\psi_1 = rF_1(\theta) = r[(A_1 + A_2\theta)\sin\theta + (A_3 + A_4\theta)\cos\theta].$$

---

[1] To illustrate this, one can add an artificial surface parameter to one of the slip models and set, say, a first-order ODE for it along the liquid-gas and liquid-solid interfaces together with an algebraic boundary condition at the contact line. Then, in the near-field asymptotics there will be two extra unknown coefficients in the expansions for this parameter and only one extra condition for them resulting from the algebraic condition at the contact line. Hence, the number of free coefficients in the local asymptotics will increase by one, that is by the order of the ODE for the surface parameter.

Boundary conditions (5.7), (5.11), (5.10) and (5.6) for F_1 become

$$F_1(\theta_d) = Q(\rho_{10}^s - \rho_G^s), \qquad F_1(0) = -Q(\rho_{20}^s - \rho_S^s),$$

$$F_1''(0) + F_1(0) = 0, \qquad F_1''(\theta_d) + F_1(\theta_d) = 0,$$

yielding

$$\psi_1 = \frac{rQ}{\sin\theta_d} \left[(\rho_{10}^s - \rho_G^s)\sin\theta + (\rho_S^s - \rho_{20}^s)\sin(\theta_d - \theta) \right]. \tag{5.24}$$

This is clearly the stream function of a uniform flow 'connecting' the two interfaces (Fig. 5.1). We will return to this type of flow in §5.6.3. The expression for ψ_1 given by (5.24) makes the right-hand sides of the radial and transversal components of equations (5.20) and (5.21) identically zero for all values of ρ_{10}^s and ρ_{20}^s, so that there is no nonintegrable pressure singularity associated with ψ_1 in §3.1.1 and hence the solution exists (requirement A of the ABC-criterion). The uniform flow (5.24) also means that the contact line is not a stagnation point for the bulk flow and hence the bulk flow is 'rolling' with no spurious low-velocity region near the contact line (requirement C). Now, we need to make sure that the pressure is regular at the contact line (requirement B), and that this is compatible with the degree of freedom our asymptotic solution should have (this is not related to the ABC-criterion but is necessary to describe the effect of 'hydrodynamic assist of wetting' to be considered later).

Consider $\psi_2 = r^2 F_2(\theta)$. A general solution for F_2 is given by (3.50), that is,

$$F_2(\theta) = B_1 + B_2\theta + B_3\sin 2\theta + B_4\cos 2\theta. \tag{5.25}$$

Boundary conditions (5.7), (5.11), (5.10), (5.6) for F_2 take the form:

$$F_2(\theta_d) = \tfrac{1}{2}Q\rho_{11}^s - \frac{g_1 Q}{2\sin\theta_d}\left[(\rho_{10}^s - \rho_G^s)\cos\theta_d - (\rho_S^s - \rho_{20}^s)\right] \equiv K_1$$

$$F_2(0) = -\tfrac{1}{2}Q\rho_{21}^s \equiv K_2$$

$$F_2''(0) = \frac{1}{Ca}\left\{ \frac{\lambda(1+4A)V^2}{\epsilon}\left[\frac{Q}{\sin\theta_d}((\rho_{10}^s - \rho_G^s) - (\rho_S^s - \rho_{20}^s)\cos\theta_d) - 1 \right] \right.$$

$$\left. -\frac{\sigma_{21}}{2} \right\} \equiv K_3$$

$$F_2''(\theta_d) = \frac{\sigma_{11}}{Ca} \equiv K_4.$$

Then,

$$B_1 = K_2 - B_4, \qquad B_2 = \frac{1}{4\theta_d}(4K_1 - 4K_2 - K_3 + K_4),$$

$$B_3 = -\frac{K_4 + 4B_4\cos 2\theta_d}{4\sin 2\theta_d}, \qquad B_4 = -\frac{K_3}{4}.$$

We require here that in a general case $\theta_d \neq \pi/2$; $\theta_d = \pi/2$ is a special case with a slightly more cumbersome asymptotics whose regularity can be demonstrated in a similar way.

A function $\psi_2 = r^2 F_2(\theta)$ makes the right-hand side of (5.21) identically zero for any F_2. Substituting ψ_2 into (5.20) we find that

$$\frac{\partial p}{\partial r} = \frac{4B_2}{r}.$$

Hence, for the pressure at the origin to be finite, $p(r, \theta) = p_0 = \text{const}$ in accordance with (5.22), we require

$$B_2(\theta_d, g_1, \rho_{10}^s, \rho_{11}^s, \rho_{20}^s, \rho_{11}^s, \sigma_{11}, \sigma_{21}) = 0. \tag{5.26}$$

We use this notation to emphasize that the constant of integration B_2 is expressed in terms of the coefficients of asymptotic expansions.

After substituting p_0 and $\psi_1 + \psi_2$ into the normal-stress condition (5.5), we have

$$Ca[p_0 + 4(B_3 \cos 2\theta_d - B_4 \sin 2\theta_d)] = -2g_1 \sigma_{10}. \tag{5.27}$$

Boundary conditions (5.8)–(5.9) give

$$\rho_{10}^s v_{11}^s + \rho_{11}^s v_{10}^s = -\epsilon^{-1}(\rho_{10}^s - \rho_G^s), \tag{5.28}$$

$$\rho_{20}^s v_{21}^s + \rho_{21}^s v_{20}^s = -\epsilon^{-1}(\rho_{20}^s - \rho_S^s), \tag{5.29}$$

$$v_{20}^s = \tfrac{1}{2}[F_1'(0) + 1] + \epsilon A[\lambda(1 + 4A)V^2]^{-1}\sigma_{21}, \tag{5.30}$$

$$\epsilon \sigma_{11} = 4\lambda V^2 [v_{10}^s - F_1'(\theta_d)]. \tag{5.31}$$

Finally, conditions at the contact line (5.14), (5.15) yield

$$\rho_{10}^s v_{10}^s + \rho_{20}^s v_{20}^s = 0 \tag{5.32}$$

$$\sigma_{10} \cos \theta_d = \sigma_{sg} - \sigma_{20}. \tag{5.33}$$

Thus, as expected, for the 15 coefficients listed in (5.23) we obtained 12 algebraic equations: (5.18) for $i = 1, 2$, $n = 0, 1$, and (6.74)–(5.33), and the problem of finding the local asymptotics is reduced to finding a three-dimensional manifold of solutions to the set of 12 algebraic equations for 15 unknowns. Numerical integration shows that this set of equations is solvable and the manifold of solutions has the required dimension. Hence the asymptotic solution indeed has the form given by (5.16), (5.17), (5.22) with no singularities at the contact line. We have also shown that no 'extra' conditions at the contact line are required.

5.1.3　Implications for numerical computations

An analysis of how the solution behaves near the contact line is important not only from the 'academic' viewpoint of the physical and mathematical criteria formulated in §2.5.2. It has immediate 'practical' implications for numerical codes dealing with problems involving moving contact lines. We will illustrate this using as an example the finite element method (Gresho & Sani 2000), which is currently the most powerful computational tool in dealing with free-surface flows. In this method, the flow domain is tessellated into finite elements and the exact solution in each element is approximated with some interpolation functions.[2] For example one can use quadratic interpolation for the velocity components and, to avoid 'locking' (Hughes 1987), linear interpolation for the pressure. Then, if this interpolation is employed for the elements incorporating the moving contact line, slip models and the interface formation theory we are examining here will lead to the following qualitatively different situations.

If in a slip model the pressure is singular (e.g., (3.55) for the Navier condition), then the finite element method will be trying to approximate the pressure which is infinite at the contact line with a piece-wise linear function that has a finite value of pressure at the contact line. As an inevitable result, the more one refines the mesh, the more evident the discrepancy between solutions for different meshes will become. This mesh-dependence of a numerical solution is clearly unacceptable since the numerical solution is supposed to uniformly approximate the exact one, which 'does not know' about any meshes. Thus, the elements incorporating the contact line have to be handled in a special way which requires knowledge of how the solution behaves there. One such way is to put the asymptotics of §3.4.1 into the code explicitly as a solution for the elements incorporating the contact line and match it with the regular numerical solution outside. An alternative and slightly more flexible way is to employ for these elements singular interpolation functions (Fix 1969, Wilson et al. 2006), which must have the same singularity at the contact line as the exact solution. Then one can proceed with the finite element method in a standard way. As is clear, in both cases the numerical method requires an analytic study of how the solution behaves as the moving contact line is approached; this information has then to be incorporated into numerics. Ironically, this additional analytical and numerical effort has to be invested to correctly describe the singularity which is physically unacceptable in the first place.

---

[2]Here the terminology has not settled yet. Interpolation functions are also referred to as 'trial functions' (Whiteman 1973). These are in their turn represented as a sum of 'basis functions' (Cuvelier, Segal & van Steenhoven 1988, Gresho & Sani 2000) also known as 'approximate functions' (Reddy 1985), or 'coordinate functions' (Lewis & Ward 1991), or even again as 'trial functions' (Strang & Fix 1973). This is not to mention 'shape functions' used by engineers.

If one chooses a slip model with the pressure singularity suppressed by imposing a high degree of slip, that is, a model satisfying (3.53), then the above difficulty will be removed at the expense of creating a different one: a spurious low-velocity region near the contact line unavoidable in all slip models (Fig. 3.29) will then be considerably expanded further into the bulk. Thus, besides replacing the right flow kinematics with a spurious one on a much larger length scale, one will face a purely numerical problem of having to resolve the low-velocity flow field in the above region and, in order to arrive at a mesh-independent streamline pattern, will have to significantly increase the accuracy of computations. As can be seen from the results of §3.4.1, in this region one can even have spurious vortices whose accurate computation is nevertheless required for one to be able to claim that the numerical solution approximates the exact one.

The situation is completely different for the interface formation model. Indeed, one can use the same regular interpolation functions for the elements incorporating the contact line as in the rest of the flow domain and proceed with the numerical solution in a standard way. Then, since both the exact solution (as shown by the above asymptotic analysis) and its numerical approximation in the finite element method are regular everywhere including the contact line itself, mesh refinement will merely give better approximation as a succession of numerical solutions uniformly converge to the exact one. In other words, unlike the case of slip models, a standard discretisation procedure of a numerical method and the subsequent computations do not give rise to any numerical problems which would have alerted the user to the necessity to handle the contact-line region analytically. This illustrates a general point that, as far as numerical algorithms are concerned, asymptotic methods as an analytic tool for tackling infinities become redundant once the unphysical infinities have been removed by the proper modelling. For a singularity-free solution the standard discretization is all one needs for a numerical code.

5.1.4   Qualitative illustration of the asymptotics

It is instructive to illustrate graphically how the interface formation model removes the singularities and at the same time preserves the correct kinematics of the flow. In order to highlight the difference between v^s and the bulk velocity at the boundary, we will represent interfaces as layers of a finite thickness. (It should be emphasized here that, in accordance with fundamentals of fluid mechanics, the model treats interfaces as mathematical surfaces with no 'interfacial thickness' present explicitly anywhere in the equations and boundary conditions; it is present implicitly in phenomenological coefficients of the boundary conditions and parameters of the surface equation of state).

Fig. 5.1 sketches the flow near the contact line at a small and large contact angle. We see that near the contact line the mass exchange associated with the interface disappearance/formation process switches on and increases as the contact line is approached. Then, the streamlines stem from the free

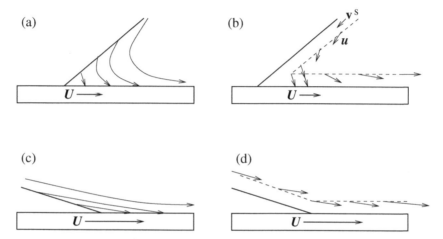

FIGURE 5.1:   Sketches of streamlines resulting from the near-field asymptotics for small (a) and large (c) contact angle. In (b) and (d) the interfaces are shown schematically as layers of a finite thickness and the arrows on the liquid-facing sides of these interfaces represent the bulk velocity $\mathbf{u}$. Mass exchange between the interfaces and the bulk in the model leads to the situation where, to leading order as $r \to 0$, one has a uniform flow near the contact line, which preserves the global 'rolling' motion without creating a low-velocity region; $\mathbf{v}^s$ is an average velocity with which ρ^s is transported along the interface; the normal component of $\mathbf{v}^s$ represents the motion of the interface itself. The numerically obtained flow field for a particular problem is given in §5.6.3.

surface and stick into the liquid-solid interface, so that near the contact line one essentially has a uniform flow described by ψ_1 (see (5.24)). The contact line is not a stagnation point for the bulk flow so that there is no low-velocity region near the contact line.

This feature highlights the crucial difference between slip models and the present one. In slip models, one always has an angle in the streamline going along the interfaces and to suppress the impact of this instantaneous change of direction of the flow velocity, one has to pose a boundary condition which would create a low-velocity region near the contact line. This way of remedying the singularity in the slip models is unavoidable. However, it is this very remedy which imposes the wrong kinematics of the flow.

In the interface formation model, the mass exchange between the interfaces and the bulk removes the above necessity since the interfaces no longer coincide with streamlines. In §5.6 we will consider the full-scale numerical solution obtained using the interface formation model and show that it indeed has the flow kinematics qualitatively described here.

An interesting qualitative feature of dynamic wetting following from the present model is that once the contact line starts moving one has $\rho_{10}^s > \rho_G^s$ and $\rho_{20}^s < \rho_S^s$, thus triggering the mass exchange described above. The first of these inequalities together with the surface equation of state (5.4) means that the surface tension in the free surface near the contact line becomes lower, although, as we will see in §5.6, very slightly, than in equilibrium. This effect of an upstream influence of the moving contact line perhaps deserves experimental research.

5.1.5   Summary of the near-field asymptotic analysis

The near-field asymptotic analysis shows that the theory formulated in chapter 4 allows one to describe flows with moving contact lines in the framework of well-posed mathematical problems whose solutions are regular everywhere, including the contact line, and predict the flow kinematics which has the same qualitative features as that observed experimentally.

The implication for numerical computations is that no special treatment of the region embracing the contact line is required in the numerical codes. The solution in this region can be discretised in the same regular way as in the rest of the flow domain and along the interfaces. No use of the asymptotic solution obtained in §5.1.2 is needed.

The degree of freedom in the local asymptotic solution indicates that, as expected from the derivation of the model, the dynamic contact angle is part of the global solution, i.e., it depends on the flow field/geometry, and that for a liquid-gas-solid system it is necessary and sufficient to have two conditions at the contact line for the surface variables.

5.2   Dynamic wetting at small capillary numbers

The importance of considering dynamic wetting at small capillary and Reynolds numbers comes from the fact that in this limiting case the contact-line motion can be analysed as a local problem whose solution can then be incorporated into a variety of global flows. This case is also the one studied in most experimental works which makes it possible to verify the results and determine phenomenological constants of the model. Below, the exposition essentially follows the analysis of Shikhmurzaev (1993, 1994).

We will consider the case where the characteristic length scale of the overall flow L is much larger than the surface-tension-relaxation length $l = U\tau$. This corresponds to an asymptotic limit $\epsilon \to 0$ in the problem (5.1)–(5.15), and localises the problem by splitting the flow domain into two asymptotic regions: an 'outer' region, where the independent variables are (r, θ) and an 'inner' region associated with the variables $(\bar{r} = r\epsilon^{-1}, \theta)$. In each of these regions we can simplify the problem further by considering the limit $Ca \to 0$. In what follows, we will be interested only in the leading-order solution in the small parameters ϵ and Ca.

In the outer region, the boundary conditions (5.5)–(5.13) to leading order in ϵ turn into the standard ones (3.5)–(3.8). As $Ca \to 0$, the normal-stress boundary condition (3.8) (i.e., (5.5)) gives that, to leading order in Ca, the free-surface curvature is zero, and hence one has a problem in a wedge. The same argument about the normal-stress boundary condition also applies to the inner region (and its asymptotic subregion we will be considering below) so that the wedge is described by $0 < \theta < \theta_d$, i.e., it is specified by the dynamic contact angle, which has yet to be determined. The solution of the outer problem (Moffatt 1964)[3] in terms of the stream function is given by (3.18), i.e.,

$$\psi = \frac{r}{\sin\theta_d \cos\theta_d - \theta_d}[(\theta - \theta_d)\sin\theta - \theta\sin(\theta - \theta_d)\cos\theta_d]. \qquad (5.34)$$

It is worth noting that the expression (5.34) has been obtained using $\psi = 0$ as a boundary condition on both interfaces. In other words, it has been assumed that there is conservation of mass in the control volume confined by the interfaces and a boundary $r = R$ lying in the far field. However, in our case conditions (5.7) and (5.11) state that in the inner region there are processes of desorption and adsorption on the interfaces so that, generally, the mass in the bulk does not conserve. Therefore, strictly speaking, the values of ψ on the two interfaces in the far field have to be different. However, this

---

[3]It is worth pointing out that, unlike the asymptotic solutions obtained in §3.4.1 for slip models, which are coordinate asymptotics valid as $r \to 0$, Moffatt's (1964) solution is valid for finite r when it is considered as the result of the parametric asymptotic limit $Ca \to 0$.

difference is of order ϵ and hence may be neglected at leading order in ϵ as $\epsilon \to 0$. Indeed, according to (5.7) and (5.11) in dimensionless terms the mass exchange between the bulk and the interfaces is proportional to Q, whereas the size of the inner region is of order ϵ. Hence the total mass exchange in the inner region is of order ϵQ and, to leading order in ϵ, can be neglected so that the solution in the outer region to this order is indeed given by (5.34).

In the outer region, both components of velocity corresponding to (5.34) are constant along the lines $\theta =$ const. In particular, at the free surface for the radial velocity one has (3.21), i.e.,

$$u_{(12)}(\theta_d, 0) = \frac{\sin \theta_d - \theta_d \cos \theta_d}{\sin \theta_d \cos \theta_d - \theta_d}. \tag{5.35}$$

In the inner asymptotic region, the boundary conditions (5.6)–(5.9) at the liquid-gas interface ($\bar{r} > 0, \theta = \theta_d$) now are given by

$$Ca \left[\frac{1}{\bar{r}} \frac{\partial u}{\partial \theta} + \bar{r} \frac{\partial}{\partial \bar{r}} \left(\frac{v}{\bar{r}} \right) \right]_{\bar{r}, \theta_d} + \lambda \frac{d\bar{\rho}^s}{d\bar{r}} = 0, \tag{5.36}$$

$$-v = Q(\rho^s - \rho_G^s), \tag{5.37}$$

$$\frac{d(\rho^s v^s)}{d\bar{r}} = -(\rho^s - \rho_G^s), \tag{5.38}$$

$$\frac{d\rho^s}{d\bar{r}} = 4V^2 \left[u(\bar{r}, \theta_d) - v^s \right]. \tag{5.39}$$

Here we used (5.4) to eliminate σ.

At the liquid-solid interface ($\bar{r} > 0, \theta = 0$), the boundary conditions (5.10)–(5.13) take the form

$$\frac{2Ca}{\lambda(1 + 4A)} \left[\frac{1}{\bar{r}} \frac{\partial u}{\partial \theta} + \bar{r} \frac{\partial}{\partial \bar{r}} \left(\frac{v}{\bar{r}} \right) \right]_{\bar{r}, 0} - \frac{d\rho^s}{d\bar{r}} = 4V^2(v^s - 1), \tag{5.40}$$

$$v = Q(\mu^s - \rho_S^s), \tag{5.41}$$

$$\frac{d(\rho^s v^s)}{d\bar{r}} = -(\rho^s - \rho_S^s), \tag{5.42}$$

$$u(\bar{r}, 0) = 2v^s - 1 + \frac{2A}{(1 + 4A)V^2} \frac{d\rho^s}{d\bar{r}}. \tag{5.43}$$

In obtaining (5.40), we eliminated u from (5.10) using (5.13), and, as in (5.36)–(5.39), σ is eliminated by using (5.4).

In connection with the boundary conditions in the inner region (5.36)–(5.43), it is important to note the following. First, as $Ca \to 0$, to leading order the tangential stresses in (5.36) and (5.40) vanish and (5.36), (5.38), (5.40), (5.42) become *ordinary* differential equations for the surface variables

$$\frac{d\rho^s}{d\bar{r}} = 0, \quad \frac{d(\rho^s v^s)}{d\bar{r}} = -(\rho^s - \rho_G^s), \quad (\bar{r} > 0, \theta = \theta_d) \tag{5.44}$$

$$\frac{d\rho^s}{d\bar{r}} = 4V^2(1-v^s), \quad \frac{d(\rho^s v^s)}{d\bar{r}} = -(\rho^s - \rho^s_S), \quad (\bar{r} > 0, \theta = 0). \quad (5.45)$$

To solve these equations, one does not require the knowledge of the bulk flow. Once their solution is found, the remaining equations (5.37), (5.39), (5.41), (5.43) provide the boundary conditions for u and v thus specifying the flow field in the bulk.

One set of the boundary conditions for (5.44), (5.45) comes from the matching of solutions in the outer and inner regions, which for the surface variables ρ^s and v^s take the form

$$\rho^s \to \rho^s_G, \quad v^s \to u_{(12)}(\theta_d, 0) \quad \text{as } \bar{r} \to \infty, \ (\theta = \theta_d), \quad (5.46)$$

$$v^s \to 1 \quad \text{as } \bar{r} \to \infty, \ (\theta = 0), \quad (5.47)$$

where $u_{(12)}(\theta_d, 0)$ is given by (5.35). (At the liquid-solid interface only one condition (5.47), or $\rho^s \to \rho^s_S$ as $\bar{r} \to \infty$, is required since $\rho^s = \rho^s_S$, $v^s = 1$ is a saddle point in the (ρ^s, v^s)-phase plane for (5.45) and the other condition follows automatically.)

The second point to be noted about (5.36)–(5.43) is that, as $Ca \to 0$, these equations are singularly perturbed and hence one has to consider an asymptotic subregion of the inner region, which we will call the 'viscous' subregion and associate with independent variables $\tilde{r} = \bar{r}Ca^{-1}$ and θ. The functions there will be also marked with a tilde. The remaining boundary conditions required for (5.44) and (5.45) are conditions of matching the solution with that in the viscous subregion.

In the variables $\tilde{r}$, θ of the viscous subregion, equations (5.36)–(5.43) immediately give that, to leading order in Ca as $Ca \to 0$, $\tilde{\rho}^s$ and $\tilde{v}^s$ are constant along both interfaces. This is understandable from the physical point of view: the surface parameters vary on the characteristic length of the surface-tension relaxation $l = U\tau$ and hence their variation is negligible on a much smaller length of the viscous subregion, which is of order $Ca\,l$ in dimensional terms.

Now, given the matching conditions

$$\lim_{\bar{r} \to 0} \varphi^s = \lim_{\tilde{r} \to \infty} \tilde{\varphi}^s$$

for the surface variables, one has that the conditions at the contact line (5.14), (5.15) can be applied to the surface parameters described by (5.44)–(5.45), i.e., for the variables in the 'outer part' of the inner region. For brevity, we will call it the 'intermediate' region. The condition (5.14) stays as it is, and (5.15), after eliminating σ using (5.4), takes the form

$$\lambda(1-\rho^s)|_{r\to 0,\,\theta=\theta_d} \cos\theta_d = \sigma_{sg} - \lambda(1-\rho^s)|_{r\to 0,\,\theta=0}. \quad (5.48)$$

Thus, to find the distribution of the surface parameters in the intermediate region *and* the dynamic contact angle θ_d we need to solve a set of ordinary differential equations (5.44), (5.45) for the surface variables in the intermediate region subject to the boundary conditions (5.46), (5.47), (5.14) and (5.48). For the constants involved in the problem formulation one has two constraints. Firstly, given that we have chosen $\rho^s_{(0)}$ and σ_e as scales for ρ^s and σ, the surface equation of state (4.61) yields

$$1 = \lambda(1 - \rho^s_G). \tag{5.49}$$

Secondly, one has the Young equation in the static situation (2.164), which can now be written down as

$$\cos\theta_s = \sigma_{sg} - \frac{1 - \rho^s_S}{1 - \rho^s_G},$$

or

$$\rho^s_S = 1 + (\cos\theta_s - \sigma_{sg})(1 - \rho^s_G). \tag{5.50}$$

Here, as before, θ_s is the static contact angle.

5.2.1 Dynamic contact angle

The equations operating along the liquid-gas interface in the intermediate region (5.44) together with the matching conditions (5.46) give that at the free surface

$$\rho^s \equiv \rho^s_G, \quad v^s \equiv u_{(12)}(\theta_d, 0) \qquad (\bar{r} > 0, \ \theta = \theta_d). \tag{5.51}$$

Then, conditions (5.14), (5.48) at the contact line for (5.45) take the form

$$\rho^s(0)v^s(0) = -\rho^s_G u_{(12)}(\theta_d, 0), \tag{5.52}$$

$$\rho^s(0) = 1 + (\cos\theta_d - \sigma_{sg})(1 - \rho^s_G). \tag{5.53}$$

Thus, for the set of nonlinear ordinary differential equations of the second order (5.45) containing an unknown constant θ_d one has 3 boundary conditions (5.47), (5.52) and (5.53). The system (5.45) is equivalent to a particular form of Abel's equation of the second kind to which no solution in quadrature is known.

The problem (5.45), (5.47), (5.52), (5.53) has been examined numerically in Shikhmurzaev (1993), and here we will consider an approximate analytical solution (Shikhmurzaev 1994) one can obtain assuming that deviations of ρ^s from ρ^s_S are small for the problem to be linearized. After integration and the use of boundary conditions (5.47), one has

$$\rho^s = \rho^s_S - C\exp(-k\bar{r}), \quad v^s = 1 - \frac{Ck}{4V^2}\exp(-k\bar{r}), \tag{5.54}$$

where

$$k = 2V(\rho^s_S)^{-1}[(V^2 + \rho^s_S)^{1/2} - V], \tag{5.55}$$

whereas condition (5.52) yields

$$C = \frac{2V(\rho_S^s + \rho_G^s u_{(12)}(\theta_d, 0))}{V + (V^2 + \rho_S^s)^{1/2}}.$$

Substituting these expressions into the Young equation (5.53) and eliminating ρ_S^s with the help of (5.50), we arrive at an algebraic equation relating the dynamic contact angle θ_d and the dimensionless contact-line speed V:

$$\cos\theta_s - \cos\theta_d = \frac{2V[\cos\theta_s - \sigma_{sg} + (1 - \rho_G^s)^{-1}(1 + \rho_G^s u_{(12)}(\theta_d, 0)]}{V + [V^2 + 1 + (\cos\theta_s - \sigma_{sg})(1 - \rho_G^s)]^{1/2}}. \quad (5.56)$$

For $V > 0$ one has $\theta_d > \theta_s$.

A point to be noted about this equation is that the bulk hydrodynamics enters the velocity dependence of the contact angle only via the solution in the *outer* asymptotic region, where one has the standard model with the no-slip boundary condition at the solid boundary and zero tangential stress at the free surface. At small *Ca*, to leading order the evolution of the surface parameters in the inner region is independent of the bulk flow. Therefore, if we take into account viscosity in the gas phase and solve the corresponding problem, the result for the contact angle dependence of the contact-line speed will differ only by the expression for $u_{(12)}$: now, instead of $u_{(12)}(\theta_d, 0)$ we have to use $u_{(12)}(\theta_d, k_\mu)$ given by (3.29), i.e.,

$$u_{(12)}(\theta_d, k_\mu) = \frac{(\sin\theta_d - \theta_d\cos\theta_d)K(\theta_2) - k_\mu(\sin\theta_2 - \theta_2\cos\theta_2)K(\theta_d)}{(\sin\theta_d\cos\theta_d - \theta_d)K(\theta_2) + k_\mu(\sin\theta_2\cos\theta_2 - \theta_2)K(\theta_d)}, \quad (5.57)$$

where k_μ is the gas-to-liquid viscosity ratio, $\theta_2 = \pi - \theta_d$ and $K(\theta) = \theta^2 - \sin^2\theta$. It is also convenient to resolve equation (5.56) with respect to V to arrive at

$$V^2 = \frac{[1 + (\cos\theta_s - \sigma_{sg})(1 - \rho_G^s)](\cos\theta_s - \cos\theta_d)^2}{4(\cos\theta_s + B)(\cos\theta_d + B)}, \quad (5.58)$$

where now

$$B = (1 - \rho_G^s)^{-1}(1 + \rho_G^s u_{(12)}(\theta_d, k_\mu)) - \sigma_{sg}, \quad (5.59)$$

and one has to take the root corresponding to $\theta_d > \theta_s$.

An important qualitative feature of equations (5.56) and (5.58) to be noted here is that the dynamic contact angle appears to be dependent not just on the contact-line speed V but also on the flow field in the bulk. The latter comes via the radial component of the bulk velocity at the free surface in the far field $u_{(12)}$. In the local problem we are considering it was assumed that there are no boundaries located close to the contact line so that, to leading order in *Ca* as *Ca* → 0, locally the fluid motion outside the intermediate region can be described as a flow in a wedge. This allowed us to use the expression (3.21) for $u_{(12)}(\theta_d, 0)$ from Moffatt's (1964) solution (or $u_{(12)}(\theta_d, k_\mu)$ if the gas viscosity is taken into account). If the outer flow is influenced, for example, by

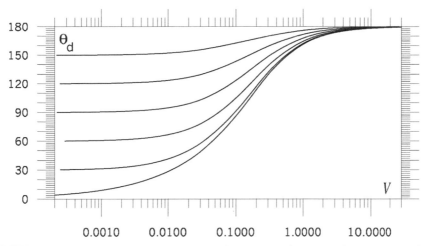

FIGURE 5.2: Dynamic contact angle versus dimensionless contact-line speed for different values of the static contact angle; $\rho_G^s = 0.9$, $\sigma_{sg} = 0$, $k_\mu = 0$.

other boundaries located sufficiently close to the contact line, then $u_{(12)}(\theta_d, 0)$ will have to be replaced by the corresponding expression for what, in the terminology of the theory of matched asymptotic expansions, is the inner limit of the outer solution. This will alter the dynamic contact angle for the *same* contact-line speed. This is exactly what we described in the experimental section of Chapter 3 as nonlocal influence of the flow field/geometry on the dynamic contact angle ('hydrodynamic assist of dynamic wetting'). We will consider this effect quantitatively in §5.6.

Figs 5.2–5.5 show the velocity dependence of the contact angle obtained using (5.58) for different values of parameters. If $\sigma_{sg} = 0$, $k_\mu = 0$, the contact angle as a function of V increases from its static value θ_s and asymptotically tends to 180° as $V \to \infty$ (Fig. 5.2). The inflection point in the semi-logarithmic plot of the velocity dependence of θ_d shifts slightly towards lower contact-line speeds as the value of the static contact angle increases.

As we discussed in §2.4.4, for a perfect solid-gas interface $\sigma_{(2)}$ and hence its dimensionless value σ_{sg} are equal to zero. Therefore, in fitting a theoretical curve to to experimental data one can view σ_{sg} as a measure of the system's imperfection, and, if the small capillary number asymptotics is used, a price for approximating a curved interface by a flat one (see Fig. 5.27). Fig. 5.3 shows that a nonzero value of σ_{sg} manifests itself only at large contact angles which in real systems correspond to finite capillary numbers.

Fig. 5.4 shows how the velocity dependence of the contact angle is influenced by the gas-to-liquid viscosity ratio k_μ. As one can see, there appears a maximum wetting speed V_* above which the algebraic equation (5.58) has no solution. For a given k_μ this speed corresponds to a certain value of $\theta_{d*} < 180°$

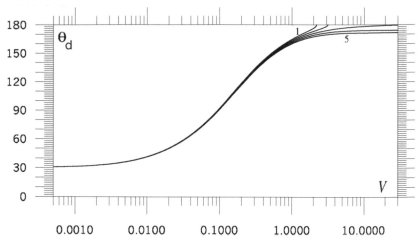

FIGURE 5.3: Dynamic contact angle versus dimensionless contact-line speed for different values of σ_{sg}. Curves 1–5 correspond to $\sigma_{sg} = -0.1$, -0.05, 0, 0.05 and 0.1, respectively; $\theta_s = 30°$, $\rho_G^s = 0.9$, $k_\mu = 0$. In the experimental data analysis, $\sigma_{sg} \neq 0$ is a measure of imperfection of the solid-gas interface as well as for approximating the curved interface with a planar one.

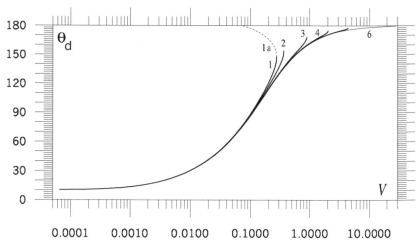

FIGURE 5.4: Dynamic contact angle versus dimensionless contact-line speed for different values of the gas-to-liquid viscosity ratio k_μ. Solid lines 1–6 correspond to $k_\mu = 2 \times 10^{-2}$, 10^{-2}, 10^{-3}, 10^{-4}, 10^{-5} and 0, respectively; $\theta_s = 10°$, $\rho_G^s = 0.9$, $\sigma_{sg} = 0$ for all curves. The dashed line 1a shows the unphysical branch of the velocity dependence of the contact angle.

and, as k_μ increases, θ_{d*} goes down. The same qualitative dependence of θ_{d*} on k_μ corresponds to what is observed in experiments for the onset of air entrainment.

A qualitative explanation of this phenomenon is very simple. The presence of a viscous gas slows down the liquid-gas interface compared to the case $k_\mu = 0$ and hence, according to (5.52), reduces the mass flux into the liquid-solid interface. This 'starvation' decreases $\rho^s(0)$ and hence, according to the surface equation of state (5.4), increases the surface tension acting on the contact line from the liquid-solid interface. Then, the force balance at the contact line (5.15) (or (5.53)) demands the contact angle to increase. This reduces the sector occupied by the (viscous) gas thus increasing the tangential stress it imposes on the liquid-gas interface and hence reducing $u_{(12)}$ even further. As V increases, the system moves towards a situation where the influence of the tangential stress from the gas on $u_{(12)}$ that slows the interface down begins to outweigh the increase in $u_{(12)}$ due to the opening up of the sector occupied by the liquid (see Fig. 3.2 in Chapter 3) so that one cannot simultaneously satisfy the mass and force balance conditions, (5.52) and (5.53). This situation corresponds to V_* and θ_{d*}.

It should be noted that here we are dealing with the leading order solution in Ca as $Ca \to 0$ and hence the interface near the contact line is planar. If the finiteness of the capillary number is taken into account, the situation remains the same qualitatively but quantitatively, due to the liquid-gas interface becoming more flexible and bending in response to the normal stresses acting on it, both the maximum speed of wetting and the maximum contact angle move towards larger values.

Finally, Fig. 5.5 shows the dependence of θ_d on V for different values of ρ_G^s. An increase in ρ_G^s towards 1 corresponds to an almost uniform shift of the curve $\theta_d = \theta_d(V)$ in semilogarithmic coordinates towards lower values of V. The solid lines correspond to the asymptotic formula (5.56) whereas the dashed ones are obtained by solving the boundary-value problem (5.45), (5.47), (5.52), (5.53) numerically. As the difference $(1 - \rho_G^s)$, which is the small parameter in the linearization of (5.45), (5.52), increases, the approximate formula (5.56) begins to slightly underpredict the contact angle. The difference, however, is well within the margin of error of standard experimental techniques so that (5.56) remains suitable for most practical purposes even for relatively large values of $(1 - \rho_G^s)$. As $\rho_G^s \to 1$, the asymptotic solution becomes indistinguishable from the numerical one.

5.2.2 Forces near the moving contact line

Consider the forces acting between the liquid and the solid in the vicinity of the moving contact line. Taking the tangential projection of (4.47), we have

$$\mathbf{n} \cdot \mathbf{P}^- \cdot (\mathbf{I} - \mathbf{nn}) = \mathbf{n} \cdot \mathbf{P}^+ \cdot (\mathbf{I} - \mathbf{nn}) + \nabla\sigma,$$

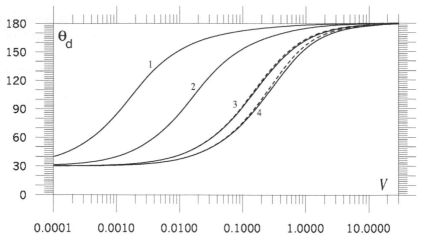

FIGURE 5.5: Dynamic contact angle versus dimensionless contact-line speed for different values of ρ_G^S. Curves 1–4 correspond $\rho_G^s = 0.999$, 0.99, 0.9 and 0.8, respectively. Solid lines are obtained using (5.56), dashed lines are solutions of the system (5.45), (5.47), (5.52), (5.53). For all curves $\sigma_{sg} = 0$, $k_\mu = 0$, $\theta_s = 30°$.

where $\mathbf{P}^-$ is the stress tensor corresponding to the force acting on the solid-facing side of the liquid-solid interface from the solid. In the inner asymptotic region associated with the limit $\epsilon \to 0$ for the force density $f = \mathbf{n} \cdot \mathbf{P}^- \cdot (\mathbf{I} - \mathbf{nn})$ in dimensionless variables we have

$$f = Ca \left[\frac{1}{\bar{r}} \frac{\partial u}{\partial \theta} + \bar{r} \frac{\partial}{\partial \bar{r}} \left(\frac{v}{\bar{r}} \right) \right]_{\bar{r},0} + \frac{d\sigma}{d\bar{r}}.$$

As $Ca \to 0$, in the 'intermediate' asymptotic region, to leading order, this expression becomes

$$\bar{f}(\bar{r}) = \frac{d\sigma}{d\bar{r}}(\bar{r}), \tag{5.60}$$

whereas in the 'viscous' region it takes the form

$$\tilde{f}(\tilde{r}) = \left[\frac{1}{\tilde{r}} \frac{\partial u}{\partial \theta} + \tilde{r} \frac{\partial}{\partial \tilde{r}} \left(\frac{v}{\tilde{r}} \right) \right]_{\tilde{r},0} + \frac{d\sigma_1}{d\tilde{r}},$$

where σ_1 corresponds to the term of $O(Ca)$ in the asymptotic expansion of σ in the viscous region. For the total force one has

$$F = \int_0^\infty \left[\bar{f}(\bar{r}) + \tilde{f}\left(\frac{\bar{r}}{Ca} \right) - \tilde{f}(\infty) \right] d\bar{r} = \sigma|_{\bar{r}=\infty} - \sigma|_{\bar{r}=0} + O(Ca)$$

$$\approx \lambda(1 - \rho_S^s) - \lambda(1 - \rho^s(0)). \tag{5.61}$$

In other words, since the force density is regular up to the contact line, the contribution from the viscous region is proportional to its size (i.e., is of $O(Ca)$) and the main contribution comes from the intermediate region, where, according to (5.60), to leading order, the force density is determined by the surface tension gradient. Now, after eliminating from (5.61) the surface densities using the Young equations (5.50) and (5.53), we arrive at

$$F = \cos\theta_s - \cos\theta_d. \tag{5.62}$$

This is a remarkable result: in the literature on wetting phenomena equation (5.62) is often postulated as an *additional* heuristic condition, whereas here we have derived it theoretically in the framework of a theory where dynamic wetting is just a particular process in a general class. The derivation makes clear the conditions of applicability attached to (5.62), first of all the importance of the capillary number being small to equate the force acting from the solid and the surface tension gradient along the liquid-solid interface. Secondly, the derivation of (5.62) implies that the surface tension in the liquid-solid interface away from the contact line is in equilibrium, i.e., that the size of this interface is large compared with the surface-tension-relaxation length. This is not the case, for example, for the initial stage of the drop spreading on a solid surface. As the drop's base starts expanding from zero (the point of contact), the whole of the solid-liquid interface is out of equilibrium, and it is only after the base sufficiently expands equation (5.62) becomes valid.

5.2.3 Flow-induced Marangoni effect

As shown in §5.2.1, in the case of small capillary numbers, it is possible to obtain the distributions of the surface parameters in the intermediate region along the liquid-gas and liquid-solid interfaces, (5.51) and (5.54), without considering the flow in the bulk. Now, once these distributions are found, equations (5.37), (5.39), (5.41) and (5.43) provide boundary conditions that specify the bulk flow:

$$u \equiv u_{(12)}(\theta_d, 0), \quad v = 0, \qquad (\bar{r} > 0, \ \theta = \theta_d),$$

$$u = 1 - \frac{Ck}{2(1 + 4A)V^2} \exp(-k\bar{r}), \quad v = -CQ\exp(-k\bar{r}), \qquad (\bar{r} > 0, \ \theta = 0).$$
$$\tag{5.63}$$

Thus, in the intermediate region we have a small degree of *apparent* slip, i.e., a difference between the tangential components of velocity on the opposite sides of the liquid-solid interface (but no *actual* slip on the solid surface) due entirely to the surface-tension gradient in the liquid-solid interface. Given that here the surface-tension gradient that causes apparent slip and hence influences the bulk velocity appears itself as a result of the flow, this can be called the *flow-induced Marangoni effect*.

The terms 'Marangoni effect' and 'Marangoni flow' conventionally refer to the fluid motion caused or influenced by the surface tension gradient (Marangoni & Stefanelli 1872, 1873). In the situations reported in the literature, the surface tension gradient appears due to nonhydrodynamic factors, such as a nonuniform temperature distribution and/or gradients in the concentration of surfactants and the dependence of the surface tension on temperature and/or the surfactant concentration (e.g., Scriven & Sterning 1960, Napolitano 1986). In the case we are considering here, the spreading of a liquid over a solid substrate continuously produces a fresh liquid-solid interface (the process of 'wetting') which leads to the appearance of the surface tension gradient along it via the boundary condition (5.63). This gradient then affects the flow that gave rise to its appearance. Thus, the essence of the flow-induced Marangoni effect is that the surface tension gradient is generated by the flow itself and then has a reverse effect on the flow. As we will see later, the flow-induced Marangoni effect, although it is not as well studied as the thermal and surfactant-driven one, is a quite common phenomenon. Its peculiarity in the moving contact-line problem is that it takes place along the liquid-*solid* interface, where, on the one hand, it is very difficult to extract it from the integral characteristics measured in real-life experiments and, on the other hand, it can lead to unexpected features in the products obtained by using coating technologies which will leave one wondering about their origin.

5.3 De-wetting and re-wetting

The processes of de-wetting and re-wetting, i.e., the spreading of a liquid over a solid surface that has previously been exposed to the same liquid, raise a number of interesting theoretical issues. As already discussed in §§3.2.2 and 3.2.3, experiments show that a receding liquid often, if not always, alters the properties of the solid substrate it leaves behind. In the absence of chemical reactions with the substrate, this influence can be described in terms of a microscopic residual film. From a theoretical viewpoint, the term 'microscopic', as before, means that the film's thickness is on a molecular scale so that in the continuum mechanics approximation it becomes zero, and hence, as well as the interfaces, the film has to be modelled as a two-dimensional 'surface phase'. Thus, the presence of a microscopic film effectively modifies the solid-gas interface. In the context of wetting phenomena, this modification is two-fold:

(i) A microscopic film alters $\sigma_{(2)}$ in the Young equation (2.164) so that the presence of such a film can alter the contact angle even in the static situation.

(ii) In the processes of de-wetting and re-wetting, the film enters the balance

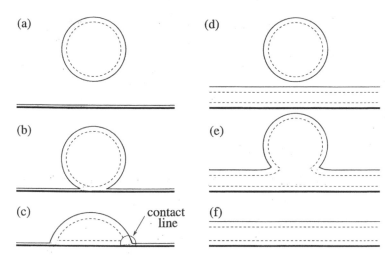

FIGURE 5.6: A sketch illustrating a qualitative difference between microscopic and macroscopic film with interfaces shown as interfacial layers of a finite thickness. A drop deposited on a microscopic film (a)–(c) assumes an equilibrium shape with a contact line and a certain contact angle, whereas a drop placed on a macroscopic film (d)–(f) coalesces with it.

of mass fluxes into and out of the contact line via condition (4.72), which now replaces (4.70), thus influencing the dynamic contact angle behaviour.

The first of these factors highlights the difference between microscopic and macroscopic films from an experimental point of view. If a film is microscopic, then one can talk of a contact line and, in general, a nonzero contact angle. Then, experimentally, if, say, a droplet is placed on such a film (Fig. 5.6a), it will evolve to an equilibrium shape with this contact angle (Fig. 5.6a–c). By contrast, a macroscopic film is effectively a layer of liquid with the corresponding bulk properties so that a drop placed on such a film (Fig. 5.6d) will simply coalesce with it (Fig. 5.6e). Then, at equilibrium one will still have a film (Fig. 5.6f), which, if not allowed to spread, will have a larger thickness. This qualitative difference between the two types of films points to a possibility of determining the range of intermolecular forces from wetting experiments.

5.3.1 Liquid spreading over a pre-wet surface

To describe the liquid spreading over a microscopically pre-wet solid surface, the only alteration we need in the problem formulation considered earlier is to replace (4.70) with (4.72), which in the dimensionless variables takes the form

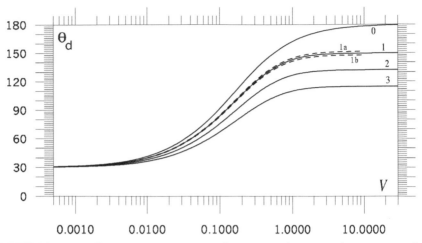

FIGURE 5.7: Dynamic contact angle versus dimensionless contact-line speed for the liquid spreading over a microscopic film. Solid lines 0–3 correspond to $\rho^s_{\text{film}} = 0$, 0.1, 0.2 and 0.3, respectively; $\theta_s = 30°$, $k_\mu = 0$, $\sigma_{sg} = 0$ for all curves. Dashed lines 1a and 1b correspond to $\sigma_{sg} = -0.1$ and 0.1, respectively; $\rho^s_{\text{film}} = 0.1$, $\theta_s = 30°$, $k_\mu = 0$.

$$(\rho^s v^s)|_{r\to 0,\,\theta=\theta_d} + (\rho^s v^s)|_{r\to 0,\,\theta=0} - \rho^s_{\text{film}} = 0. \qquad (5.64)$$

Here ρ^s_{film} is the surface density of the film scaled with $\rho^s_{(0)}$. In the process of re-wetting, a microscopic film in front of the main body of fluid is a given feature and both the tangential component of the force acting on the contact line, σ_{sg}, and ρ^s_{film} have to be prescribed.

In the case of small capillary numbers, the velocity dependence of the contact angle still has the form of equation (5.58), where this time

$$B = (1 - \rho^s_G)^{-1}[1 + \rho^s_G u_{(12)}(\theta_d, k_\mu) - \rho^s_{\text{film}}] - \sigma_{sg}. \qquad (5.65)$$

As before, $u_{(12)}(\theta_d, k_\mu)$ is given by (5.57).

Fig. 5.7 shows the velocity dependence of θ_d for different values of ρ^s_{film}. As one can see, the mass flux due to the film's presence influences this dependence in two ways: (a) the maximum contact angle value becomes significantly less than 180° and strongly dependent on the film's thickness, (b) the rate at which the contact angle increases with the contact-line speed becomes lower as the films' thickness goes up.

These features suggest a way of experimental verification and, if necessary, modification of the surface equation of state (4.61) and determining its parameters (Shikhmurzaev 1996). Broadly, the idea is to use ρ^s_{film} as an input and the measured value of θ_d as a macroscopic response to probe the microscopic physics that connects these two factors. The dimensional surface density of the film is an independent parameter, and it can be monitored or even manipulated, for example, by controlling the liquid's deposition on the solid substrate via the evaporation-condensation process. Then, using the dimensionless value of ρ_{film}^s as an adjustable parameter one can try to fit the maximum contact angle predicted by the theory to that observed in experiments. Remembering that ρ_{film}^s is scaled with $\rho_{(0)}^s$, one will then be able to determine the latter. By varying the surface density of the microscopic film, one can, in principle, determine the form of the whole function $\sigma(\rho^s)$. In the above procedure, it should be noted, however, that a variation in the film's thickness also leads to a variation in σ_{sg}, which, as we discussed earlier, could influence the maximum contact angle value. As shown in Fig. 5.7, the influence of σ_{sg} on the maximum contact angle is small compared with the role played by ρ_{film}^s and it can be easily taken into account. In order to avoid problems that might arise for large contact angles due to k_μ being nonzero, instead of the maximum contact angle, one can consider the rate at which the contact angle increases with the contact-line speed, say, at the inflection point of the semilogarithmic plot. The latter is strongly influenced by ρ_{film}^s but is practically independent of σ_{sg} and k_μ.

5.3.2 Receding contact lines

For the receding contact-line motion in the intermediate asymptotic region, instead of (5.45), (5.47), (5.52), (5.53), one has the problem

$$\frac{d\rho^s}{d\bar{r}} = -4V^2(1 + v^s), \qquad \frac{d(\rho^s v^s)}{d\bar{r}} = -(\rho^s - \rho_S^s), \qquad (\bar{r} > 0), \qquad (5.66)$$

$$v^s \to -1 \qquad \text{as } \bar{r} \to \infty, \qquad (5.67)$$

$$\rho^s(0)v^s(0) = \rho_G^s u_{(12)}(\theta_d, 0) - \rho_{\text{film}}^s(V), \qquad (5.68)$$

$$\rho^s(0) = 1 + [\cos\theta_d - \sigma_{SG}(V)](1 - \rho_G^s). \qquad (5.69)$$

An important difference between the spreading of a fluid over a pre-wet surface considered earlier and the process of de-wetting is that the former deals with a microscopic film with prescribed properties, whereas now we have the process of *formation* of the residual film behind the receding contact line. In this process, both the film's surface density ρ_{film}^s and the tangential component of the force acting on the contact line from the solid *and* the film, σ_{SG}, can be functions of the contact-line speed describing how the film is pulled out of the main body of fluid.

The problem of determining these functions theoretically is very nontrivial since, to resolve it comprehensively, one has to develop a structural model of the three-phase-interaction zone (the 'contact line') accounting for its dynamic behaviour. An alternative approach is to tackle the problem macroscopically on the basis of some sufficiently general principles and end up with a number of phenomenological parameters to be determined experimentally.

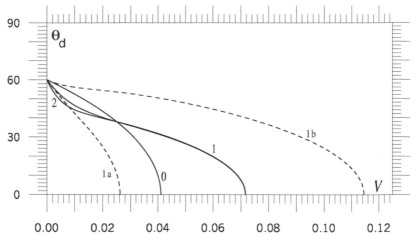

FIGURE 5.8: Receding dynamic contact angle versus dimensionless contact-line speed for different regimes of the residual film formation obtained using (5.70)–(5.72) for the following values of parameters: 0 — $\rho_\infty^s = 0$, $\sigma_\infty = 0$ (no residual film); 1 — $\rho_\infty^s = 0.4$, $\sigma_\infty = 0.2$, $V_\infty = 0.01$; 2 — $\rho_\infty^s = 0.4$, $\sigma_\infty = 0.2$, $V_\infty = 0.005$. Dashed lines 1a and 1b correspond to $\rho_\infty^s = 0$, $\sigma_\infty = 0.2$, $V_\infty = 0.01$ and $\rho_\infty^s = 0.4$, $\sigma_\infty = 0$, $V_\infty = 0.01$, respectively. For all curves $\theta_s = 60°$, $\rho_G^s = 0.9$, $\sigma_{sg} = 0$.

Here we are interested in the qualitative influence of the film on the dynamic contact angle which can be examined by using simple relaxation-type formulae. We will assume that, as the contact-line speed increases, ρ_{film}^s and σ_{SG} both tend exponentially to some limiting values ρ_∞^s and σ_∞,

$$\rho_{\text{film}}^s(V) = \rho_\infty^s[1 - \exp(-V/V_\infty)], \tag{5.70}$$

$$\sigma_{SG}(V) = \sigma_\infty[1 - \exp(-V/V_\infty)], \tag{5.71}$$

where the constants ρ_∞^s, σ_∞ and V_∞ depend on the materials of the contacting media.

The solution of the linearized problem (5.66)–(5.69) leads to the following implicit velocity-dependence of the contact angle:

$$\cos\theta_d - \cos\theta_s + \sigma_{sg} - \sigma_{SG}(V)$$

$$= \frac{2V\{\cos\theta_s - \sigma_{sg} + (1 - \rho_G^s)^{-1}[1 + \rho_G^s u_{(12)}(\theta_d, 0) - \rho_{\text{film}}^s(V)]\}}{[V^2 + 1 + (\cos\theta_s - \sigma_{sg})(1 - \rho_G^s)]^{1/2} - V}. \tag{5.72}$$

The main features of the set of equations (5.70)–(5.72) are illustrated in Fig. 5.8. The most important effect is that in both cases with or without a microscopic residual film there exists a finite maximum speed of de-wetting, V_{**}, above which there is no solution (curves 0–2, Fig. 5.8). Physically, the existence of V_{**} points to the same effects as we had with the maximum speed of wetting V_*: if a liquid is displaced by a gas from a solid surface at a speed greater than V_{**}, a macroscopic film is left behind the main body of liquid, and the actual contact line assumes a sawtooth shape emitting droplets from the trailing edges (Blake & Ruschak 1979).

Another pecularity of the velocity dependence of the contact angle given by (5.70)–(5.72) is that, in a general case, the curve $\theta_d = \theta_d(V)$ has an inflection point. This feature was observed in a number of experiments (see §3.2.3) and could not be reproduced by early theories. As shown in Fig. 5.8, the appearance of the microscopic film brings in effects acting in the opposite directions. On the one hand, the additional mass flux out of the contact line into the film slows down the rate at which the contact angle decreases and tends to increase V_{**} (dashed line 1b). In the relaxation regime of the film formation described by (5.70) this factor per se leads to an inflection point in the dependence $\theta_d = \theta_d(V)$, though this feature is not well pronounced. On the other hand, the tangential component of the force acting on the contact line, σ_{SG}, makes the dependence $\theta_d = \theta_d(V)$ steeper and reduces V_{**} (dashed line 1a in Fig. 5.8). The combined effect of these two factors leads to a well-pronounced changing sign of the second derivative in the velocity-dependence of the receding contact angle (curves 1 and 2 in Fig. 5.8), as is indeed observed in experiments (§3.2.3).

5.4   Comparison with experiments and some estimates

For a fully quantitative experimental verification of predictions for the contact angle behaviour resulting from the solution of the problem (5.45), (5.47), (5.52), (5.53) for the advancing or (5.66)–(5.69) for the receding contact angle, it is necessary to determine all parameters involved in the model from a series of independent experiments. This is a feasible experimental task since, as discussed in Chapter 4, these parameters have a clear physical meaning and, as we will see in the subsequent chapters, the model is applicable to the processes other than dynamic wetting and they could be used for measuring the constants. Below, we consider how the experimental data already available in the literature can be described using the interface formation theory.

In experimental works, the velocity dependence of the contact angle is usually presented as θ_d versus the capillary number Ca (see §3.2.3). The dimensionless speed V introduced earlier in this chapter is related to Ca by

$$V = Ca\,Sc, \qquad \text{where } Sc = \left(\frac{\sigma_e^2 \tau \beta}{\mu^2 \gamma \rho_{(0)}^s (1 + 4\alpha\beta)}\right)^{1/2} \tag{5.73}$$

is the scaling factor depending on the material properties of the fluid and the

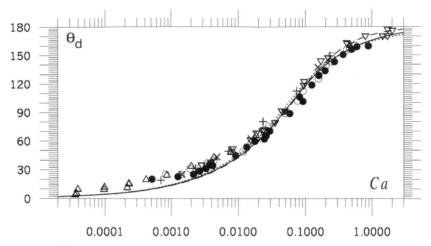

FIGURE 5.9: Comparison of the theory with experiments from Fig. 3.12. The dashed, solid and dotted lines are obtained for $\rho_G^s = 0.9$, 0.99 and 0.999 with $\log(Sc) = 0.35$, -0.55 and -1.5, respectively. For all curves $\theta_s = 0°$, $\sigma_{sg} = 0$, $k_\mu = 0$.

interfaces. In the semilogarithmic coordinates, the transition from $(\log(V), \theta_d)$ to $(\log(Ca), \theta_d)$ corresponds to a shift of the curve as a whole by $\log(Sc)$. Fig. 5.5 shows that a similar almost uniform shift results from a change in ρ_G^s. Thus, for a preliminary comparison of the theory with experiments we may fix the value of ρ_G^s and use the parameter $\log(Sc)$ as an adjustable constant trying to fit the theoretical curve to experimental data. The parameter σ_{sg} plays a role only very close to $180°$, where the experimental data themselves become unreliable, and we can set $\sigma_{sg} = 0$, whereas the gas-to-liquid viscosity ratio k_μ is known and for medium to high viscosity liquids can be neglected unless one is dealing with the contact angles very close to $180°$.

Figs 5.9–5.10 show how the predictions following from the approximating formula (5.56) compare with representative experiments reported by different authors. As one can see, the use of $\log(Sc)$ as the only adjustable constant, makes it possible to describe experimentally observed velocity dependence of the contact angle for the contact-line speed spanning over several orders of magnitude.

A more insightful way of comparing the theory with experiments is to use a series of experimental curves for the liquids whose physical properties could be correlated via some plausible physical assumptions. Ideally, this would allow one to verify the theory for a wide range of parameters, validate the physical mechanisms behind the correlating assumptions and determine at least some of the constants.

The main problem in this way is that such a procedure requires dealing with fluids which are both 'similar' and 'different'. The similarity is needed

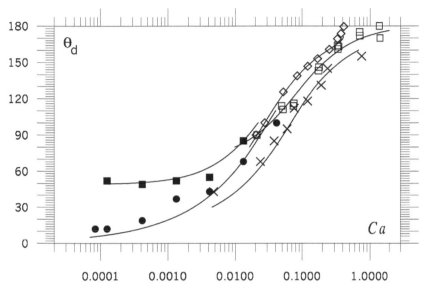

FIGURE 5.10: Comparison with some other experiments. Solid lines correspond to the theory ($\rho_G^s = 0.99$, $\sigma_{sg} = 0$ and $k_\mu = 0$ unless stated otherwise). Open squares: Hoffman (1975), Admex 760 on glass, $\theta_s = 69°$, $\log(Sc) = -0.45$; crosses: Ström et al. (1990), silocone oil 3 on untreated polystyrene, $\theta_s = 12^c icr$, $\log(Sc) = -0.6$; full circles and squares: Ström et al. (1990), paraffin oil on untreated polystyrene ($\theta_s = 0°$) and on polytetrafluoroethylene ($\theta_s = 49°$), $\log(Sc) = -0.25$ for both theoretical curves; full diamonds: Burley & Kennedy (1976b), polypropylene glycol-propyl alcohol on polyester, $\theta_s = 0°$, $\log(Sc) = -0.2$, $k_\mu = 1.8 \times 10^{-2}/46.5$.

TABLE 5.1: Parameters of the water-glycerol solutions in
Fig. 5.11 (Blake & Shikhmurzaev 2002)

% of glycerol	μ, mPa s	σ_e, dyn/cm	θ_s, °	ρ, g/cm^3	T, °C
16	1.5	69.7	72.5	1.036	22.0
43	4.2	64.9	67.0	1.104	22.0
59	10	65.3	64.5	1.150	21.1
68	19	64.8	65.5	1.175	22.0
70	23	63.5	67.0	1.181	24.0
80	58	64.5	65.2	1.208	23.1
86	104	65.8	65.0	1.224	22.3
95	672	64.5	62.0	1.248	21.5

to come up with the correlating assumptions that would bring these fluids into
the same family and minimize the number of such assumptions. On the other
hand, however, the fluids should be sufficiently different for the variation of
their parameters to be wide enough to separate the experimental trends from
experimental errors.

We will consider one step towards the realisation of this programme and use
the experimental data shown in Fig. 3.13 for a series of water-glycerol solutions
which can be regarded as 'similar' fluids with a sufficiently large range of
variation of viscosity. It should be noted here that, although these data were
reported in Blake & Shikhmurzaev (2002) together with their analysis using
the present theory, the experiments themselves had been performed before
the theory was developed and hence should be regarded as 'available in the
literature' rather than 'theory-driven'. Unlike Blake & Shikhmurzaev (2002),
where the approximate formula (5.56) was used, here we will be applying the
solution of the original problem (5.45), (5.47), (5.52), (5.53).

The data presented in Table 5.1 show that, in the series of liquids we consider, as the concentration of glycerol increases from 16 to 95 percent, the
viscosity goes up by nearly 3 orders of magnitude (about 450 times) whereas
the variation in the surface tension, static contact angle and density are relatively small. To simplify the analysis and examine only the main effect, we
will apply our theoretical model assuming that by changing the composition
of the liquid we affect only its bulk viscosity and the interfacial parameters
which depend on viscosity, and will use the mean values for the surface tension
and the static contact angle, $\sigma_e = 65.4$ dyn cm^{-1} and $\theta_s = 66.1°$, respectively.
As a result of these simplifications and the actual 10% variation of θ_s in the
table, we should not expect the theory to accurately describe the data for
very low contact-line speeds where the dynamic contact angle is close to the
static one. The system's behavior in this region may also be complicated by
the stick-slip phenomenon (see Fig. 3.16), which suggests that in this region
some 'extra' physics comes into play (Blake 1993, Shikhmurzaev 2002).

Next, we have to account for the viscosity dependence of the phenomenological parameters involved in our model, which means making assumptions

about the nature of interfaces and the microscopic physical mechanism of their formation. A discussion on the derivation of the model and the physical meaning of the coefficients in Chapter 4 suggests the following correlations.

The surface equation of state (4.61) holds for both the static and dynamic situation, and the simplest interpretation of γ is that it reflects the ability of the interfacial layer to be rarefied or compressed due the nonsymmetric action of intermolecular forces from the bulk phases. This is an equilibrium property independent of viscosity.

As discussed in §4.3.4, the analogy between the flow in an interfacial layer and a flow in a plane channel suggests that for the coefficients α and β, which describe, respectively, the effect of the surface-tension gradient and the shear stress on the velocity distribution in the interfacial layer, one has

$$\alpha \propto \frac{\ell}{\mu}, \qquad \beta \propto \frac{\mu}{\ell}. \tag{5.74}$$

Here, as before, ℓ is the characteristic thickness of the interfacial layer.

Finally, we have to consider the surface-tension-relaxation time τ, which is related to the microscopic mechanism of the interface formation. In the model's derivation, the relaxation term on the right-hand side of equations (4.58) and (4.65) results from considering the mass exchange between the interface and the bulk driven by the difference in the chemical potentials. This implies that the surface tension relaxation is of a diffusive nature and hence

$$\tau \propto \mu. \tag{5.75}$$

In our model, a theoretical curve in the (θ_d, Ca)-plane depends on three parameters, Sc, ρ_G^s, σ_{sg}, and the above assumptions mean that all these parameters are viscosity-independent. Thus, we have to describe all 8 sets of experimental data given in Fig. 3.13 with the same values of Sc, ρ_G^s and σ_{sg}, which we will use here as adjustable constants. We will pay particular attention to the high-speed end of the curves, which are more sensitive to hydrodynamic factors.

Fig. 5.11 shows that, indeed, theoretical curves generated with Sc, ρ_G^s and σ_{sg} constant do describe the data for all 8 water-glycerol solutions, where the viscosity spans nearly three orders of magnitude, over a wide range of contact-line speeds (with the anticipated discrepancy at low speeds). This can be interpreted as a qualitative argument in favour of the assumptions made about the viscosity dependence of the phenomenological coefficients of the model, which are consistent with the diffusion mechanism of interface formation. It should be emphasized that if τ were associated with some physical mechanisms unrelated to viscosity, the variation of Sc of more than an order of magnitude would have been apparent. In §5.5.1, we will show that the same conclusion about the microscopic mechanism of the interface formation follows from a completely different set of experiments, where 'similar' fluids with a wide variation of viscosity are created by the temperature variations.

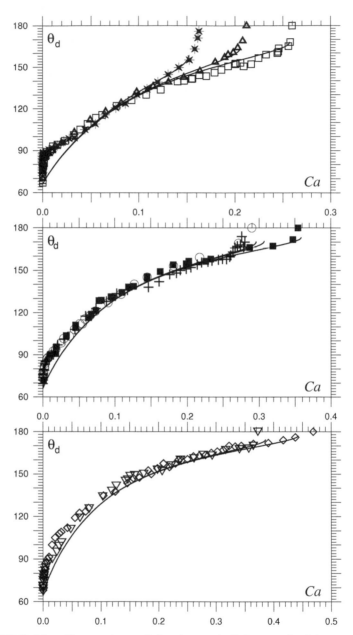

FIGURE 5.11: Comparison of the theory (solid curves) with experimental data of Fig. 3.13 for water-glycerol solutions with different concentration of glycerol. All curves correspond to the same values of nondimensional parameters: $Sc = 4.4$, $\rho_G^s = 0.55$, $\sigma_{sg} = -0.04$.

It is interesting to try to come up with an estimate for τ. In order to do this, one has to make assumptions about other phenomenological constants which are not associated with unsteadiness of the process. We will assume that both coefficients of proportionality in (5.74) are equal to 1, i.e., $\alpha = \ell\mu^{-1}$, $\beta = \mu\ell^{-1}$ and hence $\alpha\beta = 1$. It is also plausible to take, as an estimate, $\rho^s_{(0)} \approx \rho\ell$. Then, from the expression (5.73) for Sc and the equation of state (4.61) taken at equilibrium one has

$$5\mu\ell^2\rho\gamma Sc^2 = \tau\sigma_e^2, \qquad \sigma_e = \gamma\rho\ell(1 - \rho^s_G). \tag{5.76}$$

Eliminating γ from these equations, we arrive at

$$\tau = \frac{5Sc^2\mu\ell}{(1 - \rho^s_G)\sigma_e}. \tag{5.77}$$

For simple fluids we can estimate the interfacial thickness ℓ to be of order 1 to 3 nm. This range is consistent, for example, with experiments on the variation of the liquid-gas interfacial tension caused by the deposition of thin layers of one liquid on top of another (Harkins 1952). Using these values in (5.77) together with $Sc = 4.4$, $\rho^s_G = 0.55$, which we have in Fig. 5.11, one arrives at an estimate $\tau = 5 \times 10^{-9}$–1.5×10^{-8} s for $\mu = 1.5$ mPa s and $\tau = 2.2 \times 10^{-6}$–$6.6 \times 10^{-6}$ s for $\mu = 672$ mPa s. These estimates are in the range of what one would expect for simple fluids.[4]

It is worth mentioning that $\ell \sim 1$–3 nm in the second equation (5.76) leads to $\gamma = 0.5 \times 10^9$–1.5×10^9 cm^2 s^{-2}. In the surface equations of state (4.61) γ is inversely proportional to the fluid's compressibility, and the obtained estimates for γ are slightly below the square of the speed of sound ($\sim 10^{10}$ cm^2 s^{-2}), which characterizes compressibility of the liquid in macroscopic experiments. This trend is consistent with how (4.61) approximates a more general surface equation of state (4.60), as illustrated in Fig. 4.3, and the fact that, subject to strong intermolecular forces, the liquid appears to be more compressible on a microscopic length scale than in macroscopic experiments.

A slightly negative value of σ_{sg} obtained in fitting theory to experiment is a measure of error in the approximation. As one should expect, the application of asymptotic results obtained in the limit $Ca \to 0$ to the experiments where

---

[4]Here it is instructive to mention a rather eccentric view on how to estimate τ expressed by Eggers & Evans (2004). These authors suggested that relaxation of the surface tension to its equilibrium value happens... at the speed of light! Then, after dividing the typical size of a molecule of a simple liquid ($\sim 3 \times 10^{-8}$ cm) by the speed of light (3×10^{10} cm s^{-1}) they arrived at $\tau \approx 10^{-18}$ s. If correct, this estimate would have meant that the dynamic surface tension plays no role in fluid mechanics since macroscopic flows are characterized by much larger time scales. In a response to this argument (Shikhmurzaev & Blake 2004) it has been pointed out that, according to the second law of thermodynamics, relaxation of a system towards the state of thermodynamic equilibrium is a *dissipative* process, whereas the speed of light has nothing to do with dissipation, and an estimate of the rate of a dissipative process based on considering a nondissipative process is obviously incorrect.

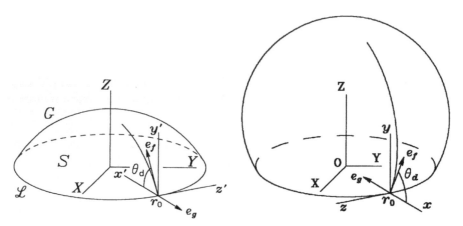

FIGURE 5.12: Sketches of a drop spreading (left) and a bubble growing (right) on a horizontal solid surface. In both cases, the contact angle θ_d is measured through the liquid. If the characteristic surface-tension-relaxation length $l = U\tau$ is small compared with the macroscopic length scale characterizing the process and $Ca \ll 1$, in the local coordinate system (x', y', z') one has the moving contact-line problem in a wedge. If $Re \ll 1$, then to leading order the inner limit of the outer flow is represented by Moffatt's (1964) solution and hence the velocity dependence of the dynamic contact angle is given by (5.58).

Ca reaches 0.45 will underpredict the value of the contact angle. As Ca increases, so does the free surface bending and an approximation with a planar free surface will lead to smaller values of the contact angle than those that would follow from the full-scale numerical solution (Fig. 5.27). The slightly negative value of σ_{sg}, which manifests itself only when θ_d is close to 180° (see Fig. 5.3) compensates for this effect.

5.5 Examples: flows in a quasi-static regime

As we have already mentioned, the results obtained for the local moving contact-line problem can be incorporated as ready-to-use elements into the models describing various global flows. Below, we illustrate this point with two closely related examples.

5.5.1 Spreading of drops on solid substrates

Consider the axisymmetric spreading of a drop of a viscous incompressible liquid deposited onto a smooth chemically homogeneous solid substrate in the presence of the gravity force. The flow domain $\mathcal{D}(t)$ is bounded by a piecewise smooth boundary consisting of a free surface G, a wetted part of the solid boundary S and the contact line $\mathcal{L}$ (Fig. 5.12, left). To describe the flow in a general case, one has to solve the Naviers-Stokes equations (4.53) in a flow domain with a varying shape subject to the boundary conditions (4.54)–(4.59), (4.61) and (4.62)–(4.66), (4.68) at the interfaces, conditions (4.70), (4.73) at the moving contact line and some initial conditions specifying the velocity at which the drop lands on the solid surface, initial distributions of velocity in the bulk and the surface density as well as the initial free surface shape. This general case is rather difficult even for a reliable computational analysis. We will consider a situation where the study can be reduced to a few simple elements (Shikhmurzaev 1997b).

In this problem, the characteristic macroscopic length scale L is now specified by the volume $\mathcal{V}$ of the drop, $L = (3\mathcal{V}/4\pi)^{1/3}$, and the characteristic velocity is given by $U = (\gamma\rho_{(0)}^s(1 + 4\alpha\beta)/(\tau\beta))^{1/2}$. The flow is controlled by the following nondimensional parameters: $\epsilon = U\tau/L$ (the dimensionless relaxation length), $Ca = \mu U/\sigma_e$ (the capillary number), $We = \rho LU^2/\sigma_e$ (the Weber number), $Bo = \rho g L^2/\sigma_e$ (the Bond number; g is the acceleration of gravity) and $\lambda = \gamma\rho_{(0)}^s/\sigma_e$. If ϵ, Ca and We are small, then the standard asymptotic analysis gives that everywhere, apart from a small 'inner' region near the contact line, the free-surface shape is not influenced by the bulk hydrodynamics and is determined by the same equation as in hydrostatics:

$$p = -\nabla \cdot \mathbf{n}. \tag{5.78}$$

The pressure inside the drop also satisfies the equation of hydrostatic equilibrium:

$$\nabla p = -Bo\,\mathbf{e}_z, \tag{5.79}$$

where $\mathbf{e}_z$ is a unit vector directed versus the gravity force. Equation (5.79) immediately yields $p = p_0 - Bo\,z$, where p_0 is a yet unknown constant of integration, and after substituting this expression into (5.78) the latter becomes an ordinary differential equation for the free surface profile:

$$\frac{ff'' - 1 - f'^2}{f(1 + f'^2)^{3/2}} = Bo\,z - p_0, \tag{5.80}$$

where $r = f(z,t)$ is the equation of the free surface and the prime denotes differentiation with respect to z. Integrating this equation subject to the boundary conditions

$$f(z_0, t) = 0, \qquad f'(z_0, t) = \infty, \tag{5.81}$$

ensuring that the drop is smooth at the top, the condition that the free surface meets the solid surface at the contact line whose location is given,

$$f(0, t) = r_0(t), \qquad (5.82)$$

and the condition that the drop's volume remains constant throughout the spreading process,

$$\int_0^{z_0} f^2 \, dz = \frac{4}{3}, \qquad (5.83)$$

allows one to find the free surface profile together with p_0 and z_0 corresponding to a given moment in time.

In other words, in the outer region at every moment the free surface shape, including the contact angle, is completely determined by the current position of the contact line. In the 'inner' region, however, one has the full-scale moving contact-line problem and the contact angle from the outer region,

$$\theta_d(t) = \frac{\pi}{2} + \arctan(f')$$

determines the contact-line speed:

$$\frac{dr_0}{dt} = V(\theta_d(t)),$$

where $V(\theta_d)$ is given by (5.58).

Thus, at every moment in time the position of the contact line determines the contact angle as part of the global (static) shape of the drop and this angle in its turn determines, via the solution (5.58) of the local moving contact-line problem, the speed with which the contact line is moving. This coupling between the static free-surface shape with the input from the full-scale dynamics of the contact line describes the so-called quasi-static evolution of the drop. A few examples illustrating this flow regime are given in Fig. 5.13.

It should be noted that, for a given Bond number, there exists a solution to the problem (5.80)–(5.83) only for r_0 greater than a certain minimum radius R_0, which is equal to 0 for $Bo = 0$ and increases with the Bond number. Physically, this means that the quasi-static regime of spreading describes the final stage of the drop evolution whereas the initial stage always depends on the details of how the drop was deposited onto the solid substrate. If $Bo = 0$ and the drop is deposited at zero velocity, this 'final' stage begins from the very moment the drop touches the solid surface. In Fig. 5.13, the calculations started from $r_0(0) = R_0$.

The solution for the drop spreading in a quasi-static regime has been verified against the experiments reported by Hocking & Rivers (1982) who investigated the evolution of droplets of molten glass of different composition and at different temperatures. In order to bring theory and experiment together,

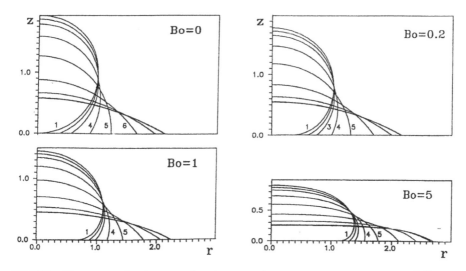

FIGURE 5.13: Evolution of a drop in the quasi-static regime at different Bond numbers (Shikhmurzaev 1997). Curves 1, 2, ... correspond to $t = 0$, 10^0, 3×10^0, 10^1, 3×10^1, ... ∞. $\theta_s = 30°$, $\rho_G^s = 0.99$, $\sigma_{sg} = 0$ for all curves.

one has to rescale the time by a factor

$$k_t = \frac{\sigma_e (\tau\beta)^{1/2}}{2\mu[\gamma\rho_{(0)}^s(1 + 4\alpha\beta)]^{1/2}},$$

and, in the cases where the beginning of the spreading in the experiment was recorded with a considerable uncertainty, to shift the time origin by δt. Some of the results are shown in Fig. 5.14.

An important qualitative result that follows from the analysis of experiments is that the surface tension relaxation time τ again turns out to be proportional to the fluid's viscosity. Indeed, assuming as before that $\alpha \propto \mu^{-1}$, $\beta \propto \mu$ and $\tau \propto \mu$, we have that k_t must be independent of μ. Then, for the same composition of the fluid and the temperature affecting only its viscosity the experimental data corresponding to different temperatures should fall into the same curve. As one can see from Fig. 5.14, this is indeed the case.

Thus, for two different ways of creating 'similar' fluids with a wide range of viscosity variation, i.e., by changing the fluid's composition, as in §5.4, and by varying the temperature of the same fluid, we have that an assumption $\tau \propto \mu$ adequately describes the experimental data. This is consistent with the diffusive mechanism of the interface formation.

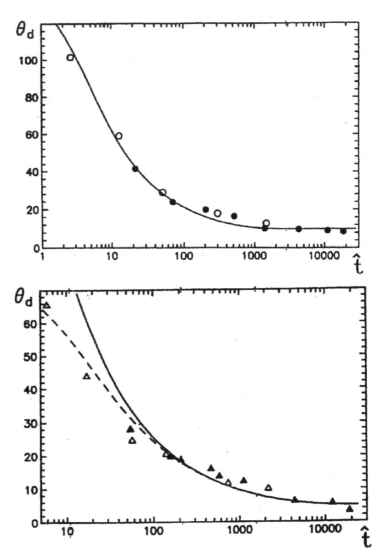

FIGURE 5.14: Experimental data of Hocking & Rivers (1982) for different types of molten glass at different temperatures versus theory (Shikhmurzaev 1997). Top: glass with addition of TiO_2 at the 5 mol % level for $T = 1010°C$, $\mu = 5600$ P (open circles) and $T = 1180°C$, $\mu = 630$ P (full circles). Theory: $\theta_s = 10°$, $\rho_G^s = 0.99$, $\sigma_{sg} = 0$; the time is rescaled using $k_t = 0.15$, $\delta t = 0$. Bottom: glass with TiO_2 at 20 mol % level for the same temperatures, $\mu = 2500$ P (open triangles), $\mu = 280$ P (full triangles). Theory: $\theta_s = 5°$ with the same ρ_G^s and σ_{sg}; the time rescaled with $k_t = 0.25$, $\delta t = 0$ (solid curve) and $\delta t = 10$ (dashed curve).

5.5.2   Bubble growth on a plate

Consider a bubble growing on a horizontal perfectly smooth solid plate
(Shikhmurzaev 1999) due to a gas injection through an infinitesimal hole so
that the bubble's volume increases linearly from some initial value b_0:[5]

$$\int_{\mathcal{B}(t)} dx\, dy\, dz = at + b_0.$$

Now, assuming that the (a priori unknown) pressure inside the bubble is
spatially uniform, one has to consider the fluid flow only outside the growing
cavity (Fig. 5.12). Since, unlike the case of the spreading drop, the volume
of the domain $\mathcal{B}$ is no longer constant, the characteristic length scale of the
process can be specified as

$$L = \left(\frac{\sigma_e}{\rho g}\right)^{1/2}, \tag{5.84}$$

thus making the Bond number equal to 1. The characteristic time scale is
now associated with the rate-of-growth of the bubble volume, $t_0 = L^3 a^{-1}$,
and $U = Lt_0^{-1}$.

The Bond number is no longer on the list of parameteres specifying the
process and instead one has to include

$$\bar{a} = \frac{\rho g a}{\sigma_e} \left(\frac{\beta \tau}{\gamma \rho_{(0)}^s (1 + 4\alpha\beta)}\right)^{1/2},$$

which is the dimensionless rate-of-growth of the bubble's volume,. and the
dimensionless initial volume $\bar{b}_0 = b_0 \rho^{3/2} g^{3/2} \sigma_e^{-3/2}$.

We can estimate the values of nondimensional parameters by considering,
for example, the experimental conditions typical of cold bubbling in water
(Lin, Banerji & Yasuda 1994). For the parameter a ranging between 5×10^{-3}
and 0.5 cm^3 s^{-1}, one has $10^{-5} \le Ca \le 10^{-3}$ and $1.8 \times 10^{-5} \le We \le 0.18$.
From (5.84) it follows that $L = 0.27$ cm (in experiments, the actual diameter
of the bubble at the moment of detachment was in the range of 0.226 to 0.58
cm), and for t_0 one has $0.04 \le t_0 \le 4$ s. Then, even using the surprisingly
high value of $\tau = 6 \times 10^{-4}$ s obtained for water by the oscillating jet method,[6]
(Kochurova & Rusanov 1981) we end up with $1.5 \times 10^{-4} \le \epsilon \le 1.5 \times 10^{-2}$.

Thus, the estimates make it reasonable to consider the limit $\epsilon \to 0$, $Ca \to 0$,
$We \to 0$, i.e., the quasi-static regime of the bubble growth. As in the case of

---

[5]Here we are not considering the problem of the bubble nucleation, which has a number of
interesting issues of its own.

[6]It should be noted, however, that this value has been obtained by 'calibrating' Bohr's (1909,
1910) original formulae and it is not supported by recent measurements that interpret a
similar phenomenon (Alakoç et al. 2004).

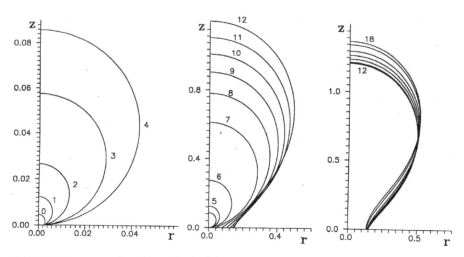

FIGURE 5.15: Profiles of a bubble growing on a solid surface in the quasi-static regime (Shikhmurzaev 1999). For all curves $\theta_s = 60°$, $\bar{a} = 10^{-2}$, $\rho_G^s = 0.99$, $\sigma_{sg} = 0$. Left: the first stage of the bubble growth where the process is dominated by the volume expansion; curves 0–4 correspond to $t = 0$, 10^{-6}, 10^{-5}, 10^{-4} and 3.3×10^{-4}, respectively. Middle: the second stage where the bubble's base expands; curves 4–12 correspond to $t = 3.3 \times 10^{-4}$, 10^{-3}, 10^{-2}, $0.1, 0.2, 0.3, 0.4, 0.5$ and 0.594, respectively. Right: the base starts to contract and the 'neck' appears; curves 12–18 correspond to $t = 0.594, 0.6, 0.62, 0.64, 0.66, 0.68$ and 0.685, respectively. For larger volumes the quasi-static regime gives way to a dynamic one.

the spreading drop, the free-surface shape is determined hydrostatically as a solution of the equation

$$\nabla \cdot \mathbf{n} = z + C$$

subject to (5.81), (5.82) and the condition

$$\int_{\mathcal{B}(t)} dx\, dy\, dz = t + \bar{b}_0,$$

which now replaces (5.83). In the inner asymptotic region, one again has the full moving contact-line problem where the value of θ_d obtained from the outer solution determines the contact-line speed.

Figs 5.15–5.16 show the profiles of a growing bubble for different values of the static contact angle and the dimensionless initial volume $\bar{b}_0 = 10^{-7}$. The initial radius of the bubble is chosen in such a way that $\theta_d = \theta_s$ at $t = 0$. Calculations show that, as one would expect, the value of $\bar{b}_0$ influences the bubble growth on a time scale comparable with $\bar{b}_0$ and becomes unimportant as the bubble volume increases.

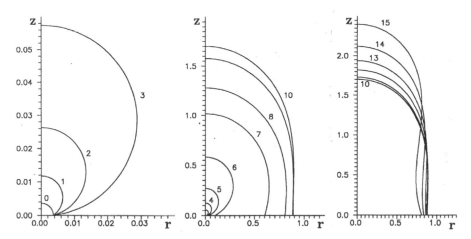

FIGURE 5.16: Profiles of a bubble growing on a solid surface for all curves $\theta_s = 90°$; other parameters are the same as in Fig. 5.15. Curves 0–10 correspond to $t = 0$, 10^{-6}, 10^{-5}, 10^{-4}, 10^{-3}, 10^{-2}, 0.1, 1, 2, 3 and 3.33, respectively. Right: the bubble base is starting to contract; curves 10–15 correspond to $t = 3.33$, 3.4, 3.6, 3.8, 4 and 4.097, respectively.

As one can see from Figs 5.15–5.16, there are three main stages of the bubble growth in a quasi-static regime. In the first stage, the main factor is the volume increase and, though the contact line is moving, the variations in its position are negligibly small compared with the bubble size. As a result, the contact line looks almost pinned (Fig. 5.15, curves 0–4; Fig. 5.16, curves 0–3). In this stage, the variations of the (hydrostatic) pressure in the liquid on the scale of the bubble are small compared with the capillary pressure, and hence the deviations of the free-surface shape from a spherical segment are relatively small. These factors result in the (receding) dynamic contact angle going down and the contact line accelerating.

As the bubble's volume increases, the relative role of its rate-of-growth decreases, and the dominating factors become the contact-line motion and the free-surface shape deformation due to the hydrostatic pressure (Fig. 5.15, curves 5–12; Fig. 5.16, curves 4–10). The bubble base expands until its radius reaches a maximum value corresponding to $\theta_d = \theta_s$. As θ_d tends to θ_s, the contact line decelerates. The free-surface shape no longer resembles a spherical cap and the bubble starts to grow in the vertical direction. In a hydraulic approach (Buyevich & Webbon 1996), this stage is modelled as the lengthening of a cylindrical column with a growing spherical cap on top.

In the third stage (Fig. 5.15, curves 12–18; Fig. 5.16, curves 10–15), as the bubble grows further in the vertical direction, a narrowing 'neck' is beginning to form making $\theta_d > \theta_s$ and hence causing the contact line to advance and the bubble base to contract. This process accelerates very rapidly and, irrespectively of the values of the small parameters, only the beginning of it takes place in the quasi-static regime. The next, fourth, stage is much shorter than the previous ones and, to describe it, one will have to include the dynamic terms neglected in the quasi-static approximation.

By manipulating the governing parameters of the problem, one can vary the shape of the bubble and the size of the base corresponding to the onset of the fourth stage where the neck connecting the departing bubble and the volume that remains attached to the solid surface. As one can see from Figs 5.15–5.16, as the solid surface wettability increases, i.e., the static contact angle goes down, both the maximum size of the bubble base and the bubble's volume corresponding to the onset of detachment become smaller. These features are in a qualitative agreement with experimental observations reported by Lin, Banerji & Yasuda (1994). These authors noted that "the diameter of the contact base increases as the [static] contact angle increases" and that "the most important factor which influences the size of bubble is the contact angle of a liquid on a surface". This is indeed what the model predicts.

5.6   Dynamic wetting at finite capillary numbers

Although in most capillary flows, especially in microfluidics, the capillary number Ca and the dimensionless relaxation length $\epsilon = U\tau/L$ are both small, there are situations where, even for numerically small Ca and ϵ, the quasi-static approximation becomes inapplicable and the problem has to be solved computationally with both Ca and ϵ regarded as finite. Then, the interface formation theory has to be used without any simplifying assumptions. One example of a flow where many qualitative features of dynamic wetting described above, in particular the influence of the flow field on the dynamic contact angle, can be illustrated in a transparent way is curtain coating in the context of microfluidics. It is also instructive from the viewpoint of numerical issues arising in implementing the theory.

5.6.1   Problem formulation

Consider a steady two-dimensional flow of an incompressible Newtonian liquid of constant density ρ and viscosity μ as a liquid sheet of an initial thickness h and a uniform velocity $\mathbf{U}_*$ parallel to the gravity force $\mathbf{g}$ impinges onto a moving solid substrate forming an angle ϕ with the horizontal (Fig. 5.17). The flow velocity $\mathbf{u}$ and pressure p in the bulk satisfy the Navier-Stokes equations (4.53) subject to the steady version of boundary conditions (4.54)–(4.59), (4.61) and (4.62)–(4.66), (4.68) on the free surfaces and the solid boundary, respectively, conditions (4.70), (4.73) at the contact line and

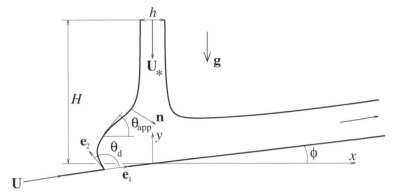

FIGURE 5.17: Definition sketch for curtain coating. θ_d is the 'actual' contact angle, i.e., the angle at which the free surface meets a solid boundary in the macroscopic fluid mechanics modelling of the flow. θ_{app} is the 'apparent' contact angle formed by the tangent to the free surface at some distance from the contact line and the solid substrate.

the following conditions specifying the flow. After introducing a Cartesian coordinate system as shown in Fig. 5.17, for the bulk flow one can set the inlet velocity and thickness of a falling liquid sheet,

$$\mathbf{u} = \mathbf{U}_* \qquad \text{for } -h/2 \leq x \leq h/2, \ y = H, \qquad (5.85)$$

where the inlet velocity $\mathbf{U}_*$ is assumed to be uniform and have only the y-component, together with boundary conditions far downstream, which we will set in a soft form:

$$\frac{\partial \mathbf{u}(x', y')}{\partial x'} \to 0 \qquad \text{as } x' \to +\infty, \ 0 < y' < \tilde{h}, \qquad (5.86)$$

where $x' = x \cos \phi + y \sin \phi$, $y' = -x \sin \phi + y \cos \phi$ and $\tilde{h}$ is to be determined. The downstream boundary condition for the bulk variables can be formulated in different equivalent ways and is primarily motivated by the convenience of its numerical implementation.

For the surface variables we will assume that at the top of the curtain the interfaces are in equilibrium, i.e.,

$$\rho^s = \rho^s_e, \quad \mathbf{v}^s = \mathbf{u} \qquad \text{for } x = \pm h/2; \ y = H, \qquad (5.87)$$

and that far downstream along the solid surface the liquid-solid interface tends to its equilibrium state:

$$\rho^s \to \rho^s_{1e} \qquad (x \to +\infty, \ y = x \tan \phi). \qquad (5.88)$$

Equations (4.53) together with conditions (4.54)–(4.59), (4.61), (4.62)–(4.66), (4.68), (4.70), (4.73), (5.85)–(5.88) fully specify the problem. In order to reduce the number of constants, we will assume, as before, that $\tau_1 = \tau$, $\alpha_1 = \alpha$ and $\beta_1 = \beta$. It is also convenient to use the equilibrium Young equation (2.164), i.e.,

$$\sigma(\rho_e^s) \cos \theta_s = \sigma_{(2)} - \sigma(\rho_{1e}^s),$$

together with the surface equations of state (4.61), (4.68) to eliminate the equilibrium surface density of the liquid-solid interface ρ_{1e}^s as a parameter of the problem replacing it with θ_s, which is relatively easy to measure experimentally.

5.6.2   The effect of the flow field/geometry on dynamic contact angle

After nondimensionalizing the problem using U, h, $\mu U h^{-1}$, the equilibrium surface tension on the free surface $\sigma_e = \gamma(\rho_{(0)}^s - \rho_e^s)$ and $\rho_{(0)}^s$ as scales for velocity, length, pressure, surface tension and the surface density, respectively, one arrives at a problem whose solution is specified by the dimensionless similarity parameters that can be divided into the following three groups. The first one comprises the Reynolds and Froude numbers, $Re = \rho U h/\mu$, $Fr = U^2/(gh)$, i.e., the parameters that characterize the bulk flow. In microfluidics, one almost invariably has creeping flows with negligible inertia. For the problem we are considering typically $h \sim 2\text{–}4~\mu m$, $U \sim 1$ cm s^{-1}, $\mu/\rho \sim 60$ cSt giving $Re \leq 10^{-3}$, so that the convective term in the Navier-Stokes equations (4.53) can be neglected. Although the ratio Re/Fr is also small ($\leq 4 \times 10^{-4}$ for our flow conditions), in the computations it is convenient to keep the body force term in the Navier-Stokes equations as a stabilizing factor for the film far downstream the solid substrate.

The second group of dimensionless parameters includes, firstly, the similarity groups formed by the material constants characterizing the contacting media,

$$\theta_s, \quad \rho_G^s = \rho_e^s/\rho_{(0)}^s, \quad A = \alpha\beta, \quad \sigma_{sg} = \sigma_{(2)}/\sigma_e,$$

and, secondly, the parameters depending on material constants and the contact-line speed only:

$$Ca = \frac{\mu U}{\sigma_e}, \quad Q = \frac{\rho_{(0)}^s}{\rho U \tau}, \quad \bar{\beta} = \frac{\beta U h}{\sigma_e}, \quad \epsilon = \frac{U\tau}{h}.$$

All these parameters remain constant for a given set of materials and a given contact-line speed.

Finally, we have three parameters,

$$\bar{U}_* = U_*/U, \quad \bar{H} = H/h, \quad \phi$$

whose variation, for a given contact-line speed and given materials of the system, leads to variation in the flow field/geometry in the vicinity of the contact line. We will consider the influence of these parameters on the dynamic contact angle θ_d as a key to the effect of 'hydrodynamic assist of dynamic wetting' we discussed in §3.2.3.

5.6.2.1 On numerical implementation of interface formation model

The main difficulty in computing the solution arises from the very nature of the physical effect we are trying to capture. The formulation (4.53), (4.54)–(4.59), (4.61), (4.62)–(4.66), (4.68), (4.70), (4.73), (5.85)–(5.88) introduces the dynamic contact angle θ_d via the Young equation (4.73) and hence makes it part of the solution dependent on the dynamic values of the surface tensions at the contact line. These values in their turn are determined by the distributions of the surface parameters along the interfaces which are linked with the bulk stress and velocity at the interfaces via (4.56), (4.57), (4.59), (4.63), (4.64), (4.66). As a result, θ_d becomes a *functional* of the flow field. This is an important point since in the process of iterations θ_d will vary in response to variations of the flow field even in the cases where the functional can be very accurately approximated by a mere function of the contact-line speed (i.e., for very small Ca and ϵ). On the other hand, however, the value of θ_d is a boundary condition for (4.55) that determines the free-surface profile and hence the shape of the flow domain, thus giving a feedback to the flow field. As a result, the bulk flow, the distributions of the surface variables and the value of the dynamic contact angle become interdependent and have to be found simultaneously. This makes even a numerical solution of the problem a formidable task since, in addition to the known difficulties of computing free-boundary flows (Tsai & Yue 1996), one has to pay particular attention to the accuracy with which the distributions of the surface parameters along the interfaces are resolved. It is their values at the contact line that, via the contact angle, have a global effect on the shape of the computational domain, hence on the bulk flow, which in its turn affects the surface distributions. Therefore, the control of accuracy with which the surface variables are computed, especially near the contact line, is the crucial element that determines the success of computations.

The problem was solved numerically in Lukyanov & Shikhmurzaev (2006, 2007a) using a combined BIE-FE algorithm (Lukyanov & Shikhmurzaev 2007b) that has the capacity of resolving the distributions of the surface parameters in the immediate vicinity of the contact line and handling the contact angle itself with sufficient accuracy (the finite element part of the method) while allowing one to describe the creeping free-surface flow away from the contact line in an efficient and flexible way (the boundary integral equation part). The computational domain had to be split into two regions (Fig. 5.18) and the solutions in these regions are matched along a separation curve. (Alternatively, the two regions can overlap; then the matching curve for each of the constitutive methods will be laying in the other method's domain.)

It should be emphasized that in implementing the interface formation model

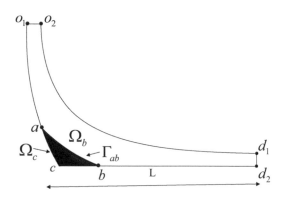

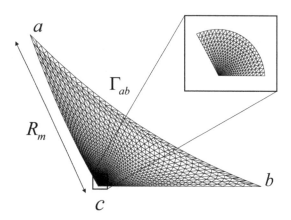

FIGURE 5.18: Top: Sketch of the computational domain for curtain coating at zero Reynolds numbers. The boundary integral element method (BIEM) ensures an efficient and accurate calculation of flow parameters in the main part Ω_b of the domain whereas the finite element method (FEM) makes it possible to compute the solution in the corner region Ω_c with the required accuracy. The two solutions are matched along a separation curve Γ_{ab}. Its position can vary with no effect on the computed flow. Bottom: Magnified view of the corner region tessellated for the finite element method. The computations are more efficient when the size of the corner region along the interfaces R_m is at least several times larger than the surface-tension-relaxation length.

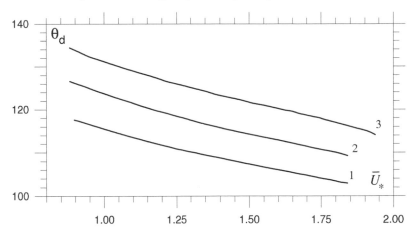

FIGURE 5.19: Dependence of the dynamic contact angle on the dimensionless inlet velocity for $\bar{H} = 10$, $\phi = 0°$ and different contact-line speeds. 1: $Ca = 0.02$, $Q = 0.04$, $\bar{\beta} = 20$, $\epsilon = 0.025$. Curves 2 and 3 are obtained by increasing the contact-line speed by 12.5 and 25%, respectively; $\theta_s = 60°$, $\rho_G^s = 0.8$, $A = 1$, $\sigma_{sg} = 0$ for all curves (Lukyanov & Shikhmurzaev 2007a).

a combined algorithm is by no means a necessity; in this particular case, it merely utilizes the fact that here we are dealing with the Stokes equations and hence can improve the accuracy by using the BIEM in a part of the computational domain. For finite Reynolds numbers the combined algorithm becomes inapplicable and one will have to use a more general method.

5.6.2.2 Mechanism of the effect

Figs 5.19–5.21 summarize the effect on θ_d of the parameters controlling the flow field. As one can see, for a given contact-line speed, parameters $\bar{U}_*$, $\bar{H}$ and ϕ do have a significant effect on the dynamic contact angle. An alternative way of interpreting these figures is that, for the same materials of the system, the dependence of θ_d on the capillary number (i.e., the dimensionless contact-line speed) is different for different flow geometries, which, in our case, are specified by $\bar{U}_*$, $\bar{H}$ and ϕ. This nonuniqueness of the velocity-dependence of the dynamic contact angle was indeed observed in experiments (§3.2.3).

As mention earlier, all elements in the model are interdependent and therefore, strictly speaking, it would be incorrect to single out direct causal links between any two of them in terms of 'causes' and 'consequences'. However, for relatively low capillary numbers, as in the flow we are considering here, one can arrive at a qualitative understanding of the mechanism by which the flow field influences the dynamic contact angle. For this purpose we will, first, examine the distributions of the surface variables ρ^s and v^s corresponding to different points along one of the curves given in Fig. 5.19. These distributions

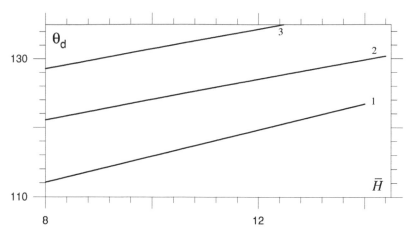

FIGURE 5.20: Dependence of the dynamic contact angle on the dimensionless height of the curtain for $\bar{U}_* = 1$, $\phi = 0°$ and different contact-line speeds with other parameters being the same as in Fig. 5.19.

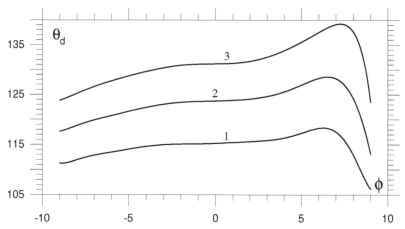

FIGURE 5.21: Dependence of the dynamic contact angle on the angle of inclination of the solid substrate for $\bar{U}_* = 1$, $\bar{H} = 10$ and different contact-line speeds as in Fig. 5.19 (Lukyanov & Shikhmurzaev 2007a).

are shown in Fig. 5.22.

As the plots of the surface density show (Fig. 5.22), on the free surface the deviations of ρ^s from its equilibrium value, being proportional to Ca, are small and hence the variations in the mass flux into the contact line, $-\mathbf{e}_1 \cdot (\rho^s \mathbf{v}^s)|_G$, that features in (4.70) are determined primarily by the variations of $v_1^s = -\mathbf{e}_1 \cdot \mathbf{v}^s|_G$. The value of v_1^s increases as the inlet velocity $\bar{U}_*$ goes up, thus increasing the mass flux into the forming liquid-solid interface. In the process of dynamic wetting, the liquid-solid interface near the contact line is always 'starving' ($\rho^s < \rho_S^s$) since it begins to form out of the liquid-gas interface that moves into the contact line with velocity lower than the velocity U of the solid substrate that drags the (solid-facing side of the) liquid-solid interface out of the contact line. The surface density carried by v_1^s is also always lower than ρ_S^s. Therefore, an increase in v_1^s due to an increase in $\bar{U}_*$ reduces 'starvation' of the liquid-solid interface, i.e., the difference between its surface density at the contact line and far away from it (see Fig. 5.22). Then, according to the Young equation (4.73), this leads to a decrease in the value of the dynamic contact angle which acts as a mechanism balancing the tangential forces exerted on the contact line by the interfaces. In other words, an increase in the mass flux into the contact line from the free surface brings the surface density of the liquid-solid interface at the contact line closer to its equilibrium value and hence drives θ_d closer to θ_s.

The above explanation also makes clear why an increase in the dimensionless curtain height $\bar{H}$ increases the dynamic contact angle (Fig. 5.20): as $\bar{H}$ goes up, the effect of the inlet velocity on the flow near the contact line diminishes, leading to lower mass flux into the contact line, thus increasing 'starvation' of the liquid-solid interface and consequently the dynamic contact angle. It should be emphasized that, unlike curtain coating in macroscopic hydrodynamics (§3.2.3), where it is driven by gravity and hence the impingement velocity increases with the curtain height, in the low Reynolds number regime considered here $\mathbf{v}_1^s$ is determined by the inlet velocity U_* with gravity playing a negligible role so that an increase in H reduces v_1^s and hence increases the contact angle.

Thus, for (numerically) small capillary numbers the mechanism of the influence of the flow field on the dynamic contact angle is relatively transparent: the contact angle responds to the influence of the flow conditions on the tangential velocity of the free surface near the contact line that controls the supply of mass into the contact line required for the formation of the liquid-solid interface. An increase in v^s reduces 'starvation' of the liquid-solid interface and hence the contact angle, thus, using the terminology of Blake et al. (1994), 'assisting' dynamic wetting.

5.6.2.3 Role of parameters

A key requirement for the 'hydrodynamic assist' to take place is that the length scale characterizing variations in the flow field must be comparable with

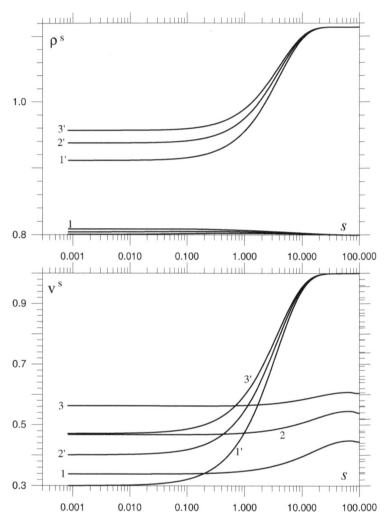

FIGURE 5.22:   Distributions of the surface parameters along the free surface and the liquid-solid interface (marked with a prime) at different points along curve 1 of Fig. 5.19. 1: $\bar{U}_* = 0.91$, 2: $\bar{U}_* = 1.38$, 3: $\bar{U}_* = 1.82$. The distance s from the contact line is scaled with $U\tau$; the data point corresponding to $s = 0$ and a few neighboring points are taken out. It is important to note that the actual distance over which the surface parameters relax to their equilibrium values is greater, often considerably, than the formally defined surface-tension-relaxation length $U\tau$.

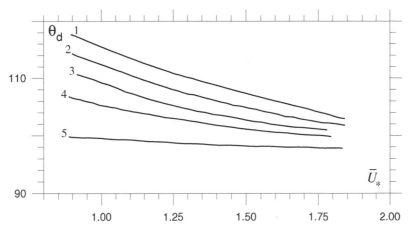

FIGURE 5.23: Dependence of the dynamic contact angle on the inlet velocity for different scales of the system (different h, the same $\bar{H}$). Curve 1 is identical to curve 1 in Fig. 5.19; 2: $\epsilon = 0.02$, $\bar{\beta} = 25$; 3: $\epsilon = 0.015$, $\bar{\beta} = 33$; 4: $\epsilon = 0.01$, $\bar{\beta} = 50$; 5: $\epsilon = 0.005$, $\bar{\beta} = 100$.

the surface-tension-relaxation length. In our model, this condition is reflected in the parameter ϵ which is exactly the ratio of the two lengths. Computations show (Fig. 5.23) that, if the system as a whole is magnified (i.e., h and H proportionally increase), the effect of 'hydrodynamic assist' diminishes and eventually disappears. (Formally, for a given system, the magnitude of the effect can be attributed to one parameter, ϵ, by eliminating h from $\bar{\beta}$, i.e., replacing it with $\epsilon^{-1}\bar{\beta}$, where $\bar{\beta} = \beta\tau U^2/\sigma_e$ is proportional to V^2 we used earlier.)

In order to understand the origin of complexity of the contact angle dependence on ϕ shown in Fig. 5.21, it is instructive to look at the role played by the upper free surface. As shown in Fig. 5.24, the dynamic contact angle reduces as this surface is moved away from the contact line, either by increasing h and H proportionally or by increasing h and keeping H constant. Then one has to conclude that, perhaps counter-intuitively, the constraint on the flow due to the presence of the free surface slows down the flow near the contact line and hence, by reducing v_1^s, increases the dynamic contact angle. It is the removal of this surface that 'assists' dynamic wetting. (It is worth emphasizing once again that here we are dealing with flows at zero Reynolds and (numerically) small capillary numbers.)

Fig. 5.24 shows that, as $\epsilon \to 0$, the dynamic contact angle tends asymptotically to a constant value slightly below the one resulting from the leading-order asymptotic solution as $Ca \to 0$ considered in §5.2.1. The reason for this effect is clear: although in the flow we are considering here the capillary number is (numerically) small, no asymptotic simplifications based on the smallness of Ca have been used and the computed surface density on the free surface

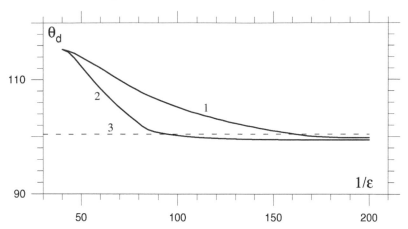

FIGURE 5.24: Dependence of the dynamic contact angle on the (inverse) dimensionless relaxation length. Curve 1: constant $\bar{H}$ ($=10$); curve 2: constant H. $Ca = 0.02$, $\bar{U}_* = 1$, $\phi = 0$ for both curves. The dashed line (3) corresponds to the asymptotic solution for low capillary numbers.

(Fig. 5.22) deviates from ρ_G^s as one approaches the contact line, whereas in the leading-order asymptotic solution in Ca as $Ca \to 0$ one obviously has $\rho^s \equiv \rho_G^s$. The slight increase in ρ^s as one approaches the contact line leads, according to (4.61), to a slight reduction in σ and hence to a lower value of θ_d needed to balance the forces at the contact line. The presence of the upper boundary also plays a role.

The dependence of θ_d on ϕ in Fig. 5.21 results from the interplay of two factors: the slowing down of the flow as its direction changes by a larger angle due to the inclination of the solid substrate (hence an increase in θ_d) and the reduction of the influence of the upper free surface as it becomes positioned further away from the contact line (hence a reduction in θ_d). Typical profiles of the free surface near the contact line corresponding to one of the curves in Fig. 5.21 are shown in Fig. 5.25.

A material-related factor that determines the magnitude of the effect of 'hydrodynamic assist' is $1 - \rho_G^s = \sigma_e/(\gamma\rho_{(0)}^s)$, which is essentially a measure of the influence of the intermolecular forces acting from the bulk phases on the interfacial layer's density. The closer ρ_G^s is to 1, the smaller is the possible amplitude of variation of the surface density and hence the stronger becomes the influence of the changes in v^s that result in variations of θ_d. This sensitivity of θ_d to $1 - \rho_G^s$ illustrated in Fig. 5.26 could be used in experiments to investigate the equation of state in the surface phase.

An important question often raising debate in the literature is whether or not the free surface experiences extremely strong bending in the immediate vicinity of the contact line. If this is the case, then the measured angle is a very poor approximation to the actual one and gives little information about the

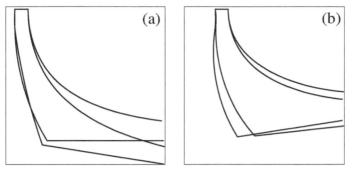

FIGURE 5.25: Curtain profiles for different points along curve 1 in Fig. 5.21: $\phi = -9°$ and $0°$ (a) and $\phi = 6.5°$ and $9°$ (b).

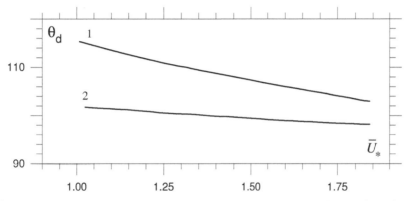

FIGURE 5.26: Dependence of the dynamic contact angle on the inlet velocity for different values of ρ_G^s. Curve 1 is identical to curve 1 in Fig. 5.19 ($\rho_G^s = 0.8$); curve 2: $\rho_G^s = 0.6$.

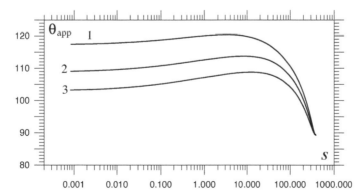

FIGURE 5.27: The free surface profiles shown as dependencies of the apparent contact angle on the distance from the contact line scaled with $U\tau$. Parameters of curves 1–3 are the same as in Fig. 5.22. The data point corresponding to $s = 0$ and a few neighboring points are removed.

wetting behavior of the system. Furthermore, if the free surface curvature is actually singular at the contact line, then the very concept of the contact angle becomes meaningless. In order to clarify this point, consider the dependence of the 'apparent' contact angle defined in Fig. 5.17 on the distance from the contact line measured in nominal relaxation lengths $U\tau$. This dependence is shown in Fig. 5.27 for different values of the inlet velocity. As one can see, even for the Stokes regime of flow we are considering here and the inlet velocities higher than the velocity of the solid substrate there is no extreme bending of the free surface. In fact, this is a mere illustration of a general premise that a physically sound mathematical theory should lead to no singularities in the calculated dynamic properties, including the capillary pressure. For a nonvanishing surface tension, the regularity of the capillary pressure obviously implies a finite surface curvature.

From a quantitative viewpoint, the obtained numerical result also has some value since the calculated profiles can be used to estimate experimental errors in determining the actual contact angle.

5.6.3   Flow field

The flow near the moving contact line has some interesting features. On a length scale large compared with the surface-tension-relaxation length the streamlines form an expected pattern (Fig. 5.28, top). Importantly, the flow is more intensive near the contact line than further afield, as is indeed observed in experiments (Clarke 1995, see §3.2.2), whereas, as we have seen in §3.4.1, the 'slip models' make the region near the contact line a stagnation zone.

After zooming in to the contact line and considering the flow on a length scale comparable with $U\tau$ (Fig. 5.28, bottom), one can see that, in accordance with (4.64), the 'starving' liquid-solid interface adsorbs the fluid so that for the bulk flow this interface is no longer a streamline. According to (4.57) and the surface density distributions shown in Fig. 5.22, one also has desorption from the liquid-gas interface, though this effect is too small numerically to be visualized here. The tangential velocity distribution on the liquid-facing side of the liquid-solid interface is very close to the distribution of v^s shown in Fig. 5.22 so that, from the viewpoint of classical fluid mechanics, there is some degree of slip. However, unlike the 'slip models', where it is the remedy for the singularity and hence has to be complete, here slip is not complete ($\mathbf{u}_{\parallel} \neq 0$ at the contact line) and its origin is the surface-tension gradient, not the tangential stress. In other words, in our case slip is a manifestation of the flow-induced Marangoni effect on the liquid-solid interface.

Computations demonstrate that in the immediate vicinity of the contact line there is a uniform flow from the free surface into the liquid-solid interface with both components of the bulk velocity being nonzero at the contact line. This confirms the results of the local asymptotic analysis of §5.1 sketched in Fig. 5.1.

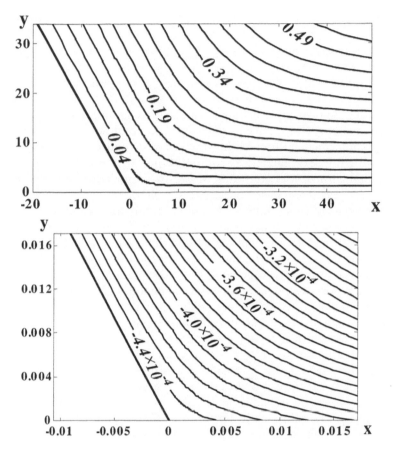

FIGURE 5.28: Typical pattern of streamlines with the distances measured in relaxation lengths $U\tau$ (Lukyanov & Shikhmurzaev 2007a). Top: The general pattern where, unlike the slip models, the region near the contact line is associated with a more intensive flow than further afield. Bottom: The magnified view of the flow near the contact line showing adsorption of the fluid into the forming liquid-solid interface.

5.7   Liquid-liquid displacement

The modelling of the liquid-liquid displacement is much more complex mathematically than the cases considered so far and its full analysis requires numerical computations. Below, we will consider a simple situation where some analytic progress can be made. We will look at the case of a steady contact line motion at small capillary numbers with the relaxation lengths in both liquids being small compared to the size of the flow domain. Our key assumption will be that there exists a solution with the surface parameter distributions close to their equilibrium values. As we will see below, for the liquid-liquid displacement this is an essential assumption, whereas in §5.2 the existence of such a solution, under certain conditions, followed from the formally applied asymptotic methods and it was confirmed by comparing the results with full-scale numerical computations. For simplicity, we will also assume that all material constants of the two liquids are of the same order and that for each component parameters γ characterizing compressibility of the fluids have the same value for the liquid-liquid and liquid-solid interfaces. These simplifications are not essential and will be used only to reduce the number of parameters. We will scale all variables and dimensional constants using parameters of the advancing fluid with the subscripts 1 and 2 referring, where necessary, to the advancing and receding fluid, respectively.

As in §5.2, for small Ca and ϵ the flow domain can be divided into three asymptotic regions: (i) the outer region where, to leading order, the standard model operates, (ii) the intermediate region where, to leading order, the surface distributions satisfy a closed set of ordinary differential equations and hence can be found, up to the constants of integration, independently of the bulk flow, and (iii) the viscous region where, again to leading order, no relaxation of the surface properties takes place. The last property makes it possible to apply the boundary conditions posed at the contact line to the inner limits of the solutions in the intermediate region. Then, these conditions together with the conditions of matching with the known (apart from the contact angle) inner limit of the outer solution fully specify the problem in the intermediate region and allow one to find the dynamic contact angle.

As in §5.3.2, the surface parameters of the receding liquid in the intermediate region satisfy (5.66), (5.67) which after linearization give

$$\rho_2^s = \rho_{S2}^s + C_2 \exp(-k_2 \bar{r}), \quad v_2^s = -1 + \frac{C_2 k_2}{4V_2^2} \exp(-k_2 \bar{r}), \qquad (5.89)$$

where $k_2 = 2V_2(\rho_{S2}^s)^{-1}[(V_2^2 + \rho_{S2}^s)^{1/2} + V_2]$, ρ_{S2}^s is the (dimensionless) equilibrium surface density in the receding liquid-solid interface and C_2 is a constant of integration.

As in §5.2, from the normal projection of (4.82) we have that, to leading order in the limit of small capillary numbers and relaxation lengths, the free surface is locally planar whereas the tangential projection yields that the surface tension in the free surface is constant. However, unlike the case of a liquid-gas interface, according to the surface equation of state (4.89) constancy of the surface tension means only that

$$\lambda_1(\bar{\rho}_{1(0)}^s - \rho_1^s) + \lambda_2(\bar{\rho}_{2(0)}^s - \rho_2^s) = 1 \qquad (5.90)$$

and does not imply that the surface densities are constant. Here and below, where necessary, the overbar is used to mark dimensionless constants; $\lambda_i = \gamma_i \rho_{(0)}^s / \sigma_e$ ($i = 1, 2$), where $\rho_{(0)}^s$ is the parameter in (4.68) for the advancing fluid we use for scaling the surface densities; σ_e is the equilibrium surface tension in the free surface.

Conditions (4.83) to leading order in Ca as $Ca \to 0$ gives that the tangential components of the bulk velocities on the opposite sides of the interface are equal and condition (4.84) together with (5.90) states that they are equal to the corresponding component of the surface velocity $\mathbf{v}^s$. The latter, according to (4.2), has only the tangential component which we will denote as v^s.

In addition to (5.90), we also have conditions (4.87), which in the intermediate region for a steady flow have the form

$$\frac{d(\rho_i^s v^s)}{d\bar{r}} = -\frac{\rho_1^s - \bar{\rho}_{1e}^s}{\bar{\tau}_{i1}} - \frac{\rho_2^s - \bar{\rho}_{2e}^s}{\bar{\tau}_{i2}}, \qquad (i = 1, 2). \qquad (5.91)$$

Here the matrix of relaxation times is scaled with the relaxation time of the advancing liquid-solid interface. The remaining boundary conditions (4.88) determine the normal components of the bulk velocity and have to be used to determine the flow field after the surface distributions are found.

In the far field one has the matching conditions

$$\rho_i^s \to \bar{\rho}_{ie}^s, \qquad v^s \to u_{(12)}(\theta_d, k_\mu), \qquad \text{as } \bar{r} \to \infty, \; i = 1, 2, \qquad (5.92)$$

where $u_{(12)}$ is given by (5.57).

The solution of the linearized equations (5.91) where the surface densities are linked by (5.90) has the form

$$\rho_1^s = \bar{\rho}_{1e}^s + \lambda_2 C \exp(-\bar{r}/(u_{(12)}\hat{\tau})), \quad \rho_2^s = \bar{\rho}_{2e}^s - \lambda_1 C \exp(-\bar{r}/(u_{(12)}\hat{\tau})), \quad (5.93)$$

$$v^s = u_{(12)}[1 + aC \exp(-\bar{r}/(u_{(12)}\hat{\tau}))] + D \qquad (5.94)$$

where

$$\hat{\tau} = \frac{\lambda_1 \bar{\rho}_{1e}^s + \lambda_2 \bar{\rho}_{2e}^s}{a_1 \bar{\rho}_{2e}^s + a_2 \bar{\rho}_{1e}^s}, \qquad a = \frac{\lambda_1 a_1 - \lambda_2 a_2}{a_1 \bar{\rho}_{2e}^s + a_2 \bar{\rho}_{1e}^s},$$

$$a_1 = \frac{\lambda_2}{\bar{\tau}_{11}} - \frac{\lambda_1}{\bar{\tau}_{12}}, \qquad a_2 = \frac{\lambda_1}{\bar{\tau}_{22}} - \frac{\lambda_2}{\bar{\tau}_{21}},$$

and C, D are constants of integration.

Since in the model τ_{ij} are expected to be much larger than τ_{ii}, one has $\hat{\tau} > 0$ and hence for the rolling motion of the advancing liquid ($u_{(12)} < 0$) the matching conditions (5.92) can be satisfied only if $C = D = 0$.

For $u_{(12)} < 0$ we also have that the surface flux of component 2 in the liquid-liquid surface phase and the flux in the receding liquid-solid interface are both directed into the contact line. Therefore, for the solution with only perturbed surface distributions to exist for the receding fluid, one needs a flux of component 2 out of the contact line. To allow for this, we will use condition (4.98), i.e.,

$$\bar{\rho}_{2e}^s u_{(12)}(\theta_d, k_\mu) + (\rho_2^s v_2^s)|_{\bar{r}=0} + \bar{\rho}_{\text{res}}^s = 0. \tag{5.95}$$

For the advancing fluid one can use the simples condition (4.97), i.e.,

$$\bar{\rho}_{1e}^s u_{(12)}(\theta_d, k_\mu) + (\rho_1^s v_1^s)|_{\bar{r}=0} = 0. \tag{5.96}$$

Thus, we have that the advancing liquid-solid interface is a two-component system and, as discussed in §4.4.1.2, assumptions must be made about the behavior of the second component. For simplicity, we will assume that the entrained molecules of fluid 2 remain in the surface phase, i.e., behave like an insoluble surfactant. Then, the surface density $\rho_{2,1}^s$ will satisfy the continuity equation $d(\rho_{2,1}^s v_1^s)/d\bar{r} = 0$ which together with the boundary condition $(\rho_{2,1}^s v_1^s)|_{\bar{r}=0} = \bar{\rho}_{\text{res}}^s$ yield

$$\rho_{2,1}^s v_1^s \equiv \bar{\rho}_{\text{res}}^s. \tag{5.97}$$

(As in §4.4.1.2, here we neglect diffusion of component 2 along the advancing liquid-solid interface, though this effect can be accounted for in a straightforward way.)

Along the advancing liquid-solid interface the distributions of parameters of component 1 satisfying the matching condition in the far field are given by (5.54), (5.55), i.e.,

$$\rho_1^s = \rho_{S1}^s - C_1 \exp(-k_1 \bar{r}), \quad v_1^s = 1 - \frac{C_1 k_1}{4V_1^2} \exp(-k_1 \bar{r}), \tag{5.98}$$

where $k_1 = 2V_1(\rho_{S1}^s)^{-1}[(V_1^2 + \rho_{S1}^s)^{1/2} - V_1]$ and C_1 is a constant of integration.

The two constants of integration, C_1 and C_2, can be determined by substituting (5.98) and (5.89) into (5.96) and (5.95), respectively:

$$C_1 = \frac{4V_1^2[\bar{\rho}_{1e}^s u_{(12)}(\theta_d, k_\mu) + \rho_{S1}^s]}{4V_1^2 + k_1 \rho_{S1}^s} \quad C_2 = \frac{4V_2^2[\bar{\rho}_{2e}^s u_{(12)}(\theta_d, k_\mu) - \rho_{S2}^s + \bar{\rho}_{\text{res}}^s]}{4V_2^2 - k_2 \rho_{S2}^s}.$$

Now, after substituting the results in the Young equation (4.96), we arrive at an algebraic equation that determines the dynamic contact angle:

$$\cos \theta_d = \lambda_2(\bar{\rho}_{(0)2}^s - \rho_{S2}^s - C_2) - \lambda_1(1 - \rho_{S1}^s + C_1 - \bar{\rho}_{\text{res}}^s(1 + k_1 C_1/(4V_1^2)). \tag{5.99}$$

Here $\bar{\rho}_{(0)2}^s$ is the constant in (4.68) for fluid 2 scaled, as all other surface densities, with the corresponding constant for fluid 1. In practice, one can almost always assume $\bar{\rho}_{(0)2}^s = \bar{\rho}_{2(0)}^s$.

It is also noteworthy that ρ_{S1}^s is the equilibrium surface density of component 1 in the advancing liquid-solid interface with the second component present as a 'surfactant' with the surface density $\bar{\rho}_{\text{res}}^s$. Therefore, it can differ from the corresponding density when the interface is not contaminated by the traces of the receding liquid.

As follows from its derivation, equation (5.99) is applicable for $u_{(12)} < 0$. By subtracting from (5.99) the Young equation for the static contact line one can obtain an equation of the form of (5.56). It should be remembered, however, that the static contact angles for a pure advancing fluid-solid interface and for the interface with traces of the receding fluid are different.

If it is the receding liquid that undergoes the rolling motion, i.e., $u_{(12)} > 0$, then $D = 0$ as before, but constant C in (5.93) and (5.94) is not necessarily zero. In this case, the surface fluxes of component 1 in both the free surface and the advancing liquid-solid interface are directed out of the contact line. Then, for a solution with only perturbed surface distributions to exist there must be a flux into the contact line from the bulk of fluid 1 that would supply mass for the outgoing fluxes. This flux and how it splits in two to feed the free surface and the forming advancing liquid-solid interface have to be specified.

More generally, conditions at the contact line with the only requirement of the physical self-consistency of the model as a whole and no particular form of the solution lead to finite deviations of the surface variables from their equilibrium values and the corresponding problem has to be solved numerically.

5.8   Summary and outstanding modelling issues

As shown in this chapter, the theory of flows with forming interfaces allows one to describe the process of dynamic wetting in a regular way, and the solutions obtained in the framework of this theory so far are in a very good agreement with all available experiments.

Importantly, the interface formation theory predicts that, for a given contact-line speed and materials of the system, the actual dynamic contact angle θ_d depends on the flow field/geometry in the vicinity of the moving contact line. This is indeed what one can observe in experiments. At present, the interface formation theory is the only one that has this property.

In the case of relatively low capillary numbers, the mechanism of the dependence of the dynamic contact angle on the flow field can be explained in terms of the influence of the flow on the tangential velocity of the free surface that determines the mass supply for the formation of the liquid-solid interface (a 'freshly wetted' solid surface area). The magnitude of the response of the contact angle to the changes in the flow field depends on the material constants specifying the equation of state in the surface phase. A way of investigating the surface equation of state experimentally, using the density of a microscopic film in front of an advancing contact line and the dynamic contact angle as a control-response pair, is described in §5.3.1.

The examples of particular flows that we considered show how the theory can be applied, including the problem formulation for finite values of all parameters, and illustrate some of the key features associated with moving contact lines.

An intrinsic limit of applicability of the theory inherent in its macroscopic nature is that the 'nominal' length of the surface-tension-relaxation 'tail' $U\tau$ must be on a macroscopic scale, i.e., one needs $U\tau \gg \ell$. This implies that the interface formation process at very low contact-line speeds is essentially described by extrapolation, and the corresponding variation of the advancing dynamic contact angle is likely to be slightly underpredicted. This limitation, however, is of little importance for most practical applications that deal with high-speed coating controlled by hydrodynamic factors.

5.8.1   Dynamic wetting of imperfect substrates

The interface formation theory provides the conceptual framework and simple mathematical models to describe dynamic wetting of 'perfect', i.e., smooth and chemically homogeneous, solid substrates. As shown earlier in this chapter, it is sufficiently accurate when applied to 'real' surfaces, provided that the influence of imperfections is in some sense small or mutually balanced. In this case, the coefficients of the models should be interpreted as 'effective', meaning that they incorporate the average effect of imperfections whereas the dominant mechanism that determines macroscopic characteristics of the flow is the interface formation. However, in many situations the solid substrate imperfections can no longer be absorbed into effective coefficients of the interface formation theory and it becomes necessary to bring them into the model in a more systematic way.

The very term 'imperfection' implies that the characteristic dimensions associated with it are much larger than the molecular scale but small compared with the characteristic length scale of the flow.[7] Imperfection of the solid substrate comes from one or more of the following sources: (a) heterogeneity of the substrate's chemical properties that are relevant to wetting, (b) variations in its profile, and (c) its porosity, or, more generally, small-scale nonuniformity in its kinematic and/or dynamic influence on the flow. The objective of modelling dynamic wetting of imperfect surfaces is to incorporate properties of the solid substrate into 'effective' boundary condition for the flow on the length large compared with the characteristic length scale associated with imperfections. In other words, the goal is to average imperfections so that the

---

[7]On the molecular scale every surface is 'imperfect' and it becomes 'perfect' after applying the continuum approximation, whereas 'imperfections' with the length scale comparable with that of the flow are simply features of the solid substrate and have to be incorporated in the boundary conditions in the usual way.

resulting model deals not with the details of the flow on their length scale but with some average characteristics of the substrate.

Classifying imperfect solid substrates according to their dominant features, we can divide imperfect surface into the following main groups:

(i) *Patterned surfaces.* These are surfaces with a regular pattern of chemical composition and/or topographic features. Examples of such substrates include solids with parallel grooves, hexagonal or square cells, concentric strips of materials with different wettability, etc.

(ii) *Rough surfaces.* Topographic features of these surfaces are random and can be characterized only statistically.

(iii) *Chemically heterogeneous surfaces.* From the viewpoint of continuum mechanics, these surfaces are smooth but their chemical properties relevant to wetting processes vary randomly (on the scale large compared with the molecular one) with the distributions characterized statistically.

(iv) *Porous substrates.* The key feature of this type of surface is that, besides spreading, the liquid can also percolate through the substrate. For the porosity of the substrate to play a role, the rate of this process must be comparable with the rate of spreading.

The general problem of averaging/modelling properties of imperfect solid substrate with respect to dynamic wetting is in many ways similar to the problem of continuum description of multiphase systems (see, e.g., Nigmatulin 1991). However, historically there is also an important difference that, perhaps, explains why dynamic wetting of imperfect substrates is considerably less developed than mechanics of multiphase media. In the latter, the microscopic dynamics that needs to be averaged or incorporated in some engineering way is known, in many cases for more than a century, like, for example, the flow around a spherical particle in suspensions. As a result, the problem of averaging appears in its classical form with difficulties coming from geometric nonlinearities. In contrast, an adequate mathematical description for dynamic wetting on the scale of imperfections, i.e., for 'perfect' surfaces, has been missing until recently, and much of research effort was directed towards by-passing this problem and using imperfection of the solid substrate as a remedy for the moving contact-line problem (e.g., Hocking 1976, Neogi & Miller 1983). The resulting models have a defect of principle: in the limit of vanishing imperfection the model break down. It should be emphasized that this limit is by no means artificial and physically unrealizable since a number of surfaces, e.g., cleft mica, satisfy all assumptions behind the concept of a 'perfect' substrate. Ideally, an adequate mathematical description must be based on the interface formation model for dynamic wetting on the scale of imperfections (and recover this model as the imperfection vanishes) whilst incorporating irregularities of the substrate in a regular way.

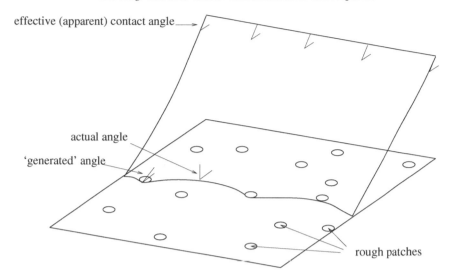

effective (apparent) contact angle

actual angle

'generated' angle

rough patches

FIGURE 5.29: Definition sketch for dynamic wetting of a flat substrate with randomly distributed rough patches.

Albeit often very challenging technically, patterned surfaces pose no diffi-culties of principle and to model them macroscopically one can employ known methods of averaging regular structures. The fundamental problems arise when microscopic properties of the substrate are random.

In the static situation, there is no essential difference between rough and porous substrates provided that the fluid is not completely absorbed by the porous medium (see below). The equilibrium 'effective' contact angle on rough surfaces has been first considered by Wenzel (1936, 1949). His idea, which later has been used in different forms in many works (e.g., Huh & Mason 1977b), was to interpret the Young equation for the actual (microscopic) con-tact angle as a relationship between surface energies of the contacting in-terfaces and relate an increase in the wetted area as the substrate becomes corrugated with a deviation of the 'effective' contact angle θ_{eff} formed by the free surface in the far field with an 'effective' smooth solid (see Figs 3.25, and 5.29) from the actual one. In other words, in a static situation, the effect of roughness is completely characterized by an increase in the surface energy of the 'effective' smooth substrate. Wenzel (1936) obtained that

$$\cos\theta_{\mathrm{eff}} = r\cos\theta_s, \qquad (5.100)$$

where r is the ratio of the actual wetted are to the effective one. This result is applicable to chemically heterogeneous surfaces if we use the averaged surface free energy of the liquid-solid interface. Equation (5.100) combined with the classical Young equation (2.164) gives a modified Young equation which shows that the effective contact angle on a rough substrate is generally smaller than the actual one.

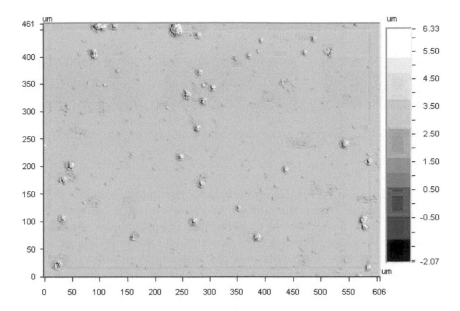

FIGURE 5.30: An example of a real solid surface (courtesy of A. Clarke).

The difficulty of combining Wenzel's approach with the interface formation theory is that for the spreading fluid the surface-tension-relaxation length becomes comparable with the size of the roughness elements and, given that, in a general case, the interface formation process is linked with the bulk flow, one will have to consider and then average the flow on the level of roughness elements.

An attempt to combine Wenzel's ideas with the methods developed in the mechanics of dilute suspensions has been made by Jansons (1985) who studied the contact-line motion over a solid surface with random sparse spots of roughness (Figs 5.29 and 5.30) at zero capillary number. The rough patches act as 'generators' of the deviation of the effective contact angle from the actual one and each of them is characterized by its 'strength' which depends on the overall position of the free surface. Mathematically, this is similar to the stress generated by the presence of slender particles in a dilute suspension (Batchelor 1971). Using the renormalization technique developed in the mechanics of suspensions, Jansons obtained an expression for the difference between the effective and the actual contact angle in terms of the concentration of rough patches per unit area and their parameters (to simplify the mathematics, rough patches with a simple structure have been considered).

It should be noted, however, that, although Jansons' analysis is intended to describe the *moving* contact line, it is not based on any specific model for dynamic wetting on the microscopic level and hence, strictly speaking, is applicable to infinitesimally slow contact-line motion. As a result, the most

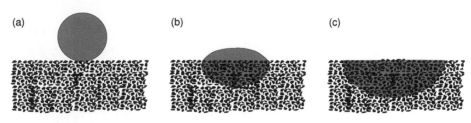

FIGURE 5.31:   Sketch of the drop spreading on an unsaturated porous substrate.

interesting dynamic effects that follow from the model, like the transition between the states where the contact line hangs on different sets of rough patches, appear to be outside the model's limits of applicability. An important advantage of Jansons' approach is that, although reliant on Wenzel's (1949) static energy balance arguments in a particular implementation, it has the potential to incorporate the microscopic contact-line dynamics in a regular way.

Chow (1998) approached the problem by assuming that the solid surface roughness can be described as a self-affine fractal. The treatment is entirely based on Wenzel's type energy arguments with the change in the contact angle now expressed in terms of the properties of correlation functions. The applicability is again limited to infinitesimally slow motion of the contact line and there is no obvious way as to how a mathematical model of dynamic wetting on the microscopic level could be incorporated.

The liquid spreading over porous substrates is a topic that covers a wide variety of situations. They range from the case where the effect of porosity is a perturbation of a regular interface-formation-based dynamic wetting to the opposite case where, pictorially, the 'effective' surface of the substrate is almost entirely made of holes so that the actual wetting takes place in the pores whereas the overall dynamics is that of the averaged flow.

Cassie & Baxter (1944) considered an equilibrium of a drop on a porous surface and suggested that the effective contact angle is related with the actual one by

$$\cos\theta_{\mathrm{eff}} = \phi_s \cos\theta_s - (1 - \phi_s), \qquad (5.101)$$

where ϕ_s is defined as the fraction of the effective substrate area on which the drop rests. This is in fact a very limited result which assumes a porous medium with no filtration through it and hence a possibility to formulate a modified Young equation for the effective contact angle by combining (5.101) with the (2.164).

A different situation illustrating the modelling difficulties is schematically shown in Fig. 5.31. As a drop placed on a porous substrate spreads, the fluid percolates through the substrate so that the equilibrium state is reached when the drop is completely absorbed by the solid. As a result, the very concept of an equilibrium effective contact angle together with all arguments based on the Young equation become meaningless, and one has to develop a completely different modelling approach to this phenomenon.

The propagation of the wetting front through the porous medium and the details of filtration are another aspect of the problem. In particular, it is necessary to analyze the role played by the unsteadiness of the process associated with the interaction of the advancing meniscus with a network of pores.

At present, the research into the modelling of the liquid spreading over porous substrates is in a very early stage. The models used so far (e.g., Davis & Hocking 1999, 2000) formally combine a slip model with a prescribed constant contact angle and the imbibition proportional to the pressure in the liquid and the substrate's porosity. These works provide a necessary reference point for more realistic theories that are now required.

Chapter 6

Cusps, corners and coalescence of drops

In this chapter, we consider singularities of the free-surface curvature that can arise in convergent flows and a related problem of coalescence of liquid volumes. Experimental implications and possible ways of using singular convergent flows for the experimental study of interfacial properties are also discussed.

6.1 Singularities of free-surface curvature in experiments

The standard model assumes that the free surface possesses the following two features:

(i) The liquid-fluid interface is smooth, i.e., the unit normal vector to it $\mathbf{n}$ can be uniquely defined at every point on the interface. It is also assumed to be differentiable and its divergence finite.

(ii) In the absence of chemical reactions and phase transitions, the fluid particles initially belonging to the free surface stay on the free surface at all times.

If the bulk parameters evaluated at the interface and the surface parameter characterizing its physical properties are smooth functions along the free surface, then the first of these assumptions allows one to formulate boundary condition in terms of differential equations to be satisfied at every point on the interface. The second assumption defines the liquid-fluid interface as a material surface. These two assumptions result in the kinematic boundary condition (2.136) specifying the normal component of the interface velocity which together with conditions for the balance of forces acting on an element of the free surface (2.154), (2.155) form the basis of the standard model in the simplest case of an interface between a viscous liquid and an inviscid gas. If both fluids are viscous, the standard model involves an additional condition for the tangential velocities (2.148) and modifications of the dynamic conditions (2.143), (2.147) with assumptions (i) and (ii) remaining intact.

(a)   (b)

FIGURE 6.1: As the angular velocity of a cylinder half immersed in a liquid increases, a rounded free surface (a) gives way to a sharp cusp (b).

TABLE 6.1: Critical angular velocities and capillary numbers for the formation of a cusp (Joseph 1992)

Fluid	Critical angular velocity, rev min^{-1}	Critical radius, cm	Ca_{cusp}
Castor oil	113	1.78	4.89
Glycerin	128	1.49	2.62
Silicone oil-500 sc	75	1.64	2.96
Silicone oil-1000 sc	41	1.40	2.75
Silicone oil-5000 sc	26	1.57	8.21

However, there are situations where both assumptions (i) and (ii) are no longer applicable and the mathematical description of experiments in the framework of the standard model becomes manifestly inadequate. One of these situations is the appearance of what looks like a two-dimensional free-surface cusp discovered experimentally by Joseph et al. (1991). This discovery and especially the subsequent theoretical work on application of the standard model to convergent flows raised a number of fundamental questions.

Joseph et al. examined fluid motion generated by a rotating cylinder half immersed in liquid with the centerline of the cylinder parallel to the free surface (Fig. 6.1). As the cylinder rotates, it drags a macroscopic liquid film out of the reservoir on one side, and, as this film plunges back into the liquid on the other side, one can observe a convergent flow. The subsequent evolution of the free surface is described as follows:

> As the rotation speed increases, the shape of the interface changes from rounded to pointed in a continuous manner, until at what we defined as the critical speed the interface appears to be a cusp. The critical values for Newtonian liquids are a little ambiguous, but did not vary by more than 5% in different determinations. Once a cusp is formed, further increase in the angular velocity will only lengthen the cusp or end in an instability characterized by fingering (Joseph 1992, p. 130).

The velocity characterizing the flow near the cusp can be defined as $U = \Omega R_c$, where Ω is the angular velocity of the cylinder and R_c is the distance between its axis and the point where the cusp is observed. Then one can introduce the capillary number $Ca = \mu U/\sigma_c$ and calculate its critical value Ca_{cusp} corresponding to the formation of a cusp. The critical capillary numbers for different Newtonian liquids in Joseph's experiments are given in Table 6.1.

The cusp as a geometric feature invalidates the assumption (i) mentioned earlier that lies behind the standard boundary conditions, and it also goes against the physical intuition which suggests that the singularity of curvature should be removed by the surface tension. However, the observations of the free-surface profile alone do not allow one to conclude that there is a cusp (or any singularity of the free-surface curvature) since they are limited by their finite spatial resolution and hence leave a possibility that on a smaller but yet macroscopic length scale the free surface is rounded and the standard boundary conditions still apply. Furthermore, if there is indeed a singularity of curvature, optical observations do not allow one to reach a conclusion about its type and distinguish between a cusp or a sharp corner. This uncertainty makes it necessary to examine the cusp phenomenon theoretically in order to bridge reliable observations of the flow on the optical scale before a cusp appears with, also relatively reliable, critical conditions for the cusp formation.

Another and by far more important aspect of the cusp problem is a qualitative change in the flow kinematics associated with the cusp formation. Jeong & Moffatt (1992), who repeated experiments of Joseph et al. (1991) using two symmetrically-placed counter-rotating cylinders and one Newtonian fluid (polybutene), described the change of kinematics in the following way:

> For very slow rotation rates, there is a stagnation line on the free surface, and in some circumstances a small rounded crest can form in the neighbourhood of this stagnation line. (...) When the rotation rate Ω is increased however, the surface dips downwards, and simple visual observation indicates the presence of a very sharp cusp on the free surface. If powder is sprinkled on the free surface, this powder is immediately swept through the cusp into the interior of the fluid. Thus, observation suggests that fluid particles on the free surface are similarly advected through the cusp into the interior (Jeong & Moffatt 1992, pp. 1–2).

Thus, the key assumption behind the kinematic boundary condition saying that a fluid particle initially belonging to the free surface will stay on the free surface at all time is no longer satisfied. Jeong & Moffatt pointed out that the kinematics observed in experiments is in conflict with the boundary condition (2.148) on the interface between a viscous liquid and a *viscous* gas which states that the tangential velocity is continuous across the interface. Hence, they reason that "if the cusp is genuine and fluid particles on the free surface do move into the interior of the fluid, then air must be entrained into the interior also. There is however no evidence in the experiments for the entrainment of air bubbles" (Jeong & Moffatt 1992, p. 2). At the same time, Jeong & Moffatt recognised that in the moving contact-line problem of dynamic wetting one has a similar situation, and hence this argument does not invalidate the possibility that the free surface indeed forms a cusp or a corner.

The kinematic aspect of cusping has important practical implications. For example, surfactants present on a smooth free surface will be swept into the bulk if a cusp (or corner) appears and the flow kinematics qualitatively changes. Then, in the subsequent motion the free surface will be surfactant-free and deform accordingly so that what looks like a local effect appears to have global consequences.

6.2   Conventional modelling

6.2.1   Solution with a presumed singularity

More than two decades before the free-surface cusps were actually discovered experimentally, Richardson (1968) had published an elegant study where, in particular, he examined the possibility of existence of a singularity of curvature at finite capillary numbers for a two-dimensional Stokes flow. He considered a steady free boundary between a viscous liquid and an inviscid gas (physically, vacuum) in the framework of the standard model. The analysis is based on the technique of analytic functions and utilizes the similarities between the mathematics of the two-dimensional Stokes flow and that of the classical theory of elasticity (Muskhelishvili 1963). Below, we reproduce Richarson's results related to the cusp problem and examine their implications.

Consider a steady two-dimensional flow satisfying the Stokes equations

$$\frac{\partial u}{\partial x} + \frac{\partial v}{\partial y} = 0, \tag{6.1}$$

$$\frac{\partial p}{\partial x} = \mu \Delta u, \qquad \frac{\partial p}{\partial y} = \mu \Delta v, \tag{6.2}$$

where (x, y) are the Cartesian coordinates, (u, v) are the corresponding components of velocity, p is the pressure, μ is the dynamic viscosity, and $\Delta = \partial^2/\partial x^2 + \partial^2/\partial y^2$. The stream function ψ introduced by

$$u = \frac{\partial \psi}{\partial y}, \qquad v = -\frac{\partial \psi}{\partial x} \tag{6.3}$$

satisfies the biharmonic equation

$$\Delta^2 \psi = 0,$$

whose solution admits a representation as either the real or the imaginary part of an expression

$$\overline{z}A_1(z) + A_2(z), \tag{6.4}$$

where $A_1(z)$ and $A_2(z)$ are analytic functions of the complex variable $z = x + iy$ and $\overline{z} = x - iy$ (Muskhelishvili 1963, p. 109; see Appendix C).

Let ψ be represented as the real part of (6.4), where, following Richardson, we will use the notation

$$\psi(x, y) = \mathrm{Re}\{\overline{z}\phi(z) + \chi(z)\}. \tag{6.5}$$

Then, for components of velocity and stress one has

$$-v + iu \equiv \frac{\partial \psi}{\partial x} + i\frac{\partial \psi}{\partial y} = \phi(z) + z\overline{\phi'(z)} + \overline{\chi'(z)}, \tag{6.6}$$

$$p_{xx} + p_{yy} \equiv -2p = 8\mu\,\mathrm{Im}\{\phi'(z)\}, \tag{6.7}$$

$$i(p_{yy} - p_{xx}) - 2p_{xy} = 4\mu(\overline{z}\phi''(z) + \chi''(z)). \tag{6.8}$$

Components of the stress tensor can be represented in terms of the Airy function $W(x, y)$ by (2.93), i.e.,

$$p_{xx} = \frac{\partial^2 W}{\partial y^2}, \quad p_{xy} = -\frac{\partial^2 W}{\partial x \partial y}, \quad p_{yy} = \frac{\partial^2 W}{\partial x^2}, \tag{6.9}$$

and, given that W is biharmonic (see (2.94)), one can write it down as the imaginary part of (6.4):

$$W = \mathrm{Im}\{\overline{z}A_1(z) + A_2(z)\}. \tag{6.10}$$

After substituting (6.10) in (6.9) and the latter in (6.7), (6.8) we get

$$A_1(z) = 2\mu\phi(z), \qquad A_2(z) = 2\mu\chi(z). \tag{6.11}$$

By examining (6.6)–(6.8) one can show that, for given velocity and pressure fields, $\phi(z)$ and $\chi(z)$ are defined uniquely up to an additive complex constant a to $\phi(z)$ and a linear function $-\overline{a}z + b$ to $\chi(z)$. Below, a and b will be specified to make two constants of integration appearing in the derivation of the boundary conditions equal to zero.

On a stationary free surface, which for the two-dimensional case considered here is a line in the z-plane (Fig. 2.8), the boundary conditions to be satisfied are zero normal velocity or, equivalently, constancy of the stream function ψ along this line, zero tangential stress and the equality of the normal stress and the capillary pressure. Using (6.5), the first of these conditions now takes the form

$$\mathrm{Re}\{\overline{z}\phi(z) + \chi(z)\} = 0, \tag{6.12}$$

where without loss of generality we have set $\psi = 0$ on the boundary.

In the absence of the surface tension gradient, the force per unit area of the free surface due to capillarity is the capillary pressure given by $-\sigma_e \mathbf{n} \nabla \cdot \mathbf{n}$ (see §2.4.2). In our case, $-\mathbf{n} \nabla \cdot \mathbf{n} = \kappa \mathbf{n} = \mathbf{t}'(s)$, where $\kappa \geq 0$ is the curvature of the line AB (Fig. 2.8). Then, in the complex form the capillary pressure can be written down as

$$\sigma_e \kappa (n_x + i n_y) = \sigma_e (t'_x(s) + i t'_y(s)) = \sigma_e z''(s),$$

and the force acting on an element ds of the line AB is given by

$$\sigma_e \kappa (n_x + i n_y)\, ds = \sigma_e \, d\left[\frac{dz}{ds}\right].$$

This force is equal in magnitude and opposite in sign to the force exerted on the boundary by the fluid. For W of the form (6.10) the latter is given by (2.98) so that, using (6.11), the force balance equation embracing both the normal and the tangential stress boundary conditions takes the form

$$2\mu\, d\{\phi(z) - z\overline{\phi'(z)} - \overline{\chi'(z)}\} = \sigma_e \, d\left[\frac{dz}{ds}\right].$$

This equation can be integrated and the constant of integration made equal to zero by the appropriate choice of the until now arbitrary constant a:

$$\phi(z) - z\overline{\phi'(z)} - \overline{\chi'(z)}\} = \frac{\sigma_e}{2\mu}\frac{dz}{ds}. \tag{6.13}$$

Differentiation of (6.12) along the boundary gives

$$\mathrm{Re}\left\{\frac{d\bar{z}}{ds}(\phi(z) + z\overline{\phi'(z)} + \overline{\chi'(z)})\right\} = 0,$$

which using (6.13) can be simplified to

$$\mathrm{Re}\left\{\frac{d\bar{z}}{ds}\phi(z)\right\} = \frac{\sigma_e}{4\mu}. \tag{6.14}$$

Multiplying (6.13) by $d\bar{z}/ds$ and eliminating $(d\bar{z}/ds)\phi(z)$ using (6.14) one obtains

$$\phi(z)\frac{d\bar{z}}{ds} + \bar{z}\phi'(z)\frac{dz}{ds} + \chi'(z)\frac{dz}{ds} = 0,$$

which after integration gives

$$\bar{z}\phi(z) + \chi(z) = 0. \tag{6.15}$$

Here the constant of integration is made equal to zero by the appropriate choice of b.

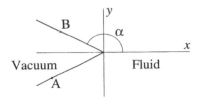

FIGURE 6.2: Sketch establishing notation for Richardson's (1968) analysis of the nature of singularity of free-surface curvature allowed by the standard model.

Now, consider the possibility of a singularity of curvature of the free surface. Assume that the free surface profile in the plane of flow has an angular point where the smooth parts of the boundary meet at an angle 2α if measured through the fluid (Fig. 6.2). Then, boundary condition (6.15) becomes

$$e^{-2i\alpha} z\phi(z) + \chi(z) = 0 \text{ on arg } z = \alpha$$
$$e^{2i\alpha} z\phi(z) + \chi(z) = 0 \text{ on arg } z = -\alpha. \tag{6.16}$$

These conditions are compatible for $\alpha = \frac{1}{2}\pi n$ for integer n and hence the singularity of curvature of the free surface must be a genuine cusp,

$$\alpha = 0, \pi.$$

Now, condition (6.16) gives that locally

$$\chi(z) = -z\phi(z). \tag{6.17}$$

Applying condition (6.13) for a part of the boundary between points A and B located on the lower and upper surface respectively (Fig. 6.2), one has

$$[\phi(z) - z\overline{\phi'(z)} - \overline{\chi'(z)}]_A^B = \pm\sigma_e/\mu, \tag{6.18}$$

where the left-hand side denoted the difference between the values of the expression in the square brackets taken at B and A. The positive and negative signs on the right-hand side correspond to $\alpha = 0$ and $\alpha = \pi$ respectively.

By combining (6.17) and (6.18) one arrives at

$$[\text{Re}\{\phi(z)\}]_A^B = \pm\sigma_e/2\mu \tag{6.19}$$

and hence for $\sigma_e \neq 0$ one has to choose the negative sign, i.e., $\alpha = \pi$. Thus, the only possibility is a cusp pointing into the fluid and, to leading order, it is modelled as a cut along the negative x-axis.

Then, it follows from (6.19) that locally

$$\phi(z) = \frac{i\sigma_e}{2\pi\mu} \log z.$$

Now, with both ϕ and χ known, one can write down the stream function which in the polar coordinates (r, θ) with $\theta = 0$ along the positive x-axis takes the form

$$\psi(r, \theta) = \frac{\sigma_e}{2\pi\mu} r \log r \sin \theta. \tag{6.20}$$

Then, for the radial normal stress component one has

$$p_{rr} = \frac{2\sigma_e}{\pi} \frac{\cos \theta}{r} \tag{6.21}$$

so that an integration around a circle with the center at the cusp produces a force equal to $2\sigma_e$ thus balancing the effect of the surface tension.

Thus, Richardson's results show that if one assumes a priori the existence of a singularity of curvature in a two-dimensional Stokes flows at finite capillary numbers and uses the standard model to describe it, then this singularity has to be a genuine cusp pointing into the fluid with the flow in its immediate vicinity described by the stream function (6.20). As emphasized by Richardson himself, this stream function corresponds to a flow in the *negative* direction, i.e., the direction against the cusp. Although this feature is not supported by experimental observations, one could still argue that this discrepancy might be attributed to a limited spatial resolution of observations and hence their inability to resolve the fine structures of the flow very close to the cusp.

More worrying is the fact that the velocity field associated with (6.20) and the stress given by (6.21) are both singular and hence the results do not satisfy the basic ABC-criterion formulated in §2.5.2 for physically realistic solutions. Furthermore, by applying the expressions from §2.3.2 for the energy dissipation due to viscosity one can show that the density of the rate of dissipation of energy in the flow described by (6.20) is nonintegrable. This means that the total rate of dissipation of energy in the vicinity of the cusp is infinite.

These obviously unphysical features of the solution indicate that at finite capillary numbers the standard model does not allow for the existence of any singularity of curvature in a steady two-dimensional creeping flow.

6.2.2 Solution with an emerging singularity

Jeong & Moffatt (1992) approached the cusp phenomenon in a different way and examined the free-surface profile in a two-dimensional convergent flow without presuming a priori that there is a free-surface curvature singularity. To qualitatively model the experiment where the flow was generated by two counter-rotating cylinders (Fig. 6.3a), they consider an idealized problem with the flow produced by a vortex dipole placed at depth $d(=1)$ below the undisturbed position of the free surface (Fig. 6.3b). The gist of Jeong & Moffatt's (1992) work is as follows.

Let the undisturbed liquid occupy a half-plane $y < 0$ with the vortex dipole of strength α placed at $z(= x + iy) = -id$. The axis of the dipole is taken to

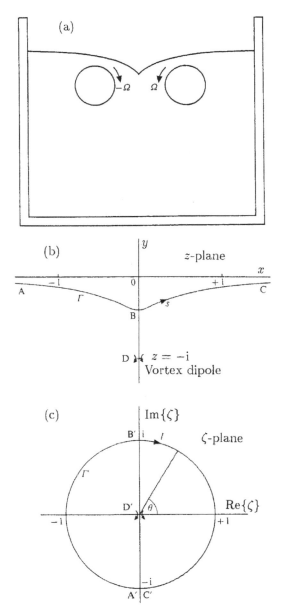

FIGURE 6.3: The flow generated by two counter-rotating cylinders (a) is modelled as produced by a dipole (b). The flow domain in the z-plane is then mapped conformally to a unit disk (b) in the ζ-plane (Jeong & Moffatt 1992; © 1992 Cambridge University Press; reproduced with permission).

be parallel to the y-axis.[1] Given that d is the only length scale in the problem, one can use it as the unit of length and hence set $d = 1$. The main physical assumption in describing the fluid motion will be that it is in the Stokes regime everywhere in the flow domain, from the high-velocity region near the dipole to the far field where the velocity vanishes. Then the stream function must obey the biharmonic equation and one can apply the technique of conformal mapping to find it. Although the flow is obviously in the Stokes regime in the far field and the biharmonic stream function associated with the vortex dipole satisfies the Stokes (and even the Navier-Stokes) equations as one approaches the dipole, one has to specify the conditions ensuring that these two regions of validity of the Stokes approximation overlap. This issue, however, is of minor importance for consideration of the cusp as a local phenomenon.

As in §6.2.1, for the Stokes flow everywhere the stream function ψ can be expressed in terms of two analytic functions $f(z)$ and $g(z)$,

$$\psi(x, y) = \text{Im}\{f(z) + \bar{z}g(z)\}, \tag{6.22}$$

so that the velocity components are given by

$$u - iv = f'(z) + \bar{z}g'(z) - \overline{g(z)}.$$

The dipole at $z = -i$ provides the condition

$$f(z) \sim \frac{i\alpha}{z+1} \qquad \text{as } z \to -i, \tag{6.23}$$

and, assuming that the fluid is at rest in the far field, i.e., $u, v \to 0$ as $|z| \to \infty$, one has

$$f \sim cz, \quad g \sim \bar{c} \qquad \text{as } |z| \to \infty, \tag{6.24}$$

where c is an arbitrary constant.

Finally, the boundary conditions on the free surface (6.14) and (6.15) take the form

$$\text{Im}\left\{\left(\frac{d\bar{z}}{ds}\right)g(z)\right\} = \frac{\sigma_e}{4\mu},$$

$$f(z) + \bar{z}g(z) = 0.$$

In order to solve the problem, one can look for a function $z = w(\zeta)$ that would map the flow domain in the z-plane to the unit disk in the ζ-plane (Fig. 6.3b,c) placing the dipole ($z = -i$) at $\zeta = 0$, the far-field points on the free surface $z = \pm\infty + i0$ at $\zeta = -i$ and the point on the free surface above the dipole ($z \in \Gamma$, $\text{Re}\{z\} = 0$) at $\zeta = i$. An analysis of the boundary conditions for the images

$$F(\zeta) = f(w(\zeta)) = f(z), \qquad G(\zeta) = g(w(\zeta)) = g(z)$$

---

[1]Later, Jeong (1999) generalized the analysis to an arbitrary angle between the axis of the dipole and the normal to the undisturbed free surface.

suggests looking for a solution within a one-parametric family of mappings

$$w(\zeta) = a(\zeta + i) + i(a + 1)\frac{\zeta - i}{\zeta + i}, \qquad (6.25)$$

where a is a real constant to be related to the capillary number. The latter is defined by

$$Ca = \frac{\mu\alpha}{d^2\sigma_e}. \qquad (6.26)$$

After a beautiful piece of analysis, Jeong & Moffatt found that

$$4\pi\,Ca = H(a), \qquad (6.27)$$

where

$$H(a) = \begin{cases} \dfrac{-a(3a+2)^2}{1+a+(-2a(a+1))^{1/2}}K(m) & \text{for } -\tfrac{1}{3} < a \le 0, \\[3mm] \dfrac{-a(3a+2)^2}{((a+1)(3a+1))^{1/2}}K(m'), & \text{for } a \ge 0, \end{cases} \qquad (6.28)$$

$$m = \frac{2}{(-2a/(a+1))^{1/4} + ((a+1)/(-2a))^{1/4}}, \qquad m' = \left(\frac{2a}{3a+1}\right)^{1/2}, \qquad (6.29)$$

and K is the complete elliptic integral of the first kind:

$$K(m) = \int_0^{\pi/2} \frac{d\theta}{(1 - m^2 \sin^2\theta)^{1/2}} = \int_0^1 \frac{dx}{[(1 - x^2)(1 - m^2 x^2)]^{1/2}}. \qquad (6.30)$$

$H(a)$ is a strictly monotonically decreasing smooth function on $-\tfrac{1}{3} < a < \infty$, and hence, for a given Ca, equation (6.27) determines a uniquely. As $a \to -\tfrac{1}{3}$, one has $Ca \to \infty$.

The free-surface profile in the z-plane is the image of the circumference $|\zeta| = 1$, and substitution $\zeta = e^{i\theta}$ into (6.25) gives this profile in a parametric form:

$$x = a\cos\theta + (a+1)\frac{\cos\theta}{1 + \sin\theta}, \qquad y = a(1 + \sin\theta).$$

After eliminating θ from the above expressions one arrives at the equation of the free-surface profile as a cubic curve:

$$x^2 y = (2a - y)(y + a + 1)^2. \qquad (6.31)$$

For $a > -\tfrac{1}{3}$ near the point $(0, 2a)$ the curve is a parabola

$$x^2 \approx -(2a)^{-1}(3a + 1)^2(y - 2a),$$

and its radius of curvature at $(0, 2a)$ is given by

$$R = -(4a)^{-1}(3a + 1)^2. \qquad (6.32)$$

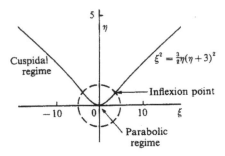

FIGURE 6.4:   As $Ca \to \infty$, one has $a = -\frac{1}{3} + \epsilon$, $\epsilon \to 0$ and, to leading order in ϵ the curve (6.31) in similarity variables $\xi = \epsilon^{-3/2}x$, $\eta = \epsilon^{-1}(y - 2a)$ takes a universal form $\xi^2 = \frac{3}{2}\eta(\eta + 3)^2$. The inflexion points at $\xi = \pm 2\sqrt{6}$, $\eta = 1$ correspond to the transition from the cuspidal to parabolic regime (Jeong & Moffatt 1992; © Cambridge University Press; reproduced with permission).

Since

$$H(a) \sim \tfrac{1}{4} \log \frac{32}{3(3a + 1)} \qquad \text{as } a \to -\tfrac{1}{3}, \tag{6.33}$$

one has from (6.27) that $Ca \to \infty$ as $a \to -\frac{1}{3}$ and then from (6.32) and (6.33) that in dimensional form

$$R/d \sim \tfrac{256}{3} \exp(-32\pi \, Ca). \tag{6.34}$$

This is a very important result describing analytically how the radius of curvature of the free surface at a point above the dipole evolves as the capillary number increases. The asymptotic estimate (6.34) shows that, strictly speaking, at a finite capillary number the free surface is always rounded and a genuine cusp appears only in the limit $Ca \to \infty$ (Fig. 6.4). Furthermore, the very formulation of the problem dictates that the free surface is always a streamline, and hence there is no transition to a different type of kinematics that the authors observed in their experiment.

Setting the experimental issues aside, it is interesting to consider Jeong & Moffatt's result from the viewpoint of the limits of applicability of the model in the framework of which this result has been obtained. The authors' estimates give that "for $Ca = 0.25$, $R/d \approx 10^{-9}$, and for $Ca = 1$, we have $R/d \approx 1.87 \times 10^{-42}$" (Jeong & Moffatt 1992, p. 11). Thus, regardless of what was observed in a particular experiment, the exact solution indicates that, at finite values of characteristic parameters, the radius of curvature of the free surface, which must be, by definition, macroscopic, can reach a submolecular scale. In other words, a continuous variation of parameters within a finite range drives the model outside its limits of applicability.

This point can be illustrated by considering the critical conditions in Joseph et al. (1991) experiments. Using that for the same flow Jeong & Moffatt's capillary number is, according to their assessment, 16 times smaller than that of Joseph et al. one has that for d of the order of centimeters the radius of curvature of the free surface corresponding to the experimental capillary numbers of cusping, Ca_cusp, given in Table 6.1 is on the molecular scale or even less. Hence, as the capillary number increases towards Ca_cusp, the model clearly falls outside its limits of applicability since its description of the liquid-gas interface as a mathematical surface with a constant surface tension relies on an assumption that the interfacial layer's thickness is much smaller than the radius of curvature of the free surface. Recognizing this fact, Jeong & Moffatt (1992, p. 11) conclude that "from a continuum point of view, it seems fair to state that when d is of order 1 m or less, a cusp does indeed form when $Ca \geq 0.25$".

This conclusion leaves open the question of how to actually describe what happens *after* the standard model becomes inapplicable. Indeed, if one formally uses Jeong & Moffatt mathematical description of the flow, this would not only mean application of the standard model outside its limits of applicability but it would also contradict experimental observations. In particular, one would have the kinematic picture of the flow with a stagnation line on the free surface, whereas in experiments there is a flux through the cusp into the bulk of the fluid. One implication of this discrepancy is that, if this flow is used as a background for calculating the surfactant concentration distribution, the solution (Antanovskii 1994) predicts that the surfactant will remain on the free surface at all time, whereas the true kinematics will immediately sweep the surfactant through the cusp into the interior leading to a surfactant-free liquid-gas interface.

An alternative approach could be to impose an external physical criterion and assume that, as the radius of curvature reaches a certain preset value, one has to introduce a line where the free-surface curvature is singular whereas up to this line it must be finite. However, if this singularity is to be described in the framework of the standard model, then, according to Richardson's (1968) solution, it must be a genuine cusp, and we come back to the difficulties considered in the previous section: by switching to this solution one will impose a sudden change of the flow direction, which is now against the cusp, and, more importantly, imply an infinite rate of dissipation of energy in the fluid which is no more physical than the radius of curvature of the free surface on the submolecular scale.

6.2.3 The paradox of cusping

Thus, by examining two exact analytical solutions (Richardson 1968, Jeong & Moffatt 1992), which describe convergent flows and exhaust all possibilities for the standard model, we arrived at a paradox:

- If no singularity is presumed a priori, then gradual increase in the capillary number leads to an exponentially vanishing radius of curvature of the free surface and hence drives the model outside its limits of applicability as the radius of curvature reaches molecular and then submolecular scales.

- If a free-surface singularity is presumed from the start or introduced when the singularity-free approach moves the solution outside its limits of applicability, then the standard model dictates that this singularity has to be a genuine cusp and the corresponding solution gives rise to an infinite rate of dissipation of energy which is also unphysical.

It should be emphasized here that these solutions per se indicate a problem with the standard model. The experiments simply confirm that the problem is a real one.

It is also important to note that the inapplicability of the Navier-Stokes equations together with the standard boundary conditions on the free surface to the description of flows with steady two-dimensional cusps does not at all mean the breakdown of the continuum approximation itself: to describe the macroscopically observed cusp/corner on the macroscopic (hydrodynamic) length scale with a qualitatively correct kinematics of fluid motion one should include in the corresponding mathematical model the macroscopic 'outcome' of those physical mechanisms that become important when a rounded free surface transforms into a piece-wise smooth one. More specifically, one should expect the appearance of a concentrated nonhydrodynamic force borne and maintained by the external flow, which will balance the capillary force acting on the line where the free-surface curvature is singular, thus allowing the hydrodynamic flow field to be regular.

6.3 'Missing' physics

6.3.1 Essential factors and additional influences

The simplest idea that comes to mind in connection with the above paradox of modelling is to suggest that this problem is irrelevant and in reality Jeong & Moffatt's sharpening tip is followed by air entrainment once the tip radius becomes sufficiently small (but yet remains macroscopic) so that, in practice, a true singularity of curvature never forms. One can then calculate the conditions required for the lubricating air between free surfaces to produce the required instability (Eggers 2001). A similar way of 'sweeping the problem under the carpet' is known in dynamic wetting where one can assume that there is no actual contact line and the spreading liquid is lying on an air cushion dragged by the solid substrate (Bourgin & Saintlos 2001). However, this argument falls apart once the air above the cusping liquid is removed. Indeed, as in dynamic wetting, the air is a physical factor *additional* to the essential hydrodynamics of the convergent flow, and its influence can be removed, for example, by placing the experimental setup with a low-volatility fluid in a low-pressure chamber. Then one will have almost exactly the situation modelled by Jeong & Moffatt's solution and hence the fully-blown paradox we identified earlier. Experiments of this type carried out in the context of dynamic wetting show that air is indeed a factor that comes *on top* of the physics of wetting, and the onset of air entrainment can be manipulated by reducing the air pressure (Benkreira, Khan & Patel 2004).

An important indicator that air is not associated with the 'missing' physics of cusping is the fact that in experiments reported in the literature there is no consistency with respect to air entrainment. Joseph et al. (1991) observed what looked like a stable two-dimensional cusp and reported that the instability appears only after a further increase in the capillary number. Jeong & Moffatt (1992) also found "no evidence in the experiment for the entrainment of air bubbles", leave alone a sheet of air. On the other hand, Lorenceau, Restagno & Quéré (2003) reported that in their experiments there was no actual cusp or corner and sharpening of the tip (to the radius of about 10 μm) was followed by the entrainment of an air film into the liquid. A possible explanation for the discrepancy is that a cusp/corner itself is a local phenomenon whereas the viscous effect of air depends on the flow configuration on a much greater scale where one essentially needs a sufficiently long narrow channel (with its width being on a macroscopic, not molecular scale) between free surfaces. In the photographs presented by Jeong & Moffatt one can see a fast opening free-surface profile whereas Lorenceau et al. report a small aspect ratio in their flow configuration with the depth of immersion of the tip on the scale of several millimeters and typical radius of the tip of less than 100 μm.

As we emphasized earlier, the solutions obtained by Richardson (1968) and Jeong & Moffatt (1992) *per se* pose a fundamental theoretical problem as they fall outside the limits of applicability of the underlying model.

6.3.2   Singularity of curvature as a contact line in the interface disappearance process

In addressing the problem, we have to resolve two issues:

(a) What 'extra' physics should be incorporated to allow the model to describe a singularity of curvature in a regular way?

(b) What is the type of this singularity, i.e., is it a cusp or a corner?

The key to the physics missing in the conventional model is suggested by the flow kinematics observed in experiments. A quotation from Jeong & Moffatt given in §6.1 describes a qualitative change in the flow associated with the formation of a singularity: as the singularity, which the authors, following Joseph and co-workers, describe as a cusp, appears, the fluid particles belonging to the free surface no longer stay there at all time, as assumed in the standard model. Instead, they go through the singularity line into the interior

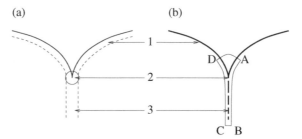

FIGURE 6.5: Schematic illustration of the physical picture of the flow near a singularity of free-surface curvature. In the continuum approximation, the interfacial layers 1(a) are modelled as sharp interfaces 1(b), the transition region 2(a) is seen as a contact line 2(b), and the surface-tension-relaxation tail 3(a) becomes a gradually disappearing internal interface 3(b). For the total force acting on the control volume ABCD to be zero, the surface tensions acting on AD can only be balanced by shear stress from the fluid outside acting on AB and CD, thus suggesting the existence of the surface-tension gradient along the internal interface.

to eventually become 'ordinary' bulk particles. Hence, here we have another case of the interface disappearance/formation process or, more precisely, the interface disappearance process. Indeed, in macroscopic terms the liquid-gas interface disappears as it goes through the singularity line and evolves into an 'internal interface', i.e., a 'tail' stretching from the free surface into the interior of the fluid where the former interface eventually looses its specific 'surface' properties, such as the surface tension (Fig. 6.5). Given that in mechanical processes the most important surface parameter is the surface tension, this tail can be referred to as 'the surface-tension-relaxation tail'.

One can describe the process in other words as 'dynamic wetting of a mathematical surface' where the equilibrium values of the surface parameters correspond to zero surface tension. This is also suggested by an obvious similarity between the kinematics observed by Jeong & Moffatt (1992) and 'rolling motion' we discussed in Chapter 3. The line along which this mathematical surface ('internal interface') meets the free surface is completely analogous to the contact line we had in dynamic wetting. Hereafter, for brevity we will refer to this line as the contact line.

The 'equilibrium contact angle' associated with the equilibrium state of the internal interface is obviously equal to 90°, i.e., in the situation where there is no surface tension relaxation along the internal interface one has a conventional smooth free surface.

The idea of an internal interface, or the surface-tension-relaxation tail, suggests a simple and natural mechanism for balancing the forces acting on the contact line and hence removing the paradox we described earlier. Indeed, the concentrated force from the surface tensions of the two parts of the free surface that meet at the contact line is balanced by another concentrated force, i.e., a force from the surface-tension-relaxation tail so that, as we will show later, the flow field in the vicinity of the contact line remains regular.

It is instructive to illustrate the above argument, as well as the solutions described in §§6.2.1 and 6.2.2, in different way. Consider the control volume ABCD (Fig. 6.5b) which comprises the contact line and has sides AB and CD lying just outside the range of influence of intermolecular forces while AD and BC are at a finite macroscopic distance from the contact line. Then, all boundaries of this volume are located in the bulk where only the physical mechanisms incorporated in the classical hydrodynamics are involved.[2] The surface tensions acting on AD can be compensated either by the normal stress applied to BC and/or by the tangential stress acting on AB and CD. The standard boundary conditions on the free surface and along the plane of symmetry in the bulk imply zero tangential stress so that the tangential forces on AB and CD are negligible, and the only possibility is that the capillary forces are balanced by the normal stress imposed on BC. If one neglects the length scale associated with intermolecular forces compared to the characteristic length scale of the flow (this, as we remember, is the limit of the continuum approximation), then the surface tension acting on AD remain finite whereas the size of BC becomes infinitesimal. Hence, for the total force acting on BC to remain finite, the normal stress distribution should have a singularity. Hence one arrives at Richardson's (1968) solution considered in §6.2.1, which deals with a genuine cusp at finite capillary numbers and therefore has to produce a finite total force as a result of the action of the bulk stress on a line. The same conclusion follows from Jeong & Moffatt's (1992) solution, where the stress density increases exponentially with the capillary number.

The control volume consideration applied to the singularity of curvature described as an interface disappearance process suggests a different balance. As the liquid-gas interface goes through the contact line into the bulk, it gradually looses its surface tension leading to the appearance of the surface tension gradient along the internal interface. Then, as is always the case in the Marangoni flows, this gradient has to be balanced by the shear stress from the liquid. Hence the concentrated forces acting on the side AD of the control volume will be balanced by shear stress acting on *lateral* sides AB and CD, whose length remains macroscopic in the continuum limit. This allows for balancing the capillary forces without giving rise to singularities in the flow field. Similarly to the situation we considered in the previous chapter, one will have a reverse influence on the flow of the surface-tension gradient caused by this flow, i.e., the flow-induced Marangoni effect.

---

[2]In this situation, the intermolecular forces, being internal with respect to volume ABCD, cannot balance those from the surface tension acting on AD, and hence the arguments in favour of intermolecular forces as an explanation for the cusp suggested in Joseph (1992) become irrelevant.

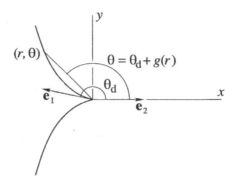

FIGURE 6.6:   Definition sketch for the free-surface curvature singularity problem as a flow with disappearing interface.

Thus, the cusp/corner phenomenon turns out to be a particular case of the interface formation/disappearance process, and to describe the flow in the vicinity of the contact line one can apply, without any ad hoc alterations, the theory of such processes described in Chapter 4. Then, it is the mathematics of the model that should decide whether the singularity of curvature is a corner or a genuine cusp. In the next section, we will address the problem following the analysis in Shikhmurzaev (2005).

6.4   Singularity-free solution: cusp or corner?

6.4.1   Problem formulation

Consider a symmetric flow domain bounded by an interface between an incompressible Newtonian liquid and an inviscid gas (i.e., physically a vacuum). Consider a general case where the interface forms a corner coinciding with the origin of a Cartesian coordinate frame (x, y) with an angle $2\theta_d$ if measured through the liquid (Fig. 6.6). A cusp will then be a particular case corresponding to $\theta_d = \pi$. Owing to the symmetry of the problem, we may consider the region $y \geq 0$ and in what follows will assume that the flow is steady and the associated Reynolds number, based on the characteristic distance from the contact line and velocity in the far field, is small. Then the flow in the vicinity of the contact line is described by the Stokes equations

$$\nabla \cdot \mathbf{u} = 0, \quad \nabla p = \mu \Delta \mathbf{u}. \tag{6.35}$$

Since for the symmetric case the tangential force on the plane of symmetry is zero, it is convenient to model the internal interface as composed of two free surfaces glued together. Then, as the free surface is swept into the interior, this results in a change of the equilibrium state towards which the internal interface has to relax. Thus, the same boundary conditions (4.54), (4.56)–(4.59) apply both at the free surface ($x < 0$, $y = h(x)$) and at the internal interface ($x > 0$, $y = 0$). For a steady flow it follows from (4.54) that $\mathbf{v}^s \cdot \mathbf{n} = 0$, and conditions (4.56)–(4.59) take the form

$$\mu \mathbf{n} \cdot [\nabla \mathbf{u} + (\nabla \mathbf{u})^*] \cdot (\mathbf{I} - \mathbf{nn}) + \nabla \sigma = 0, \tag{6.36}$$

$$\rho \mathbf{u} \cdot \mathbf{n} = \frac{\rho^s - \rho_{eq}^s}{\tau}, \tag{6.37}$$

$$\nabla \cdot (\rho^s \mathbf{v}^s) = -\frac{\rho^s - \rho_{eq}^s}{\tau}, \tag{6.38}$$

$$(1 + 4\alpha\beta)\nabla \sigma = 4\beta(\mathbf{v}_{\|}^s - \mathbf{u}_{\|}), \tag{6.39}$$

$$\sigma = \gamma(\rho_{(0)}^s - \rho^s), \tag{6.40}$$

where

$$\rho_{eq}^s = \begin{cases} \rho_e^s, & \text{for } x < 0, \ y = h(x) \\ \rho_{(0)}^s, & \text{for } x > 0, \ y = 0 \end{cases}.$$

Thus at equilibrium the surface tension on the free surface $\sigma_e = \gamma(\rho_{(0)}^s - \rho_e^s) > 0$, whereas on the internal boundary one has $\sigma(\rho_{(0)}^s) = 0$.

The normal stress boundary condition (4.55), where we put $p_g = 0$, applies only on the free surface

$$-p + \mu \mathbf{n} \cdot [\nabla \mathbf{u} + (\nabla \mathbf{u})^*] \cdot \mathbf{n} = \sigma \nabla \cdot \mathbf{n}, \qquad \text{for } x < 0, \ y = h(x), \tag{6.41}$$

and determines the free-surface profile $y = h(x)$.

At the contact line one has the mass and momentum balance conditions similar to those we discussed in §4.3.5. In the simplest case, they are given by

$$(\rho^s \mathbf{v}^s)|_{x \to 0-, y=h(x)} \cdot \mathbf{e}_1 + (\rho^s \mathbf{v}^s)|_{x \to 0+, y=0} \cdot \mathbf{e}_2 = 0, \tag{6.42}$$

$$\sigma|_{x \to 0-, y=h(x)} \cos \theta_d + \sigma|_{x \to 0+, y=0} = 0, \tag{6.43}$$

where $\mathbf{e}_1$ and $\mathbf{e}_2$ are unit vectors normal to the contact line and tangential to the free surface and the internal interface, respectively (Fig. 6.6).

To complete the problem formulation, one has to prescribe the velocity distribution in the bulk and the values of the surface parameters in the far field.

6.4.1.1 Nonsymmetric flow

If the flow is not symmetric with respect to a plane passing through the contact line, one will have to model the internal interface as a whole (not as two sublayers) with the surface equation of state for this interface taking the form

$$\sigma = 2\gamma(\rho_{(0)}^s - \rho^s),$$

which will replace (6.40). Then, since the shape of the internal interface is not known a priori and this interface experiences tangential stress on both its sides, condition (6.36) has to be replaced by (4.47), and instead of (6.39) one has to use (4.37) and (4.48). Condition (6.37) now applies on both sides of the internal interface with a negative sign if **n** is an external normal. Finally, the boundary conditions at the contact line (6.42), (6.43) have to be replaced with

$$(\rho^s \mathbf{v}^s)|_{G+} \cdot \mathbf{e}_1 + (\rho^s \mathbf{v}^s)|_{G-} \cdot \mathbf{e}_3 + (\rho^s \mathbf{v}^s)|_I \cdot \mathbf{e}_2 = 0,$$

$$\sigma|_{G+} \cos\theta_{d+} + \sigma|_{G-} \cos\theta_{d-} + \sigma|_I = 0, \tag{6.44}$$

$$\sigma|_{G+} \sin\theta_{d+} = \sigma|_{G-} \sin\theta_{d-}, \tag{6.45}$$

where I marks the parameters of the internal interface; $G\pm$, $\theta_{d\pm}$ denote the upper/lower branches of the free surface and the angles they form with the internal interface; $\mathbf{e}_3$ is tangential to the lower branch of the free surface. In addition to (6.44) expressing the balance of projections of forces on the direction tangential to the internal interface at the contact line, now we also need condition (6.45) for the normal force balance.

The relaxation of the surface tension to zero along the internal interface gradually turns the latter into an ordinary streamline.

6.4.2   Near-field asymptotics

The analysis of the asymptotic behaviour of the solution in the near field of the singularity of curvature is similar to that given in §5.1.

Let

$$\theta = \theta_d + g(r), \quad \theta \neq \tfrac{\pi}{2}, \ g(r) \to 0 \text{ as } r \to 0, \tag{6.46}$$

be the equation of the profile of the upper branch of the free surface in the vicinity of the contact line in polar coordinates shown in Fig. 6.6 and u, v denote the radial and transversal components of the bulk velocity. For the near-field asymptotic analysis it is convenient to use the formulation in terms of the stream function ψ introduced by

$$u = \frac{1}{r}\frac{\partial\psi}{\partial\theta}, \quad v = -\frac{\partial\psi}{\partial r}$$

and reduce the Stokes equations (6.35) to a biharmonic equation

$$\Delta^2\psi = 0 \tag{6.47}$$

for ψ. After scaling the velocities, lengths, pressure, surface tension and the surface density respectively with a characteristic velocity U, $L = U\tau$, $\mu U/L$, σ_e and $\rho^s_{(0)}$ we get that at the free surface the boundary conditions (6.41), (6.36)–(6.39) take the form

$$Ca\, p_{nn} = \frac{\sigma(2g' + rg'' + r^2 g'^3)}{(1 + r^2 g'^2)^{3/2}}, \tag{6.48}$$

$$Ca\,p_{n\tau} + \frac{1}{(1+r^2g'^2)^{1/2}}\frac{d\sigma}{dr} = 0, \tag{6.49}$$

$$\frac{rg'u - v}{(1+r^2g'^2)^{1/2}} = Q(\rho^s - \bar{\rho}_e^s), \tag{6.50}$$

$$\frac{1}{(1+r^2g'^2)^{1/2}}\frac{d(\rho^s v^s)}{dr} = -(\rho^s - \bar{\rho}_e^s), \tag{6.51}$$

$$\frac{1}{(1+r^2g'^2)^{1/2}}\frac{d\sigma}{dr} = 4\lambda V^2\left[v^s - \frac{u + rg'v}{(1+r^2g'^2)^{1/2}}\right], \tag{6.52}$$

whereas at the internal interface one has

$$Ca\,p_{r\theta} + \frac{d\sigma}{dr} = 0, \tag{6.53}$$

$$v = Q(\rho^s - 1), \tag{6.54}$$

$$\frac{d(\rho^s v^s)}{dr} = 1 - \rho^s, \tag{6.55}$$

$$\frac{d\sigma}{dr} = 4\lambda V^2(v^s - u) \tag{6.56}$$

with σ and ρ^s related in both sets of the boundary conditions by the surface equation of state (6.40), i.e., in a dimensionless form

$$\sigma = \lambda(1 - \rho^s). \tag{6.57}$$

Here

$$Ca = \frac{\mu U}{\sigma_e}, \quad Q = \frac{\rho^s_{(0)}}{\tau\rho U}, \quad V^2 = \frac{\tau\beta U^2}{\gamma\rho^s_{(0)}(1 + 4\alpha\beta)}, \quad \lambda = \frac{\rho^s_{(0)}\gamma}{\sigma_e},$$

$$\bar{\rho}_e^s = \frac{\rho^s_e}{\rho^s_{(0)}} < 1 \tag{6.58}$$

and the expressions for the stresses are given by

$$p_{n\tau} \equiv \tfrac{1}{2}(p_{rr} - p_{\theta\theta})\sin 2\varphi - p_{r\theta}\cos 2\varphi,$$

$$p_{nn} \equiv \tfrac{1}{2}(p_{rr} + p_{\theta\theta}) + \tfrac{1}{2}(p_{\theta\theta} - p_{rr})\cos 2\varphi - p_{r\theta}\sin 2\varphi,$$

$$p_{rr} = -p + 2L(\psi), \quad p_{\theta\theta} = -p - 2L(\psi), \quad p_{r\theta} = M(\psi),$$

where

$$L = \frac{1}{r}\frac{\partial^2}{\partial r\partial\theta} - \frac{1}{r^2}\frac{\partial}{\partial\theta}, \quad M = \frac{1}{r^2}\frac{\partial^2}{\partial\theta^2} - \frac{\partial^2}{\partial r^2} + \frac{1}{r}\frac{\partial}{\partial r},$$

$$\sin 2\varphi = \frac{2rg'}{1 + r^2g'^2}, \quad \cos 2\varphi = \frac{1 - r^2g'^2}{1 + r^2g'^2}.$$

The bulk pressure p is related to the stream function via the equations of motion (6.35), which in a scalar form after nondimensionalisation become

$$\frac{\partial p}{\partial r} = \left(\frac{1}{r}\frac{\partial^3}{\partial r^2 \partial \theta} + \frac{1}{r^3}\frac{\partial^3}{\partial \theta^3} + \frac{1}{r^2}\frac{\partial^2}{\partial r \partial \theta}\right)\psi, \tag{6.59}$$

$$\frac{\partial p}{\partial \theta} = -\left(r\frac{\partial^3}{\partial r^3} + \frac{\partial^2}{\partial r^2} + \frac{1}{r}\frac{\partial^3}{\partial r \partial \theta^2} - \frac{1}{r}\frac{\partial}{\partial r} - \frac{2}{r^2}\frac{\partial^2}{\partial \theta^2}\right)\psi. \tag{6.60}$$

We will consider whether the problem (6.47)–(6.57), (6.42), (6.43) leads to a physically meaningful solution and, if this is the case, what type of the free-surface curvature singularity it corresponds to. In accordance with the ABC-criterion of §2.5.2, by a 'physically meaningful solution' we will understand a solution having finite velocity and pressure distributions in the flow domain and at its boundary as well as finite values of the surface variables along the interfaces. Then, as $r \to 0$ for the stream function we can use a general power coordinate expansion beginning with a linear term corresponding to a finite velocity distribution,

$$\psi(r, \theta) = \psi_1(r, \theta) + \sum_{n=1}^{\infty} \psi_{q_n}(r, \theta) \equiv rF_1(\theta) + \sum_{n=1}^{\infty} r^{q_n} F_{q_n}(\theta), \tag{6.61}$$

whereas for the pressure distribution one has

$$p(r, \theta) = p_0 + \sum_{n=1}^{\infty} p_n(r, \theta), \tag{6.62}$$

$p_1 \to 0$, $p_{n+1}/p_n \to 0$ $(n = 1, \ldots \infty)$ as $r \to 0$. Conditions (6.51), (6.55) give that as $r \to 0$ for the surface variables the coordinate expansions have the form

$$\rho^s(r) = \rho_{i0}^s + r\rho_{i1}^s + \sum_{n=2}^{\infty} a_n(r)\rho_{in}^s, \quad v^s(r) = v_{i0}^s + rv_{i1}^s + \sum_{n=2}^{\infty} b_n(r)v_{in}^s, \tag{6.63}$$

where $i = 1$ for the free surface and $i = 2$ for the internal interface and $a_2 = o(r)$, $b_2 = o(r)$, $a_{n+1}/a_n \to 0$, $b_{n+1}/b_n \to 0$ $(n = 2, \ldots \infty)$ as $r \to 0$. Then, from (6.57) one has that

$$\sigma(r) = \sigma_{i0} + r\sigma_{i1} + \sum_{n=2}^{\infty} a_n(r)\sigma_{in},$$

where

$$\sigma_{i0} = \lambda(1 - \rho_{i0}^s), \quad \sigma_{in} = -\lambda\rho_{in}^s \quad (i = 1, 2; n = 1, \ldots \infty). \tag{6.64}$$

Finally,

$$g(r) = \sum_{n=0}^{\infty} f_n(r), \tag{6.65}$$

where, given (6.46), $f_0 \to 0$ and $f_{n+1}/f_n \to 0$ $(n = 0, \dots \infty)$ as $r \to 0$.

The leading-order term in the expansion for ψ satisfying the biharmonic equation (6.47) has the form

$$\psi_1 \equiv rF_1(\theta) = r[(A_1 + A_2\theta)\sin\theta + (A_3 + A_4\theta)\cos\theta], \qquad (6.66)$$

where the constants of integration A_i $(i = 1, \dots, 4)$ are determined by the conditions

$$F_1(\theta_d) = Q(\rho_{10}^s - \bar{\rho}_e^s), \qquad F_1(0) = -Q(\rho_{20}^s - 1), \qquad (6.67)$$

$$F_1''(0) + F_1(0) = 0, \qquad F_1''(\theta_d) + F_1(\theta_d) = 0, \qquad (6.68)$$

that follow from (6.50), (6.54), (6.53) and (6.49). The resultant stream function,

$$\psi_1 = \frac{rQ}{\sin\theta_d} \left[(\rho_{10}^s - \bar{\rho}_e^s)\sin\theta + (1 - \rho_{20}^s)\sin(\theta_d - \theta) \right], \qquad (6.69)$$

describes a uniform flow directed from the free surface to the internal interface. This feature together with the flux through the contact line allowed by (6.42) correctly reproduce the flow kinematics observed experimentally by Jeong & Moffatt (1992). The stream function (6.69) obviously makes the right-hand sides of (6.59) and (6.60) identically zero.

Examine whether there is a fractional-power term in the expansion for ψ, i.e., consider $\psi_q = r^q F_q(\theta)$, $1 < q < 2$. The biharmonic equation for ψ_q yields

$$F_q(\theta) = C_1 \sin q\theta + C_2 \cos q\theta + C_3 \sin(q - 2)\theta + C_4 \cos(q - 2)\theta,$$

and boundary conditions (6.53) and (6.54) immediately give that $C_2 = C_4 = 0$. From (6.50) it follows that $f_0(r) = r^{q-1}g_0$ and conditions (6.49), (6.50) take the form

$$C_1 q \sin q\theta_d + C_3(q - 2)\sin(q - 2)\theta_d = 0, \qquad (6.70)$$

$$C_1 q \sin q\theta_d + C_3 q \sin(q - 2)\theta_d = -(q - 1)g_0 F_1'(\theta_d). \qquad (6.71)$$

Then, after substitution of ψ_q together with (6.62) into (6.59), (6.60) we arrive at $C_3 = 0$ so that for a nontrivial solution equation (6.70) requires $q = \pi/\theta_d$ and hence from (6.71) one has that $g_0 F_1'(\theta_d) = 0$. Given $C_3 = 0$ and $q = \pi/\theta_d$, the normal-stress boundary condition (6.48) takes the form

$$C_1 = (2Ca)^{-1}\sigma_{10}g_0. \qquad (6.72)$$

Equation (6.43) together with (6.64) ensure that $F_1'(\theta_d) \neq 0$: otherwise $\bar{\rho}_e^s = 1$ which would contradict (6.58). Then $g_0 F_1'(\theta_d) = 0$ means $g_0 = 0$ and hence from (6.72) one has $C_1 = 0$.

Thus, we have shown that there is no fractional power term $r^q F_q(\theta)$, $1 < q < 2$ in the expansion for the stream function. Consequently, the expansion for $g(r)$ begins, at least, with the linear term $f_1(r) = rg_1$. The latter means that the free-surface curvature remains finite up to the contact line, which is what is implied when, as always in fluid mechanics, the interface is modelled as a mathematical surface of zero thickness. (This requirement, i.e., $f_0(r) \equiv 0$, $f_1(r) = rg_1$, can be used as an alternative to (6.62): it turns (6.70), (6.71) into a homogeneous system and replaces (6.72) with $C_1 + C_3 \cos 2\theta_d = 0$ thus making $C_1 = C_3 = 0$ the only solution and hence leads to the expansion (6.62) for p.)

Consider $\psi_2 = r^2 F_2(\theta)$. The general solution for F_2 is given by

$$F_2(\theta) = B_1 + B_2\theta + B_3 \sin 2\theta + B_4 \cos 2\theta, \tag{6.73}$$

and the boundary conditions (6.50), (6.54), (6.53), (6.49) for F_2 take the form:

$$F_2(\theta_d) = \tfrac{1}{2}Q\rho_{11}^s - \frac{g_1 Q}{2 \sin \theta_d}\left[(\rho_{10}^s - \bar{\rho}_e^s)\cos\theta_d - (1 - \rho_{20}^s)\right] \equiv K$$

$$F_2(0) = -\tfrac{1}{2}Q\rho_{21}^s, \quad F_2''(0) = -\sigma_{21}/Ca, \quad F_2''(\theta_d) = \sigma_{11}/Ca.$$

Then,

$$B_1 = -\tfrac{1}{2}Q\rho_{21}^s - \frac{\sigma_{21}}{4Ca}, \quad B_2 = \frac{1}{4Ca\theta_d}(4CaK + 2CaQ\rho_{21}^s + \sigma_{11} + \sigma_{21}),$$

$$B_3 = -\frac{\sigma_{11} + \sigma_{21}\cos 2\theta_d}{4Ca \sin 2\theta_d}, \quad B_4 = \frac{\sigma_{21}}{4Ca}.$$

After substituting ψ_2 and the expansion (6.62) for p into (6.60), (6.59) we arrive at

$$B_2(\theta_d, g_1, \rho_{10}^s, \rho_{11}^s, \rho_{20}^s, \rho_{11}^s, \sigma_{11}, \sigma_{21}) = 0. \tag{6.74}$$

The above notation emphasizes that the constant of integration B_2 is a function of the coefficients of asymptotic expansions, and hence (6.74) provides a constraint on them.

After substituting p_0 and $\psi_1 + \psi_2$ into the normal-stress condition (6.48), we obtain

$$Ca[p_0 + 4(B_3 \cos 2\theta_d - B_4 \sin 2\theta_d)] = -2g_1\sigma_{10}. \tag{6.75}$$

For the remaining coefficients in the expansions (6.62), (6.63) one has

$$\rho_{10}^s v_{11}^s + \rho_{11}^s v_{10}^s = -(\rho_{10}^s - \bar{\rho}_e^s), \tag{6.76}$$

$$\rho_{20}^s v_{21}^s + \rho_{21}^s v_{20}^s = -(\rho_{20}^s - 1), \tag{6.77}$$

$$\sigma_{11} = 4\lambda V^2[v_{10}^s - F_1'(\theta_d)], \tag{6.78}$$

$$\sigma_{21} = 4\lambda V^2(v_{20}^s - F_1'(0)) \tag{6.79}$$

following from (6.51), (6.55), (6.52), (6.56) and two equations

$$\rho_{10}^s v_{10}^s + \rho_{20}^s v_{20}^s = 0, \tag{6.80}$$

$$\sigma_{10} \cos \theta_d + \sigma_{20} = 0, \tag{6.81}$$

following from (6.42) and (6.43).

The set of algebraic equations (6.64) for $i = 1, 2$ and $n = 0, 1$, (6.74) (6.81) provide 12 constraints on 15 coefficients of the asymptotic expansions

$$p_0, \ \theta_d, \ g_1, \ \sigma_{in}, \ \rho^s_{in}, \ v^s_{in} \quad (i = 1, 2; \ n = 0, 1).$$

The degree of freedom for these coefficients, which, importantly, include θ_d, is the same as in the case of dynamic wetting and they are determined by the outer flow. The higher-power terms in the asymptotic expansions are obviously regular at $r = 0$ and do not interest us here.

Thus, we have shown that the flow field near the contact line is regular with the finite pressure and reproduces the kinematics observed in experiments, as required by the ABC-criterion of §2.5.2. Now, we can consider a more interesting question of its *type*.

6.4.2.1 Type of singularity

Assume that at finite values of all parameters one has $\theta_d = \pi$ and hence the singularity of curvature is a genuine cusp. Then, it follows from (6.81) that $\sigma_{10} = \sigma_{20}$ and, given that, according to (6.64), $\sigma_{i0} = \lambda(1 - \rho^s_{i0})$ for $i = 1, 2$, one has $\rho^s_{10} = \rho^s_{20}$. Then, since $\bar{\rho}^s_e < 1$, the expression in the square brackets on the right-hand side of (6.69) is not identically zero. Also for any finite U the parameter $Q \neq 0$ whereas for $\theta_d = \pi$ in the denominator on the right-hand side of (6.69) one has $\sin \theta_d = 0$. Thus, our initial assumption that $\theta_d = \pi$ makes the solution (6.69) meaningless. For ψ_1 to remain meaningful, one must have $\theta_d < \pi$ and hence the singularity of the free-surface curvature is always a *corner*, not a cusp.

Alternatively, one can set $\theta_d = \pi$ from the start, i.e., in the boundary conditions (6.67), (6.68). Then, given that (6.81) and (6.64) still yield $\rho^s_{10} = \rho^s_{20}$, from (6.67) and (6.68) it follows that in (6.66) $A_2 = 0$ whereas

$$A_4 = \pi^{-1}Q(\bar{\rho}^s_c - 1) \neq 0.$$

Now, after substituting (6.66) in the radial projection of the equation of motion (6.59) we obtain that as $r \to 0$

$$p = \frac{2A_4 \cos \theta}{r} + O(1),$$

so that the pressure field is (nonintegrably) singular and hence one cannot satisfy the normal-stress boundary condition (6.48) for $g(r) \to 0$ as $r \to 0$. Thus, to have a regular solution described earlier, one must again have $\theta_d < \pi$, i.e., a corner, not a cusp. It should be noted that this conclusion follows directly from the leading-order (linear) term in the expansion (6.61) for ψ, i.e., regardless of the higher orders in the near-field asymptotics.

On a qualitative level, the obtained result is easy to understand. As the fluid composing the liquid-gas interface passes through the contact line into

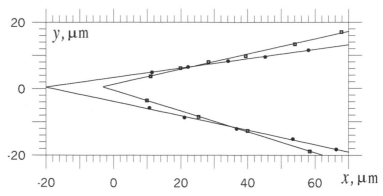

FIGURE 6.7: The measured values of the cusp tip for 500 cs silicone oil (circles) and castor oil (squares) reported by Joseph et al. (1991, Fig. 17b,c) show a piecewise linear profile of the free surface with the two branches forming a corner, not a cusp (Shikhmurzaev 2005a).

the bulk, it finds itself in a nonequilibrium state as an 'internal interface'. In its relaxation towards equilibrium, this internal interface absorbs mass from the bulk thus generating the nonzero component of the bulk velocity normal to itself. If there is a genuine cusp, then the surface density has to be continuous across the contact line whereas the equilibrium surface density is obviously discontinuous. As a result, there must be a discontinuity in the normal component of the bulk velocity at the contact line and hence a nonintegrable stress singularity which is physically unacceptable (and does not allow one to satisfy the normal-stress condition (6.48)). Equivalently, one can see this situation as the geometric incompatibility of the uniform flow directed from the free surface to the internal interface, where the normal component of the bulk velocity at both interfaces is nonzero, with a cusp. It is only once the mass exchange between the interfaces and the bulk is neglected in the boundary condition for the bulk velocity (i.e., $Q = 0$), the component of the mass flux normal to the interfaces disappears, and one has a cusp at finite velocities with the leading-order stream function representing a uniform flow parallel to it (Shikhmurzaev 1998).

Our conclusion that the singularity of curvature in convergent flows is a corner and not a cusp is supported by available experimental data. Fig. 6.7 shows that in experiments reported by Joseph et al. (1991) the profiles of the two branches of the free surface are very accurately approximated by straight lines forming a well-defined corner. A possibility that the free surface is rounded on a length scale below the spatial resolution of the measurements is ruled out by the flow kinematics which is incompatible with a stagnation line always present on a smooth free surface in convergent flows. On the other hand, there is no experimental evidence suggesting that on this 'invisible' length scale the free surface forms a cusp.

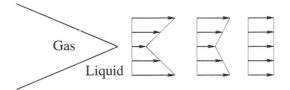

FIGURE 6.8: The surface-tension-relaxation tail can be detected by its effect on the velocity profile near the singularity of free-surface curvature.

6.4.3 Numerical and experimental implications

The main conclusion that follows from the local analysis for numerical computations of flows with singularities of the free-surface curvature is that these singularities, or more precisely the contact lines formed on the free surface, once described in the framework of the theory of flows with forming/disappearing interfaces (Chapter 4) do not require any special treatment. Given that the flow field in the vicinity of contact lines is regular and the angle formed at the contact line is well defined, one can discretise the whole problem, including the region near the contact line, in the same way. The regularity of solution means that there is no need to incorporate the local asymptotics into a code. Away from the contact line, as the surface tension in the internal interface vanishes, the latter becomes an ordinary streamline and there is no necessity to compute its surface properties anymore. It should be emphasized that, no matter how small the parameter Q is for a particular situation, it must be taken into account since, as is clear from (6.69), $1/\sin\theta_d$ acts as a magnifying factor making the role played by the mass exchange between the interfaces and the bulk in the boundary conditions (5.8), (6.54) for the normal component of the bulk velocity important in the vicinity of the contact line.

From an experimental point of view, the flow with a steady contact line on the free surface offers some interesting possibilities. This is a situation where one has the process of relaxation of the surface parameters which is, on the one hand, steady and hence free from complications unavoidable in most existing techniques of measuring the dynamic surface properties. On the other hand, this flow configuration does not involve a solid substrate with all the difficulties associated with controlling its state (roughness, chemical heterogeneity, etc.). Given that the instability of the contact line can be postponed almost indefinitely by removing the air (e.g., a vacuum chamber), the length of the surface-tension-relaxation tail can be driven well within the range of the spatial resolution of existing experimental techniques. A distinctive feature that allows one to detect the presence of surface tension gradient from the observations of the macroscopic flow is the discontinuity in the tangential stress in the bulk (Fig. 6.8).

Another experimental possibility arises from the similarity between the corner formed on the free surface in covergent flows and the situation in dynamic wetting with the nonlocal influence of the flow field/geometry on the dynamic contact angle ('hydrodynamic assist of dynamic wetting'). An important difference is that, unlike the case of dynamic wetting, the asymptotic limit $Ca \to 0$ for the problem (6.35)–(6.43) to leading order yields a trivial solution with $\theta_d = \pi/2$. This actually means that, unlike dynamic wetting where the overall flow is an *additional* influence coming on top of the velocity-dependence of the contact angle, here the velocity near the contact line is part of the global flow and hence θ_d is *completely* determined by the global influence. This makes θ_d a sensitive macroscopic indicator linking the overall flow with the properties of and processes in the interfaces which can be used in experimental investigations of dynamics of interfaces.

6.4.4   Summary

The paradox arising in the conventional modelling of steady two-dimensional convergent flows near a free surface described in §6.2.3 is resolved when these flows are considered as a particular case of fluid motion with forming/disappearing interfaces. Then, the singularity of curvature becomes a particular case of the contact line formed at the intersection of interfaces. The theory of Chapter 4 can then be applied, without any adhoc alterations, to describe the phenomenon. The local analysis of the flow near the contact line shows that

- The flow parameters and the free-surface curvature (where it is defined) remain finite up to the contact line formed at the intersection of two parts of the free surface.

- The line singularity of the free-surface curvature is always a corner, not a cusp.

The flow kinematics in the theory allows the material points on the free surface to go through the contact line into the interior of the fluid, which is the defining feature in the transition from a conventional smooth free surface to the one with a singularity of curvature.

The fact that all parameters in the local asymptotics for the flow near the contact line are regular means that, in developing numerical codes dealing with flows where the free surface is not smooth, one does not need to incorporate the local asymptotics into the code and can discretize the equations and the boundary conditions near the contact line in the same regular way as in the rest of the flow domain.

As described in §6.4.3, the convergent flow forming a corner on the free surface offers remarkable experimental opportunities for studying interfacial phenomena. The contact angle formed by the two parts of the free surface is determined by the flow field in a way similar to the nonlocal influence of the flow on the dynamic contact angle in dynamic wetting described in §3.2.3.

However, unlike dynamic wetting, in convergent flows there is no involvement of the solid surface, with its complex and poorly definable properties (roughness, heterogeneity, etc.), and no contact-line speed that dominates dynamic wetting and makes nonlocal effects only an *additional* factor (hence the term 'hydrodynamic *assist* of dynamic wetting'). In the convergent flow configuration, the nonlocal influence of the flow field becomes the main effect, and hence one has a multiparametric system of controls with the contact angle becoming a macroscopic response. These two elements of the control-response pair are linked by the physics of interface formation, in the simplest case represented by the model of Chapter 4, that allows one to study dynamic interfacial properties. Importantly, the absence of the contact-line speed in this system reduces bending of the free surface near the corner thus making it easier to measure the contact angle (see Fig. 6.7). On the other hand, however, the very nature of the nonlocal influence of the flow means that there can be no 'magic' algebraic formula, similar to what one can derive in dynamic wetting for small capillary and Reynolds numbers in the absence of 'hydrodynamic assist', that would allow one to describe and interpret experimental data. As a result, a numerical solution of the full hydrodynamic problem becomes a necessity. Although time-consuming, this problem poses no difficulties of principle given that the solution is regular everywhere and the effect of interest is associated with low Reynolds numbers.

6.5   Coalescence of drops

A phenomenon closely related with what we for brevity will call the problem of cusping is coalescence of liquid volumes. This is a particular case from a wide class of fluid motion where the flow domain undergoes a topological transition in a finite time. Other flows from this class include breakup of liquid threads leading to drop formation, rupture of free liquid films, in particular, associated with coalescence of bubbles and evolution of foams, disintegration of liquid films on solid substrates, including pattern formation on solid surfaces of varying wettability, nucleation and collapse of bubbles, breakup of gas jets in an ambient fluid, and many others.

The coalescence of fluid bodies, especially on a small length scale, is a very interesting phenomenon often involving quite unexpected dynamic effects. An example illustrating its complexity is the coalescence cascade (Fig. 6.9) reported by Thoroddsen & Takehara (2000). The post-coalescence dynamics is also very nontrivial (e.g., Anilkumar, Lee & Wang 1991).

From the viewpoint of mathematical modelling, the essence of the coalescence problem is to describe the transition from the initial contact of two fluid bodies, i.e., the moment $t = 0$ when they first touch, to a stage where one has

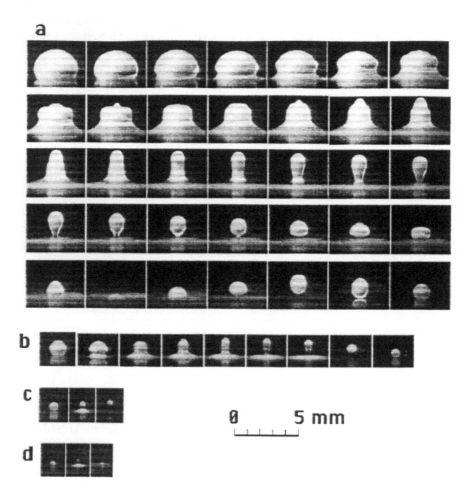

FIGURE 6.9: Coalescence cascade for a water drop (Thoroddsen & Take-hara 2000; © 2000 American Institute of Physics; reproduced with permission). The drop merges with a pool of liquid via several coalescence-breakup cycles (a–d) each time losing part of its volume to the pool.

a single body of fluid with a smooth boundary whose subsequent evolution can be modelled in the standard way. It should be emphasized that our definition of coalescence, which focuses on the transition in the topology of the flow domain, differs from a broad qualitative definition used in the experimental literature where 'coalescence' means the whole process of transition from one equilibrium state (two drops) to another (one larger drop in equilibrium). The second definition allows one to talk about 'late stages' of coalescence where, from a mathematical view point, one has simply a deformed large drop.

To address the problem of coalescence, the first question one has to ask is whether or not coalescence, as we defined it, takes place instantly, i.e., the *whole* process for $t > 0$ can be described in the framework of the standard model in a physically acceptable way. This question has been completely elucidated for plane two-dimensional flows, and it is instructive to begin with this case.

6.5.1 Coalescence in the standard model

Plane two-dimensional Stokes flows resulting from the coalescence of two cylinders of equal and unequal diameters, a cylinder and a half-space as well as the coalescence of several touching cylinders initiated by injection of fluid into one of them were studied in a series of works by Hopper (1984, 1990, 1991, 1992, 1993a,b) and Richardson (1992, 1997). The motion is assumed to be driven by a constant surface tension and the ambient fluid is neglected. Hopper and Richardson employed the technique of conformal mapping and for each of the above situations found an exact analytical solution. Fig. 6.10 shows the evolution of the free-surface profile for different diameter ratios of coalescing cylinders.

These remarkable results, being exact solutions, completely exhaust the coalescence problem for the standard model in a plane two-dimensional case. According to them, a piecewise smooth free surface present at $t = 0$ becomes smooth for $t > 0$ and the flow evolves in accordance with the standard model. Then, the initial free surface configuration is the limit as $t \to 0+$ for a continuous succession of smooth profiles, and the free-surface curvature κ, defined at every point of the free surface for $t > 0$, tends to infinity as $t \to 0+$. The capillary pressure $\sigma_e \kappa$ and hence the normal component of stress in the fluid, which is equal to $\sigma_e \kappa$, the pressure and velocity are all unbounded as $t \to 0+$. Therefore, as one approaches the starting point of the process, the solution falls outside the limits of applicability of the model in the framework of which it was obtained. For example, infinite stresses at the initial stage of the process are incompatible with the physical assumption of incompressibility of the fluid as well as its Newtonian rheology, and the assumption of a constant surface tension breaks down for infinitely large curvatures. Some of these deficiencies of the standard model have been discussed by Hopper (1990, 1992, 1993a), who wrote, in particular:

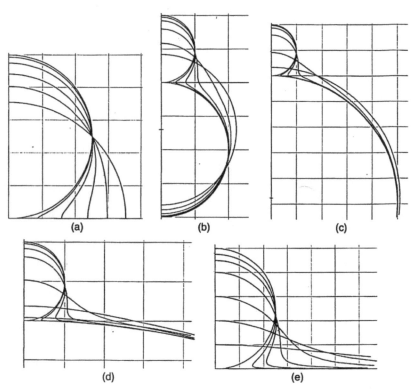

FIGURE 6.10: Free-surface profiles for the coalescence of liquid cylinders at zero Reynolds number (Hopper 1993b; © 1993; American Ceramic Society; reproduced with permission). The diameter ratios D and times t are: (a) $D = 1$, $t = 0.032.0.141, 0.325, 0.647, 1.120, \infty$; b) $D = 2$, $t = 0.0224, 0.170, 0.350, 0.703, 1.529, \infty$; (c) $D = 5$, $t = 0.0172, 0.127, 0.286, 0.632, 1.642, \infty$; (d) $D = 20$, $t = 0.0199, 0.149, 0.345, 0.603, 1.349, 3.519, \infty$; (e) $D = \infty$ (cylinder and a half-space), $t = 0.035, 0.248, 0.390, 0.631, 1.107, 1.749, 3.207$.

At extremely high stresses < ... >, the material may be non-Newtonian (Hopper 1993a, p. 2948). The surface tension of a real liquid depends on the local curvature if it is large enough, and on the separation of two surfaces if they are close enough. Primarily for this reason, the situation near the neck surface may in the very earliest stages of coalescence fail to satisfy the model (Hopper 1992, p. 172).[3]

It should be noted here that for unsteady flows, like that in the coalescence process, the model of an incompressible fluid can be inadequate due to the role played by the compressibility effects. This is the case, for example, where one is dealing with high-speed collisions of drops and should expect shock waves propagating away from the point of contact (Lesser 1983). In the situations we are considering here the compressibility effects can also play a role so that one may expect a spatially localized singularity in the time-dependence of the pressure. Then, for the results to be physically meaningful and practically applicable this singularity must be integrable. The requirement of finiteness of the fluid's velocity remains unchanged.

Thus, Hopper and Richardson have demonstrated that, in the plane two-dimensional case, the assumption inherent in the standard model that coalescence is instant leads to physically unacceptable consequences. This outcome is similar to the conclusion that follows from Jeong & Moffatt's (1992) solution, where as $Ca \to \infty$ the radius of curvature tends to zero thus driving the solution outside the limits of applicability of the model.

In a three-dimensional case, in particular when coalescence is axisymmetric, the situation becomes more complex. There are no exact analytical results exploring the fact that now, although both principal curvatures tend to infinity as $t \to 0+$, they have the opposite signs so that, at least in principle, one could think of a singularity-free scenario of coalescence for the standard model. For this scenario to materialize, the neck connecting the merging volumes in the early stages has to evolve asymptotically as a catenoid thus making the capillary pressure in the neck finite.

The first study of the axisymmetric coalescence of spherical drops was reported by Frenkel (1945), who examined this problem in connection with the so-called "cold welding" of crystalline bodies. This work considers the process in an integral form by equating the total work per unit time done by the surface tension on minimization of the free-surface area with the total dissipation of energy due to viscous friction. Both quantities are calculated approximately using the following approach. Assuming that at a moment $t > 0$ the two drops of initial radii R_0 form a single body consisting of two spherical segments of radii $R(t)$ joint over the area of a circle of radius $r_m(t)$

---

[3]This statement should perhaps refer to the model failing to describe the reality in the earliest stages of the process, i.e., when coalescence, as we defined it, is actually taking place, rather than to the reality failing to satisfy the model.

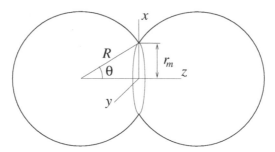

FIGURE 6.11: A definition sketch for Frenkel's (1945) semi-qualitative analysis.

(Fig. 6.11), one has that, for the volume of each drop to remain constant,

$$\tfrac{1}{3}\pi R^3(2 + 3\cos\theta - \cos^3\theta) = \tfrac{4}{3}\pi R_0^3, \qquad (6.82)$$

where the angle $\theta(t)$ is defined by $\sin\theta = r_m/R$. The corresponding variation in the free-surface area S is given by

$$S_0 - S = 8\pi R_0^2 - 4\pi R^2(1 + \cos\theta). \qquad (6.83)$$

For the initial stage of the process, i.e., as $\theta \to 0+$, one has from (6.82) that $R = R_0 + O(\theta^4)$ and hence, to leading order in θ, equation (6.83) becomes

$$S_0 - S = 2\pi R_0^2 \theta^2.$$

Then, during the initial stage of the coalescence the work of the surface tension per unit time, $-\sigma_e(dS/dt)$, is given by

$$-\sigma_e\frac{dS}{dt} = 2\pi R_0^2 \sigma_e \frac{d\theta^2}{dt}. \qquad (6.84)$$

This expression has an asymptotic accuracy as $\theta \to 0+$.

The rate of dissipation of energy in the fluid due to viscosity is given by $\int_{V(t)} \boldsymbol{\sigma} : \mathbf{E}\, dV$, where $V(t)$ is the space occupied by the two drops at a given moment $t > 0$. In order to calculate it, one has to solve the hydrodynamic problem and, in particular, resolve the modelling issues (later) noted by Hopper. Instead, Frenkel considers an extensional flow,

$$u_x = -\tfrac{1}{2}\alpha x, \quad u_y = -\tfrac{1}{2}\alpha y, \quad u_z = \alpha z, \qquad (6.85)$$

as an 'effective representation' of the actual fluid motion in the coalescence process. On the one hand, this simplification allows him to find the functional dependences between some key characteristics of the phenomenon in a simple way. On the other hand, however, the application of (6.85), which is asymptotically accurate only near the point of the initial contact of the drops, to the whole body of fluid reduces the analysis to a semi-qualitative level.

The effective extension (or more precisely, contraction) is defined as the variation of the distance between the center of each drop and their surface of contact, $R(1 - \cos\theta) = \frac{1}{2}R_0\theta^2 + O(\theta^4)$ as $\theta \to 0$, so that the absolute value of the extension coefficient α can be estimated as the ratio of the rate of extension $d(\frac{1}{2}R_0\theta^2)/dt$ to R_0. Using α as a characteristic value of $\nabla\mathbf{u}$ in (2.131), Frenkel obtains the following approximation for the rate of energy dissipation:

$$\int_V \boldsymbol{\sigma} : \mathbf{E}\, dV \approx \frac{16}{3}\pi R_0^3 \mu \left[\frac{d}{dt}\left(\frac{\theta^2}{2}\right)\right]^2. \tag{6.86}$$

Now, by equating the right-hand sides of (6.84) and (6.86) one arrives at an ordinary differential equation

$$\frac{d\theta^2}{dt} = \frac{3}{2\pi}\frac{\sigma_e}{\mu R_0},$$

which, after integration subject to an initial condition $\theta(0) = 0$, yields

$$\theta^2 = \frac{3}{2\pi}\frac{\sigma_e}{\mu R_0}t.$$

Given that $r_m = R\sin\theta \approx R_0\theta$, it is now possible to express the minimal radius of the neck r_m as

$$r_m = \left(\frac{3}{2\pi}\right)^{1/2}\left(\frac{R_0\sigma_e}{\mu}\right)^{1/2}t^{1/2}. \tag{6.87}$$

Here the first factor on the right-hand side depends on the exact specification of the 'effective' extensional flow used in the analysis and can be replaced with a geometry-dependent calibration coefficient. The second factor shows that, in the modelling scheme used by Frenkel, the characteristic value of the fluid's velocity is given by $U = \sigma_e/\mu$, which is associated with $Ca = 1$. The time-dependence of r_m results from the quadratic approximation of the shape at the point of contact.

Frenkel's analysis has an important implication. If the fluid-gas interface is assumed to be a material surface with the fluid particles staying there at all time, then the velocity of these particles at the neck is equal to the velocity at which the neck's radius increases, $u_m \equiv \dot{r}_m \sim t^{-1/2}$, and hence it is singular as $t \to 0+$. Since this result follows from a very general approach which provides an integral frame for the underlying dynamics, one has to conclude that in the initial stage of coalescence either some additional mechanisms of dissipation come into play to slow the process down or the fluid particles do not follow the free surface thus producing a kinematic pattern than cannot be reduced to an effective extensional flow. Then a more detailed analysis of the flow dynamics is needed and, due to the mathematical complexity of the problem, it has to be carried out numerically.

One of the first numerical studies of the axisymmetric coalescence was reported by Hiram & Nir (1983), who used the boundary integral equations method based on the integral representation for the Stokes flow. In this work, the initial configuration is assumed to be two spherical drops joint by a small-diameter smooth neck, i.e., the computations start after the coalescence, as we defined it earlier, has already taken place. Nevertheless, one can still infer from this study some information about how the standard model performs as the initial diameter of the neck decreases. As noted by the authors, "at the initial steps the numerical scheme is inadequate to describe the evolutionary process" and "the exact features of the inner region (i.e., the region including the neck where initially the free-surface curvature and the normal velocity are extremely high) grossly depend on the relative position of the droplets when coalescence started" (Hiram & Nir 1983, p. 467). These features are typical of a regular numerical approximation of a singular exact solution.[4]

In the subsequent numerical works (e.g., Lafaurie 1994, Martínez-Herrera & Derby 1995, Eggers, Lister & Stone 1999, Menchaca-Rocha et al. 2001), the initial free-surface shape is also prescribed as the after-coalescence configuration, sometimes either explicitly using Hopper's analytical results (Martínez-Herrera & Derby 1995) or assuming the asymptotic behaviour of the free-surface profile that makes the three-dimensional flow at small times similar to Hopper's solution for two dimensions (Eggers et al. 1999). In the latter case,

$$r_m \sim -\pi^{-1} t \ln t$$

and hence

$$u_m \equiv \dot{r}_m \sim -\pi^{-1} \ln t \to \infty \qquad \text{as } t \to 0 + .$$

In the rare studies addressing the influence of the initial configuration on numerical computations, the authors report grid-dependence of the results (e.g., Menchaca-Rocha et al. 2001), thus again indicating the presence of a singularity in the exact solution the numerical one is trying to approximate.

Oğuz & Prosperetti (1989) considered the evolution of two drops of an incompressible *inviscid* fluid joint initially by a small bridge with the free surface being smooth everywhere, as in all previous studies. They found numerically that in this simple system, where one has only inertial and capillary effects, the evolution of the free-surface shape leads to entrapment of a number of toroidal bubbles which are then likely to break up into spherical ones. In principle, this feature suggests an experimental way of finding out the radius of the initial bridge and hence the scale on which the actual coalescence took place. Also for the case of an inviscid fluid Eggers et al. (1999) offered a 'scaling law': by equating the characteristic value of the capillary pressure at the

---

[4]Similar problems with accuracy in the initial stage have been reported in the case of array of cylinders (Ross, Miller & Weatherly 1981), where later Hopper and Richardson have obtained exact (singular) solutions.

neck $\sigma_e/(r_m^2 R_0)$ driving the flow and the characteristic kinetic energy there, $\frac{1}{2}\rho(\dot{r}_m)^2$, one arrives at a differential equation which after integration yields

$$r_m \propto \left(\frac{\sigma_e R_0}{\rho}\right)^{1/4} t^{1/2}, \qquad (6.88)$$

i.e., the same dependence of r_m on t as in (6.87) with the same singularity at $t \to 0+$. It should be noted, however, that, unlike (6.87), the scaling (6.88) explicitly relies on the neck being smooth, thus implying that the coalescence has already taken place and one is dealing with a single body of fluid which can be described in the framework of the standard (inviscid) model. As we will see below, an important assumption behind (6.88) that the process is driven *locally* helps to elucidate the nature of the singularity at the onset of coalescence.

Experimental information about coalescence is very limited. Most studies consider the post-coalescence evolution of a single body of fluid towards its equilibrium shape (e.g., Menchaca-Rocha et al. 2001, Wu, Cubaud & Ho 2004, Thoroddsen, Takehara & Etoh 2005). For low-viscosity fluids the dependence of r_m on t and R_0 for the post-coalescence evolution of the drops given by the scaling (6.88) has been confirmed within the accuracy of the experiments by Wu et al. (2004). Here it should be pointed out that, as has been shown in a number of experiments (e.g., Rogers 1858, Thomson & Newall 1885, Dooley et al. 1997), the process of coalescence is associated with the appearance of vorticity in the fluid whereas in the inviscid theories the flow is assumed to be potential.

The most recent and accurate experiments to date (Thoroddsen et al. 2005) show "slight, but significant deviations" from the known scaling law which, according to the authors, are "most probably caused by the finite initial contact radius". Thoroddsen et al. remark that in their experiments, where the spatial resolution was about 1 μm and the temporal resolution could reach 1 million frames per second,

> The drops do not touch at a point, but along a finite ring, having a diameter of the order of 100 μm. The most likely explanation is the presence of air, as has been studied, among others, by Jones & Wilson (1978) and Rother et al. (1997). No matter how slowly the two surfaces meet, the lubrication pressure will eventually become strong enough to overcome the capillary pressure and deform he drops. (Thoroddsen et al. 2005, pp. 88–89).

This observation highlights an important point associated with an additional physical factor, i.e., the presence of air. However, it leaves open the question of how the actual merger of two fluid volumes takes place whether the drops first touch at a point or along a finite area. This process remains inaccessible to straightforward experimentation based on direct optical observations and

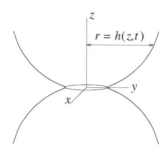

FIGURE 6.12: Definition sketch for the coalescence problem in the standard formulation.

further progress could be made via more focussed theory-driven measurements with a clear idea of what should be confirmed or ruled out.

From a theoretical viewpoint, the essence of the problem is to make a mathematical model of coalescence which at most has an integrable singularity in the time-dependence of the bulk pressure at $t = 0$. The first step in developing such a model is to try to understand the origin of singularities in the standard model.

6.5.2 Local self-similarity and the origin of the singularity

In order to elucidate the nature of the singularity inherent in the standard formulation consider the evolution of the free surface of a thin liquid bridge formed after the onset of coalescence of two liquid volumes as a local problem. For simplicity, we will assume that the bridge is surrounded by an inviscid dynamically-passive gas, and the liquid is incompressible Newtonian with a uniform density ρ and a constant viscosity μ. The surface tension of the liquid-gas interface σ_e is also assumed to be constant. To nondimensionalize the lengths, time, velocity, and pressure, one can use the following 'natural scales':

$$l_\mu = \frac{\mu^2}{\sigma_e \rho}, \quad t_\mu = \frac{\mu^3}{\sigma_e^2 \rho}, \quad U_\mu = \frac{l_\mu}{t_\mu} \equiv \frac{\sigma_e}{\mu}, \quad P_\mu = \frac{\sigma_e}{l_\mu} \equiv \frac{\sigma_e^2 \rho}{\mu^2}. \tag{6.89}$$

These scales remove all material constants from the Navier-Stokes equations and standard boundary conditions on the free surface.

We will use cylindrical coordinates (r, φ, z) with the origin at the point of the initial contact of the fluid volumes and the z-axis directed along the axis of symmetry (Fig. 6.12). It is assumed that all functions are independent of φ and the velocity $\mathbf{u}$ has only the radial and axial components, u and w. In the bulk, the distributions of the flow parameters are described by the Navier-Stokes equations which in the nondimensional variables take the form:

$$\frac{1}{r}\frac{\partial(ru)}{\partial r} + \frac{\partial w}{\partial z} = 0, \tag{6.90}$$

$$\frac{\partial u}{\partial t} + u\frac{\partial u}{\partial r} + w\frac{\partial u}{\partial z} = -\frac{\partial p}{\partial r} + \frac{1}{r}\frac{\partial}{\partial r}\left(r\frac{\partial u}{\partial r}\right) + \frac{\partial^2 u}{\partial z^2} - \frac{u}{r^2}, \qquad (6.91)$$

$$\frac{\partial w}{\partial t} + u\frac{\partial w}{\partial r} + w\frac{\partial w}{\partial z} = -\frac{\partial p}{\partial z} + \frac{1}{r}\frac{\partial}{\partial r}\left(r\frac{\partial w}{\partial r}\right) + \frac{\partial^2 w}{\partial z^2}, \qquad (6.92)$$

where u and w are the radial and axial components of velocity and p is the pressure.

On the free surface given by $r = h(z,t)$ one has the standard kinematic condition

$$u - w\frac{\partial h}{\partial z} = \frac{\partial h}{\partial t}, \qquad (6.93)$$

the zero tangential-stress condition

$$2\frac{\partial h}{\partial z}\left(\frac{\partial u}{\partial r} - \frac{\partial w}{\partial z}\right) + \left[1 - \left(\frac{\partial h}{\partial z}\right)^2\right]\left(\frac{\partial w}{\partial r} + \frac{\partial u}{\partial z}\right) = 0, \qquad (6.94)$$

and the normal-stress condition (p_0 is the pressure in the surrounding gas)

$$p - p_0 - \frac{2}{1 + \left(\dfrac{\partial h}{\partial z}\right)^2}\left[\frac{\partial u}{\partial r} + \left(\frac{\partial h}{\partial z}\right)^2\frac{\partial w}{\partial z} - \frac{\partial h}{\partial z}\left(\frac{\partial w}{\partial r} + \frac{\partial u}{\partial z}\right)\right]$$

$$= \frac{1}{h\left[1 + \left(\dfrac{\partial h}{\partial z}\right)^2\right]^{1/2}} - \frac{\dfrac{\partial^2 h}{\partial z^2}}{\left[1 + \left(\dfrac{\partial h}{\partial z}\right)^2\right]^{3/2}}. \qquad (6.95)$$

To complete the problem formulation, one also needs some boundary conditions in the far field as well as initial condition which we will not discuss here.

The standard problem formulation given above implies that the initial stage of coalescence (i.e. as $t \to 0+$) is driven locally due to the singularly high free-surface curvature and hence singular capillary pressure. Mathematically, this means that in the initial stage the flow in the neck between the two volumes is only weakly dependent on the boundary conditions in the far field. One can also see that the length and time scales characterizing the flow near the point of the initial contact are separated from those characterizing the overall flow. Indeed, the expansion of the initial infinitesimally-thin bridge is associated with the length measured from zero and hence the characteristic length scale of the local flow is infinitesimal compared with the initial radii of curvature of the coalescing volumes. The time of the process, which is also measured from zero, is characterized by the scale infinitesimal compared with the radius of curvature divided by the initial speed with which the two volumes approached each other or any other macroscopic time scale. To put it pictorially, in the initial stage of the process the length has to be measured with time. These arguments suggest that for the earliest stage of coalescence in the immediate vicinity of the point of the initial contact, to leading order as $t \to 0+$, the solution will be locally self-similar. Consider some implications of this assumption.

After introducing independent variables

$$\xi = \frac{r}{t^\alpha}, \qquad \zeta = \frac{z}{t^\beta} \qquad (\alpha, \beta > 0) \tag{6.96}$$

one has that, if the solution is self-similar, $h(z,t) = t^\alpha H(\zeta)$, and the free-surface shape in the $\chi\zeta$-plane is given by $\xi = H(\zeta)$. The kinematic condition (6.93) together with the continuity equation (6.90) immediately give that the velocities have to have the form

$$u(r,z,t) = t^{\alpha-1}U(\xi,\zeta), \qquad w(r,z,t) = t^{\beta-1}W(\xi,\zeta). \tag{6.97}$$

A simple way to illustrate this is to consider the minimum cross-section where $\partial h/\partial z = 0$ and hence the two remaining terms in (6.93) must be of the same order in t as $t \to 0$. This gives the first equality in (6.97). Then, using this equality together with (6.96) and equation (6.90), where there are only two terms which then also must be of the same order, one arrives at the second equality in (6.97). In other words, the kinematic condition (6.93) and the continuity equation (6.90) couple the scales for lengths and velocities.

To avoid an overdetermined set of equations in the bulk, one has to keep the term with the pressure gradient at least in one of the bulk equations to leading order in t as $t \to 0+$. The left-hand sides of equations (6.91) and (6.92) have the order of $t^{\alpha-2}$ and $t^{\beta-2}$, respectively. The viscous terms are proportional to $t^{-\alpha-1}$ and $t^{\alpha-2\beta-1}$ in (6.91) and to $t^{-\beta-1}$ and $t^{\beta-2\alpha-1}$ in (6.92). Therefore, if

$$p - p_0 = t^\gamma P(\xi,\zeta), \tag{6.98}$$

then, after substituting (6.98) into (6.91) and (6.92), we find that to keep the terms with the pressure gradient in (6.91) and/or (6.92), one needs

$$\gamma = \min(2\alpha - 2, -1, 2\alpha - 2\beta - 1) \quad \text{or/and} \quad \gamma = \min(2\beta - 2, -1, 2\beta - 2\alpha - 1). \tag{6.99}$$

Thus, in both cases

$$\gamma \leq -1. \tag{6.100}$$

As a result one has that the absolute value of $p \sim 1/t$ as $t \to 0+$, i.e., p is nonintegrably singular as a function of t.

The only assumptions that have lead us to (6.99) and hence (6.100) are:

(a) The Navier-Stokes equations (6.90)–(6.92) in the bulk;

(b) The kinematic boundary condition (6.93) everywhere on the free surface at $t > 0$;

(c) The assumption that locally the process of the topological transition is self-similar.

Even though no use has been made yet of the dynamic boundary conditions (6.94) and (6.95), we have already arrived at a nonintegrable singularity! In order to proceed in a physically meaningful way, it becomes necessary to relax at least one of conditions (a)–(c).

The arguments behind the assumption of local self-similarity have been outlined earlier, and essentially it follows from the separation of scales for the flow in the neck and that in the far field. Hence the local self-similarity cannot be relaxed since this separation of scales is a defining feature of the geometry and kinematics of the problem.

The Navier-Stokes equations define the fluid we are dealing with, i.e., they specify its rheology and state that the compressibility effects are unimportant on the time and length scale these equations are applied. In other words, they are the equations for which the boundary conditions must be set up. If any other physical mechanism becomes important in the process in question, in a macroscopic (fluid-mechanical) model it must be accounted for in the boundary conditions.

The vulnerable element in (a)–(c) appears to be (c), i.e., the only boundary condition used so far. The condition (6.93) everywhere on the free surface implies that the liquid-gas interface is smooth for $t > 0$ and its physical meaning is that the fluid particles at every point of the free surface travel with it. This requirement together with the continuity equation (6.90) and the assumption of local self-similarity of the flow couples the scales for lengths and velocities. As we remember, the assumption that the fluid particles travel with the free surface results in a singular velocity field in Frenkel's (1945) integral analysis, which gives a qualitative frame for the standard model.

As discussed earlier in this chapter, there are situations where the free surface is not smooth and the fluid's kinematics differs from that of the standard model. If this is the case in the coalescence process, then the standard kinematic boundary condition (6.93) is no longer valid at every point of the free surface and the scales for lengths and velocities are no longer coupled. Thus, by relaxing (b) and invoking the problem of cusping one can proceed with the modelling of the coalescence phenomenon.

6.5.3 Singularity-free description

Consider the main assumption we made in trying to describe the coalescence of drops in the framework of the standard formulation. Given that the two drops that are approaching each other have smooth boundaries at $t < 0$ (Figs 6.13a and 6.14a), when at $t = 0$ they touch along a flat section of their free surfaces (Fig. 6.13b) or at a point (Fig. 6.14b), the resultant liquid-gas interface is invariably *not* smooth: it has a cusp. Then it is a very strong assumption that this singularity of curvature disappears instantly so that

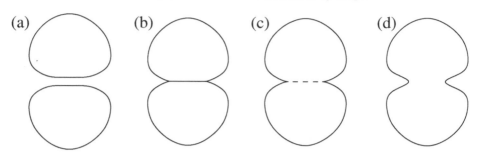

FIGURE 6.13:   Sketch of the coalescence process with the initial contact along a finite area. In this case, a part of the free surface becomes 'trapped' between the bulk phases, so that coalescence essentially becomes the process of disappearance of the 'surface properties' of the former free surface.

the free surface of the forming single volume of fluid becomes smooth for $t > 0$. It is this assumption that brings in the standard boundary conditions, couples the scales for lengths and velocities thus leading to the nonintegrable singularity in the flow field. One should also note that this assumption implies the use of the standard normal-stress boundary condition in the situation where the radii of curvature of the free surface are infinitesimal whereas the model considering the surface tension as a material constant relies on an assumption that they are large compared with the interfacial layers' thickness.

In order to understand what actually happens at the onset of coalescence, it is instructive to look first at the situation where the two drops, deformed by the air resistance as they approach each other, touch along a finite area. This scenario is suggested by some experiments (e.g., Thoroddsen et al. 2005) and shown schematically in Fig. 6.13. In this case, at $t = 0$ a part of the drops' free surfaces becomes sandwiched between the bulk phases (Fig. 6.13b). This area of the former free surfaces no longer separates two different bulk phases and hence the influence of intermolecular forces from them is no longer asymmetric. In macroscopic terms, this means that the *equilibrium* surface parameters of this area now correspond to zero surface tension since in equilibrium this area will be made of 'ordinary' fluid particles that form the rest of the bulk of the fluid. Thus, at $t > 0$ the 'trapped' section of the free surface will begin to relax to its equilibrium state gradually (though, in physical terms, very quickly) losing its surface tension and surface energy (Fig. 6.13c). This process will lead to a gradual disappearance of the concentrated force acting on the contact line formed at the intersection of the free surfaces and the trapped ('internal') interface. The contact line will then move outwards and the contact angle formed by the free surfaces will open up. When the relaxation process is complete, one ends up with a smooth boundary confining a single body of fluid which will then be describable by the standard model (Fig. 6.13d). In terms of the definition we gave earlier, at this point the coalescence process is complete and the subsequent motion is simply an evolution

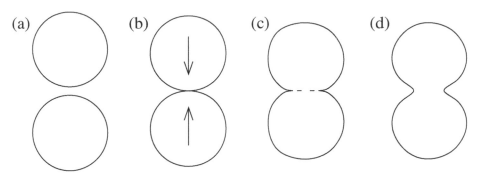

FIGURE 6.14: Sketch of the coalescence process with the initial contact at a point. In this situation, for coalescence to take place the drops have to be pressed against each other by the inertia of their preceding motion and/or external forces.

of a single drop.

If the two drops touch at a point (Fig. 6.14a), then the 'trapping' of the free surface in the bulk, the relaxation of its properties towards the equilibrium and the outward motion of the contact line take place simultaneously, as opposed to the scenario sketched in Fig. 6.13, where 'trapping' occurs as a separate stage at $t = 0$. However, the essence of the process remains the same. If the drops approach each other at a finite velocity, then the initial creation of the contact area is associated with the contact-line speed proportional to $t^{1/\alpha-1}$, where α is the degree of touching of the two free surfaces ($\alpha = 2$ for spherical drops). It should be noted here that, unlike the scheme of Fig. 6.13, in terms of macroscopic fluid mechanics, the initial position where the drops touch at a point is a position of equilibrium. For the coalescence to begin the drops must be pressed against each other by the inertia of their preceding motion or/and external forces. These factors, no matter how small they are, will create a finite area of contact and trigger the relaxation process. It is also necessary to emphasize that the velocity of the contact line, which in a general case is singular at $t = 0$, is not equal to the fluid's velocity: as in the problem of cusping we considered earlier, the fluid particles belonging to the interface travel across the contact line, or, in other words, the contact line propagates with respect to the fluid particles.

Thus, the coalescence of liquid volumes appears to be another case of the interface disappearance/formation process, and for its mathematical description we can again use the model formulated in Chapter 4, as we did for the problem of cusping. An essentially new element now is that the process is unsteady and the 'internal interface' starts developing when the singularity of the free-surface curvature (i.e., a cusp formed when the drops touch) is already in place. What we need to show is that the scenario outlined above makes it possible to describe the process without singularities in the flow field.

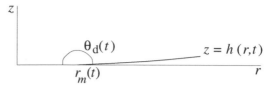

FIGURE 6.15:   Definition sketch for the onset of coalescence of two drops of equal radii.

6.5.3.1   Initial stage of coalescence

Consider two spherical drops, for simplicity of equal radii R_0, that at $t = 0$ are in contact at a point (Fig. 6.14). Then, if they are pressed against each other due to some external factors, e.g., their inertia or external forces, there appears an area of contact, the system finds itself out of equilibrium, and the coalescence process begins. We will assume that the external factors that create the area of contact to be sufficiently small so that the compressibility effects in the liquid can be neglected. A review of the issues brought in by the fluid's compressibility in the drop collision problems can be found in Lesser (1983).

We will use cylindrical coordinates (r, φ, z) with the origin at the point of the initial contact of the drops and the z-axis directed along their centerline. Due to the symmetry of the problem, it is sufficient to consider one drop whose shape near the origin is $z = h(r, t)$. The contact line confining the contact area propagates outwards leaving behind a gradually disappearing internal interface (Fig. 6.15). It is convenient to nondimensionalize the problem using the natural scales (6.89) so that the bulk flow will again be described by (6.90)–(6.92). As before, we will use the notation u and w for the radial and axial components of the fluid's velocity and the corresponding components of the surface velocity will have a superscript s.

At the free surface $z = h(r, t)$, the boundary conditions (4.54)–(4.59) take the form:

$$\frac{\partial h}{\partial t} + u^s \frac{\partial h}{\partial r} = w^s, \tag{6.101}$$

$$(u - u^s)\frac{\partial h}{\partial r} - (w - w^s) = -Q(\rho^s - \rho_e)\left[1 + \left(\frac{\partial h}{\partial r}\right)^2\right]^{1/2}, \tag{6.102}$$

$$2\frac{\partial h}{\partial r}\left(\frac{\partial w}{\partial z} - \frac{\partial u}{\partial r}\right) + \left[1 - \left(\frac{\partial h}{\partial r}\right)^2\right]\left(\frac{\partial w}{\partial r} + \frac{\partial u}{\partial z}\right) - \lambda\left[1 + \left(\frac{\partial h}{\partial r}\right)^2\right]^{1/2}\frac{\partial \rho^s}{\partial r} = 0, \tag{6.103}$$

$$p - 2\left[1 + \left(\frac{\partial h}{\partial r}\right)^2\right]^{-1}\left[\left(\frac{\partial h}{\partial r}\right)^2\frac{\partial u}{\partial r} + \frac{\partial w}{\partial z} - \frac{\partial h}{\partial r}\left(\frac{\partial w}{\partial r} + \frac{\partial u}{\partial z}\right)\right]$$

$$= \lambda(1 - \rho^s) \left\{ \frac{1}{r} \frac{\partial h}{\partial r} \left[1 + \left(\frac{\partial h}{\partial r} \right)^2 \right]^{-1/2} + \frac{\partial^2 h}{\partial r^2} \left[1 + \left(\frac{\partial h}{\partial r} \right)^2 \right]^{-3/2} \right\}, \quad (6.104)$$

$$\frac{\partial \rho^s}{\partial t} + \left[1 + \left(\frac{\partial h}{\partial r} \right)^2 \right]^{-1} \left[\frac{\partial(\rho^s u^s)}{\partial r} + \frac{\partial h}{\partial r} \frac{\partial(\rho^s w^s)}{\partial r} \right] + \frac{\rho^s u^s}{r} = -\epsilon^{-1}(\rho^s - \rho_e^s),$$
$$(6.105)$$

$$\epsilon \frac{\partial \rho^s}{\partial r} = 4V^2 \left[(u - u^s) + (w - w^s) \frac{\partial h}{\partial r} \right], \quad (6.106)$$

where

$$Q = \frac{\mu \rho_{(0)}^s}{\rho \tau \sigma_e}, \quad \lambda = \frac{\gamma \rho_{(0)}^s}{\sigma_e}, \quad \epsilon = \frac{\rho \tau \sigma_e^2}{\mu^3}, \quad V^2 = \frac{\beta \tau \sigma_e^2}{\mu^2 \rho_{(0)}^s \gamma (1 + 4\alpha\beta)},$$

and pressure is measured with respect to a constant pressure in the ambient gas. For convenience we have eliminated σ from the boundary conditions above and below using the simplest linear surface equation of state (4.61). After scaling, this equation gives an obvious link between λ and ρ_e^s:

$$\lambda(1 - \rho_e^s) = 1. \quad (6.107)$$

At the internal interface, which is not moving in the vertical direction ($w^s \equiv 0$), one has the boundary conditions (4.56)–(4.59):

$$w = Q(\rho^s - 1), \quad (6.108)$$

$$\frac{\partial w}{\partial r} + \frac{\partial u}{\partial z} - \lambda \frac{\partial \rho^s}{\partial r} = 0, \quad (6.109)$$

$$\frac{\partial \rho^s}{\partial t} + \frac{1}{r} \frac{\partial}{\partial r} (r \rho^s u^s) = -\epsilon^{-1}(\rho^s - 1), \quad (6.110)$$

$$\epsilon \frac{\partial \rho^s}{\partial r} = 4V^2 (u - u^s). \quad (6.111)$$

Finally, at the contact line $r = r_m(t)$, $z = 0$ formed at the intersection of the free surfaces of the two drops and the internal interface ($h(r_m(t), t) \equiv 0$), in the simplest case, one can apply conditions (6.42), (6.43):

$$\left\{ \rho^s \left(u^s - \frac{dr_m}{dt} + w^s \frac{\partial h}{\partial r} \right) \left[1 + \left(\frac{\partial h}{\partial r} \right)^2 \right]^{-1/2} \right\}_G = \left[\rho^s \left(u^s - \frac{dr_m}{dt} \right) \right]_I,$$
$$(6.112)$$

$$(1 - \rho^s)_G \cos \theta_d + (1 - \rho^s)_I = 0, \quad (6.113)$$

where

$$\theta_d(t) = \pi - \arctan \left[\frac{\partial h}{\partial r} (r_m(t), t) \right] \quad (6.114)$$

is the contact angle formed by the free surface with the internal interface
(Fig. 6.15) and the subscripts G and I refer to the limits $r \to r_m+$, $z = h(r,t)$
and $r \to r_m-$, $z = 0$, respectively. For convenience we will eliminate θ_d from
(6.113) and (6.114) so that the second condition at the contact line will take
the form

$$(1 - \rho^s)_G \left[1 + \left(\frac{\partial h}{\partial r}\right)^2\right]^{-1/2} = (1 - \rho^s)_I. \qquad (6.115)$$

Conditions at the axis of symmetry can be formulated as

$$u = u^s = \frac{\partial w}{\partial r} = 0 \qquad (r = 0, z > 0, t > 0). \qquad (6.116)$$

We are interested in an early stage of the process where, for the reasons
discussed earlier, the solution is going to be locally self-similar, and the appropriate boundary conditions in the far field will be formulated later.

To consider the onset of coalescence in the vicinity of the origin, we will
introduce an asymptotically small time scale δ^2 ($\delta \to 0$) and use the corresponding variable

$$\bar{t} = \frac{t}{\delta^2}. \qquad (6.117)$$

The corresponding scaling suggested by the initial shape of the drops near the
point of their contact has the form

$$r = \delta\bar{r}, \quad z = \delta\bar{z}. \qquad (6.118)$$

Now we can formulate boundary conditions in the far field. The scales
(6.117), (6.118) on which we now consider the problem near the origin are
infinitesimal compared with the time and length scale of the global flow. This
separation of scales suggests that, as boundary conditions in the far field,
one can use the requirement that, in the rescaled variables, the flow becomes
independent of the distance from the origin as this distance tends to infinity:

$$\frac{1}{(\bar{r}^2 + \bar{z}^2)^{1/2}} \left(\bar{r}\frac{\partial u}{\partial \bar{r}} + \bar{z}\frac{\partial u}{\partial \bar{z}}\right) \to 0, \qquad (6.119)$$

$$\frac{1}{(\bar{r}^2 + \bar{z}^2)^{1/2}} \left(\bar{r}\frac{\partial w}{\partial \bar{r}} + \bar{z}\frac{\partial w}{\partial \bar{z}}\right) \to 0 \qquad \text{as } \bar{r}^2 + \bar{z}^2 \to \infty. \qquad (6.120)$$

Essentially, this is the condition allowing for the local self-similarity of the
initial stage solution. For the surface density we have

$$\frac{\partial \rho^s}{\partial \bar{r}} \to 0 \qquad \text{as } \bar{r} \to \infty, \qquad (6.121)$$

which is compatible with the initial condition $\rho^s(\bar{r}, 0) \equiv \rho_e^s$.

The form of asymptotic expansions for the dependent variables is as follows.
Since the expansions for ρ^s on both the free surface and the internal interface obviously begin with ρ_e^s (formally, this is a trivial consequence of (6.105),

(6.110), (6.112) and the initial state of the free surface), the kinematic condition (6.108) suggests

$$(u, u^s, w, w^s) = \sum_{n=0}^{\infty} \delta^n (u_n, u_n^s, w_n, w_n^s) \qquad \text{as } \delta \to 0. \tag{6.122}$$

Then, from (6.106), (6.111), (6.118) and (6.122) it follows that

$$\rho^s = \rho_e^s + \sum_{n=1}^{\infty} \delta^n \rho_n^s \tag{6.123}$$

on both interfaces, and the kinematic condition (6.101) yields the asymptotic form for the free-surface profile

$$h = \sum_{n=0}^{\infty} \delta^{2+n} h_n \qquad \text{as } \delta \to 0. \tag{6.124}$$

Finally, to keep p to leading order in δ in the equations of motion (6.91), (6.92), we need

$$p = \sum_{n=0}^{\infty} \delta^{n-1} p_n \qquad \text{as } \delta \to 0. \tag{6.125}$$

Now, the free surface is given by $\bar{z} = \delta(h_0 + \delta h_1 + \dots)$ and, to leading order in δ, we have a flow domain $\bar{r} > 0$, $\bar{z} > 0$, where u_0, w_0 and p_0 satisfy the unsteady Stokes equations

$$\frac{1}{\bar{r}} \frac{\partial(\bar{r} u_0)}{\partial \bar{r}} + \frac{\partial w_0}{\partial \bar{z}} = 0, \tag{6.126}$$

$$\frac{\partial u_0}{\partial \bar{t}} = -\frac{\partial p_0}{\partial \bar{r}} + \frac{1}{\bar{r}} \frac{\partial}{\partial \bar{r}} \left(\bar{r} \frac{\partial u_0}{\partial \bar{r}} \right) + \frac{\partial^2 u_0}{\partial \bar{z}^2} - \frac{u_0}{\bar{r}^2}, \tag{6.127}$$

$$\frac{\partial w_0}{\partial \bar{t}} = -\frac{\partial p_0}{\partial \bar{z}} + \frac{1}{\bar{r}} \frac{\partial}{\partial \bar{r}} \left(\bar{r} \frac{\partial w_0}{\partial \bar{r}} \right) + \frac{\partial^2 w_0}{\partial \bar{z}^2}. \tag{6.128}$$

Condition (6.108) and as its consequence condition (6.109) take a simple form

$$w_0 = Q(\rho_e^s - 1) \qquad \frac{\partial u_0}{\partial \bar{z}} = 0, \qquad (\bar{r} < \bar{r}_m(\bar{t}), \bar{z} = 0), \tag{6.129}$$

whereas (6.103) and (6.104) give

$$\frac{\partial w_0}{\partial \bar{r}} + \frac{\partial u_0}{\partial \bar{z}} = 0, \qquad p_0 - 2\frac{\partial w_0}{\partial \bar{z}} = 0 \qquad (\bar{r} > \bar{r}_m(\bar{t}), \bar{z} > 0). \tag{6.130}$$

The symmetry conditions (6.116) and conditions (6.119), (6.120) in the far field for u_0, w_0 have the same form as for u and w. Thus, the flow field to leading order can be determined from (6.126)–(6.130), (6.116), (6.119), (6.120) independently of the surface variables. The requirement of self-similarity of the solution replaces the initial conditions. In the similarity variables

$$u_0(\bar{r}, \bar{z}, \bar{t}) = U_0(\xi, \zeta), \quad w_0(\bar{r}, \bar{z}, \bar{t}) = W_0(\xi, \zeta), \quad p_0(\bar{r}, \bar{z}, \bar{t}) = \bar{t}^{-1/2} P_0(\xi, \zeta),$$
(6.131)

where

$$\xi = \frac{\bar{r}}{\bar{t}^{1/2}}, \quad \zeta = \frac{\bar{z}}{\bar{t}^{1/2}},$$
(6.132)

the problem becomes two-dimensional. The form of (6.131) and (6.132) is dictated by the requirement that in the original variables r, z, t the first terms in the expansions (6.122), (6.125) for u, w and p must be independent of the artificially introduced parameter δ.

The solution to the problem is given by $U_0 = 0$, $W_0 = -Q(1 - \rho_e^s)$, $P_0 = 0$, i.e.,

$$u_0 = p_0 = 0, \quad w_0 = -Q(1 - \rho_e^s) < 0.$$
(6.133)

This solution shows that the potentially singular first term in the expansion (6.125) for p is identically zero. This is the main outcome we were looking for since it indicates that the model allows one to proceed from the first contact of the drops to the situation where they have a contact *area* without singularities in the flow parameters. Then one has a propagating and gradually disappearing singularity of curvature, similar to that considered earlier in this chapter for the problem of steady cusping, and we already know that the flow field in the vicinity of this singularity remains regular. Here we will make a few more steps to consider some features of the initial stage of coalescence.

Now, once the bulk flow has been found, we can obtain the surface variables. To leading order (6.105), (6.106) and (6.110), (6.111) have the same form on both interfaces:

$$\frac{\partial \rho_1^s}{\partial \bar{t}} + \frac{\rho_e^s}{\bar{r}} \frac{\partial (\bar{r} u_0^s)}{\partial \bar{r}} = 0, \quad \epsilon \frac{\partial \rho_1^s}{\partial \bar{r}} = -4V^2 u_0^s.$$
(6.134)

From (6.115) it follows that ρ_1^s is continuous across the contact line and then from (6.112) one has that u_0^s is continuous as well. These conditions together with (6.116), (6.121) and the initial condition $\rho_1^s(\bar{r}, 0) = 0$ lead to a trivial solution $\rho_1^s = u_0^s = 0$.

Finally, the kinematic condition (6.102) yields $w_0^s = w_0$ so that, taking into account (6.133), equation (6.101) becomes

$$\frac{\partial h_0}{\partial \bar{t}} = -Q(1 - \rho_e^s).$$
(6.135)

The initial condition is provided by the free-surface profile at $t = 0$:

$$h_0(\bar{r}, 0) = \frac{\bar{r}^2}{2R},$$
(6.136)

where $R = R_0 \sigma_e \rho / \mu^2$ is the dimensionless radius of the drop. The solution to (6.135), (6.136) is given by

$$h_0(\bar{r}, \bar{t}) = -Q(1 - \rho_e^s)\bar{t} + \frac{\bar{r}^2}{2R} \tag{6.137}$$

and from $h_0(\bar{r}_*, \bar{t}) = 0$ one finds the position of the contact line (to leading order in δ) as a function of time:

$$\bar{r}_m(\bar{t}) = [2RQ(1 - \rho_e^s)]^{1/2} \bar{t}^{1/2}. \tag{6.138}$$

The contact angle θ_d formed by the free surface and the internal interface (i.e., the plane of symmetry) is given by (6.114) so that, taking into account (6.137), (6.138) and (6.107), we have

$$\theta_d = \pi - \delta \left(\frac{2Q}{\lambda R}\right)^{1/2} \bar{t}^{1/2} = \pi - \left(\frac{2Q}{\lambda R}\right)^{1/2} t^{1/2}.$$

Thus, the free surface shape evolves continuously from a cusp formed when the two drops touched ($\bar{t} = 0$, $\theta_d = \pi$) towards a smooth free surface (i.e., $\theta_d = \pi/2$), when the process of coalescence will be complete. Close to the initial moment one has that $\pi - \theta_d \propto t^{1/2}$. In other words, for $t > 0$ the singularity of curvature is a corner, not a cusp, and this corner gradually opens up as the area of contact widens. This is qualitatively different from the scenario described by the standard model where one has a jump from $\theta_d = \pi$ at $t = 0$ to $\theta_d = \pi/2$ at $t = 0+$.

The algorithm for obtaining higher-order terms in the asymptotic expansions (6.122)–(6.125) is the same as for the leading terms: one finds the flow parameters in the bulk that depend on the previous-order solution for the bulk and surface variables and then, using these solutions, obtains the surface variables and a correction to the free-surface profile and the position of the contact line. These higher-order terms are regular at $t = 0$ so that their analysis only refines the leading-order solution we considered earlier. We will illustrate the procedure by considering the next step.

For the bulk variables at the next order we again have the unsteady Stokes equations. Since $\rho_1^s \equiv 0$, the boundary conditions obtained from (6.103), (6.108) and (6.109) are homogeneous. The normal-stress condition (6.104) now takes the form

$$p_1 - 2\frac{\partial w_1}{\partial \bar{z}} = \frac{2}{R}.$$

These conditions together with the homogeneous conditions (6.116), (6.119), (6.120) and the requirement of self-similarity lead to a trivial solution

$$u_1 = w_1 = 0, \quad p_1 = \frac{2}{R},$$

and, since it follows from (6.102) that $w_1^s = w_1$, equation (6.101) yields $h_1 = 0$. The remaining step is to find u_1^s and ρ_2^s on the free surface and the internal interface from equations (6.105), (6.106) and (6.110), (6.111), respectively, subject to conditions (6.112), (6.115), (6.116), (6.121) and the requirement of self-similarity. After eliminating u_1^s from (6.105) and (6.110) using (6.106) and (6.111), respectively, the problem for ρ_2^s in the similarity variables $\rho_2^s = \bar{t} F(\xi)$ can be reduced to the following one:

$$\xi F_i'' + (1 + \frac{a}{2}\xi^2)F_i' - a\xi F_i = -ab_i\xi \quad (i = 1, 0 < \xi < \xi_*; \; i = 2, \xi_* < \xi < \infty),$$
$$(6.139)$$
$$F_1(\xi_*) - F_2(\xi_*) = A, \quad F_1'(\xi_*) - F_2'(\xi_*) = B, \quad (6.140)$$
$$F_1'(0) = 0, \quad F_2'(\infty) = 0, \quad (6.141)$$

where

$$\xi_* = [2RQ(1 - \rho_e^s)]^{1/2}, \quad a = \frac{4V^2}{\epsilon\rho_e^s}, \quad b_1 = \epsilon^{-1}(1 - \rho_e^s), \quad b_2 = 0,$$

$$A = \frac{Q(1 - \rho_e^s)^2}{R}, \quad B = \frac{4V^2Q^{3/2}(1 - \rho_e^s)^{3/2}(2\rho_e^s - 1)}{(2R)^{1/2}\rho_e^s\epsilon}.$$

The solution of (6.139)–(6.140) is given by

$$F_1(\xi) = C_1(2 + \frac{a}{2}\xi^2) + b_1, \quad F_2(\xi) = C_2(2 + \frac{a}{2}\xi^2) \int_{\sqrt{a/2}\,\xi}^{\infty} \frac{\exp(-x^2/2)\,dx}{x(x^2 + 2)^2},$$

where

$$C_1 = \frac{A - b_1}{2 + \eta_*^2} + JC_2, \quad C_2 = \frac{B}{S}\left(\frac{2}{a}\right)^{1/2} - \frac{2\eta_*(A - b_1)}{S(2 + \eta_*^2)},$$

$$\eta_* = \left(\frac{a}{2}\right)^{1/2}\xi_*, \quad J = \int_{\eta_*}^{\infty} \frac{\exp(-x^2/2)\,dx}{x(x^2 + 2)^2}, \quad S = \frac{\exp(-\eta_*^2/2)}{\eta_*(\eta_*^2 + 2)}.$$

Now, using equations (6.111) and (6.106), we can find u_1^s on both interfaces:

$$u_1^s = \bar{t}^{1/2}G(\xi) = -\frac{\epsilon}{4V^2}\frac{\partial\rho_2^s}{\partial\bar{t}} = -\frac{\epsilon}{4V^2}\bar{t}^{1/2}F'(\xi).$$

Here $F = F_1$, $G = G_1$ on the internal interface and $F = F_2$, $G = G_2$ on the free surface.

As we can see, at this order the surface density and hence the surface tension are no longer uniform along the interfaces. As the fluid particles of the former free surface find themselves between the bulk phases, their surface properties start evolving towards new equilibrium values thus creating a nonuniform distribution of the surface density along the internal interface. Due to the spatial ellipticity of the problem, this affects the fluid particles that are yet

on the free surface and makes their surface properties deviate from the equilibrium ones, though the magnitude of this deviation is exponentially small.
However, it is only at the next order in δ that the gradient of the surface
density/tension manifests itself in the flow field due to the Marangoni effect.
The kinematic boundary condition on the internal interface (6.108) becomes
modified as well. For u_2, w_2 and p_2 one again has the unsteady Stokes equations and the boundary conditions (6.108), (6.109), (6.103) and (6.104) take
the form:

$$w_2 = Q\rho_2^s, \qquad \frac{\partial w_2}{\partial \bar{r}} + \frac{\partial u_2}{\partial \bar{z}} = \lambda \frac{\partial \rho_2^s}{\partial \bar{r}} \qquad (\bar{r} < \bar{r}_m, \ \bar{z} = 0),$$

$$\frac{\partial w_2}{\partial \bar{r}} + \frac{\partial u_2}{\partial \bar{z}} = \lambda \frac{\partial \rho_2^s}{\partial \bar{r}}, \qquad p_2 - 2\frac{\partial w_2}{\partial \bar{z}} = 0 \qquad (\bar{r} > \bar{r}_m, \ \bar{z} = 0).$$

The flow field satisfying these conditions together with the symmetry conditions (6.116) and the far-field condition (6.119), (6.120) in the similarity
variables $u_2 = \bar{t}U_2(\xi, \zeta)$, $w_2 = \bar{t}W_2(\xi, \zeta)$, $p_2 = \bar{t}^{1/2}P_2(\xi, \zeta)$ can be found numerically. Then, once these terms are known, one can obtain ρ_3^s and h_2 from
(6.105), (6.106), (6.110), (6.111) and (6.101) with conditions (6.112), (6.115)
at the contact line, conditions (6.116) and (6.121) and initial conditions in the
same way as we obtained ρ_2^s and h_1. All these terms are obviously regular.

Thus, we have shown that the coalescence problem modelled in the framework of the interface disappearance/formation theory allows one to describe
the flow without singularities in the flow parameters.

6.5.4   Computational and experimental aspects

The main conclusion for numerical computations that follows from the
asymptotic analysis of the previous section is that the problem of coalescence does not need any special treatment if it is considered in the framework
of the theory described in Chapter 4. By introducing an internal interface
at the first time step, when the area of contact is on the scale of the spatial
resolution of the code, one avoids the necessity to consider high (in the limit,
infinitely high) curvatures unavoidable in the standard model. Since the flow
parameters remain finite throughout the process, the flow field can be dis-
cretized in a regular way. The boundary condition (6.108) on the internal
interface that comes into play once we have a finite (though however small)
area of contact will generate the flow, and this flow will force the contact line
to propagate and the corner formed by the free surfaces of the touching drops
to evolve towards a smooth surface (i.e., $\theta_d = \pi/2$).

The main problem in the experimental studies of coalescence is that coalescence as a transition in the topology of the flow domain takes place well below
the temporal and spatial resolution of currently used experimental techniques.
One way of alleviating the problem is to use fluids with very high viscosity
which would relax the limitations on the temporal resolution. In this connec-

tion, it is instructive to consider the evolution of the radius of the contact area as it follows from the asymptotic analysis of the previous section.

Using (6.107) and the definitions of the dimensionless parameters, we can write down equation (6.138) in the dimensional form as

$$r_m(t) = \left(\frac{2R_0\sigma_e}{\rho\tau\gamma} \right)^{1/2} t^{1/2}. \tag{6.142}$$

This expression is quite remarkable. As we discussed in Chapter 5, experiments show that $\tau \propto \mu$ and hence equation (6.142) gives the same dependence of the radius of the minimal cross section on t, R_0, σ_e *and* μ as Frenkel's formula (6.87), which was obtained on the basis of completely different arguments. An important difference is that (6.142) exhibits also a dependence on γ, which is inversely proportional to the fluid's compressibility, and ρ.

A potentially useful scaling argument that follows from (6.142) is that the initial stage of coalescence is associated with a characteristic velocity $U = \sigma_e/(\rho\tau\gamma)$ and, given that this scale comes from (6.108), one should expect the same scale for the velocity field at the onset of coalescence also in the situation where the two drops first touch along a finite area.

A promising approach that allows one to overcome the limitations associated with the spatial resolution of optical techniques is to use electrical measurements where the contact of the two drops closes an electric circuit. Given the sensitivity of electrical measurements, it would allow one to obtain the time-dependence of the contact area with a very high accuracy. An important advantage of this method is also that it measured the actual area of contact and hence takes into account that, due to the pre-flattening of the free surfaces, there could be bubbles trapped between the drops, the flow can lose its symmetry, etc. This type of experiment for the capillary breakup problem has been reported recently by Burton, Rutledge & Taborek (2004).

6.5.5   Summary

The mathematical modelling of coalescence in the framework of the standard approach implies that the change in the topology of the flow domain takes place instantly as the two fluid volumes touch. This approach unavoidably leads to unphysical singularities in the flow field at the point of coalescence and hence, by continuity, near this point both in space and time. The singularities occur due to the coupling of scales for lengths, times and velocities dictated by the standard kinematic boundary condition, which prescribes that the fluid particles that initially belong to the free surface stay there at all time.

Experiments, in particular those where the free surfaces of the contacting volumes touch along a finite area, indicate that the process of coalescence is a particular case of fluid motion with disappearing interfaces, similar to that observed in steady convergent flows ('the phenomenon of cusping'). The asymptotic analysis of the onset of coalescence shows that this phenomenon can be described in the framework of the interface disappearance/formation theory without singularities in the flow field and any adjustments to the theory. This result ensures that the whole coalescence flow can be modelled numerically in a regular way, without any special treatment of the early stages of the process.

Chapter 7

Breakup of jets and rupture of films

In this chapter, we examine the modeling issues arising in the mathematical description of disintegration of liquid volumes. The two representative examples to be considered from this general class are the breakup of axisymmetric liquid threads and the rupture of films. Singular solutions obtained in the framework of the standard model and a regular way of describing the breakup phenomenon are discussed and compared with experimental observations. We conclude by giving a brief overview of the modeling difficulties emerging when one applies the continuum approach on mesoscopic length scales and some results obtained on this way.

7.1   Background

The dynamics of drop formation is a topic that combines industrial relevance, experimental challenges and fundamental theoretical issues. Formation of microscopic droplets and their subsequent manipulation are key element in many technologies, ranging from the established ones, such as ink-jet printing (Le 1998) and powder manufacturing, to bioengineering and various medical applications (Schena et al. 1998, Northrup, Jensen & Harrison 2003). Tiny amounts of fluid used in these technologies must be controlled with high precision which requires understanding and accurate description of capillary effects that are dominant on such small scales. The main difficulty in applications is to ensure uniformity of the forming drops and minimize parasitic effects such as the appearance of secondary "satellite" droplets and a secondary breakup of the primary ones. These phenomena are in the focus of ongoing research effort where the principal goals are to investigate the free-boundary flow leading to the formation of drops and to understand the process by which drops pinch off (see Barrero & Loscertales 2007 and Villermaux 2007 for broad-ranging reviews).

Experiments on drop formation have a long history. In the late seventeenth century, Mariotte (1686) mentioned the decay into drops of a stream of water flowing from an orifice in the bottom of a container and made an attempt to explain the phenomenon by attributing it to the physical factors, such as

Breakup of jets and rupture of films

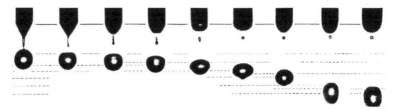

FIGURE 7.1: One of the first series of photographs of drop formation (Lenard 1887).

gravity, that were known to researchers at the time. Savart (1833) carried out a study of the breakup of liquid jets and, by using direct naked-eye observations, was able to determine the main qualitative features of the phenomenon, namely that the breakup results from the growth of disturbances of the free-surface profile and that the primary drops are almost invariably accompanied by much smaller satellite droplets.

Plateau (1843) was the first to recognize the significance of the surface tension in the drop formation and produced detailed sketches of the process highlighting its remarkable complexity and a wide variation of characteristic length scales involved. The advent of photography made it possible for researchers to document their observations (Lenard 1887, Rayleigh 1891; see Figs 7.1 and 7.4) and focus on analysing fine structures appearing in the free-surface evolution. Further progress of direct optical observations was brought by the use of high-speed cameras (Hauser et al. 1936, Edgerton, Hauser & Tucker 1937) which allowed the experimenters to capture and analyze the dynamics of the process in time. This technique, with much increased and continuously improving temporal and spatial resolution, remains the main investigative tool in experimental research to the present day.

Recently, it has been demonstrated that one can achieve a very high resolution of the measurements via an indirect method based on using the electric circuit disconnected as a result of the breakup (Burton, Rutledge & Taborek 2004). For medium and high conductivity fluids, this technique promises substantial advantages and is expected to be widely used in the near future.

The first detailed experiments on the dynamics of jets stimulated intensive theoretical research into the nature of instabilities that eventually lead to the change in the topology of the flow domain (Plateau 1849, 1873, Rayleigh 1879a,b, 1892). Rayleigh has shown that a cylindrical column of fluid is linearly unstable with respect to long-wave disturbances as the cross-sectional curvature, which is higher than the longitudinal one, drives the jet away from the equilibrium. This effect for nonaxisymmetric disturbances allowed Bohr (1909, 1910) to develop a technique of 'jet vibration' for measuring the surface tension of a freshly-formed free surface. This technique as well as the methods based on the jet instability analysis are used to the present day (Kochurova & Rusanov 1981, Alakoç et al. 2004).

For the most part of the last century, research on the jet dynamics has been focussed primarily on the stability issues, developing and extending Rayleigh's ideas to weakly and then strongly nonlinear regimes of flow (see Lin & Reitz 1998, Lin 2003 and references therein). Recently, the emphasis shifted towards the studies of the dynamics of the breakup itself, i.e., the topological transition leading to the drop formation, and the post-breakup evolution of the jet, including satellite formation, secondary breakup of the main drop and related issues. As we will see below, even in the simplest case of a Newtonian liquid thread in vacuum the breakup is far from being a trivial phenomenon to model and the solutions obtained in the framework of the standard model exhibit unphysical singularities whose origin must be understood in order to propose a singularity-free mathematical description of the process.

7.2 Drop formation: emerging singularity

It is convenient to begin with the simplest though fully representative situation where the pre-breakup evolution of a liquid thread can be described in the framework of the slender-jet (long-wavelength) approximation. This regime has been intensively studied in the past two decades and its analysis helps to identify the main problems associated with the modelling of the breakup phenomenon.

7.2.1 Slender-jet approximation

Consider a cylindrically-symmetric thread of viscous Newtonian fluid surrounded by an inviscid dynamically-passive gas. If $(u, 0, w)$ are the components of velocity in the cylindrical coordinates (r, φ, z) with the z-axis directed along the axis of symmetry of the thread, then, after making the Navier-Stokes equations and the standard boundary conditions dimensionless using the natural scales (7.1), i.e.,

$$l_\mu = \frac{\mu^2}{\sigma_e \rho}, \quad t_\mu = \frac{\mu^3}{\sigma_e^2 \rho}, \quad U_\mu = \frac{l_\mu}{t_\mu} \equiv \frac{\sigma_e}{\mu}, \quad P_\mu = \frac{\sigma_e}{l_\mu} \equiv \frac{\sigma_e^2 \rho}{\mu^2}, \quad (7.1)$$

and assuming that all variables are independent of φ, we arrive at the problem (6.90)–(6.95), i.e., equations

$$\frac{1}{r}\frac{\partial(ru)}{\partial r} + \frac{\partial w}{\partial z} = 0, \quad (7.2)$$

$$\frac{\partial u}{\partial t} + u\frac{\partial u}{\partial r} + w\frac{\partial u}{\partial z} = -\frac{\partial p}{\partial r} + \frac{1}{r}\frac{\partial}{\partial r}\left(r\frac{\partial u}{\partial r}\right) + \frac{\partial^2 u}{\partial z^2} - \frac{u}{r^2} \quad (7.3)$$

$$\frac{\partial w}{\partial t} + u\frac{\partial w}{\partial r} + w\frac{\partial w}{\partial z} = -\frac{\partial p}{\partial z} + \frac{1}{r}\frac{\partial}{\partial r}\left(r\frac{\partial w}{\partial r}\right) + \frac{\partial^2 w}{\partial z^2} \qquad (7.4)$$

in the bulk together with boundary conditions

$$u - w\frac{\partial h}{\partial z} = \frac{\partial h}{\partial t} \qquad (7.5)$$

$$2\frac{\partial h}{\partial z}\left(\frac{\partial u}{\partial r} - \frac{\partial w}{\partial z}\right) + \left[1 - \left(\frac{\partial h}{\partial z}\right)^2\right]\left(\frac{\partial w}{\partial r} + \frac{\partial u}{\partial z}\right) = 0, \qquad (7.6)$$

$$p - \frac{2}{1 + \left(\frac{\partial h}{\partial z}\right)^2}\left[\frac{\partial u}{\partial r} + \left(\frac{\partial h}{\partial z}\right)^2\frac{\partial w}{\partial z} - \frac{\partial h}{\partial z}\left(\frac{\partial w}{\partial r} + \frac{\partial u}{\partial z}\right)\right]$$

$$= \frac{1}{h\left[1 + \left(\frac{\partial h}{\partial z}\right)^2\right]^{1/2}} - \frac{\frac{\partial^2 h}{\partial z^2}}{\left[1 + \left(\frac{\partial h}{\partial z}\right)^2\right]^{3/2}} \qquad (7.7)$$

on the free surface $z = h(r,t)$. For convenience, the pressure p is measured with respect to a constant pressure in the ambient gas.

The slender-jet approximation of (7.2)–(7.7) is based on the assumption that the ratio of characteristic length scales in the r and z directions, denoted hereafter as ϵ, is small. Then, as follows from (7.2), the ratio of characteristic velocities in the radial and axial directions will also be of order ϵ, thus making the axial flow the dominant one. To specify the slender-jet approximation, we will make two assumptions. First, we will assume that $u = O(1)$ as $\epsilon \to 0$, i.e., assume that U, which corresponds to the capillary number $Ca = \mu U/\sigma_e = 1$, is the scale for the radial velocity. The second assumption is that in the axial projection of the momentum balance equation (7.4) the convective and longitudinal viscous term are of the same order,

$$w\frac{\partial w}{\partial z} \sim \frac{\partial^2 w}{\partial z^2} \qquad \text{as } \epsilon \to 0.$$

These assumptions yield the following asymptotic scaling as $\epsilon \to 0$:

$$r = \epsilon^2 \bar{r}, \quad z = \epsilon \bar{z}, \quad t = \epsilon^2 \bar{t}, \quad h(z,t) = \epsilon^2 \bar{h} = \epsilon^2(h_0 + \dots), \qquad (7.8)$$

$$u = \bar{u} = u_0 + \epsilon^2 u_1 + \dots, \quad w = \epsilon^{-1}\bar{w} = \epsilon^{-1}(w_0 + \epsilon^2 w_1 + \dots),$$

$$p = \epsilon^{-2}\bar{p} = \epsilon^{-2}(p_0 + \epsilon^2 p_1 + \dots).$$

Then, equations (7.2)–(7.4) take the form

$$\frac{1}{\bar{r}}\frac{\partial(\bar{r}\bar{u})}{\partial \bar{r}} + \frac{\partial \bar{w}}{\partial \bar{z}} = 0, \qquad (7.9)$$

$$\frac{\partial \bar{u}}{\partial \bar{t}} + \bar{u}\frac{\partial \bar{u}}{\partial \bar{r}} + \bar{w}\frac{\partial \bar{u}}{\partial \bar{z}} = -\frac{1}{\epsilon^2}\frac{\partial \bar{p}}{\partial \bar{r}} + \frac{1}{\epsilon^2}\left[\frac{1}{\bar{r}}\frac{\partial}{\partial \bar{r}}\left(\bar{r}\frac{\partial \bar{u}}{\partial \bar{r}}\right) - \frac{\bar{u}}{\bar{r}^2}\right] + \frac{\partial^2 \bar{u}}{\partial \bar{z}^2} \tag{7.10}$$

$$\frac{\partial \bar{w}}{\partial \bar{t}} + \bar{u}\frac{\partial \bar{w}}{\partial \bar{r}} + \bar{w}\frac{\partial \bar{w}}{\partial \bar{z}} = -\frac{\partial \bar{p}}{\partial \bar{z}} + \frac{1}{\epsilon^2}\frac{1}{\bar{r}}\frac{\partial}{\partial \bar{r}}\left(\bar{r}\frac{\partial \bar{w}}{\partial \bar{r}}\right) + \frac{\partial^2 \bar{w}}{\partial \bar{z}^2}, \tag{7.11}$$

and the boundary conditions (7.5)–(7.7) at $\bar{r} = \bar{h}(\bar{z},\bar{t})$ are given by

$$\bar{u} - \bar{w}\frac{\partial \bar{h}}{\partial \bar{z}} = \frac{\partial \bar{h}}{\partial \bar{t}} \tag{7.12}$$

$$2\epsilon^2 \frac{\partial \bar{h}}{\partial \bar{z}}\left(\frac{\partial \bar{u}}{\partial \bar{r}} - \frac{\partial \bar{w}}{\partial \bar{z}}\right) + \left[1 - \epsilon^2\left(\frac{\partial \bar{h}}{\partial \bar{z}}\right)^2\right]\left(\frac{\partial \bar{w}}{\partial \bar{r}} + \epsilon^2 \frac{\partial \bar{u}}{\partial \bar{z}}\right) = 0, \tag{7.13}$$

$$\bar{p} - \frac{2}{1+\epsilon^2\left(\dfrac{\partial \bar{h}}{\partial \bar{z}}\right)^2}\left[\frac{\partial \bar{u}}{\partial \bar{r}} + \epsilon^2\left(\frac{\partial \bar{h}}{\partial \bar{z}}\right)^2\frac{\partial \bar{w}}{\partial \bar{z}} - \frac{\partial \bar{h}}{\partial \bar{z}}\left(\frac{\partial \bar{w}}{\partial \bar{r}} + \epsilon^2 \frac{\partial \bar{u}}{\partial \bar{z}}\right)\right]$$

$$= \frac{1}{\bar{h}\left[1+\epsilon^2\left(\dfrac{\partial \bar{h}}{\partial \bar{z}}\right)^2\right]^{1/2}} - \frac{\epsilon^2 \dfrac{\partial^2 \bar{h}}{\partial \bar{z}^2}}{\left[1+\epsilon^2\left(\dfrac{\partial \bar{h}}{\partial \bar{z}}\right)^2\right]^{3/2}}. \tag{7.14}$$

Equation (7.11) to order ϵ^{-2} gives that a solution regular at $\bar{r} = 0$ has the form

$$w_0 = w_0(\bar{z},\bar{t}), \tag{7.15}$$

that is to leading order we are dealing with a plug flow in the z-direction. Given this, from (7.9) one has

$$u_0(\bar{r},\bar{z},\bar{t}) = -\frac{\partial w_0(\bar{z},\bar{t})}{\partial \bar{z}}\frac{\bar{r}}{2}. \tag{7.16}$$

This makes the square bracket in equation (7.10) to leading order identically zero and hence this equation to order ϵ^{-2} yields $p_0 = p_0(\bar{z},\bar{t})$. Then, given that both p_0 and w_0 are independent of $\bar{r}$, one can partially separate the variables in equation (7.11) by writing it, to order $O(1)$, in the form of two equations:

$$\frac{\partial w_0}{\partial \bar{t}} + w_0\frac{\partial w_0}{\partial \bar{z}} + \frac{\partial p_0}{\partial \bar{z}} - \frac{\partial^2 w_0}{\partial \bar{z}^2} = C_1(\bar{z},\bar{t}) = \frac{1}{\bar{r}}\frac{\partial}{\partial \bar{r}}\left(\bar{r}\frac{\partial w_1}{\partial \bar{r}}\right), \tag{7.17}$$

where $C_1(\bar{z},\bar{t})$ is the (yet unknown) separation function. Integrating the second of these two equations with respect to $\bar{r}$ and assuming that w_1 is regular at $\bar{r} = 0$, we obtain

$$w_1 = C_1(\bar{z},\bar{t})\frac{\bar{r}^2}{4} + C_2(\bar{z},\bar{t}), \tag{7.18}$$

that is a parabolic correction to the plug flow.

Now we can turn to the boundary conditions. Substituting (7.18) into the tangential-stress boundary condition (7.13), which to leading order has the form

$$2\frac{\partial h_0}{\partial \bar{z}}\left(\frac{\partial u_0}{\partial \bar{r}} - \frac{\partial w_0}{\partial \bar{z}}\right) + \frac{\partial w_1}{\partial \bar{r}} + \frac{\partial u_0}{\partial \bar{z}} = 0 \qquad \text{at } \bar{r} = h_0, \qquad (7.19)$$

and using (7.16) for u_0, we find the separation function in terms of h_0 and w_0

$$C_1(\bar{z}, \bar{t}) = \frac{\partial^2 w_0}{\partial \bar{z}^2} + \frac{6}{h_0}\frac{\partial h_0}{\partial \bar{z}}\frac{\partial w_0}{\partial \bar{z}}. \qquad (7.20)$$

Condition (7.14) to leading order expresses p_0 also in terms of h_0 and w_0:

$$p_0(\bar{z}, \bar{t}) = \frac{1}{h_0} + 2\frac{\partial u_0}{\partial \bar{r}} = \frac{1}{h_0} - \frac{\partial w_0}{\partial \bar{z}}. \qquad (7.21)$$

Now, substituting (7.20), (7.21) into the first equation (7.17) and (7.16) into (7.12), we arrive at the slender-jet approximation for the original problem

$$\frac{\partial w_0}{\partial \bar{t}} + w_0\frac{\partial w_0}{\partial \bar{z}} = -\frac{\partial}{\partial \bar{z}}\left(\frac{1}{h_0}\right) + \frac{3}{h_0^2}\frac{\partial}{\partial \bar{z}}\left(h_0^2\frac{\partial w_0}{\partial \bar{z}}\right), \qquad (7.22)$$

$$\frac{\partial h_0}{\partial \bar{t}} + w_0\frac{\partial h_0}{\partial \bar{z}} + \frac{h_0}{2}\frac{\partial w_0}{\partial \bar{z}} = 0. \qquad (7.23)$$

It is noteworthy that, to leading order in ϵ, in the slender-jet approximation the longitudinal curvature of the free surface is negligible: as is clear from (7.14), it would be asymptotically inconsistent to artificially keep it in the equations.[1] The driving force of the process is the first term on the right-hand side of (7.22), i.e., the spatially nonuniform (z-dependent) cross-sectional curvature, which creates the pressure gradient pushing the fluid out of the minimal cross-section (the 'neck').

If the assumptions we made in deriving (7.22), (7.23) remain satisfied, these equations can be used to describe the process of *thinning* of the thread as it evolves towards the breakup. The question now is whether or not they remain valid up to the *breakup itself* and what information one can obtain for the post-breakup retraction of the tips of the disintegrated thread.

---

[1] If the formal parametric asymptotic analysis is replaced with mathematical manipulations of unknown/undefinable asymptotic accuracy (Markova & Shkadov 1972, Eggers 1993, Eggers & Dupont 1994), then the longitudinal curvature can pop up at leading order even in the slender-jet approximation, where, as one can see from (7.14), it does not belong.

7.2.2   Similarity solutions

As the moment of pinch-off t_0 approaches, the characteristic time and length scales for the local flow in the thinning neck of the liquid thread become $t_0 - t$ and the distance from the point where the thread will break up (we will take this point as the origin of our coordinate system). These are infinitesimal scales so that the flow in the immediate vicinity of the breakup point becomes separated in scale from that in the far field. Then, if the process of pinching off is driven locally, it becomes independent of the initial conditions and the flow in the far field, which are specified via the corresponding initial and boundary conditions. Then one may assume that the process of breaking up of the liquid thread is locally self-similar and look for a similarity solution of (7.22), (7.23). It should be emphasized here that, even given the obvious separation of scales between the local process and the global flow, the local self-similarity of a solution is an *assumption* which must be verified.

The first similarity solution of (7.22), (7.23) has been proposed by Eggers (1993, 1995), who used the similarity variables

$$h_0(\bar{z},\bar{t}) = \hat{t}\phi(\xi), \quad w_0(\bar{z},\bar{t}) = \hat{t}^{-1/2}\psi(\xi), \tag{7.24}$$

where $\xi = \bar{z}\hat{t}^{-1/2}$ and $\hat{t} = \bar{t}_0 - \bar{t}$ is the time to the breakup scaled with t_μ. The substitution (7.24) turns (7.22), (7.23) into a set of nonlinear ordinary differential equations

$$\frac{\psi}{2} + \frac{\xi\psi'}{2} + \psi\psi' = \frac{\phi'}{\phi^2} + 3\psi'' + \frac{6\psi'\phi'}{\phi}, \quad \phi' = \phi\frac{1 - \psi'/2}{\phi + \xi/2} \tag{7.25}$$

and makes h and w to leading order independent of ϵ. The boundary conditions at infinity come from the requirement that far away from the pinch-off point the time-dependence of h_0 and w_0 dies out. In the similarity variables this gives

$$\phi(\xi) \to \xi^2, \quad \psi(\xi) \to \xi^{-1} \qquad \text{as } \xi \to \pm\infty. \tag{7.26}$$

Given that $\psi(\xi)$ is bounded, there exists $\xi = \xi_0$ that makes the denominator on the right-hand side of the second equation (7.25) equal to zero. For ϕ to remain regular at this point, the numerator must be also zero, so that

$$\psi'(\xi_0) = 2. \tag{7.27}$$

The similarity functions ϕ and ψ satisfying (7.25)–(7.27) have been obtained numerically by the shooting method (Eggers 1993, 1995). The results are shown in Fig. 7.2.

The solution given by (7.24) is obviously singular. The axial velocity diverges as $\hat{t} \to 0$ and, using (7.21) to calculate p_0, we find that

$$p_0(\bar{z},\bar{t}) = \frac{1}{h_0} - \frac{\partial w_0}{\partial\bar{z}} = \frac{1 - \phi(\xi)\psi'(\xi)}{\hat{t}\phi(\xi)}, \tag{7.28}$$

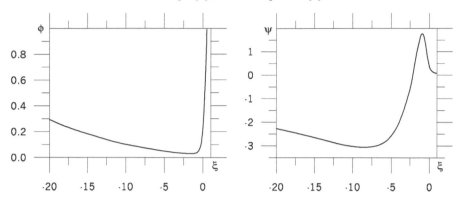

FIGURE 7.2:   Functions ϕ and ψ in the similarity solution (7.24) obtained by Eggers (1993, 1995).

i.e., the bulk pressure is nonintegrably singular at $\hat{t} = 0$. In dimensional variables (marked with an asterisk) the solution (7.24), (7.28) has the form:

$$h^*(z^*, t^*) = \frac{\sigma_e}{\mu}(t_0^* - t^*)\phi(\xi), \quad w^*(z^*, t^*) = \frac{\nu^{1/2}\psi(\xi)}{(t_0^* - t^*)^{1/2}}, \tag{7.29}$$

$$p^*(z^*, t^*) = \frac{\mu[1 - \phi(\xi)\psi'(\xi)]}{(t_0^* - t^*)\phi(\xi)}, \tag{7.30}$$

where

$$\xi = \frac{z^*}{\nu^{1/2}(t_0^* - t^*)^{1/2}}, \tag{7.31}$$

and $\nu = \mu/\rho$ is the kinematic viscosity.

The singularities in the time-dependence of w^* and p^* at $t^* = t_0^*$ indicate that the above similarity solution is not valid at and, by continuity, near this moment. This solution can be regarded as describing the process of *thinning* of the liquid thread and becomes inapplicable in some finite time interval before the breakup.[2] Some 'extra' physics not accounted for in the standard model must step in as the breakup is approached, and this physics must be incorporated into the model to remove the singularities and allow the thread to go through the topological transition with meaningful values of the flow parameters. It is the nature of this 'extra' physics and how it handles the topological transition that determine the initial conditions for the post-breakup retraction of the ends of the disintegrated thread when, after some period following the breakup, the system will again be describable in the framework of the standard model.

_____

[2] As we discussed in §2.1.3, all time intervals that one is dealing with in continuum mechanics are, by definition, macroscopic.

However, Eggers argued that his analysis based on the standard model, with no 'extra' physics incorporated to remove the singularity, is "universal" in linking the pre-breakup and post-breakup states of the thread and predicted (Eggers 1995) that the recoil velocity of the ends of the broken thread, which in a sense is the macroscopic 'summary' of how the pinch-off has occurred, must be described by

$$v_{\text{recoil}} = \tfrac{1}{2}\xi_{\text{neck}}\hat{t}^{-1/2}, \tag{7.32}$$

where $\xi_{\text{neck}}/2 \approx 8.7$ and $\hat{t} = t - t_0$ is the time after the breakup scaled with t_μ. In dimensional terms, this means

$$v^*_{\text{recoil}} = \frac{\xi_{\text{neck}}}{2} \frac{\nu^{1/2}}{(t^* - t_0^*)^{1/2}}, \tag{7.33}$$

i.e., the retraction velocity is infinite at the moment of pinching off and for $t^* > t_0^*$ depends only on the time from the pinch-off and, counter-intuitively, on the kinematic viscosity of the fluid. Furthermore, more viscous fluids are expected to recoil *faster* than less viscous ones. As we will see in §7.3, these predictions disagree with experiments both quantitatively and qualitatively.

Following on from Eggers' work a group of researchers produced, in the words of Lister et al. (1997), "a zoo of singularity structures" corresponding to different balances in the long-wavelength approximation (Papageorgiou 1995a,b, Brenner, Lister & Stone 1996, Lister & Stone 1998, Zhang & Lister 1999) and later also by relaxing this approximation (Sierou & Lister 2003). Some mathematical aspects of the axisymmetric long-wavelength approximation have been clarified by Renardy (2005).

The obtained solutions can be regarded as continuing the direction of research into the dynamics of liquid jets initiated by Rayleigh's (1879a,b) seminal papers, and they uncovered a number of interesting features related to the evolution of liquid threads *before* the physics of breakup kicks in to take the system through the topological transition. The interested reader can find reviews of some of these as well as earlier results in Eggers (1997) and Lister et al. (1997).

A question that suggests itself here is whether it is the long-wavelength approximation that gives rise to the singularities in the solutions and what would happen if this approximation is abandoned and the process is considered as a fully two-dimensional axisymmetric flow. In fact, we already know the answer to this question: the analysis of the similarity scaling of (6.90)–(6.95) in the case of coalescence (§6.5.2) remains valid for the breakup problem, and it shows that a locally self-similar solution of this problem in the framework of the standard model is bound to have, at least, a nonintegrable singularity in the bulk pressure. As we remember, this outcome is a direct consequence of (a) the Navier-Stokes equations in the bulk, (b) the standard kinematic boundary condition everywhere on the free surface (thus presuming that the free surface is smooth), and (c) the assumption of local self-similarity of the flow that follows from the separation of scales.

In the case of coalescence, the initial state of the system, i.e., the flow parameters and the free-surface shape at the moment $t = 0$ when the two fluid volumes are brought into contact, is prescribed independently, and the free surface has a singularity of curvature, which suggests how to resolve the problem. In this respect, the breakup problem is very different since the breakup is the culmination of the preceding dynamics governed by the very equations that drive the solution into a singularity, i.e., outside their own limits of applicability. Will the singularity of curvature, similar to that in the coalescence process, develop as the breakup is approached? If not, what is the physics that switches on close to the breakup and what are the conditions for it to come into play? In order to have a broader view of the situation, we will look first at some of the numerical results obtained without a presumed self-similarity of the process and then, having at our disposal all what the standard model has to offer, examine the experimental data related to the breakup phenomenon.

7.2.3 Numerical results

The dynamics of drop formation, the evolution of liquid jets and bridges have been studied numerically by many authors for potential flows (e.g., Chen & Steen 1997, Day, Hinch & Lister 1998, Leppinen & Lister 2003), the creeping (i.e., zero-Reynolds-number) flow regime (Gaudet, McKinley & Stone 1996, Pozrikidis 1999) and by considering the full Navier-Stokes equations (e.g., Ashgriz & Mashayek 1995, Wilkes, Phillips & Basaran 1999, Notz, Chen & Basaran 2001, Chen, Notz & Basaran 2002, Ambravaneswaran, Wilkes & Basaran 2002, Notz & Basaran 2004). Some results have also been obtained for three-dimensional creeping flows (e.g., Cristini, Blawzdziewicz & Loewenberg 1998)

The power of computational methods makes it possible to address a wide range of issues associated with the regular flow, from the global behaviour of an axisymmetric or a fully three-dimensional fluid body and conditions for the primary and the satellite drop formation (e.g., Ambravaneswaran et al. 2002) to the accuracy of various asymptotic approximations and transitions between the corresponding flow regimes (e.g., Notz et al. 2001). What we are interested in here is what happens as an axisymmetric liquid jet or a bridge connecting two solid bodies approaches the breakup.

It should be emphasized that numerical computations in the framework of the standard model per se do not allow one to go through the topological transition. The method used to mimic the breakup is to stop computations when the diameter of the neck becomes smaller than a certain preset value. Then the free surface is reconnected as if the breakup has taken place and the computations are restarted with this new 'post-breakup' initial configuration (Ashgriz & Mashayek 1995). This procedure obviously introduces a finite discontinuity in the time-dependence of the free-surface curvature and hence creates an instant jump in the bulk pressure. An alternative way used in

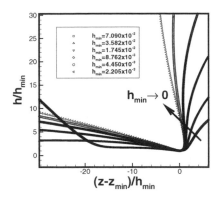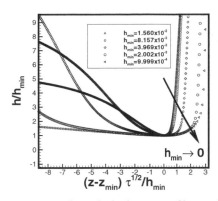

FIGURE 7.3: Computations of variations of scaled drop profiles with scaled axial coordinate for low-viscosity (left) and high-viscosity (right) fluids as the neck approached pinch-off (Chen, Notz & Basaran 2002; © 2002 American Physical Society; reproduced with permission). The parameters of the low-viscosity fluid correspond to water and those of high-viscosity fluid to a 83% glycerol-water solution; τ is the time to the expected pinch-off; h_{min} is the dimensionless diameter of the neck.

some works is to link the pre-breakup and the post-breakup states via some singular self-similar solution which amounts to an assumption of unphysical behaviour of the system in between these states. Both these ways of an effective numerical 'cut-off' have a finite influence on the distributions of the flow parameter in the near field but, for sufficiently large systems, introduce a small error for the far-field flow and hence allow one to make sufficiently accurate predictions for global characteristics of the process. However, as one moves towards smaller scales, in particular the scales typical for applications in microfluidics, the necessity to have a physically justifiable way of modelling the topological transition becomes unavoidable because the 'far field' is simply not far enough for the dynamics of the breakup to be neglected.

Computations performed for the nondimensional parameters of the flow varying over a wide range have shown that for a sufficiently thin thread, as the thread's diameter in the minimal cross-section decreases, the free-surface profile evolves in a self-similar way, i.e., if suitably scaled, it becomes time-independent. This confirms one of the key assumptions used in the asymptotic analysis of the breakup. Fig. 7.3 shows the scaled free-surface profiles for the pinching of low- and high-viscosity liquid threads computed numerically and confirmed experimentally (Chen et al. 2002).

For the low-viscosity fluid with the parameters corresponding to water, the free-surface profile of the departing drop overturns so that the drop and the thread form two co-axial cones that are joint *smoothly* near the neck (Fig. 7.3 left). This double-cone structure has been computed by several authors for

FIGURE 7.4:   Evolution and breakup of a jet (Rayleigh 1891).

potential flow (Chen & Steen 1997, Day et al. 1998) and the full-scale Navier-Stokes equations (Wilkes et al. 1999).

For a high-viscosity fluid, for example, a 83% glycerol-water solution with the 85 cP viscosity (Fig. 7.3 right), there develops an elongated filament (a micro-thread) with a decreasing aspect ratio. Computations confirm that this filament is accurately described by the slender-jet approximation (Chen et al. 2002).

Thus, for the commonly used and sufficiently general initial conditions the full-scale numerical computations do not indicate any tendency of the free-surface shape to develop cusp-like features, which would allow one to remove the unphysical singularities in the flow field in a way similar to that used in Chapter 6 for the coalescence problem. (This is still a possibility for some exotic initial data, but we will not consider such cases here.) As we will see below, experiments confirm this conclusion and pose some additional and very intriguing questions.

7.3   Experiments on capillary pinch-off

Experiments have revealed that the dynamics of an axisymmetric liquid jet leading to the drop formation is remarkably complex. In general terms, the process can be described as follows. As the jet emerges out of an orifice, its free surface starts to develop a wave pattern that evolves into a distinct structure comprising bulges, which will later become primary drops, and threads that connect them. As the connecting (primary) thread breaks up close to its front end, the main drop is released, whereas the breakup at the rear end turns the former thread into a satellite droplet. Since the thread detached at one or both ends finds itself in a very nonequilibrium state, a very fast recoil of its ends takes place generating a wave motion along the thread. Very often the result of this is a series of secondary breakups leading to the appearance of multiple satellite droplets. This scenario, that has already been known to Plateau (1849) and documented by Lenard (1887) and Rayleigh (1891), is illustrated in Fig. 7.1 and Fig. 7.4.

More detailed research into the capillary pinch-off discovered a number of interesting features. It has been found that, for sufficiently viscous fluids

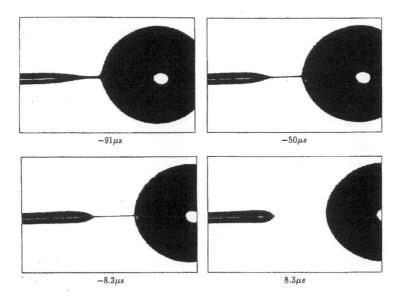

FIGURE 7.5: The evolution of the free-surface profile close to departing drop for a fluid of kinematic viscosity, $\nu = 45$ mm^2/s and $\sigma_e = 30.5$ mN/m (Kowalewski 1996; © 1996 Elsevier; reproduced with permission). The time is given relative to the moment of the neck rupture. The width of one frame is 1 mm.

in the stage immediately preceding the breakup, there appears a secondary filament ('micro-thread') between the primary thread and the departing drop (Fig. 7.5). This phenomenon has been observed first for a drop falling from a faucet (Shi, Brenner & Nagel 1994) and then for a jet (Kowalewski 1996). The formation of this structure and the conditions for the subsequent pinch-off have become the subject of several investigations.

Kowalewski (1996) carried out a detailed study of various aspects of the capillary breakup for a range of fluids with viscosities spanning more than 2 orders of magnitude ($\nu = 1$–320 mm^2/s, see Table 7.1). Surprisingly, it has been found that the minimal diameter of the micro-thread $d_{\min}$ at which the actual pinch-off is triggered remains approximately the same (about 1 μm, see Table 7.2) for fluids of different viscosities. Discussing this result and ruling out the possibility that there might be a tertiary filament of a macroscopic length, Kowalewski writes:

> We believe this, because the micro-thread can be well seen both shortly before and also after the pinch-off. Hence, except for an infinitesimal region around the breakup point, the rest of the micro-thread to be visible must preserve its final diameter of the order of 1 μm (Kowalewski 1996, p. 130).

TABLE 7.1: Parameters of the fluids used in Kowalewski's (1996) experiments

Notation	Content	ν	σ_e	ρ	l_μ	t_μ
		mm^2/s	dyn/cm	g/cm^3	μm	μs
W	W	1.0	72.6	1.000	0.014	0.0002
A	A	1.48	22.5	0.803	0.036	0.001
MIXD	G-W	11.24	65.6	1.161	2.23	0.44
MIXE	G-A	45.0	30.5	1.081	71.8	114.5
MIXG	G-W	46.0	65.8	1.209	38.8	32.7
GLY1	G-W	120	64.0	1.220	274.5	627.9
GLY2	G-W	180	64.0	1.230	622.7	2154
GLY3	G-W	320	63.8	1.260	2022	12780

A: ethyl alcohol, W: water, G: glycerin; l_μ and t_μ are natural scales defined by (7.1). (Kowalewski 1996; © 1996 Elsevier; reproduced with permission).

TABLE 7.2: Typical characteristics of the jet breakup within a time interval Δt before (after) the rupture (Kowalewski 1996)

Liquid	r_j	F_0	Δt	L_{max}	L_{max}/l_μ	d_{min}	v_{recoil}
	μm	Hz	μs	μm		μm	m/s
W_{FT}	48	–	5.5	< 1	–	< 1	6.5
A	98.5	7840	1.6	< 1	–	< 1	10.0
MIXD	98.5	4337	2.66	45	20.1	0.7	4.9
MIXE	140	1391	8.3	294	4.09	0.7	24
MIXG	37.5	8625	1.34	220	5.67	< 1.2	30
$MIXG_{FT}$	37.5	7723	5.55	220	5.67	–	25
MIXG	98.5	1401	8.28	250	6.44	0.7	10.6
$MIXG_{FT}$	98.5	2590	5.55	380	9.8	–	8.8
$MIXG_{FT}$	190	775	19.45	400	10.3	–	5.5
$MIXG_{FT}$	250	2852	5.55	380	9.8	–	5.0
MIXG	650	73	–	350	9.02	0.7	–
$MIXG_{FT}$	650	75	19.45	250	6.44	–	8.0
GLY1	195	380	15.1	860	3.13	0.7	4.0
GLY2	900	85	137	1420	2.28	0.7	4.3
GLY3	195	603	10.97	1750	0.86	0.7	–

Here r_j is the initial radius of the jet, F_0 is the frequency, L_{max} is the maximum length of the micro-thread, d_{min} is its minimum diameter, and v_{recoil} is its recoil velocity. The subscript FT corresponds to the data obtained using the Frame Transfer technique. (Kowalewski 1996; © 1996 Elsevier; reproduced with permission).

An illustration of the process is given in Figs 7.6 and 7.7, which present the snapshots of the free-surface profile before, at the moment of, and shortly after the breakup. The fact that d_{min} is independent of the fluid's viscosity for viscosities varying in such a wide range is in stark conflict with the similarity solution (7.29)–(7.30), which expects the thread's diameter to be a strong function of the fluid's viscosity. Hence one has to conclude that even the scaling of the similarity solution (7.29)–(7.30) is not valid up to the breakup. At the same time, the slender-jet approximation is certainly applicable and the similarity solution seems to be working when the jet's diameter becomes sufficiently small and until it reaches the order of d_{min}. Then the departure of the observed behaviour of the system from the solution that uses the standard model as the jet diameter goes below the order of d_{min} indicates that, when the breakup is approached, some new physics kicks in and the micro-thread ruptures *locally* with the rest of it having the diameter on the scale of d_{min}.

Another important result obtained by Kowalewki is that, contrary to Eggers' prediction (7.33), the recoil velocity of the ends of the micro-thread turns out to be independent of ν and, instead, depends on σ_e, ρ and d_{min}:

$$v^{*}_{\text{recoil}} \propto \left(\frac{\sigma_e}{\rho d_{min}} \right)^{1/2}. \tag{7.34}$$

Commenting on the failure of Eggers' prediction, Kowalewki writes:

> The comparison of both diverging ends of the thread with experimental observations shows severe discrepancies. Also the kinematics of the thread contraction after the pinch-off shows for higher viscosities discrepancies difficult to interpret. Predicted increase of the retraction velocity in function of viscosity could not be confirmed and rather opposite relation was observed (Kowalewski 1996, p. 142).

Thus, the qualitative discrepancy between the prediction (7.33) made on the basis of the similarity solution and the experiments completely undermines the idea of using singular solutions to 'link' the pre-breakup and post-breakup states of the system regardless of what in reality happens in between.

Interestingly, the empirically discovered scaling (7.34) coincides exactly with what follows from Keller's (1983) elementary analysis, where he considered how a blob at the end of a cylindrical thread would recoil under the action of the surface tension. This match of Keller's analysis and (7.34) indicates, firstly, that the recoil is driven *from the end* of the micro-thread and this end must be *rounded* to produce the required capillary pressure, i.e., it must have the characteristic radii of curvature on the scale of the thickness of the micro-thread. This is completely opposite to what is described by (7.33), i.e., a thread with "a sharp front at the end" (Eggers 1995, p. 950) which remains self-similar as it withdraws due to the nonuniform *cross-sectional* curvature *along* it.

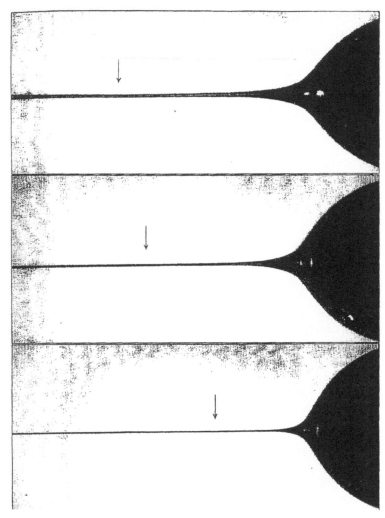

FIGURE 7.6: Process of the drop separation for a high-viscosity fluid ($\nu =$ 320 mm^2/s, $\sigma_e = 63.8$ mN/m) with the arrows indicating the minimal cross-section (Kowalewski 1996; © 1996 Elsevier; reproduced with permission). Time (from the top) relative to the moment of breakup: -1053 μs, -614.6 μs, -175.6 μs; the frame width corresponds to 2 mm.

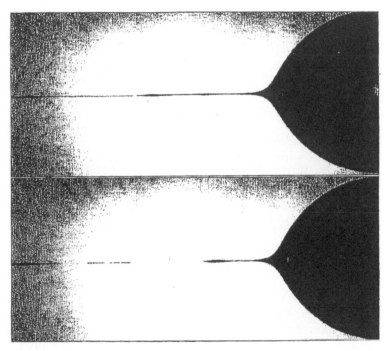

FIGURE 7.7:   Continuation of the sequence shown in Fig. 7.6. Top: the moment of breakup. Bottom: 44 μs later (Kowalewski 1996; © 1996 Elsevier; reproduced with permission).

Secondly, the scaling (7.34) and its simple explanation in Keller (1983) support Kowalewski's claim that in his experiment there was no further necking of the micro-thread, at least on the scale of observations. Additional experiments performed in a low-pressure chamber have shown that the presence of air does not change the observed scenario of the pinch-off.

Henderson, Pritchard & Smolka (1997) carried out a series of experiments on the formation of drops as the fluid flowed under gravity from a reservoir. They found that the structure consisting of the primary thread and the micro-thread attached to the drop is well reproducible (the lengths of the two threads before the breakup are reproducible within 3% and 10% respectively). It is noteworthy that neither Kowalewski nor Henderson et al. observed the appearance of a micro-thread at the rear end of the primary thread nor a cascade of threads at the front, whereas Shi et al. (1994) reported both of these features for the flow configuration close to that of Henderson et al. (1997).

Confirming Kowalewski's result, which is illustrated in Figs 7.6 and 7.7, for different flow conditions, Henderson et al. point out that

> The actual pinch-off does not occur at the point of attachment between the secondary thread and the drop. Instead, it occurs between the disturbances of the secondary thread. After the initial pinch-off, additional breaks occur between the disturbances, resulting in several secondary satellite drops with a broad distribution of sizes (Henderson et al. 1997, p. 3188).

The measurements of the recoil velocity as a function of time (Fig. 7.8) again demonstrate severe discrepancies between the predictions based on the supposed 'universality' of the singular similarity solution and the measured values.[3] Together with Kowalewski's experiments, this result indicates that the specific 'physics of breakup' does affect the initial conditions for the post-breakup evolution of the thread that determine the recoil velocity.

The process of thinning of the liquid thread as the system evolves towards the breakup, i.e., the dependence of the thread's diameter, d^*, on the time to the pinch-off, $t_0^* - t^*$, is another topic that received considerable attention in the past decade. Kowalewski (1996) found that, until the thread's diameter reaches $d_{\min}$, for fluids of high viscosity (μ =146–403 cP, $\sigma_e \approx 64$ dyn/cm, $\rho = 1.22$–1.26 g/cm^3) the dependence is linear, $d^* \sim (t_0^* - t^*)$, as it follows from (7.29), whereas for low-viscosity fluids (μ =13–55 cP, σ_e =30.5–65.8 dyn/cm, ρ =1.16–1.2 g/cm^3) one observes a departure from this behaviour. Rothert, Richter & Rehberg (2001), who investigated a water-glycerin mixture

---

[3]Discussing the discrepancies between his similarity solution and experiments, Eggers (1997) attributed them to the drag from the ambient air and dismissed Kowalewski's experiments in the low-pressure chamber, where this explanation had been investigated and ruled out, on the grounds that "the breakup was quite unsteady in this particular series of experiments". The *qualitative* disagreement between (7.33) and (7.34) is simply not mentioned.

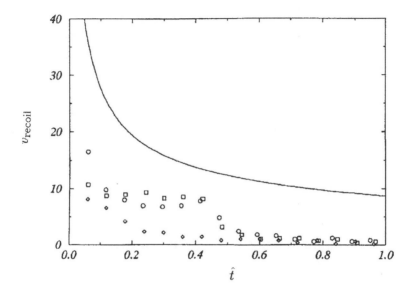

FIGURE 7.8: Recoil velocity of the ends of the micro-thread vs time from pinching off for three representative experiments (Henderson, Pritchard & Smolka 1997; © 1997 American Institute of Physics; reproduced with permission). Circles: $\nu = 96.6$ mm^2/s; squares: $\nu = 96.0$ mm^2/s; diamonds: $\nu = 96.2$ mm^2/s. The solid line corresponds to the prediction (7.33) made by Eggers (1995).

($\mu = 99$ cP, $\sigma_e = 64.6$ dyn/cm, $\rho = 1.25$ g/cm^3) and measured the thread's diameters down to about 10 μm, observed a transition between two linear dependencies that takes place at $t_0^* - t^* \approx 10 t_\mu$. The results are interpreted as a transition from the viscous (Papargeorgiou 1995a) to the inertial-viscous regime described by (7.29), (7.30). Chen et al. (2002) reported the transition from the inviscid regime, $d^* \propto (t_0^* - t^*)^{2/3}$, to the inertial-viscous one for a 83% glycerol-water solution ($\mu = 85.4$ cP, $\sigma_e = 63.1$ dyn/cm, $\rho = 1.21$ g/cm^3) as they observed the thread's thinning down to the diameter of about 7 μm and described this transition using the full-scale Navier-Stokes numerical computations. The computed free-surface profiles are given in Fig. 7.3 in the similarity variables.

The constraints of optical instrumentation regarding the spatial resolution of the measurements and the inability to see the neck of the thread when the free-surface profile overturns (Fig. 7.3) have been overcome by Burton et al. (2004). They investigated the breakup of a drop of mercury ($\mu = 1.526$ cP, $\sigma_e = 485.5$ dyn/cm, $\rho = 13.59$ g/cm^3) as it disconnects an electric circuit. It was found that, in this system, down to $d^* \approx 2.7$ nm and $t_0^* - t^* \approx 0.18$ ns, the thread's diameter obeys the inviscid scaling, $d^* \propto (t_0^* - t^*)^{2/3}$. This regime has already been investigated numerically (Chen & Steen 1997, Day et al. 1998) who found no evidence of the formation of a singularity of free-surface curvature before the breakup.

Thus, experiments and their analysis in terms of solutions obtained for the standard model give no indication of a singularity in the free-surface curvature prior to the breakup. The experimental results for the post-breakup recoil of the ends of the micro-thread are in a qualitative and quantitative disagreement with the singular solution put forward for linking the post-breakup with the pre-breakup state, although this solution adequately describes the process of thinning of the liquid thread until its diameter reaches a certain critical value. This is an indication that the proper singularity-free modelling of the process of pinching off is important for the correct description of the post-breakup evolution of the system.

7.4   'Missing' physics and its qualitative verification

A brief review of theoretical and experimental results shows that the jet breakup phenomenon presents us with a double paradox. Firstly, the solutions obtained in the framework of the standard model, being realistic and, as experiments show, accurate for the process of thinning of the liquid thread, develop a singularity and hence lead outside the limits of applicability of the underlying model as the pinch-off is approached. This is typical of most paradoxes of modelling — provided that a solution exists at all. Secondly, although on a qualitative level the breakup can be perceived as a phenomenon broadly opposite to coalescence, it is not obvious how to make this connection in terms of mathematical modelling.

We will start from this general perception and, given that our analysis of coalescence in Chapter 6 has shown it to be essentially a process of the interface disappearance, will look at the interface *formation* in the pre-breakup thinning of a liquid thread. For this purpose, it is convenient to examine the slender-jet approximation whose properties (e.g., the type of singularity in the bulk pressure) are representative of the general case.

Consider the relative rate-of-change of the free-surface area S^* in the process of thinning of the micro-thread. This quantity can be defined by

$$\frac{1}{S^*}\frac{dS^*}{dt^*} \equiv \nabla \cdot \mathbf{u}_\parallel + (\mathbf{u} \cdot \mathbf{n})\nabla \cdot \mathbf{n}, \tag{7.35}$$

where the first term on the right-hand side is the standard divergence of a vector field on a manifold and the second one corresonds to the creation (reduction) of the free-surface area due to the motion of a curved surface in the normal direction.[4] Then, using (7.29), we obtain

$$\frac{1}{S^*}\frac{dS^*}{dt^*} = \frac{1}{h^*}\left(\frac{\partial h^*}{\partial t^*} + w^*\frac{\partial h^*}{\partial z^*}\right) + \frac{\partial w^*}{\partial z^*} = \frac{F(\xi)}{t_0^* - t^*}, \tag{7.36}$$

where

$$F(\xi) = \frac{\xi\phi'}{2\phi} - 1 + \frac{\psi\phi'}{\phi} + \psi'. \tag{7.37}$$

Thus, as $t^* \to t_0^*$, one has $(1/S^*)(dS^*/dt^*) \to \infty$.

This outcome can also be illustrated in an elementary way by noticing that, firstly, at leading order the velocity profile is given by (7.15), i.e., it is the profile of a plug flow, $w_0 = w_0(\bar{z}, \bar{t})$. Secondly, one has that in the slender-jet approximation the longitudinal curvature is negligible compared with the cross-sectional one so that, essentially, we have the stretching of a liquid cylinder. If L^* is the length of this cylinder in the z-direction, then since the volume $V^* = \pi(h^*)^2 L^*$ remains constant (plug flow), for the lateral area S^* one has $S^* = 2\pi h^* L^* = 2V^*/h^* \to \infty$ as $h^* \to 0$. The relative rate-of-change of the surface area, $(1/S^*)dS^*/dt^*$, also tends to infinity if h^* goes down faster than exponentially, in particular if h^* reaches zero in a finite time.

As we can see now, the breakup phenomenon is indeed the process of the interface formation, and in this sense it is opposite to coalescence where one has the disappearance of the liquid-gas interface trapped between the coalescing bodies. Similarly to the situation we had in coalescence, where the inherent

---

[4]In more general terms, the relative rate-of-change of the surface area is the divergence of the surface velocity, see Appendix A.

— and admissible — singularity of the free-surface curvature at $t = 0$ indicated the way to resolve the problem, the breakup is also associated with an 'additional' singularity, this time it is an emerging singularity in the rate-of-creation of the free-surface area, that gives a key to the modelling.

This 'additional' singularity suggests the following physical picture of the process. As the micro-thread evolves towards the breakup, the rate at which the fresh free surface area is created in the standard model tends to infinity and hence it will inevitably become greater than any finite rate at which the newly created free surface acquires its dynamic 'surface' properties, such as the surface tension. The latter is broadly characterized by the inverse surface-tension-relaxation time, $1/\tau$, which is a measurable finite quantity. This suggests that when the rate-of-stretching of the free surface becomes high enough, i.e.,

$$\frac{1}{S^*}\frac{dS^*}{dt^*} \sim \frac{1}{\tau}, \tag{7.38}$$

the liquid-gas interface will no longer be in equilibrium. The surface tension in this interface will start evolving as the interface formation mechanisms, which will be activated to restore the thermodynamic equilibrium, interact with the flow. Then, the capillary pressure, which in the standard model is the only driving force of the breakup, will become dependent not only on the free-surface curvature but also on the *variable* surface tension. As a result, the capillary pressure and hence the pressure in the bulk no longer have to become singular at the pinch-off.

The second dynamic consequence of the increasing rate-of-change of the free-surface area is that, being spatially nonuniform, this process will give rise to the surface tension gradient along the free surface. This gradient and the tangential stress at the free surface it generates will lead to the Marangoni effect driving the liquid away from the neck and becoming a mechanism of the breakup competing with the capillary pressure.

Thus, in order to describe the breakup process we may again use the theory of flows with forming interfaces of Chapter 4. No modifications are required; we only need to use the surface equation of state in its more general form (4.60), instead of its linear approximation (4.61) applied previously. This will allow us to model the evolution of the surface tension at high rates of variation of the free-surface area.

As a preliminary step, it is interesting to examine whether the very idea of considering the pinch-off as an interface formation process can make sense of the surprising experimental results reported by Kowalewski's (1996) who observed that the minimal diameter of the micro-thread, when the breakup is triggered, is independent of the fluid's viscosity.

In terms of the similarity solution (7.29), condition (7.38), which determines the time $t_{\text{new regime}}$ when the mechanisms of interface formation come into

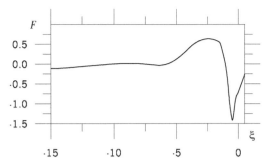

FIGURE 7.9: The dimensionless factor $F(\xi)$ defined by (7.37) in the expression (7.36) for the rate of change of the free-surface area.

play, has the form

$$\frac{\max F(\xi)}{t_0^* - t_{\text{new regime}}^*} \sim \frac{1}{\tau},$$

where $F(\xi)$ is given by (7.37) and plotted in Fig. 7.9. We have already shown that, according to experiments on dynamic wetting, $\tau \propto \mu$ so that one can write this proportionality down as $\tau = \mu\tau_\mu$, where τ_μ is a dimensional factor independent of viscosity. Then, using the expression for h^* from (7.29), we obtain

$$d_{\min} \equiv 2h_{\min} \approx \frac{2\sigma_e}{\mu}(t_0^* - t_{\text{new regime}}^*) \sim 2\sigma_e\tau_\mu \max F(\xi), \qquad (7.39)$$

so that $d_{\min}$ is indeed indepedent of the fluid's viscosity. We can take one step further and actually estimate $d_{\min} \sim 2\sigma_e\tau_\mu$ using the estimates for τ obtained in §5.4 ($\max F(\xi)$ is of order one; see Fig. 7.9). According to these estimates,

$$\tau_\mu \sim 3 \times 10^{-7}\text{--}10^{-6} \text{ s/P} \qquad (7.40)$$

so that, taking $\sigma_e \sim 30\text{--}60$ dyn/cm, as in the experiments, we arrive at

$$d_{\min} \sim 2 \times 10^{-5}\text{--}1.2 \times 10^{-4} \text{ cm}, \qquad (7.41)$$

which is exactly the range reported by Kowalewski (see Table 7.2). This is a remarkable result, especially given the crudeness of assumptions made in §5.4 to estimate τ.

The simple explanation we obtained for the viscosity independence of $d_{\min}$ and the closeness of the estimate (7.41) to the experimental data are encouraging, and in the next section we will consider the features that the model described in Chapter 4 predicts for the breakup phenomenon.

7.5 Axisymmetric capillary pinch-off: singularity-free solution

To describe the breakup process, consider the solution of the axisymmetric Navier-Stokes equations (7.2)–(7.4) subject to the boundary conditions (4.54)–(4.60). We will nondimensionalize these boundary conditions using the same natural scales (7.1) together with a scale for the surface density $\rho_{(0)}^s = a/b$, where a and b are parameters in the surface equation of state (4.60). Below we write down the boundary conditions in a scalar form for the axisymmetric flow and for clarity highlight where they differ from (7.12)–(7.14).

Instead of the kinematic boundary condition (7.12) one now has (4.54) and (4.57), that is

$$u^s - w^s \frac{\partial h}{\partial z} = \frac{\partial h}{\partial t} \qquad \text{at } r = h, \tag{7.42}$$

$$(u - u^s) - \frac{\partial h}{\partial z}(w - w^s) = -Q_1(\rho^s - \rho_G^s)\left[1 + \left(\frac{\partial h}{\partial z}\right)^2\right]^{1/2} \qquad \text{at } r = h, \tag{7.43}$$

where

$$Q_1 = \frac{\mu \rho_{(0)}^s}{\rho \sigma_e \tau} = \frac{\mu a}{\rho \sigma_e \tau b}, \qquad \rho_G^s = \frac{\rho_e^s}{\rho_{(0)}^s} = \rho_e^s a^{-1} b.$$

The tangential and normal-stress boundary conditions (7.13) and (7.14) are now replaced with (4.56) and (4.55), where the surface tension is a variable:

$$2\frac{\partial h}{\partial z}\left(\frac{\partial u}{\partial r} - \frac{\partial w}{\partial z}\right) + \left[1 - \left(\frac{\partial h}{\partial z}\right)^2\right]\left(\frac{\partial w}{\partial r} + \frac{\partial u}{\partial z}\right) = \left[1 + \left(\frac{\partial h}{\partial z}\right)^2\right]^{1/2} \frac{\partial \sigma}{\partial z}, \tag{7.44}$$

$$p - \frac{2}{1 + \left(\frac{\partial h}{\partial z}\right)^2}\left[\frac{\partial u}{\partial r} + \left(\frac{\partial h}{\partial z}\right)^2 \frac{\partial w}{\partial z} - \frac{\partial h}{\partial z}\left(\frac{\partial w}{\partial r} + \frac{\partial u}{\partial z}\right)\right]$$

$$= \frac{\sigma}{h\left[1 + \left(\frac{\partial h}{\partial z}\right)^2\right]^{1/2}} - \frac{\sigma \frac{\partial^2 h}{\partial z^2}}{\left[1 + \left(\frac{\partial h}{\partial z}\right)^2\right]^{3/2}}. \tag{7.45}$$

The surface equation of state (4.60) in the dimensionless variables takes the form

$$\sigma = \bar{a}\rho^s(1 - \rho^s), \tag{7.46}$$

where $\bar{a} = 1/(\rho_G^s(1 - \rho_G^s))$.

Finally, equations (4.58) and (4.59) become

$$\frac{\partial \rho^s}{\partial t} + \left[1 + \left(\frac{\partial h}{\partial z}\right)^2\right]^{-1} \left[\frac{\partial h}{\partial z}\frac{\partial (\rho^s u^s)}{\partial z} + \frac{\partial (\rho^s w^s)}{\partial z}\right] + \frac{\rho^s u^s}{h} = -Q_2(\rho^s - \rho_G^s),$$

$$(7.47)$$

$$\frac{\partial \sigma}{\partial z} = 4\bar{\beta}\left[\frac{\partial h}{\partial z}(u^s - u) + (w^s - w)\right] \qquad \text{at } r = h, \qquad (7.48)$$

where

$$Q_2 = \frac{\mu^3}{\rho\tau\sigma_e^2}, \qquad \bar{\beta} = \frac{\beta\mu}{\rho\sigma_e(1 + 4\alpha\beta)}.$$

Thus, to describe the whole process, from the conventional evolution of the thread to the breakup, one has to find the solution of (7.2)–(7.4) subject to the boundary conditions (7.42)–(7.48) together with initial conditions and some conditions in the far field specifying a particular flow.

Clearly, the above formulation includes the conventional one: when the rate-of-change of the surface area is low, for the dimensionless surface tension one has $\sigma \approx 1$. Then, the process is driven solely by the capillary pressure which for a long-wave disturbance of the free surface will be higher in the minimal cross-section and increase as the radius of the cross-section decreases. This self-accelerating mechanism inherent in the standard formulation would lead to the singularities in the solution we discussed in §7.2.2. However, as the rate-of-change of the surface area increases, the surface tension starts to deviate from its equilibrium value causing the surface tension gradient and the associated Marangoni effect. Ultimately, the surface tension in the minimal cross-section will be lower than σ_e and the process will be driven by a combined action of the (finite) capillary pressure and the surface tension gradient with their relative roles varying as the breakup is approached.

We will consider the final stage of the process in the framework of the slender-jets approximation which in the present context should be seen as a mathematically transparent 'test case'.

7.5.1 Slender-jet approximation

Now one has to repeat the slender-jet approximation derivation and incorporate additional boundary conditions on the free surface that involve the variable surface tension. As before, we will use (7.8) to rescale the variables and expand them in asymptotic series in ϵ as $\epsilon \to 0$. In addition, the following expansions for the surface quantities will be required:

$$\sigma = \bar{\sigma} = \sigma_0 + \epsilon^2\sigma_1 + \dots, \qquad \rho^s = \bar{\rho}^s = \rho_0^s + \epsilon^2\rho_1^s + \dots,$$

$$u^s = \bar{u}^s = u_0^s + \epsilon^2 u_1^s + \dots, \qquad w^s = \epsilon^{-1}\bar{w}^s = \epsilon^{-1}(w_0^s + \epsilon^2 w_1^s + \dots), \qquad \text{as } \epsilon \to 0.$$

$$(7.49)$$

The derivation of §7.2.1 up to and including (7.18) remains intact. Then, equation (7.44), which replaced (7.6) of the standard model, will take the form

$$2\epsilon^2 \frac{\partial \bar{h}}{\partial \bar{z}} \left(\frac{\partial \bar{u}}{\partial \bar{r}} - \frac{\partial \bar{w}}{\partial \bar{z}} \right) + \left[1 - \epsilon^2 \left(\frac{\partial \bar{h}}{\partial \bar{z}} \right)^2 \right] \left(\frac{\partial \bar{w}}{\partial \bar{r}} + \epsilon^2 \frac{\partial \bar{u}}{\partial \bar{z}} \right)$$

$$= \epsilon^2 \left[1 + \epsilon^2 \left(\frac{\partial \bar{h}}{\partial \bar{r}} \right)^2 \right]^{1/2} \frac{\partial \bar{\sigma}}{\partial \bar{z}} \quad \text{at } \bar{r} = \bar{h},$$

or, using that $w_0 = w_0(\bar{z}, \bar{t})$, to leading order one has

$$2 \frac{\partial h_0}{\partial \bar{z}} \left(\frac{\partial u_0}{\partial \bar{r}} - \frac{\partial w_0}{\partial \bar{z}} \right) + \frac{\partial w_1}{\partial \bar{r}} + \frac{\partial u_0}{\partial \bar{z}} = \frac{\partial \sigma_0}{\partial \bar{z}} \quad \text{at } \bar{r} = h_0,$$

instead of (7.19). Using that, to leading order in ϵ as $\epsilon \to 0$, the condition that σ is independent of the coordinate normal to the interface is given by $\partial \sigma_0 / \partial \bar{r} = 0$, we arrive at the following expression for the separation function

$$C_1(\bar{z}, \bar{t}) = \frac{\partial^2 w_0}{\partial \bar{z}^2} + \frac{6}{h_0} \frac{\partial h_0}{\partial \bar{z}} \frac{\partial w_0}{\partial \bar{z}} + \frac{2}{h_0} \frac{\partial \sigma_0}{\partial \bar{z}}, \tag{7.50}$$

instead of (7.20).

Equation (7.45), which replaced (7.7), to leading order yields

$$p_0(\bar{z}, \bar{t}) = \frac{\sigma_0}{h_0} - \frac{\partial w_0}{\partial \bar{z}}, \tag{7.51}$$

instead of (7.21).

Now, after substituting (7.50), (7.51) into the first equation (7.17), we arrive at

$$\frac{\partial w_0}{\partial \bar{t}} + w_0 \frac{\partial w_0}{\partial \bar{z}} = \frac{1}{h_0^2} \frac{\partial}{\partial \bar{z}} (h_0 \sigma_0) + \frac{3}{h_0^2} \frac{\partial}{\partial \bar{z}} \left(h_0^2 \frac{\partial w_0}{\partial \bar{z}} \right) \tag{7.52}$$

instead of (7.22). It is important to note that the only difference between (7.52) and (7.22) is in the first term on the right-hand side: now $\sigma_0 \neq 1$ and it appears under the sign of differentiation. This term is the driving-force of the process where now the gradients of h_0 and σ_0 can compete or act together depending on the circumstances.

The first of the two kinematic boundary conditions (7.42) and (7.43) remains unchanged in the slender-jet approximation,

$$u_0^s - w_0^s \frac{\partial h_0}{\partial \bar{z}} = \frac{\partial h_0}{\partial \bar{t}} \quad \text{at } \bar{r} = h_0, \tag{7.53}$$

whilst the second can be slightly simplified and, after using (7.8), (7.49) and (7.16), to leading order one arrives at

$$\frac{h_0}{2} \frac{\partial w_0}{\partial \bar{z}} + u_0^s + \frac{\partial h_0}{\partial \bar{z}} (w_0 - w_0^s) = Q_1(\rho^s - \rho_G^s). \tag{7.54}$$

After using (7.8) and (7.49), the surface mass balance equation (7.47) becomes

$$\frac{\partial \bar{\rho}^s}{\partial \bar{t}} + \left[1 + \epsilon^2 \left(\frac{\partial \bar{h}}{\partial \bar{z}}\right)^2\right]^{-1} \left[\epsilon^2 \frac{\partial \bar{h}}{\partial \bar{z}} \frac{\partial (\bar{\rho}^s \bar{u}^s)}{\partial \bar{z}} + \frac{\partial (\bar{\rho}^s \bar{w}^s)}{\partial \bar{z}}\right]$$

$$+ \frac{\bar{\rho}^s \bar{u}^s}{\bar{h}} = -\epsilon^2 Q_2 (\rho^s - \rho_G^s). \tag{7.55}$$

To leading order as $\epsilon \to 0$, the right-hand side of equation (7.55) as well as the second term on its left-hand side can be neglected, and, using (7.53) to express u_0^s via w_0^s and h_0, we can write (7.55) down as

$$\frac{\partial}{\partial \bar{t}} (h_0 \rho_0^s) + \frac{\partial}{\partial \bar{z}} (h_0 \rho_0^s w_0^s) = 0. \tag{7.56}$$

Finally, expansions (7.49) turn equation (7.48) into

$$\frac{\partial \bar{\sigma}}{\partial \bar{z}} = 4 \bar{\beta} \left[\epsilon^2 \frac{\partial \bar{h}}{\partial \bar{z}} (\bar{u}^s - \bar{u}) + (\bar{w}^s - \bar{w})\right],$$

or, to leading order,

$$\frac{\partial \sigma_0}{\partial \bar{z}} = 4 \bar{\beta} (w_0^s - w_0). \tag{7.57}$$

The surface equation of state (7.46) to leading order remains unchanged:

$$\sigma_0 = \bar{a} \rho_0^s (1 - \rho_0^s). \tag{7.58}$$

Thus, the slender-jet approximation for our model to leading order consists of 6 equations (7.52)–(7.54), (7.56)–(7.58) for 6 unknowns w_0, h_0, σ_0, u_0^s, w_0^s and ρ_0^s. It is noteworthy that in (7.55) the small parameter appears in front of the source term on the right-hand side so that, to leading order, the mass exchange between the surface and the bulk phase may be neglected and the surface mass balance equation takes the form of a conservation law (7.56). At the same time, the term describing this mass exchange is still present in the boundary condition for the normal component of the bulk velocity (7.54), and this fact, together with the finiteness of $\bar{\beta}$ in (7.57), shows that, generally, in the slender-jet approximation both components of the surface velocity, u_0^s and w_0^s, differ from the corresponding components of the bulk velocity, u_0 and w_0, evaluated at the free surface. To make further simplifications we need to consider estimates for Q_1 and $\bar{\beta}$.

7.5.2   Further simplifications: estimates for Q_1 and $\bar{\beta}$

An estimate for Q_1 can be obtained in a straightforward way. Assuming that $\rho_{(0)}^s \sim \rho \ell$, where, as in Chapter 4, ℓ is the characteristic thickness of the interfacial layer, and using the representation $\tau = \mu \tau_\mu$, one has

$$Q_1 \equiv \frac{\mu \rho_{(0)}^s}{\rho \sigma_e \tau} \sim \frac{\ell}{\sigma_e \tau_\mu}. \tag{7.59}$$

Then, for $\ell \sim$ 1–3 nm, $\sigma_e \sim$ 30–60 dyn/cm and τ_μ given by (7.40) we arrive at

$$Q_1 \sim 2 \times 10^{-3}\text{–}3 \times 10^{-2}. \tag{7.60}$$

Given this estimate, an asymptotic limit $Q_1 \to 0$ becomes a meaningful one, and to leading order in this limit we can neglect the right-hand side of (7.54). Thus, in the process of breakup the normal to the free surface component of the surface and bulk velocities are, to leading order, equal irrespective of the liquid's velocity.

Now, we can turn to $\bar{\beta}$. The physical meaning of the coefficients α and β following from the derivation of the model (Chapter 4) together with the analysis of experiments on dynamic wetting (Chapter 5) suggest that $\beta \propto \mu/\ell$, $\alpha \propto \ell/\mu$. Assuming for the estimates that $\beta \approx \mu/\ell$, $\alpha \approx \ell/\mu$ and using, as before, $\ell \sim$ 1–3 nm, we find that

$$\bar{\beta} \approx \frac{\mu^2}{5\rho\sigma_e\ell} \sim \mu^2 \times 10^4\text{–}\mu^2 \times 10^5 \text{ P}^{-2}.$$

Then, for low-viscosity liquids, like water, one has $\bar{\beta} \sim$ 1–10 and hence, according to (7.57), the difference between the axial component of the surface velocity, w_0^s, and the corresponding component of the bulk velocity, w_0, can be significant.[5] At the same time, given that $\bar{\beta} \propto \mu^2$, even for moderately viscous liquids one has $\bar{\beta} \gg 1$ and hence can consider equations (7.52)–(7.54), (7.56)–(7.58) in the asymptotic limit $\bar{\beta} \to \infty$. Then, it follows from (7.57) that in this limit to leading order the axial components of the surface and bulk velocities are equal, $w_0^s = w_0$. (It should be noted that, from a physical viewpoint, as $\rho^s \to 0$ both α and β can no longer be regarded as constants, so that, strictly speaking, the assumption that $\bar{\beta} \gg 1$ will eventually become inapplicable. However, here we are interested in the qualitative mechanism of the breakup, and it is convenient to use this simplifying assumption instead of going into the details of rheology of a rarefied interface, which is a separate and rather complicated topic.)

Thus, in the limit $Q_1 \to 0$, $\bar{\beta} \to \infty$ the surface velocity is equal to the bulk velocity evaluated at the interface so that: (a) the two kinematic conditions (7.53) and (7.54) now collapse into the standard one,

$$\frac{\partial h_0}{\partial \bar{t}} + w_0 \frac{\partial h_0}{\partial \bar{z}} + \frac{h_0}{2} \frac{\partial w_0}{\partial \bar{z}} = 0, \tag{7.61}$$

and (b) the surface density will now be carried with w_0 and hence, using (7.61), one can write down equation (7.56) in the form

$$\frac{\partial \rho_0^s}{\partial \bar{t}} + w_0 \frac{\partial \rho_0^s}{\partial \bar{z}} + \frac{\rho_0^s}{2} \frac{\partial w_0}{\partial \bar{z}} = 0. \tag{7.62}$$

---

[5]Here we must remember that for low-viscosity liquids the very use of the slender-jet approximation becomes questionable.

Thus, the evolutions of h_0 and ρ_0^s are governed by identical equations (7.61) and (7.62). Initial conditions for h_0 and ρ_0^s in a general case can of course be different.

7.5.3 Flow regimes

7.5.3.1 Pre-breakup evolution

For medium to high-viscosity liquids in the slender-jet approximation we have 4 unknowns w_0, h_0, σ_0 and ρ_0^s satisfying 3 differential equations (7.52), (7.61) and (7.62) together with the algebraic equation of state (7.58) relating σ_0 and ρ_0^s:

$$\frac{\partial w_0}{\partial \bar{t}} + w_0 \frac{\partial w_0}{\partial \bar{z}} = \frac{1}{h_0^2} \frac{\partial}{\partial \bar{z}} (h_0 \sigma_0) + \frac{3}{h_0^2} \frac{\partial}{\partial \bar{z}} \left(h_0^2 \frac{\partial w_0}{\partial \bar{z}} \right), \tag{7.63}$$

$$\frac{\partial h_0}{\partial \bar{t}} + w_0 \frac{\partial h_0}{\partial \bar{z}} + \frac{h_0}{2} \frac{\partial w_0}{\partial \bar{z}} = 0, \tag{7.64}$$

$$\frac{\partial \rho_0^s}{\partial \bar{t}} + w_0 \frac{\partial \rho_0^s}{\partial \bar{z}} + \frac{\rho_0^s}{2} \frac{\partial w_0}{\partial \bar{z}} = 0, \tag{7.65}$$

$$\sigma_0 = \bar{a} \rho_0^s (1 - \rho_0^s). \tag{7.66}$$

Equations (7.65), (7.64) and with the help of the latter equation (7.63) can be written down as conservation laws so that (7.63)–(7.66) take the form

$$\frac{\partial (h_0^2 w_0)}{\partial \bar{t}} + \frac{\partial (h_0^2 w_0^2)}{\partial \bar{z}} = \frac{\partial (h_0 \sigma_0)}{\partial \bar{z}} + 3 \frac{\partial}{\partial \bar{z}} \left(h_0^2 \frac{\partial w_0}{\partial \bar{z}} \right), \tag{7.67}$$

$$\frac{\partial h_0^2}{\partial \bar{t}} + \frac{\partial (h_0^2 w_0)}{\partial \bar{z}} = 0, \tag{7.68}$$

$$\frac{\partial (\rho_0^s)^2}{\partial \bar{t}} + \frac{\partial \left((\rho_0^s)^2 w_0 \right)}{\partial \bar{z}} = 0, \tag{7.69}$$

$$\sigma_0 = \bar{a} \rho_0^s (1 - \rho_0^s) \tag{7.70}$$

convenient for numerical integration.

Before the flow enters the slender-jet regime described by these equations, one has $w = O(1)$ and the surface-tension-relaxation mechanism makes $\rho^s = \rho_G^s$ and hence $\sigma = 1$. Then, as the thread gets thinner, the scales gradually change to (7.8), (7.49) where $w = O(\epsilon^{-1})$ and the relaxation term with Q_2 vanishes so that the surface density and tension become variables whose evolution is coupled with the evolution of the flow. The transition to the slender-jet regime is, of course, asymptotic, and the initial conditions for (7.67)–(7.70) come from matching with the outer solution, that is depend on the evolution of the thread before it enters the regime described by the slender-jet approximation. Therefore, in a general case the initial profile of the thread is asymmetric in the z-direction, w_0 is nonzero and ρ_0^s (and hence σ_0) is z-dependent. Here,

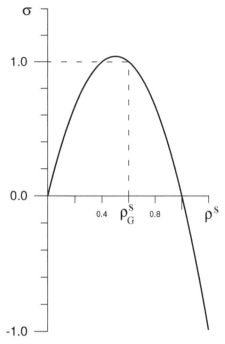

FIGURE 7.10: Dependence of the surface tension on the surface density according to the surface equation of state (4.60); the surface tension and surface density are scaled with σ_e and $\rho^s_{(0)} = a/b$, respectively. As the rate at which the fresh free-surface area is created becomes higher than the rate of the surface tension relaxation back to equilibrium, the surface density will deviate from ρ^s_G. For a spatially nonuniform process this will give rise to variation in the surface-tension gradient and the associated Marangoni effect.

our goal is to illustrate how the physical mechanism of the breakup as an interface formation process works and for this purpose we will consider a model case of a liquid cylinder subject to an initial symmetric disturbance given by

$$h_0(\bar{z}, 0) = 1 - A \exp(-B\bar{z}^2) \tag{7.71}$$

with initial distributions

$$\rho^s_0(\bar{z}, 0) \equiv \rho^s_G, \qquad w_0(\bar{z}, 0) \equiv 0 \tag{7.72}$$

and boundary conditions for w_0 given by

$$w_0(0, \bar{t}) = 0, \qquad \frac{\partial w_0}{\partial \bar{z}} \to 0 \quad \text{as } \bar{z} \to \infty. \tag{7.73}$$

For the equilibrium surface density we will use $\rho^s_G = 0.6$ (Fig. 7.10), which is consistent with the estimates obtained by analysing experiments on dynamic

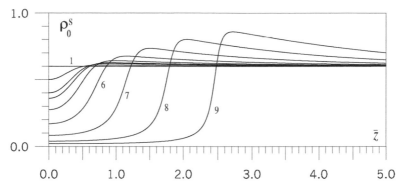

FIGURE 7.11: Profiles of the surface density at different times (Shikhmurzaev 2005b). Curves 1–9 correspond to $\bar{t} = 0$, 5.90, 8.37, 9, 10, 11, 12, 13, 14, respectively.

wetting of Chapter 5. It should be emphasized here that the illustration we are going to consider deals with the late stage of thinning and the breakup itself. In other words, we consider the stage where the rate-of-change of the free surface area is already so high that the surface tension relaxation mechanisms have no time to restore the equilibrium state of the interface and hence the mass exchange between the surface phase and the bulk is negligible both in the surface continuity equation and in the boundary condition for the normal component of the bulk velocity. In a general case, one has to compute the whole process, from the thinning using the full interface-formation model to the final stage we are examining here.

The results of numerical integration of equations (7.67)–(7.70) subject to initial conditions (7.71), (7.72) and boundary conditions (7.73) are shown in Figs 7.11–7.16 for $A = 0.1$, $B = 10$ (Shikhmurzaev 2005). It is instructive to begin their description from the profiles of ρ_0^s and σ_0 at different times shown in Figs 7.11 and 7.12, that is from the point of view of the driving mechanism. The corresponding profiles of the free-surface shape are given in Fig. 7.13. Qualitatively, the process looks as follows.

As the thread gets thinner in the minimal cross-section (the 'neck') due to the action of capillary pressure, the rate at which the surface area is created increases and the surface density in the neck starts to decrease (curve 2, Fig. 7.11) since the time scale we are dealing with is too short for the relaxation mechanism to restore the equilibrium. Then, according to the surface equation of state (7.70) sketched in Fig. 7.10, the surface tension in the neck increases slightly compared to its equilibrium value (curve 2, Fig. 7.12). As a result, on the one hand, the capillary pressure in the neck becomes slightly higher than what one would have in the standard model ($\sigma_0 \equiv 1$), but on the other hand the appearing surface-tension gradient is directed, and hence generates the flow, *towards* the neck, thus hindering the (increased) capillary pressure in squeezing the liquid out of the neck. Curves 2 in Figs 7.11 and 7.12 correspond

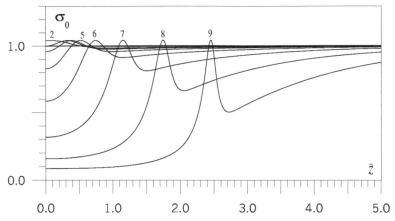

FIGURE 7.12: Profiles of the surface tension for the same times as in Fig. 7.11. The surface tension in front of the propagating wave is lower than in equilibrium as a result of the surface density being higher than ρ_G^s.

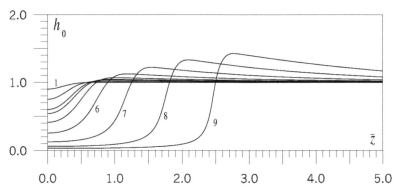

FIGURE 7.13: Profiles of the thread's radius for the same times as in Fig. 7.11. Curves 8 and 9 show the emerging structure consisting of (i) a departing blob, which will be macroscopically seen as a recoiling tip of the micro-thread, and (ii) a vanishing 'residual' thread.

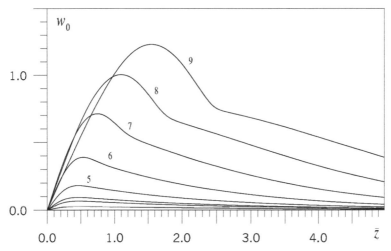

FIGURE 7.14: Distributions of the axial velocity for the same times as in Fig. 7.11. Curves 8 and 9 exhibit a different trend compared to 1–7 due to the onset of formation of the structure comprising the departing macroscopic blob and the vanishing residual thread (see curves 8 and 9 in Fig. 7.13).

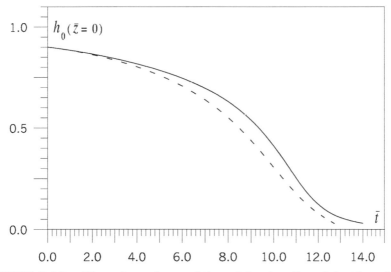

FIGURE 7.15: Time-dependence of the minimal radius of the thread. The solid line correspond to the interface-formation model and the dashed line to the standard one ($\sigma_0 \equiv 1$).

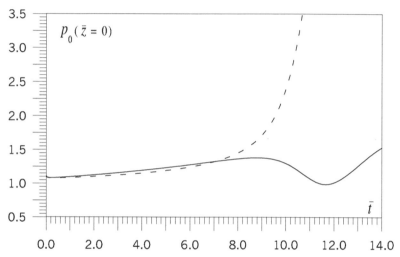

FIGURE 7.16:   Time-dependence of the pressure in the neck. The solid line corresponds to the interface-formation model, where the pressure is calculated using (7.51), and the dashed line is for the standard model with p_0 obtained from (7.21). Initially, the two models give close results as the process is driven by the capillary pressure in the neck. Then, the variation of the surface tension in the interface-formation model starts to play a role, reducing the capillary pressure and, instead, removing the fluid from the neck via the Marangoni effect. The limiting value of the pressure is given by the self-similar solution of §7.5.3 and it is finite.

to the moment when σ_0 reaches its maximum value in the neck ($\rho_0^s = 0.5$, see Fig. 7.10). After that, further decrease in ρ_0^s will make σ_0 in the neck go down, the profile of the surface tension becomes nonmonotone and the surface-tension gradient starts to *support* the capillary pressure in removing liquid from the minimal cross-section. Curves 3 in Figs 7.11 and 7.12 correspond to the moment when the surface tension in the neck reaches the equilibrium value once again, this time at a lower surface density ($\rho_0^s = 0.4$, see Fig. 7.10). Now, further decrease in ρ_0^s (curves 4–9, Fig. 7.11) drives σ_0 below its equilibrium value (curves 4–9, Fig. 7.12) thus reducing the capillary pressure compared to the standard model, whereas the role of the surface tension gradient increases.

As we can see, the capillary pressure and the Marangoni effect due to the flow-induced surface-tension gradient act as two balancing factors of the process: the capillary-pressure-driven flow ultimately reduces the surface tension in the neck and hence creates the surface tension gradient which, on the one hand, supports the capillary pressure in removing the fluid from the neck but, on the other hand, leads to further reduction of the surface tension in the neck and hence reduces the capillary pressure that gave rise to it.

An interesting feature of the flow is that, since for the surface density in our case one has a conservation law (7.69), a decrease in ρ_0^s in the neck is accompanied by its *increase* further down the thread (Fig. 7.11). According to the surface equation of state (7.70), this leads to a decrease in the surface tension so that one ends up with a propagating 'wave' in the surface tension distribution (Fig. 7.12). Obviously, this 'wave' will disappear, physically on the time scale τ, once the relaxation mechanism is taken into account.

The profiles of h_0 and the longitudinal velocity w_0 as functions of the distance $\bar{z}$ from the pinch-off point for the same times as the distributions of ρ_0^s and σ_0 in Figs 7.11 and 7.12 are given in Figs 7.13 and 7.14. They have a number of interesting features. In Fig. 7.13 one can see that, as the thread gets thinner in the neck, it develops a distinct structure comprising the following two main elements: (i) a receding blob, which will be macroscopically seen as a recoiling tip of the micro-thread, and (ii) an elongating and gradually vanishing 'residual' thread.

The question to be addressed now is whether the emerging structure that one can clearly see in curves 8 and 9 of Fig. 7.13 corresponds to a transition to a new regime of flow.

The first indication that this is indeed the case comes from Fig. 7.14, where curves 8 and 9 also exhibit a different trend compared with 1–7: $\partial w_0 / \partial \bar{z}$ at the origin starts to decrease. If we look at the evolution of the minimal radius of the thread as a function of time (Fig. 7.15), the indications pointing to a transition become clearer. One can see that the solid line representing $h_0(0, \bar{t})$ for the interface-formation model behaves differently from the dashed one corresponding to the standard one ($\sigma_0 \equiv 1$) as the minimal radius decreases.

Finally, to make it completely obvious that there is indeed a transition to another regime of flow, we consider the temporal evolution of the pressure in the neck, $p_0(0, \bar{t})$. This is given in Fig. 7.16, where the solid line corresponds to the interface formation model with the pressure calculated using (7.51) and the dashed line describes the pressure of the standard model obtained from (7.21). One can see now that in the standard model the pressure shoots to infinity without exhibiting any features which might suggest a different flow regime taking over from the one considered in §7.2.2, whereas the situation with the interface-formation model is qualitatively different. The interplay between the capillary pressure with varying surface tension and the Marangoni effect described earlier in this section results in the pressure growing very slowly and then starting to decrease. At a certain moment, $p_0(0, \tilde{t})$ reaches its minimum, which can be regarded as the onset of formation of the residual thread, and then starts to increase again. In order to determine its limit and the evolution of parameters associated with the residual thread we have to consider the asymptotics of (7.67)–(7.70) as the radius of the residual thread tends to zero. It is important to note here that the residual thread emerges as a result of the evolution of *any* initial disturbance so that one would expect that this thread corresponds to a self-similar regime of flow.

7.5.3.2   Self-similar regime

In order to study the flow in the residual thread we introduce the following scales and asymptotic expansions:

$$\tilde{t} = \delta^2 \bar{t}, \qquad \tilde{z} = \delta^2 \bar{z}, \qquad h_0 = \delta(h_{00} + \delta h_{01} + \dots), \qquad (7.74)$$

$$w_0 = w_{00} + \delta w_{01} + \dots, \qquad \rho_0^s = \delta(\rho_{00}^s + \delta \rho_{01}^s + \dots), \qquad \sigma_0 = \delta(\sigma_{00} + \delta \sigma_{01} + \dots),$$

as $\delta \to 0$. Here the dimensionless characteristic thickness of the residual thread is used for a small parameter δ. Then the scales for ρ_0^s and consequently σ_0 must also be of $O(\delta)$ whereas the corresponding length in the z-direction for the cylindrical geometry of the thread has the order of δ^{-2}. By balancing the orders of the terms on the left-hand side of (7.67) with the first term on the right-hand side one has δ^{-2} and 1 as the scales for time and w_0, respectively. Then, from (7.67)–(7.70) for leading order terms h_{00}, w_{00}, ρ_{00}^s and σ_{00} we arrive at

$$\frac{\partial(\rho_{00}^s)^2}{\partial \tilde{t}} + \frac{\partial\left((\rho_{00}^s)^2 w_{00}\right)}{\partial \tilde{z}} = 0, \qquad (7.75)$$

$$\frac{\partial h_{00}^2}{\partial \tilde{t}} + \frac{\partial(h_{00}^2 w_{00})}{\partial \tilde{z}} = 0, \qquad (7.76)$$

$$\frac{\partial(h_{00}^2 w_{00})}{\partial \tilde{t}} + \frac{\partial(h_{00}^2 w_{00}^2)}{\partial \tilde{z}} = \frac{\partial(h_{00}\sigma_{00})}{\partial \tilde{z}}, \qquad (7.77)$$

$$\sigma_{00} = \bar{a} \rho_{00}^s. \qquad (7.78)$$

Now we need only the first boundary condition (7.73), which remains unchanged:

$$w_{00}(0, \tilde{t}) = 0. \qquad (7.79)$$

We will look for a self-similar solution of (7.77)–(7.79) and the scales (7.74) together with (7.78) suggest that it must have the form

$$w_{00} = W(\zeta), \quad h_{00} = \tilde{t}^{-1/2} H(\zeta), \quad \sigma_{00} = \bar{a}\rho_{00}^s = \bar{a}\tilde{t}^{-1/2} R(\zeta), \quad (\zeta = \tilde{z}/\tilde{t}). \tag{7.80}$$

Then, equation (7.78) is satisfied automatically and (7.75)–(7.77), (7.79) take the form

$$[(W - \zeta)R^2]' = 0, \tag{7.81}$$

$$[(W - \zeta)H^2]' = 0, \tag{7.82}$$

$$[(W - \zeta)WH^2]' = \bar{a}(RH)', \tag{7.83}$$

$$W(0) = 0, \tag{7.84}$$

where the prime denotes differentiation with respect to ζ. Equations (7.82) and (7.81) together with the boundary condition (7.84) immediately give $W(\zeta) = \zeta$. Then, it follows from (7.83) that $RH = \text{const}$. This is the only constraint imposed on R and H by (7.82)–(7.83) which means that the self-similar regime preserves the profiles of ρ_{00}^s and h_{00} with which the flow enters it. Given the results of the present section, we have that both R and H are constant, $R(\zeta) \equiv R_0$, $H(\zeta) \equiv H_0$. This is what one would expect since the evolution of both ρ_0^s and h_0 is described by the same equation and the velocity distribution tends to level out the profile of the initial disturbance. Thus, as the radius of the residual thread tends to zero, we have:

$$h_0 = \frac{H_0}{\sqrt{\tilde{t}}}, \quad w_0 = \frac{\tilde{z}}{\tilde{t}}, \quad \sigma_0 = \bar{a}\rho_0^s = \frac{\bar{a}R_0}{\sqrt{\tilde{t}}} \quad \text{as } \tilde{t} \to \infty, \tag{7.85}$$

where constants R_0 and H_0 (as well as the time shift corresponding to the onset of the self-similar regime) are determined by the previous evolution of the flow. It should be emphasized that the large time scale specified by (7.74) takes place *within* the short-time limit specified by (7.8), i.e., $\epsilon\delta^{-3} \to 0$ as $\epsilon \to 0$, $\delta \to 0$, so that, physically, the whole process of breakup occurs on a time scale short compared with the surface-tension-relaxation time τ. It should be also noted that the limit $\epsilon\delta^{-3} \to 0$ implies that, in terms of the dimensional length scales (7.1), the length of the residual thread is infinitesimal.

Using (7.51) and (7.85) we can now find the pressure in the residual thread as the thread's diameter tends to zero:

$$p_0(\tilde{z}, \tilde{t}) = \frac{\sigma_0}{h_0} - \frac{\partial w_0}{\partial \tilde{z}} = \frac{\bar{a}R_0}{H_0} - \frac{1}{\tilde{t}} \leq \frac{\bar{a}R_0}{H_0}. \tag{7.86}$$

Thus, unlike the standard model, in the interface-formation theory pressure remains finite throughout the breakup.

Another important feature of the interface formation theory in the application to the jet breakup problem is that it predicts a finite recoil velocity. Indeed, the velocity with which the visible macroscopic 'blob' recoils in Fig. 7.13

is actually the elongation rate of the residual thread. Given that the similarity variable is $\zeta = \bar{z}/\bar{t}$, we have that for the length of the residual thread $l_z \propto \bar{t}$, so that $dl_z/d\bar{t} = O(1)$. On the contrary, in the standard model the similarity variable is given by (7.31) so that for the retraction velocity one has (7.33), that is $v_{\text{recoil}} \propto \hat{t}^{-1/2}$ as $\hat{t} \to 0$, thus leading to a retraction velocity infinite at breakup.

It is noteworthy that the self-similar solution (7.85) makes the term $\partial(h\sigma)/\partial z$ in the equation of motion identically zero. Before the flow entered the self-similar regime, this term provided the driving mechanism, and hence, in the self-similar stage when it vanishes, the evolution of the residual thread is driven entirely by the departing ends of the main thread. In other words, there is no influence of the self-similarly evolving residual thread on the ends of the main thread and hence there is no influence of these ends on each other. In effect, this is what 'breakup' means in dynamic terms.

It should be emphasized here that when a continuum mechanics description produces a residual thread with a diameter tending to zero, this *per se* does not mean disappearance of this thread in physical terms: physically, the thread reaches a molecular scale and hence becomes 'invisible' to continuum mechanics as a bulk phase. The thread's physical rupture on this scale is, from the viewpoint of continuum mechanics, an *assumption*. An alternative way of describing this thread formally as a 'line phase' makes little sense since this phase has no dynamic properties that one would need to take into account.

Since the thinning of the residual thread described above in the framework of continuum mechanics leads to no singularities in the solution, in practice, as one computes the global flow numerically, the residual thread can be neglected once its thickness becomes smaller than the spatial resolution of the numerical code. Unlike the "cut-off" used in the singular solutions, this will not lead to a singularity and hence an unphysical impact on the post-breakup evolution of the thread's ends. In the next section, we will see what this difference means from a molecular viewpoint.

7.6 Pinch-off from a molecular viewpoint

As was shown in the previous section, the interface formation theory makes it possible to describe the breakup of a liquid thread without singularities in the distributions of the flow parameters, predicts finiteness of the retraction velocity of the recoiling filaments after the thread is broken and explains the viscosity-independence of the minimal diameter of the thread observed in experiments.

In should be also noted that, if viewed in macroscopic terms, the mechanism of the final (molecular) stage of the breakup process is in fact a natural

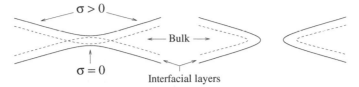

FIGURE 7.17: Schematic representation of the breakup on the molecular scale. In the place where the opposite interfacial layers begin to overlap (left), the surface tension vanishes whereas elsewhere it remains finite. The resulting Marangoni effect pulls the thread apart.

continuation of the macroscopic mechanism we have considered. Indeed, once the radius of the residual thread becomes comparable with the thickness of the interfacial layers, the surface tension becomes dependent on, and reduces with, the radius of curvature (Tolman 1949, Koenig 1950). Then, as the opposite interfacial layers begin to overlap, they cease to be 'interfaces' since they no longer separate two 'bulk' phases (Fig. 7.17). As the asymmetry in the intermolecular forces acting from the two bulk phases disappears, so does the surface tension given that it is this asymmetry that is the physical reason for its existence. Consequently, as the opposite interfacial layers merge, the surface tension and hence the capillary pressure in the place where the interfaces are merging both vanish, whereas elsewhere, i.e., away from the region of merger, the interfaces are still separated by the 'bulk' and hence the surface tension remains finite. As a result of this situation, there have to be surface-tension gradients directed away from the place where the interfacial layers are merging. These gradients will literally pull the liquid thread apart. Thus, pinch-off at a microscopic level, which we describe here in *macroscopic* terms, appears to be due entirely to the Marangoni effect and not to capillary pressure.

It should be emphasized that in the continuum modelling all molecular length scales are zero (thermodynamic limit) so that the breakup on the molecular scale takes place when, from the viewpoint of continuum mechanics, the process is already over and the neck is reduced to a point. Hence the macroscopic continuum-mechanical theory has to hand the matters over to a molecular-dynamic one with *finite* values of macroscopic parameters (pressure, velocity). This ensures meaningful values of the underlying microscopic ones whose subsequent evolution (on the molecular scale) will then be described by a molecular dynamics theory. The interface formation theory allows for this 'takeover' since its (macroscopic) parameters remain finite at all time.

As mentioned in §7.2.3, numerical simulations often use a so-called 'cut-off' diameter to stop the computed (macroscopic) parameters from blowing up. As all lengths in continuum mechanics, this diameter is, by definition, macroscopic, not molecular, and it becomes an artificial parameter that takes away the model's predictive power. Sometimes, it is suggested that this 'cut-off' di-

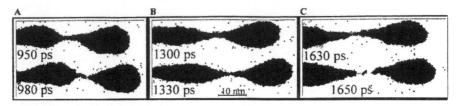

FIGURE 7.18: The final stage of rupture of a nanojet (Moseler & Landmann 2000; © 2000 AAAS; reproduced with permission).

ameter should be on the molecular length scale (Eggers 1997). This approach would introduce a molecular length scale into the continuum mechanics modelling thus going against the very fundamentals of the field (thermodynamic limit, continuum approximation, etc.).

An insight into the process on the molecular scale and how it should be modelled macroscopically could be offered by molecular dynamics simulations. Compared with the case of dynamic wetting, molecular dynamics simulations of the breakup phenomenon are, on the one hand, relatively straightforward since one does not have to deal with the issues associated with the modelling of the solid substrate and an extremely large number of molecules to produce a fully representative 'bulk' phase. On the other hand, however, it is extremely difficult to extract meaninful macroscopic parameters since the breakup process is intrinsically time dependent and leaves no room for long averaging intervals. Nevertheless, the results obtained by now are helpful, at least, as a qualitative indicator of the main effects on the molecular level.

Koplik & Banavar (1993) carried out a detailed study of the breakup and found that, for systems of about 20,000 Lennard-Jones molecules, "the Eulerian velocity and stress fields are essentially unmeasurable — dominated by thermal noise" (p. 521). This disappointing outcome is useful in one respect: it completely rules out the suggestion that the macroscopic flow is (nearly) singular when a 'molecular cut-off' takes place. The simulations only indicate that, just before the breakup, "the fluid is moving away from the point where the rupture occurs" (ibid, p. 527). An interesting observation is also that

> The rupture process itself seems to consist of a gradual withdrawal of molecules from a thinning neck region, leaving a thin string of mutually attracted molecules in a background of vapor or a second fluid. This thin string is somewhat persistent on the molecular time scale, although of course quite evanescent on laboratory time scale (Koplik & Banavar 1993, p. 535).

This feature is qualitatively reminiscent of the vanishing 'residual' thread described in the previous section.

Some information about the pinch-off on the molecular level can be derived from the molecular dynamics simulations of nanojets. Moseler & Landman (2000) observed that for the simulated nanojet of propane molecules

The neck radius decreases steadily to its final value of one or two molecular diameters, and a smooth separation of a few remaining molecules in a double-cone apex completes the pinch-off process (Moseler & Landman 2000, p. 1167).

In other words, again there is no indication of singularly high velocities in the final stage of the breakup that would follow from the standard model (§7.2.2). The simulated profiles are shown in Fig. 7.18. The authors note that their MD simulations are indeed "in obvious disagreement" with the results produced by the standard model and propose to remedy the situation by adding Gaussian white-noise forcing to the right-hand side of equation (7.22).[6] Although even the initial diameter of a nanojet in the simulations is comparable with the thickness of the interfacial layer and in the process of pinching off it goes down to the scale of a couple of molecular diameters, it is assumed in the continuum modelling that the surface tension remains constant throughout the process. (We will discuss this issue in §7.7.2.) Setting aside the shortcomings of the proposed theory, one may conclude that the MD-simulated behaviour of nanojets is consistent with the scenario we outlined earlier.

7.7 Rupture of films

The rupture of a liquid film is both similar to and different from the jet breakup process. Both these phenomena belong to the same general class of flows characterized by disintegration of a liquid volume and hence should be describable by a mathematical model incorporating broadly the same physics. However, there is also an essential difference. A cylindrical jet is linearly unstable with respect to long-wave disturbances so that, once perturbed, it is driven towards the breakup by the capillary pressure due to the cross-sectional curvature being higher than the longitudinal one. In the case of a film, the capillary pressure at the point where the film's thickness has its minimum is always a restoring force so that there is no intermediate stage of capillarity-driven thinning of the film that would eventually trigger some breakup mechanism. In other words, it is the external disturbances coupled with the flow conditions that have to trigger this mechanism. Below, we discuss a simple situation where a possible macroscopic mechanism of rupture have been examined. Then, we will consider the modelling difficulties arising

---

[6]Following Eggers (1993), the authors included the term with the longitudinal curvature on the right-hand side of this equation, which, as we have shown earlier, is asymptotically incompatible with the slender-jet approximation.

when the film's thickness becomes comparable with the length scale of van der Waals forces.

7.7.1 Macroscopic mechanism

A purely macroscopic mechanism similar to that described in §7.5 was proposed in Shikhmurzaev (2005c). Its essence is that, if a film is sufficiently thin, then its perturbation by external disturbances will lead to the rate of creation of a fresh free-surface area becoming comparable with the inverse surface-tension-relaxation time thus driving the surface tension out of equilibrium. Then, the Marangoni effect will act as a film rupture mechanism. Unlike the jet breakup problem, this mechanism is not supported by the capillary pressure so that, if it is eliminated by the surface-tension-relaxation process that acts to restore the equilibrium surface tension, the capillary pressure, then becoming the only force, will bring the free surface shape back to equilibrium. In other words, for the film to rupture, it must be sufficiently thin and the external disturbance sufficiently strong so that the film disintegrates before the interface formation process restores the thermodynamic equilibrium of the interface.

We will illustrate this mechanism of rupture by considering a free film in the situation where it is easy to highlight the similarities and differences with the jet breakup process. In a general case, one has a three-dimensional problem (4.53), (4.54)–(4.60), i.e., the Navier-Stokes equations

$$\nabla \cdot \mathbf{u} = 0, \quad \rho \left(\frac{\partial \mathbf{u}}{\partial t} + \mathbf{u} \cdot \nabla \mathbf{u} \right) = -\nabla p + \mu \nabla^2 \mathbf{u} \tag{7.87}$$

together with the boundary conditions

$$\frac{\partial f}{\partial t} + \mathbf{v}^s \cdot \nabla f = 0, \tag{7.88}$$

$$-p + \mu \mathbf{n} \cdot [\nabla \mathbf{u} + (\nabla \mathbf{u})^*] \cdot \mathbf{n} = \sigma \nabla \cdot \mathbf{n}, \tag{7.89}$$

$$\mu \mathbf{n} \cdot [\nabla \mathbf{u} + (\nabla \mathbf{u})^*] \cdot (\mathbf{I} - \mathbf{n}\mathbf{n}) + \nabla \sigma = 0, \tag{7.90}$$

$$\rho(\mathbf{u} - \mathbf{v}^s) \cdot \mathbf{n} = \frac{\rho^s - \rho_e^s}{\tau}, \quad \frac{\partial \rho^s}{\partial t} + \nabla \cdot (\rho^s \mathbf{v}^s) = -\frac{\rho^s - \rho_e^s}{\tau}, \tag{7.91}$$

$$(1 + 4\alpha\beta)\nabla \sigma = 4\beta(\mathbf{v}^s - \mathbf{u}) \cdot (\mathbf{I} - \mathbf{n}\mathbf{n}), \quad \sigma = a\rho^s - b(\rho^s)^2, \tag{7.92}$$

on an a priori unknown free surface $f(\mathbf{r}, t) = 0$ with the inward normal $\mathbf{n} = \nabla f / |\nabla f|$, and some conditions in the far field. This problem can be considerably simplified in the thin-film approximation with the ratio of characteristic length scales in the directions normal and tangential to the film ϵ as a small parameter. As a further simplification, we will consider this approximation for a plane two-dimensional flow and the perturbation symmetric with respect to the centerline, i.e., for the film's shape given by $y = \pm h(x, t)$ in a suitably chosen Cartesian coordinate frame. Using u, u^s, v, v^s to denote the x and y components of the bulk and surface velocities, respectively, the same natural scales (7.1) to nondimensionalise the variables and the same asymptotic expansions as in §7.5, i.e.,

$$x = L\epsilon\bar{x}, \quad y = L\epsilon^2\bar{y}, \quad t = t_\mu\epsilon^2\bar{t}, \quad hL^{-1} = \sum_{n=0}^{\infty} \epsilon^{2(n+1)}h_n, \qquad (7.93)$$

$$(u, u^s)U^{-1} = \sum_{n=0}^{\infty} \epsilon^{2n-1}(u_n, u_n^s), \quad (v, v^s)U^{-1} = \sum_{n=0}^{\infty} \epsilon^{2n}(v_n, v_n^s),$$

$$pP^{-1} = \sum_{n=0}^{\infty} \epsilon^{2(n-1)}p_n, \quad (\sigma\sigma_e^{-1}, \rho^s ba^{-1}) = \sum_{n=0}^{\infty} \epsilon^{2n}(\sigma_n, \rho_n^s), \qquad \text{as } \epsilon \to 0,$$

after the standard asymptotic analysis to leading order one arrives at the following set of equations:

$$\frac{\partial u_0}{\partial \bar{t}} + u_0\frac{\partial u_0}{\partial \bar{x}} = \frac{1}{h_0}\frac{\partial \sigma_0}{\partial \bar{x}} + \frac{4}{h_0}\frac{\partial}{\partial \bar{x}}\left(h_0\frac{\partial u_0}{\partial \bar{x}}\right), \quad v_0 = -\bar{y}\frac{\partial u_0(\bar{x}, \bar{t})}{\partial \bar{x}}, \qquad (7.94)$$

$$\frac{\partial h_0}{\partial \bar{t}} + u_0^s\frac{\partial h_0}{\partial \bar{x}} = v_0^s, \quad \frac{\partial \rho_0^s}{\partial \bar{t}} + \frac{\partial(\rho_0^s u_0^s)}{\partial \bar{x}} = 0, \qquad (7.95)$$

$$\frac{\partial \sigma_0}{\partial \bar{x}} = 4\bar{\beta}(u_0^s - u_0), \quad (u_0 - u_0^s)\frac{\partial h_0}{\partial \bar{x}} - (v_0 - v_0^s) = \bar{\chi}(\rho_0^s - \bar{\rho}_e^s), \qquad (7.96)$$

$$\sigma_0 = a\rho_0^s - b(\rho_0^s)^2. \qquad (7.97)$$

Here $\bar{\beta} = \beta\mu\rho^{-1}\sigma^{-1}(1 + 4\alpha\beta)^{-1}$, $\bar{\chi} = a\mu(b\rho\sigma_e\tau)^{-1}$ and $\bar{\rho}_e^s = \rho_e^s ba^{-1}$. In obtaining (7.94)–(7.97) it was assumed that all nondimensional parameters appearing after the use of scaling (7.93) are of $O(1)$ as $\epsilon \to 0$. In particular, this means that $\epsilon^2 t_\mu\tau^{-1} \to 0$ as $\epsilon \to 0$ so that, physically, the process described by (7.94)–(7.97) takes place on a time scale small compared with τ and, according to the second equation (7.95), the relaxation mechanisms have no time to restore the equilibrium surface density (and hence the equilibrium surface tension).

As in §7.5, equations (7.94)–(7.97) can be simplified further for medium to high-viscosity fluids by using the estimates for material constants of the model obtained from experiments on dynamic wetting (Chapter 5). These estimates show that for such fluids $\bar{\chi} \ll 1$ and $\bar{\beta} \gg 1$ and hence, to leading order in these parameters, it follows from (7.96) that one can neglect the difference between (u_0^s, v_0^s) and (u_0, v_0). Then, after simple algebra we arrive at an initial-value problem for the following set of nonlinear equations:

$$\frac{\partial u}{\partial t} + u\frac{\partial u}{\partial x} = \frac{1}{h}\frac{\partial \sigma}{\partial x} + \frac{4}{h}\frac{\partial}{\partial x}\left(h\frac{\partial u}{\partial x}\right), \qquad (7.98)$$

$$\frac{\partial F}{\partial t} + \frac{\partial(Fu)}{\partial x} = 0, \quad F = (h, \rho^s); \quad \sigma = a\rho^s - b(\rho^s)^2. \qquad (7.99)$$

(Hereafter for brevity we drop the overbar and the subscript 0.)

In (7.98), the surface tension gradient on the right-hand side appears as the only driving force, unlike the case of the breakup of a cylindrical jet (§7.5), where the Marangoni effect and the capillary pressure due to the cross-sectional curvature play comparable roles and support each other as the breakup mechanisms.

As already mentioned, the scaling (7.93) implies, in particular, that equations (7.98), (7.99) operate on a time scale small compared with τ, and, with no relaxation in (7.99), $u \equiv \hat{u} = $ const, $h \equiv \hat{h} = $ const, $\rho^s \equiv \hat{\rho}^s = $ const is a solution of (7.98), (7.99) for arbitrary values of $\hat{u}$, $\hat{h}$ and $\hat{\rho}^s$. A linear stability analysis of this solution leads to a dispersion relationship of the form

$$\omega = k\hat{u} - 2ik^2 \left[1 \pm \left(1 + \frac{\hat{\rho}^s \lambda}{4k^2 \hat{h}} \right)^{1/2} \right], \qquad \text{where } \lambda = \frac{d\sigma}{d\rho^s}(\hat{\rho}^s), \qquad (7.100)$$

and ω and k are the angular frequency and wavenumber, respectively. Hence the solution is stable for $\lambda < 0$, which for the surface equation of state given by the second equation (7.99) corresponds to $\hat{\rho}^s > a/(2b)$ and unstable otherwise. In other words, it is stable if the rarefaction of the surface phase increases the surface tension, which then contracts the surface and replenishes ρ^s. On the other hand, if the surface tension decreases with decrease in ρ^s, then a local rarefaction of the surface phase due to an external disturbance leads to a local reduction in the surface tension whose gradient then acts to pull the film apart and reduce ρ^s even further. The dispersion relationship (7.100) also indicates that, unlike Rayleigh's instability of a cylindrical jet, the disturbances become more destabilizing as the wavelength becomes shorter. Therefore, to find the most destabilizing mode one has to abandon the thin-film (i.e., long-wavelength) approximation and consider a two-dimensional stability problem.

In order to illustrate how the Marangoni effect incorporated in (7.98)–(7.99) leads to the rupture of a free film in the nonlinear regime, consider the evolution of a film after an external finite-amplitude disturbance makes $\rho^s \equiv \hat{\rho}^s$, where $\hat{\rho}^s$ is in the unstable zone ($\lambda \geq 0$). It is instructive to look at the borderline case where $\lambda = 0$, i.e., $\hat{\rho}^s = a/(2b)$.

Consider a small disturbance $\rho^s = a/(2b)[1 - A \exp(-x^2/l^2)]$, where A is the relative amplitude and l is the width of the disturbance. Fig. 7.19 shows the film's profile at various times obtained via numerical integration of (7.98) and (7.99) for $A = 0.1$ and $l = 0.5$. As one can see, the initial disturbance of the surface tension indeed leads to the local thinning of the film. As the minimum thickness decreases, the profile develops qualitatively the same distinct structure as in the jet breakup phenomenon, namely a combination of (a) a gradually vanishing 'residual' film and (b) the main film whose rim moves away at a constant speed. (In reality, the transition zone between the two elements is not as steep as shown in Fig. 7.19 given that the scales in the x and y directions are different.) The distributions of the surface parameters, the x-component of the fluid's velocity and the pressure in the minimum cross-

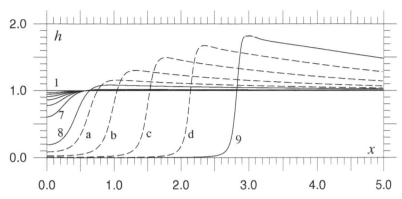

FIGURE 7.19: Profiles of the free surface at different moments in time (Shikhmurzaev 2005c). Solid lines 1–9 are obtained for $t = 0$, 10, 15, 20, 25, 30, 35, 40 and 45, respectively. Dashed lines a–d correspond to $t = 41$, 42, 43 and 44.

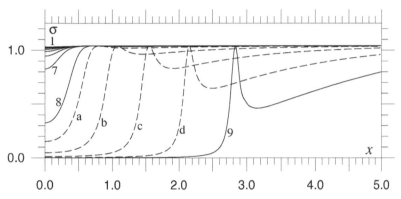

FIGURE 7.20: Profiles of the surface tension for the same moments as in Fig. 7.19.

section are also qualitatively the same as the corresponding distributions in the jet-breakup problem (see Figs 7.20, 7.22–7.24).

An initial disturbance of the surface tension creates a surface-tension gradient that starts to pull the fluid out of the cross-section with the minimum surface tension due to the Marangoni effect. The resulting creation of a fresh free-surface area accelerates as the film gets thinner, and, given that the surface density is in the unstable zone, this leads to a further decrease in the surface tension in the place of the initial disturbance and hence an increase in the surface-tension gradient. The increased surface-tension gradient leads to a stronger Marangoni effect so that the process of thinning of the film becomes self-perpetuating (curves 1–8 in Fig. 7.19).

The above process has an obvious limit: no positive rate-of-creation of a

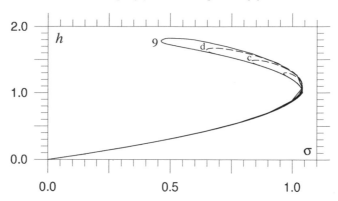

FIGURE 7.21: Correlation between h and σ in the process of rupture following from Fig. 7.19 and Fig. 7.20.

fresh free-surface area can make the surface tension negative, and, as the surface tension approaches zero (curves a–b in Fig. 7.20), the reduction of the surface tension in the minimum cross-section slows down. Then, the surface tension gradient that has its maximum along the transition zone between the residual and the main film (see Figs 7.19, 7.20) becomes a force acting on the lip of the main film that will drive it away from the place of the initial disturbance (curves c, d, 9 in Figs 7.19 and 7.20).

After eliminating x from the dependencies $h = h(x, t)$ and $\sigma = \sigma(x, t)$ shown in Figs 7.19 and 7.20 one can see the intrinsic link between h and σ developing as t increases in the form of a limit cycle (Fig. 7.21). The low branch ($\sigma < 1$) in Fig. 7.21 corresponds to the transition zone and the residual film.

An interesting feature of the plots in Fig. 7.20 corresponding to the upper branch ($\sigma > 1$) in Fig. 7.21 is a propagating and deepening local minimum in the surface tension distribution ahead of the recoiling lip. The origin of this minimum is different from the lowering of the surface tension in the residual film. As illustrated in Fig. 7.22, the latter is a consequence of the surface phase rarefaction caused by the stretching of the film in the place of the initial disturbance whereas the former comes from the surface phase becoming less rarefied as a result of the nonuniformity of the longitudinal velocity distribution (Fig. 7.23): the film is at rest in the far field whilst its lip is recoiling.

The distributions of the longitudinal velocity at different times (Fig. 7.23) show the same change in the trend as in Fig. 7.14, and, as the two-element structure forms, the slope of the velocity profile becomes less steep at the origin whereas the maximum velocity continues to increase.

Given that in the thin-film approximation $p = -2\partial u/\partial x$ and, as t increases, the pressure in the minimum cross-section goes through a global minimum (Fig. 7.24). This minimum can be used as a formal indicator of a transition to a new dynamic regime. Since the (almost constant) thickness of the macrofilm

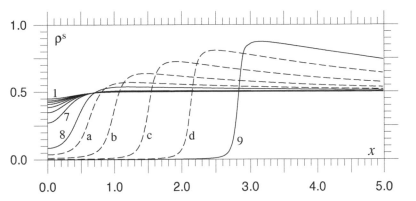

FIGURE 7.22: Distributions of the surface density behind the surface tension profiles given in Fig. 7.20.

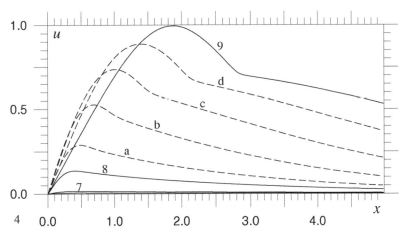

FIGURE 7.23: Distributions of the longitudinal velocity corresponding to the profiles given in Fig. 7.19.

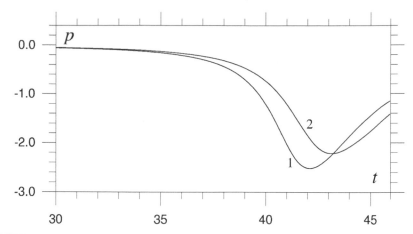

FIGURE 7.24:   The pressure in the minimum cross-section vs time for $l = 0.5$ (curve 1) and $l = 0.7$ (curve 2).

and the (infinitesimal) thickness of the residual film are separated in scale, to describe the late stage of the evolution of the residual film one can look for a self-similar solution. This solution has the form

$$u = x(t - t_0)^{-1}, \quad (h, \rho^s, \sigma) = (H, R, aR)(t - t_0)^{-1},$$

where t_0, H and R are constants determined by the preceding evolution of the film. In this solution, both terms on the right-hand side of (7.98) become separately zero thus indicating that the process is driven by the departing ends of the macroscopic film whereas the residual film simply follows and gets thinner accordingly. This means that, from a dynamic viewpoint, the lips of the macroscopic film behave independently so that, macroscopically, the film can be regarded as disintegrated. However, from a physical point of view, the residual film simply reaches a molecular scale and its subsequent evolution has to be considered. What is important to us from the viewpoint of macroscopic modelling is that at this stage the residual film is associated with the vanishing surface tension and regular values of all other parameters. Then, since the solution does not develop singularities as the breakup is approached, the subsequent evolution of the residual film can be handed over to a model describing the flow on a scale comparable with the range of intermolecular forces. We will discuss the issues arising in developing such a model in the next section. In practice, however, in many cases for computing the global flow numerically, the residual film can be neglected once its thickness becomes smaller than the spatial resolution of the code. Then, the surface of the lip of the macroscopic film will acquire its equilibrium surface tension on a time scale of order τ and the lip will recoil as a result of the action of the capillary pressure, i.e., driven from the end, as is the case with the recoil of a broken cylindrical thread.

7.7.2   Mesoscopic films

In this section, we consider the issues arising when the film's thickness becomes comparable with the range of intermolecular forces and discuss the modeling implications for the film rupture problem.

7.7.2.1   Fluid mechanics and intermolecular forces

When the thickness of a film becomes comparable with what we generally referred to as 'molecular scales', the film becomes 'microscopic' from the viewpoint of classical fluid mechanics in a sense that it has to be modelled as a 'surface phase'. In order to continue treating such a film as a 'bulk' phase, we need to revisit the fundamental assumptions made in Chapter 2 where we introduced the concepts and formulated the equations of fluid mechanics.

The continuum approximation, as the 0th-order approximation in the limit (2.2), i.e.,

$$\frac{\ell}{L} \to 0, \quad \frac{t_{\mathrm{mol}}}{t_{\mathrm{macro}}} \to 0 \qquad (\ell = \max(\ell_1, \ell_2)), \tag{7.101}$$

leads to a number of important simplifications. Firstly, in this limit a discrete system of molecules can be represented in terms of smooth functions describing its averaged macroscopic properties. Secondly, intermolecular forces and momentum fluxes due to thermal molecular motion become reduced to distributed surface forces (stresses) that, in simple fluids, are considered to be dependent locally on other macroscopic characteristics. Thirdly, interfacial layers between 'bulk' phases have to be modelled as 'interfaces' of zero thickness ('surface phases') possessing their own dynamic and thermodynamic properties. Finally, if we need to consider, for example, the free energy of the fluid, then, within this approximation, it can be treated as an additive quantity.

As the length scale L on which macroscopic properties are described becomes smaller, we have to make a step from the 0th-order approximation in ℓ/L, $t_{\mathrm{mol}}/t_{\mathrm{macro}}$ towards accounting for the finiteness of molecular scales. For condensed matter, one has $\ell_2 \gg \ell_1$, i.e., the range of intermolecular forces ℓ_2 is much greater than intermolecular distances ℓ_1, so that we can relax the 0th-order approximation by taking into account the finiteness of the range of long-range intermolecular forces ($\ell_2 \le L$) whilst still treating the matter as a continuum ($\ell_1 \ll L$). This range of length scales is often referred to as 'mesoscopic'.

From a methodological viewpoint, the consequences of this step are as follows. The physical mechanisms responsible for long-range intermolecular forces must now be taken out of the stress tensor and explicitly represented macroscopically in terms of field variables and the corresponding equations relating them. These mechanisms must be also taken out of the variables representing properties of interfaces as geometric surfaces so that these interfaces now become 'diffuse' with the mechanisms generating long-range intermolecular forces determining their structure. The field equations for intermolecular forces will also imply, in particular, nonadditivity of the medium's free energy. A classical example of implementation of this scheme (on a macroscopic scale) is electrohydrodynamics, where electrostatic forces are taken out of the stress tensor with their 'generator', the electric field, described macroscopically by Maxwell's equations. Below, we briefly review some results obtained on this way for electrically neutral fluids in the context of the ultrathin film modelling.

The physical nature and form of long-range intermolecular forces ('action at a distance') remained elusive for a long time.[7] The quantum origin of these forces was uncovered by London (1930), who applied the second-order perturbation technique to the electrostatic interaction between two oscillating dipoles. He found that the force between the dipoles results from the matching of their phases with the energy of interaction being inversely proportional to the sixth power of the separation distance R. This result holds for the separation distances much less than the length λ of the wave generated by the dipole. For $R \geq \lambda$ one has to take into account the effect of 'retardation', i.e., the finiteness of time needed for a signal transmitted at the speed of light to travel the distance between the dipoles and the phase lag that occurs during this time.

This effect has been calculated by Casimir & Polder (1946, 1948), who used the fourth-order perturbation technique. One of the consequences of Casimir & Polder's analysis is that, in the limiting case $R \gg \lambda$, the interaction potential appears to be proportional to R^{-7}. Experiments performed by Derjaguin and Abrikosova (Derjaguin & Abrikosova 1951, 1957, Abrikosova & Derjaguin 1953, 1958) and later measurements reported by Kitchener & Prosser (1957) and Black et al. (1960) broadly confirmed Casimir & Polder's theoretical results. A much improved experimental setup used by Tabor and co-workers (Tabor & Winterton 1968, 1969, Israelachvili & Tabor 1972) allowed them not only to widen the range of distances at which intermolecular forces could be studied but also to show that the transition between the normal and retarded force occurs as the separation distance increases from about 100 Å to about 200 Å. This is consistent with the signal propagating at the speed of light.

The problem arises when one tries to represent the long-range forces macroscopically. Direct integration of intermolecular interactions using either London's potential (Kallmann & Willstaetter 1932, Bradley 1932, Hamaker 1937) or Casimir & Polder's results implies that pairs of molecules interact independently. However, in a condensed matter the electronic envelope of every molecule and hence its properties relevant to intermolecular interactions are significantly influenced by the neighboring molecules. A way forward based on calculating collective effects encounters formidable mathematical difficulties

---

[7] A comprehensive account of the history of research into the forces of cohesion can be found in Rowlinson (2002).

which, in the case of liquids, are compounded by the necessity to calculate also the molecular distribution functions.

An alternative and purely macroscopic approach has been developed by Dzyaloshinskii, Lifshitz and Pitaevskii (Lifshitz 1954, 1956, Dzyaloshinskii & Pitaevskii 1959, Dzyaloshinskii, Lifshitz & Pitaevskii 1960, see Dzyaloshinkii, Lifshitz & Pitaevskii 1961 for a review). The idea is to consider long-wavelength ($\lambda \gg \ell_1$) fluctuations of the electromagnetic field in a condensed medium caused by thermal fluctuations and specify their contribution to macroscopic thermodynamic properties in terms of the complex dielectric permittivity ε of the medium. Using the methods of quantum field theory, Dzyaloshinskii & Pitaevskii (1959) developed a general framework for calculating the parameters of fluctuating electromagnetic field in the state of thermodynamic equilibrium and their contribution to the stress tensor and the free energy in an inhomogeneous condensed medium. It has been shown that the temperature Green's function for the photon $\mathbf{G}(\mathbf{r}, \mathbf{r}'; \omega_n)$ satisfies the equation

$$\left[\varepsilon(\mathbf{r}; i\omega_n)\omega_n^2 - \nabla^2\right]\mathbf{G}(\mathbf{r}, \mathbf{r}'; \omega_n) + \nabla[\nabla \cdot \mathbf{G}(\mathbf{r}, \mathbf{r}'; \omega_n)] = -4\pi\delta(\mathbf{r} - \mathbf{r}')\mathbf{I}, \quad (7.102)$$

where the dielectric permittivity for an imaginary frequency $\varepsilon(i\omega_n)$ $(\omega_n > 0)$ is real and related to the imaginary part $\mathrm{Im}[\varepsilon]$ of the dielectric permittivity for real frequencies by (Landau & Lifshitz 1960, §58)

$$\varepsilon(i\omega_n) = 1 + \frac{2}{\pi}\int\limits_0^\infty \frac{\omega\,\mathrm{Im}[\varepsilon(\omega)]}{\omega^2 + \omega_n^2}\,d\omega.$$

The function $\mathbf{G}$ is analogous to the vector potential in the classical case; the derivative in (7.102) is taken with respect to $\mathbf{r}$ whereas $\mathbf{r}'$ is a parameter.

In order to calculate the contribution to the stress tensor from the van der Waals forces we need to introduce two functions $\mathbf{G}^{\mathrm{E}}$ and $\mathbf{G}^{\mathrm{H}}$ by

$$G_{ik}^{\mathrm{E}}(\mathbf{r}, \mathbf{r}'; \omega_n) = -\omega_n^2 G_{ik}(\mathbf{r}, \mathbf{r}'; \omega_n), \quad (7.103)$$

$$G_{ik}^{\mathrm{H}}(\mathbf{r}, \mathbf{r}'; \omega_n) = \epsilon_{ilp}\epsilon_{kmq}\nabla^l\nabla'^m G^{pq}(\mathbf{r}, \mathbf{r}'; \omega_n), \quad (7.104)$$

where ϵ_{ijk} are components of the Levi-Civita pseudo-tensor (see Appendix A). Since $\mathbf{G}^{\mathrm{E}}$ and $\mathbf{G}^{\mathrm{H}}$ are obviously divergent at $\mathbf{r}' = \mathbf{r}$, one has to subtract the contribution corresponding to a homogeneous space, i.e., to consider, in addition to (7.102), an equation for the Green's function $\overline{\mathbf{G}}(\mathbf{r} - \mathbf{r}', \omega_n; \mathbf{r}_0)$ for an unbounded homogeneous medium with the dielectric permittivity equal to that of the inhomogeneous medium under consideration at any given point $\mathbf{r}_0$:

$$\left[\varepsilon(\mathbf{r}_0; i\omega_n)\omega_n^2 - \nabla^2\right]\overline{\mathbf{G}}(\mathbf{r} - \mathbf{r}', \omega_n; \mathbf{r}_0) + \nabla[\nabla \cdot \overline{\mathbf{G}}(\mathbf{r} - \mathbf{r}', \omega_n; \mathbf{r}_0)] = -4\pi\delta(\mathbf{r} - \mathbf{r}')\mathbf{I}.$$

Then, following Dzyaloshinskii & Pitaevskii, one can introduce $\overline{\mathbf{G}}^{\mathrm{E}}$ and $\overline{\mathbf{G}}^{\mathrm{H}}$ as in (7.103) and (7.104), define the regular at $\mathbf{r}' = \mathbf{r}$ functions

$$\mathbf{D}^{\mathrm{E}}(\mathbf{r}, \mathbf{r}; \omega_n) = \lim_{\mathbf{r}' \to \mathbf{r}}[\mathbf{G}^{\mathrm{E}}(\mathbf{r}, \mathbf{r}'; \omega_n) - \overline{\mathbf{G}}^{\mathrm{E}}(\mathbf{r} - \mathbf{r}', \omega_n; \mathbf{r})], \quad (7.105)$$

$$\mathbf{D}^{\mathrm{H}}(\mathbf{r}, \mathbf{r}; \omega_n) = \lim_{\mathbf{r}' \to \mathbf{r}} [\mathbf{G}^{\mathrm{H}}(\mathbf{r}, \mathbf{r}'; \omega_n) - \overline{\mathbf{G}}^{\mathrm{H}}(\mathbf{r} - \mathbf{r}', \omega_n; \mathbf{r})] \qquad (7.106)$$

and show that the van der Waals part of the stress tensor is given by

$$\mathbf{P}_{\mathrm{vdW}} = \frac{\hbar T}{2\pi} \sum_{n=0}^{\infty} {}' \Big[\frac{1}{2}\Big(\varepsilon - \rho\frac{\partial \varepsilon}{\partial \rho}\Big) \mathbf{I} : \mathbf{D}^{\mathrm{E}}(\mathbf{r}, \mathbf{r}; \omega_n)\, \mathbf{I} - \varepsilon \mathbf{D}^{\mathrm{E}}(\mathbf{r}, \mathbf{r}; \omega_n)$$

$$+ \tfrac{1}{2}\mathbf{I} : \mathbf{D}^{\mathrm{H}}(\mathbf{r}, \mathbf{r}; \omega_n)\, \mathbf{I} - \mathbf{D}^{\mathrm{H}}(\mathbf{r}, \mathbf{r}; \omega_n) \Big], \qquad (7.107)$$

where $\omega_n = 2\pi n T/\hbar$, $\hbar$ is Planck's constant, and the prime on the summation sign indicates that term with $n = 0$ is taken with a factor $\frac{1}{2}$.

Thus, in equilibrium for a particular configuration of condensed media (solid bodies, bodies separated by a film, etc.) it is necessary to find $\mathbf{G}$ that satisfies (7.102) together with the boundary conditions of continuity of the tangential (with respect to the first index) components of $\mathbf{G}^{\mathrm{E}}$ and $\mathbf{G}^{\mathrm{H}}$, which correspond to the continuity of the tangential components of the electric and magnetic fields. Then, after using (7.105) and (7.106) to regularize $\mathbf{G}^{\mathrm{E}}$ and $\mathbf{G}^{\mathrm{H}}$, one can calculate the van der Waals part of the stress tensor from (7.107). As shown by Dzyaloshinskii et al. (1961, p. 196–198), in the appropriate limits the macroscopic theory allows one to recover the results of London (1930) and Casimir & Polder (1948) mentioned earlier. Importantly, the macroscopic approach does not imply additivity of intermolecular interactions but at a price of having to specify the macroscopic function ε.

The general framework allowed Dzyaloshinskii et al. (1960) to consider a model problem of the equilibrium of a liquid film of mesoscopic thickness h between a flat solid surface and a vacuum. After reducing the problem to the one considered earlier by Lifshitz (1956), who calculated the van der Waals forces between two solid bodies in a vacuum, they found that the force of attraction acting on the free surface per unit area depends on the film's thickness and is given by

$$F(h) = \frac{\hbar}{2\pi^2 c^3} \int\limits_0^{\infty} \int\limits_1^{\infty} p^2 \xi^3 \varepsilon^{3/2} \left\{ \left[\frac{(s_1 + p)(s_2 + p)}{(s_1 - p)(s_2 - p)} \exp(2hp\xi\varepsilon^{1/2}c^{-1}) - 1 \right]^{-1} \right.$$

$$+ \left. \left[\frac{(s_1 + p\varepsilon_1/\varepsilon)(s_2 + p/\varepsilon)}{(s_1 - p\varepsilon_1/\varepsilon)(s_2 - p/\varepsilon)} \exp(2hp\xi\varepsilon^{1/2}c^{-1}) - 1 \right]^{-1} \right\} dp\, d\xi, \qquad (7.108)$$

where

$$s_1 = (\varepsilon_1/\varepsilon - 1 + p^2)^{1/2}, \quad s_2 = (1/\varepsilon - 1 + p^2)^{1/2},$$

the dielectric permittivities of the solid ε_1 and the liquid ε are functions of imaginary frequency $\omega = i\xi$, and c is the speed of light. For the values of h small compared with the wave lengths that are important in the absorption spectrum of the two media, it follows from (7.108) that $F \propto h^{-3}$, whereas for large h one has $F \propto h^{-4}$.

Taking into account that in equilibrium the chemical potential is constant, Dzyaloshinskii et al. (1960) identify its part dependant on the fluctuating electromagnetic field with $F(h)$. Then, after formally defining the surface tension $\sigma(h)$ for the whole film in terms of the chemical potential (as in the adsorption theory), one has that

$$h = - \left(\frac{\partial \sigma}{\partial F} \right)_T,$$

and, after integrating this equation, arrives at the following expression for the surface tension

$$\sigma(h) = \int_h^\infty h \frac{dF}{dh} \, dh + \sigma_{1e} + \sigma_e. \tag{7.109}$$

Here the last two terms specify the constant of integration from the condition that as $h \to \infty$ the formally (macroscopically) defined surface tension of the film will tend to the sum of the surface tension of the liquid-solid boundary σ_1 and the equilibrium surface tension on the free surface σ_e. Since $dF/dh < 0$, one has that the sum of the surface tensions on the two interfaces decreases as the film becomes thinner. It should be noted, of course, that both $F(h)$ and $\sigma(h)$ are divergent as $h \to 0$ thus indicating a limit of applicability of (7.108), (7.109). It is also necessary to point out that, as experiments show (Derjaguin, Churaev & Muller 1987, ch. 7), the distribution of the fluid's density in the direction normal to the solid surface is nonuniform and, in quantifying the dependence of the surface tension on the film's thickness, this effect should be taken into account.

7.7.2.2   Implications for modelling

An important general fact illustrated by the results we have briefly reviewed is that, once the two interfaces start 'feeling' each other's presence via fluctuating electromagnetic field, i.e., via van der Waals forces, the (macroscopically defined) surface tensions of these interfaces start deviating from their equilibrium values. The words 'macroscopically defined' are important since (7.109) interprets the analysis of *mesoscopic* effects in *macroscopic* terms using a macroscopic picture of the film as a body of fluid confined by interfaces with all interfacial effects attributed to the surface tensions. It should also be noted that σ_1 and σ_e in (7.109) include contributions from (7.107), and it is only the interactions between interfaces that are taken out and represented explicitly. Once the interfaces are considered on a mesoscopic scale, where the long-range intermolecular forces appear explicitly and $\mathbf{P}_{\mathrm{vdW}}$ given by (7.107) is nonzero near each interface, the surface tensions that remain attributed to these interfaces are no longer the ones considered in macroscopic fluid mechanics. In other words, if one takes some intermolecular forces out of the stress tensor and represents them explicitly, then these forces should also be taken out of the surface tensions, thus, in the case of a film, making the latter, in general, dependant on the free-surface profile.

Another consequence of taking long-range intermolecular forces out of the stress tensor is that, strictly speaking, the constitutive equations for the 'remaining' stress tensor will not be the same as for the macroscopic one. For the model to be self-consistent it is also necessary now to take into account thermal fluctuations of the thermodynamic parameters of the fluid and represent them as field variables, as in the description of van der Waals interactions outlined earlier.

The general framework developed by Dzyaloshinskii, Lifshitz and Pitaevskii made a connection between the van der Waals forces on the molecular level (London 1930, Casimir & Polder 1946, 1948) and their continuum description on the mesoscopic scale. It also gives an idea of the level of difficulties that stand on the way of a systematic and self-consistent extension of classical fluid mechanics to the mesoscopic scale.

Subsequent research into mesoscopic films focussed primarily on their stability and the evolution of disturbances under different conditions. It has been shown that the destabilizing role of long-range intermolecular forces can be qualitatively understood by considering the free energy of the film postulated in such a way that its dependence on the film's thickness broadly reflects the known properties of the potential of intermolecular interactions (Vrij 1966; see Sheludko 1967 for a review of related works). Then, small-amplitude disturbances of the film's thickness with wavelengths larger than a certain critical wavelength can lead to a decrease of the free energy of the film and hence grow, so that the film appears to be linearly unstable. The mechanism is similar to the phase separation in a single phase fluid by spinodal decomposition (Cahn 1965).

A step from the static analysis of the energy balance towards the full dynamic description of the film's evolution is nontrivial, even if one adopts a simplified phenomenological treatment of the process. An approach that gained popularity, especially in the engineering community, is to 'augment' the Navier-Stokes equations with the gradient of some potential energy which, after integration across the film, results in a force per unit area inversely proportional to h^3, as it follows from (7.108) in the limit of small h or from the direct integration of London's (1930) potential (Ruckenstein & Jain 1974, Gumerman & Homsy 1975, Williams & Davis 1982, Sharma & Ruckenstein 1986, Kheshgi & Scriven 1991, Erneux & Davis 1993, Chou & Kwong 2007). This phenomenological approach is obviously ad hoc as it relies on the quasi one-dimensional geometry of a film, where 'integration across' is well defined, and additional simplifications can be used by employing the lubrication approximation that makes it sufficient to have only the cross-sectional integral of the force. For a two- or three-dimensional problem one will have to go back to the general framework described above. The relative simplicity of the 'augmented' Navier-Stokes equations in the lubrication approximation makes it possible to capture the main mechanism that *initiates* the process of rupture when the film's thickness is on the mesoscopic scale. It should be noted, however, that this approach fails to describe the process down to the microscopic scale since, in the model, the force that squeezes the film is singular at $h = 0$ and hence needs regularization at small h.

Another issue with this approach is that the surface tension on the free surface is always assumed to be constant, which is in conflict with (7.109). This drawback is relatively easy to address, and the inclusion of the surface tension dependence on h could explain at least some of quantitative discrepancies between theoretical and experimental values for the critical film thickness reported recently (Coons et al. 2005). Since for large h one has $F \propto h^{-4}$ whereas equation (7.109) gives $\sigma \propto h^{-3}$, it is in the range of thicknesses where the effect of retardation becomes important that one should expect a relatively stronger influence of the surface-tension gradient. It should be noted, however, that equation (7.109) was obtained for the state of a thermodynamic equilibrium and for the moving fluid will need a suitable generalization.

As the film's thickness evolves through the mesoscopic down to the microscopic scale, a number of physical factors, such as thermal noise, evaporation, electrical double layer etc., become increasingly important. Their role, open theoretical questions and possible experiments that might elucidate various aspects of the ultrathin film kinetics receive regular attention in the literature (e.g., Churaev 2003, Thiele 2003). However, there is still very little understanding of how a systematic approach to their modelling which would allow one to describe the film's evolution down to the microscopic scale in a singularity-free way could be developed.

7.8 Summary

It has been shown that, unlike the standard model, the interface formation theory provides a unified approach to the problems of disintegration of liquid volumes and ensures regularity of the resulting solutions. In particular, it correctly identifies the conditions for the onset of the transition to the final stage of breakup of axisymmetric liquid threads and indicates a macroscopic mechanism that can rupture a liquid film. Importantly, the theory does not need any ad hoc alterations so that the material constants determined from unrelated experiments on capillary flows with forming interfaces can be used in the breakup problem and vice versa.

Although the results obtained to date in the modeling of capillary breakup are encouraging, they are only a starting point for research into this type of flows. Many issues are still unresolved, especially in the dynamics of thin films. In particular, the nonlinear stability analysis of films in the situations where the rate-of-change of the free surface area triggers interface formation mechanisms is lacking and will require considerable research effort. An important aspect of this problem is the question of two-dimensional disturbances of the film's free surface and their subsequent evolution. This question arises in many practical applications where the interplay of rupture mechanisms and self-healing properties of films leads to a complex picture of possible outcomes.

Appendix A

Elements of vector and tensor calculus

The summation convention. In the tensor algebra, it is convenient to use the following summation convention: if an object has a repeated upper and lower index, then, unless explicitly stated otherwise, it is understood that there is summation over this index from 1 to 3 (if we are dealing with a 3-dimensional space). For example,

$$a^i b_i = \sum_{i=1}^{3} a^i b_i, \qquad A_k^k = \sum_{k=1}^{3} A_k^k.$$

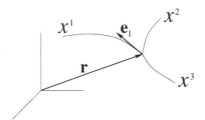

FIGURE A.1:   Definition sketch for basis vectors

Basis vectors. If (x^1, x^2, x^3) are coordinates in space, which hereafter we will assume to be Euclidean, and $\mathbf{r}$ is the radius-vector (or position vector) defined at every point by the coordinates of this point, then the basis vectors $\mathbf{e}_i$ $(i = 1, 2, 3)$ can be defined by

$$\mathbf{e}_i = \frac{\partial \mathbf{r}}{\partial x^i} \qquad i = 1, 2, 3. \tag{A.1}$$

Clearly, these vectors are not necessarily unit vectors and both their directions and lengths can vary from one point to another.

Covariant and contravariant. The basis vectors and components of vectors and tensors (to be introduced later) with low indices are called *covariant*;

those with upper indices are called *contravariant*. As it follows from (A.1), contravariant coordinates correspond to covariant basis vectors.

Transformation of coordinates. If we have two coordinate systems, (x^1, x^2, x^3) and (y^1, y^2, y^3), then

$$\mathbf{e}'_i = \frac{\partial \mathbf{r}}{\partial y^i} = \frac{\partial \mathbf{r}}{\partial x^k} \frac{\partial x^k}{\partial y^i} = \mathbf{e}_k \frac{\partial x^k}{\partial y^i},$$

and hence the covariant basis vectors in (y^1, y^2, y^3) and (x^1, x^2, x^3) are related by

$$\mathbf{e}'_i = \mathbf{e}_k \frac{\partial x^k}{\partial y^i} \qquad \text{and} \qquad \mathbf{e}_i = \mathbf{e}'_k \frac{\partial y^k}{\partial x^i}.$$

So far, we had only the radius-vector and basis vectors; now we can introduce simply a vector.

Vector. A vector $\mathbf{a}$ is an object invariant with respect to the transformation of coordinates which in a given coordinate frame (x^1, x^2, x^3) is characterized by its components (a^1, a^2, a^3) in the following way:

$$\mathbf{a} = a^i \mathbf{e}_i.$$

Therefore

$$\mathbf{a} = a^i \mathbf{e}_i = a^i \underbrace{\mathbf{e}'_k \frac{\partial y^k}{\partial x^i}}_{=\mathbf{e}_i} = \underbrace{a^i \frac{\partial y^k}{\partial x^i}}_{=a'^k} \mathbf{e}'_k = a'^k \mathbf{e}'_k,$$

and hence the transformation of contravariant components of a vector when we go from one coordinate frame to another is given by

$$a'^k = a^i \frac{\partial y^k}{\partial x^i} \qquad \text{and} \qquad a^i = a'^k \frac{\partial x^i}{\partial y^k}.$$

Tensor. A tensor $\mathbf{T}$ of rank n is defined as an invariant object related to the basis vectors of a given coordinate system through its components by

$$\mathbf{T} = T^{i_1 i_2 i_3 \ldots i_n} \mathbf{e}_{i_1} \mathbf{e}_{i_2} \mathbf{e}_{i_3} \ldots \mathbf{e}_{i_n}.$$

The polyadic products $\mathbf{e}_{i_1} \mathbf{e}_{i_2} \mathbf{e}_{i_3} \ldots \mathbf{e}_{i_n}$ $(i_1, i_2, \ldots i_n = 1, 2, 3)$ are formal objects introduced to make $\mathbf{T}$ invariant with respect to transformations of coordinates. They extend the concept of basis vectors for the case of tensor space and are linearly independent by definition.

If the coordinate system changes $(x^1, x^2, x^3) \mapsto (y^1, y^2, y^3)$, the contravariant components of a tensor in the old and new coordinate systems will be related by

$$T'^{i_1 i_2 \ldots i_n} = T^{j_1 j_2 \ldots j_n} \frac{\partial y^{i_1}}{\partial x^{j_1}} \frac{\partial y^{i_2}}{\partial x^{j_2}} \cdots \frac{\partial y^{i_n}}{\partial x^{j_n}} \tag{A.2}$$

and

$$T^{j_1 j_2 \ldots j_n} = T'^{i_1 i_2 \ldots i_n} \frac{\partial x^{j_1}}{\partial y^{i_1}} \frac{\partial x^{j_2}}{\partial y^{i_2}} \cdots \frac{\partial x^{j_n}}{\partial y^{i_n}}. \tag{A.3}$$

Symmetric and antisymmetric. A tensor of the second rank is called symmetric, if $T^{ij} = T^{ji}$ for all values of i and j. If for all i and j one has $T^{ij} = -T^{ji}$, the tensor is called antisymmetric. The definition can be extended to any pair of indices of a tensor of any rank.

Clearly, any tensor can be split into a symmetric and antisymmetric parts by

$$T^{ij} = \frac{1}{2}(T^{ij} + T^{ji}) + \frac{1}{2}(T^{ij} - T^{ji}).$$

Metric tensor. To measure distances in a given coordinate system, we will introduce the scalar product of basis vectors[1]

$$\mathbf{e}_i \cdot \mathbf{e}_j = g_{ij},$$

that is a set of numbers $\{g_{ij}\}$ corresponding to every point in space. The matrix $\{g^{ij}\}$ is defined as inverse to $\{g_{ij}\}$, that is $g^{ik}g_{kj} = \delta^i_j$, where δ^i_j is the Kronecker symbol. Then one can introduce a tensor

$$\mathbf{I} = g^{ij}\mathbf{e}_i\mathbf{e}_j,$$

which is referred to as the metric tensor; $\{g_{ij}\}$ and $\{g^{ij}\}$ are, by definition, its covariant and contravariant components.

Covariant components of vectors and tensors. Using the metric tensor, one can introduce contravariant basis vectors and the corresponding covariant coordinates by

$$\mathbf{e}^i = g^{ik}\mathbf{e}_k \quad \text{and} \quad x_i = x^j g_{ij}. \tag{A.4}$$

(The geometric meaning of $\mathbf{e}^i$ will be explained later.) Covariant components of vectors as well as covariant and mixed components of tensors are introduced in a similar way

$$a_i = a^k g_{ik}, \qquad T^i{}_j = T^{ik}g_{kj}, \quad T_i{}^j = T^{kj}g_{ki}, \quad T_{ij} = T^{kl}g_{ki}g_{lj}.$$

The tensor itself remains, of course, invariant, and, using the above definitions, one can show that

$$\mathbf{T} = T^{ij}\mathbf{e}_i\mathbf{e}_j = T^i{}_j\mathbf{e}_i\mathbf{e}^j = T_i{}^j\mathbf{e}^i\mathbf{e}_j = T_{ij}\mathbf{e}^i\mathbf{e}^j,$$

and hence that the mixed and covariant components are transformed as

$$T'^i{}_j = T^\alpha{}_\beta \frac{\partial y^i}{\partial x_\alpha}\frac{\partial x^\beta}{\partial y^j} \quad \text{and} \quad T'_{ij} = T_{\alpha\beta}\frac{\partial x^\alpha}{\partial y^i}\frac{\partial x^\beta}{\partial y^j}. \tag{A.5}$$

---

[1]Strictly speaking, one can introduce a metric without defining a scalar product, and the resulting metric space will not necessarily be generalizable to a Hilbert space, where the scalar product is defined, or even a Banach space, where one has a norm. Here our aim is to introduce the notions that will eventually give us a Hilbert space suitable for describing real-life objects.

Obviously, $\mathbf{e}^i \cdot \mathbf{e}^j = g^{ij}$ and $g^i_j = \delta^i_j$.

<u>Orthogonal coordinates.</u> The coordinate systems in which $g_{ij} = 0$ for $i \neq j$ at every point in space are called orthogonal. An orthogonal coordinate system where $g_{ii} = 1$ for $i = 1, 2, 3$ is called Cartesian. Clearly, for orthorgonal coordinates systems covariant and contravariant basis vectors are parallel and for Cartesian coordinates they are identical. It follows from the way the contravariants components of the metric tensor were introduced that the mixed components of $\mathbf{I}$ are Kronecker's symbols δ^i_j.

<u>Examples of metric tensors.</u> Cartesian coordinates $x^1 = x$, $x^2 = y$, $x^3 = z$:

$$g_{11} = g_{22} = g_{33} = 1, \qquad g_{ij} = 0 \text{ if } i \neq j. \tag{A.6}$$

Cylindrical coordinates $x^1 = r$, $x^2 = \varphi$, $x^3 = z$:

$$g_{11} = 1, \quad g_{22} = r^2, \quad g_{33} = 1, \qquad g_{ij} = 0 \text{ if } i \neq j.$$

Spherical coordinates $x^1 = r$, $x^2 = 0$, $x^3 = \varphi$:

$$g_{11} = 1, \quad g_{22} = r^2, \quad g_{33} = r^2 \sin^2 \theta, \quad g_{ij} = 0 \text{ if } i \neq j.$$

In the above coordinates, it is convenient to use the notation g_{xx}, g_{yy}, g_{zz}; $g_{rr}, g_{\theta\varphi}, g_{zz}$; $g_{rr}, g_{\theta\theta}, g_{\varphi\varphi}$, respectively.

<u>Scalar product.</u> The scalar product (or 'dot product') is introduced via the already known scalar product of the basis vectors by

$$\mathbf{a} \cdot \mathbf{b} = (a^i \mathbf{e}_i) \cdot (b^j \mathbf{e}_j) = (\mathbf{e}_i \cdot \mathbf{e}_j) a^i b^j = g_{ij} a^i b^j = a_i b^i = a^i b_i.$$

It is also convenient to use the dot notation for convolution of tensors of the second rank. Then,

$$\mathbf{A} \cdot \mathbf{B} \equiv A^{ik} B_k{}^j \mathbf{e}_i \mathbf{e}_j \qquad \text{and} \qquad \mathbf{A} : \mathbf{B} \equiv A^{ij} B_{ij},$$

so that one can operate with tensors rather than their components.

<u>Length.</u> Using the definition of the scalar product, one can find that the length of a vector, which is its only invariant, is given by

$$|\mathbf{a}|^2 \equiv \mathbf{a} \cdot \mathbf{a} = g_{ij} a^i a^j.$$

<u>Scalar invariants of tensors.</u> The fundamental role in physical applications of tensors is played by functions of their *components* which remain invariant with respect to transformations of coordinates,

$$f(T'^{ij}, T'^i_j, T'_{ij}) = f(T^{ij}, T^i_j, T_{ij}).$$

For a vector $\mathbf{a}$ it can be shown that all its invariants can be expressed as functions of one, $a^i a_i$, or its square root, that is the absolute value of the vector. For a tensor of the second rank $\mathbf{T}$ there are 3 independent invariants:

$$I_1 = T^i{}_i, \quad I_2 = T^i{}_j T^j{}_i, \quad I_3 = T^i{}_j T^j{}_n T^n{}_i.$$

Obviously, I_1, I_2 and I_3 are functionally independent and independent of the coordinate system. One can show that all other invariants are functions of these three.

Principal axes and principal components. It can be shown that for a symmmetric tensor of the second rank **T** at every point M in space there are orthogonal axes in which $T^{ij} = 0$ for $i \neq j$. In these axes, we also have $g^{ij} = g_{ij} = \delta^i_j$ and hence there is no difference between covariant, contravariant and mixed components of **T**. These axes are called the principal axes and the diagonal components are called the principal components of tensor **T** at point M. In other words, for a given symmetric tensor of the second rank at every point one can introduce a Cartesian coordinate system in which this tensor will have a diagonal form. In order to find the principal axes one has to consider the eigenvalues of the matrix $\{T^i_j\}$ and the corresponding eigenvectors.

Differentiation of vectors and tensors.

$$\frac{\partial \mathbf{a}}{\partial x^i} = \frac{\partial(a^k \mathbf{e}_k)}{\partial x^i} = \frac{\partial a^k}{\partial x^i}\mathbf{e}_k + a^k \frac{\partial \mathbf{e}_k}{\partial x^i} = \frac{\partial a^k}{\partial x^i}\mathbf{e}_k + a^k \Gamma^\alpha_{ki}\mathbf{e}_\alpha$$

$$= \left(\frac{\partial a^k}{\partial x^i} + a^\alpha \Gamma^k_{\alpha i}\right)\mathbf{e}_k = (\nabla_i a^k)\mathbf{e}_k. \tag{A.7}$$

Here we have introduced the following notation. The coefficients Γ^i_{jk} defined by

$$\frac{\partial \mathbf{e}_j}{\partial x^k} = \frac{\partial^2 \mathbf{r}}{\partial x^k \partial x^j} = \Gamma^i_{jk}\mathbf{e}_i \tag{A.8}$$

are known as the Christoffel symbols. The operation

$$\nabla_i a^k = \frac{\partial a^k}{\partial x^i} + a^\alpha \Gamma^k_{\alpha i},$$

is called covariant differentiation, and its result is a covariant derivative. For the covariant derivative of a covariant component one has

$$\nabla_i a_k = \frac{\partial a_k}{\partial x^i} - a_\alpha \Gamma^\alpha_{ik}.$$

Similarly, for a tensor of the second rank **T**

$$\nabla_k T^{ij} = \frac{\partial T^{ij}}{\partial x_k} + T^{\alpha j}\Gamma^i_{\alpha k} + T^{i\alpha}\Gamma^j_{\alpha k},$$

$$\nabla_k T_{ij} = \frac{\partial T_{ij}}{\partial x_k} - T_{\alpha j}\Gamma^\alpha_{ik} - T_{i\alpha}\Gamma^\alpha_{jk},$$

$$\nabla_k T^i_j = \frac{\partial T^i_j}{\partial x_k} + T^\alpha_j \Gamma^i_{\alpha k} - T^i_\alpha \Gamma^\alpha_{jk}.$$

The idea of covariant differentiation is that covariant derivatives of the tensor components, unlike their partial derivatives, are also components of a tensor, that is, they transform according to (A.2), (A.3) and (A.5) from one coordinate system to another. In particular, since for Cartesian coordinates

$$\nabla_i g_{jk} = 0 \quad \text{and} \quad \nabla_i g^{jk} = 0, \qquad (i, j, k = 1, 2, 3),$$

this is true for all other coordinate systems as well. Obviously, one also has $\nabla_i \mathbf{e}_k = 0$, $\nabla_i \mathbf{e}^k = 0$ for all i and k and hence

$$\nabla_i \mathbf{a} = \nabla_i (a^k \mathbf{e}_k) = (\nabla_i a^k) \mathbf{e}_k = \frac{\partial \mathbf{a}}{\partial x^i}.$$

An operator ∇^i is defined as $\nabla^i \equiv g^{ij} \nabla_j$.

Christoffel symbols. Using the definition of the Christoffel symbols (A.8) it can be demonstrated that they are related to components of the metric tensor by

$$\Gamma^i_{jk} = \frac{g^{is}}{2} \left(\frac{\partial g_{ks}}{\partial x^j} + \frac{\partial g_{js}}{\partial x^k} - \frac{\partial g_{jk}}{\partial x^s} \right), \qquad (A.9)$$

and, since $g_{ij} = g_{ji}$, one has $\Gamma^i_{jk} = \Gamma^i_{kj}$. Given the expressions for g_{ij} in the Cartesian coordinates (A.6), it follows from (A.9) that in these coordinates $\Gamma^i_{jk} \equiv 0$, whilst in curvilinear coordinates $\Gamma^i_{jk} \neq 0$. Hence Γ^i_{jk} are not components of a tensor.

Using (A.9), one can show that Christoffel symbols $\{\Gamma^i_{jk}\}$ in the coordinate system (x^1, x^2, x^3) and $\{\Gamma'^i_{jk}\}$ corresponding to (y^1, y^2, y^3) are related by

$$\Gamma'^i_{jk} = \left(\Gamma^\alpha_{\beta\gamma} \frac{\partial x^\beta}{\partial y^j} \frac{\partial x^\gamma}{\partial y^k} + \frac{\partial^2 x^\alpha}{\partial y^j \partial y^k} \right) \frac{\partial y^i}{\partial x^\alpha}.$$

It also follows from (A.9) that in orthogonal coordinates, where $g_{ij} = 0$ if $i \neq j$, one has (no summation here!)

$$\Gamma^\alpha_{\alpha\beta} = \frac{1}{2} g^{\alpha\alpha} \frac{\partial g_{\alpha\alpha}}{\partial x^\beta}, \qquad (\alpha \neq \beta)$$

$$\Gamma^\alpha_{\beta\beta} = -\frac{1}{2} g^{\alpha\alpha} \frac{\partial g_{\beta\beta}}{\partial x^\alpha},$$

$$\Gamma^\alpha_{\alpha\alpha} = \frac{1}{2} g^{\alpha\alpha} \frac{\partial g_{\alpha\alpha}}{\partial x^\alpha},$$

$$\Gamma^\alpha_{\beta\gamma} = 0 \qquad (\alpha \neq \beta \neq \gamma \neq \alpha).$$

If $g = \det \| g_{ij} \|$, then it can be shown that (here we use the summation convention again)

$$\Gamma^\alpha_{\alpha i} = \frac{1}{2g} \frac{\partial g}{\partial x^i} = \frac{1}{\sqrt{g}} \frac{\partial \sqrt{g}}{\partial x^i}. \qquad (A.10)$$

Gradient. Using the notation ∇_i $(i = 1, 2, 3)$ and ∇^j $(j = 1, 2, 3)$ as covariant and contravariant components of a vector, one can introduce a notation

$$\nabla \equiv \mathbf{e}_i \nabla^i \equiv \mathbf{e}^i \nabla_i.$$

A vector defined by

$$\operatorname{grad} f \equiv \nabla f \equiv \frac{\partial f}{\partial x^i} \mathbf{e}^i = \frac{\partial f}{\partial x^i} g^{ik} \mathbf{e}_k.$$

is known as the gradient of a scalar function f. Note that the components $\partial f/\partial x^i$ are covariant.

Divergence. Using (A.10), we can introduce the divergence of a vector by

$$\operatorname{div} \mathbf{a} \equiv \nabla \cdot \mathbf{a} \equiv \nabla_i a^i = \frac{\partial a^i}{\partial x^i} + a^k \Gamma^i_{ki} = \frac{\partial a^i}{\partial x^i} + \frac{a^k}{\sqrt{g}} \frac{\partial \sqrt{g}}{\partial x^k} = \frac{1}{\sqrt{g}} \frac{\partial (a^i \sqrt{g})}{\partial x^i}. \quad (A.11)$$

Laplacian. An operator

$$\nabla^2 \equiv \nabla \cdot \nabla \equiv \nabla_i \nabla^i \equiv \Delta$$

is known as the Laplacian. Note that, unlike the case of scalar functions, the Laplacian of a vector, in general, involves nondifferential terms due to the Christoffel simbols.

In orthogonal coordinates, the Laplacian of a scalar function is given by

$$\nabla^2 f = \frac{1}{\sqrt{g}} \left[\frac{\partial}{\partial x^1} \left(\frac{\sqrt{g}}{g_{11}} \frac{\partial f}{\partial x^1} \right) + \frac{\partial}{\partial x^2} \left(\frac{\sqrt{g}}{g_{22}} \frac{\partial f}{\partial x^2} \right) + \frac{\partial}{\partial x^3} \left(\frac{\sqrt{g}}{g_{33}} \frac{\partial f}{\partial x^3} \right) \right],$$

where $g = g_{11}g_{22}g_{33}$.

Vector product. To calculate vector products in an arbitrary coordinate system we introduce the Levi-Civita (pseudo)tensor $\boldsymbol{\epsilon} = \epsilon^{ijk} \mathbf{e}_i \mathbf{e}_j \mathbf{e}_k$ by

$$\epsilon^{ijk} = \begin{cases} \dfrac{1}{\sqrt{g}}, & \text{if } i,j,k \text{ is an even permutation of } 1,2,3; \\[2mm] -\dfrac{1}{\sqrt{g}}, & \text{if } i,j,k \text{ is an odd permutation of } 1,2,3; \\[2mm] 0, & \text{if } i,j,k \text{ are not all different.} \end{cases} \quad (A.12)$$

If we change the coordinate system $\{x^i\} \longmapsto \{y^i\}$, then

$$\epsilon'^{ijk} = \frac{J}{|J|} \epsilon^{\alpha\beta\gamma} \frac{\partial y^i}{\partial x^\alpha} \frac{\partial y^j}{\partial x^\beta} \frac{\partial y^k}{\partial x^\gamma},$$

where

$$J = \frac{\partial(y^1, y^2, y^3)}{\partial(x^1, x^2, x^3)}$$

is the Jacobian of transformation. For the coordinate transformations where $J > 0$, components of the Levi-Civita pseudotensor $\{\epsilon^{ijk}\}$ behave like components of a tensor.

We can introduce the vector product by

$$\mathbf{c} = \mathbf{a} \times \mathbf{b} = \epsilon^{ijk} a_j b_k \mathbf{e}_i, \quad (A.13)$$

that is,

$$c^i = \frac{1}{\sqrt{g}}(a_j b_k - a_k b_j),$$

where i, j, k form a circle permutation of 1, 2, 3.

The vector product defines a normal direction: vector $\mathbf{c}$ in (A.13) is said to be pointing in the positive normal direction with respect to a plane defined by $\mathbf{a}$ and $\mathbf{b}$. Using (A.4), (A.12) and (A.13) one can show that

$$\mathbf{e}^i = \frac{\mathbf{e}_j \times \mathbf{e}_k}{\sqrt{g}},$$

where i, j, k form a circular permutation of $1, 2, 3$.

Curl. The curl

$$\boldsymbol{\Omega} = \operatorname{curl} \mathbf{u} \equiv \nabla \times \mathbf{u}$$

is introduced by

$$\boldsymbol{\Omega} = \epsilon^{ijk} \nabla_j u_k \mathbf{e}_i.$$

One can verify that

$$\epsilon^{ijk} \epsilon_{kmn} = \delta^i_m \delta^j_n - \delta^i_n \delta^j_m. \tag{A.14}$$

This identity makes it easy to obtain the following ones known in the vector calculus:

$$\mathbf{a} \times (\mathbf{b} \times \mathbf{c}) = \mathbf{b}(\mathbf{a} \cdot \mathbf{c}) - \mathbf{c}(\mathbf{a} \cdot \mathbf{b}),$$

$$\nabla \times (\nabla \times \mathbf{a}) = \nabla(\nabla \cdot \mathbf{a}) - \nabla^2 \mathbf{a},$$

$$\nabla \times (\mathbf{a} \times \mathbf{b}) = \mathbf{a}(\nabla \cdot \mathbf{b}) - \mathbf{a} \cdot \nabla \mathbf{b} - \mathbf{b}(\nabla \cdot \mathbf{a}) + \mathbf{b} \cdot \nabla \mathbf{a}.$$

Physical components of vectors and tensors. For a curvilinear coordiante system the basis vectors defined by (A.1) are not necessarily unit vectors. As a result, covariant and contravariant components of vectors and tensors in such a coordinate frame in a general case will differ from one another and, if the vector or tensor describes a certain physical quantity, the dimensions of its components will not necessarily be the dimensions of this quantity. In applications of the tensor analysis, it is convenient to introduce physical components of vectors and tensors which are associated with the unit basis vectors:

$$\mathbf{a} = \sum_i a^i \mathbf{e}_i = \sum_i \underbrace{a^i \sqrt{g_{ii}}}_{=\bar{a}^i} \underbrace{\frac{\mathbf{e}_i}{\sqrt{g_{ii}}}}_{=\hat{\mathbf{e}}_i} = \bar{a}^i \hat{\mathbf{e}}_i.$$

Here summation is shown explicitly rather than via the summation convention, $\bar{a}^i = a^i \sqrt{g_{ii}}$ (no summation here!) are the contravatiant physical components of $\mathbf{a}$ and $\hat{\mathbf{e}}_i = \mathbf{e}_i / \sqrt{g_{ii}}$ are the unit basis vectors. Physical components of tensors of a higher rank are introduced in a similar way. With respect to transformations of coordinates the physical components do not behave as components of a tensor.

In orthogonal coordinate systems, one has $\hat{\mathbf{e}}_i = \hat{\mathbf{e}}^i$, $\bar{g}^{ij} = \bar{g}_{ij} = \delta^i_j$ and there is no difference between covariant and contravariant physical components of vectors and tensors. The divergence (A.11) of a vector $\mathbf{a}$ in terms of the physical components of the latter in orthogonal coordinates (x^1, x^2, x^3) is given by

$$\nabla \cdot \mathbf{a} = \frac{1}{\sqrt{g}} \frac{\partial}{\partial x^i} \left(\bar{a}^i \sqrt{\frac{g}{g_{ii}}} \right). \tag{A.15}$$

For the vector product one has

$$\mathbf{a} \times \mathbf{b} = \begin{vmatrix} \hat{\mathbf{e}}_2 \times \hat{\mathbf{e}}_3 & \bar{a}_1 & \bar{b}_1 \\ \hat{\mathbf{e}}_3 \times \hat{\mathbf{e}}_1 & \bar{a}_2 & \bar{b}_2 \\ \hat{\mathbf{e}}_1 \times \hat{\mathbf{e}}_2 & \bar{a}_3 & \bar{b}_3 \end{vmatrix} = \pm \begin{vmatrix} \hat{\mathbf{e}}_1 & \bar{a}_1 & \bar{b}_1 \\ \hat{\mathbf{e}}_2 & \bar{a}_2 & \bar{b}_2 \\ \hat{\mathbf{e}}_3 & \bar{a}_3 & \bar{b}_3 \end{vmatrix},$$

where the signs $+$ and $-$ correspond to the right and left orientation of $\hat{\mathbf{e}}_1, \hat{\mathbf{e}}_2, \hat{\mathbf{e}}_3$, respectively.

'Surface vector'. If S is a smooth surface, and $\mathbf{n}$ is a unit normal to it, then $\mathbf{a}^s$ will be called a 'surface vector' if $\mathbf{n} \cdot \nabla \mathbf{a}^s = 0$ everywhere on S. By definition, $\mathbf{n} \cdot \nabla \mathbf{n} = 0$.

Divergence of a 'surface vector'. For the divergence of the 'surface vector' $\mathbf{a}^s$ on a surface S whose orientation is characterized by a unit normal vector $\mathbf{n}$ one has:

$$\nabla \cdot \mathbf{a}^s = \nabla \cdot [\mathbf{a}^s_\parallel + (\mathbf{a}^s \cdot \mathbf{n})\mathbf{n}] = \nabla \cdot \mathbf{a}^s_\parallel + (\mathbf{a}^s \cdot \mathbf{n})\nabla \cdot \mathbf{n} + \mathbf{n} \cdot \nabla(\mathbf{a}^s \cdot \mathbf{n})$$

$$= \nabla \cdot \mathbf{a}^s_\parallel + (\mathbf{a}^s \cdot \mathbf{n})\nabla \cdot \mathbf{n} + \underbrace{(\mathbf{n} \cdot \nabla \mathbf{a}^s)}_{=0} \cdot \mathbf{n} + \underbrace{(\mathbf{n} \cdot \nabla \mathbf{n})}_{=0} \cdot \mathbf{a}^s$$

$$= \nabla \cdot \mathbf{a}^s_\parallel + (\mathbf{a}^s \cdot \mathbf{n})\nabla \cdot \mathbf{n}.$$

Here $\mathbf{a}^s_\parallel = \mathbf{a}^s \cdot (\mathbf{I} - \mathbf{nn})$ is a tangential to the surface component of $\mathbf{a}^s$. In order to calculate $\nabla \cdot \mathbf{a}^s_\parallel$, one can use (A.11) or (A.15), where $\{x^i\}$ are now the coordinates parametrizing the surface S, and g_{ii} are components of the corresponding metric tensor.

Appendix B

Equations of fluid mechanics in curvilinear coordinates

Cylindrical coordinates r, φ, z are related with the Cartesian coordinates x, y, z by $x = r\cos\varphi$, $y = r\sin\varphi$, $z = z$. The square of an elementary distance is then given by $ds^2 = dr^2 + r^2 d\theta^2 + dz^2$, so that

$$g_{rr} = 1, \ g_{\varphi\varphi} = r^2, \ g_{zz} = 1,$$

whilst other components $g_{ij} = 0$. The only nonzero Christoffel symbols are given by

$$\Gamma^r_{\varphi\varphi} = -r, \ \Gamma^\varphi_{r\varphi} = \Gamma^\varphi_{\varphi r} = \frac{1}{r}.$$

The Laplacian acting on a *scalar* function is given by

$$\Delta = \frac{1}{r}\frac{\partial}{\partial r}\left(r\frac{\partial}{\partial r}\right) + \frac{1}{r^2}\frac{\partial^2}{\partial\varphi^2} + \frac{\partial^2}{\partial z^2}. \tag{B.1}$$

The continuity equation written in terms of the physical components of velocity takes the form

$$\frac{\partial\rho}{\partial t} + \frac{1}{r}\frac{\partial(\rho u_r r)}{\partial r} + \frac{1}{r}\frac{\partial(\rho u_\varphi)}{\partial\varphi} + \frac{\partial(\rho u_z)}{\partial z} = 0.$$

The pressure gradient is expressed in terms of its physical components by

$$\nabla p = \frac{\partial p}{\partial r}\hat{\mathbf{e}}_r + \frac{1}{r}\frac{\partial p}{\partial\varphi}\hat{\mathbf{e}}_\varphi + \frac{\partial p}{\partial z}\ddot{\mathbf{e}}_z.$$

For the physical components of acceleration one has

$$a_r = \frac{\partial u_r}{\partial t} + u_r\frac{\partial u_r}{\partial r} + \frac{u_\varphi}{r}\frac{\partial u_r}{\partial\varphi} + u_z\frac{\partial u_r}{\partial z} - \frac{u_\varphi^2}{r},$$

$$a_\varphi = \frac{\partial u_\varphi}{\partial t} + u_r\frac{\partial u_\varphi}{\partial r} + \frac{u_\varphi}{r}\frac{\partial u_\varphi}{\partial\varphi} + u_z\frac{\partial u_\varphi}{\partial z} + \frac{u_r u_\varphi}{r},$$

$$a_z = \frac{\partial u_z}{\partial t} + u_r\frac{\partial u_z}{\partial r} + \frac{u_\varphi}{r}\frac{\partial u_z}{\partial\varphi} + u_z\frac{\partial u_z}{\partial z}.$$

In cylindrical coordinates, the physical components of the rate-of-strain tensor are expressed via the physical components of velocity in the following way

$$e_{rr} = \frac{\partial u_r}{\partial r}, \ 2e_{r\varphi} = \frac{1}{r}\frac{\partial u_r}{\partial\varphi} + r\frac{\partial}{\partial r}\left(\frac{u_\varphi}{r}\right), \ 2e_{rz} = \frac{\partial u_z}{\partial r} + \frac{\partial u_r}{\partial z},$$

$$e_{\varphi\varphi} = \frac{1}{r}\frac{\partial u_\varphi}{\partial \varphi} + \frac{u_r}{r}, \quad 2e_{\varphi z} = \frac{\partial u_\varphi}{\partial z} + \frac{1}{r}\frac{\partial u_z}{\partial \varphi}, \quad e_{zz} = \frac{\partial u_z}{\partial z}.$$

The Navier-Stokes equations for incompressible fluid with constant viscosity in the absence of body forces are given by

$$\rho a_r = -\frac{\partial p}{\partial r} + \mu \left(\Delta u_r - \frac{u_r}{r^2} - \frac{2}{r^2}\frac{\partial u_\varphi}{\partial \varphi} \right),$$

$$\rho a_\varphi = -\frac{1}{r}\frac{\partial p}{\partial \varphi} + \mu \left(\Delta u_\varphi - \frac{u_\varphi}{r^2} + \frac{2}{r^2}\frac{\partial u_r}{\partial \varphi} \right),$$

$$\rho a_z = -\frac{\partial p}{\partial z} + \mu \Delta u_z,$$

where Δ is defined by (B.1).

Spherical coordinates r, θ, φ are related with the Cartesian coordinates x, y, z by

$$x = r \sin \theta \cos \varphi, \quad y = r \sin \theta \sin \varphi, \quad z = r \cos \theta.$$

The square of an elementary distance is then given by $ds^2 = dr^2 + r^2 \, d\theta^2 + r^2 \sin^2 \theta \, d\varphi^2$, so that

$$g_{rr} = 1, \quad g_{\theta\theta} = r^2, \quad g_{\varphi\varphi} = r^2 \sin^2 \theta.$$

For the Christoffel symbols one has

$$\Gamma^\theta_{r\theta} = \Gamma^\theta_{\theta r} = \Gamma^\varphi_{r\varphi} = \Gamma^\varphi_{\varphi r} = \frac{1}{r}, \quad \Gamma^\varphi_{\varphi\theta} = \Gamma^\varphi_{\theta\varphi} = \cot\theta,$$

$$\Gamma^r_{\theta\theta} = -r, \quad \Gamma^r_{\varphi\varphi} = -r\sin^2\theta, \quad \Gamma^\theta_{\varphi\varphi} = -\sin\theta\cos\theta,$$

with all other Γ^i_{jk} equal to zero.

The Laplacian of a scalar function f is given by

$$\Delta f = \frac{1}{r^2} \left[\frac{\partial}{\partial r}\left(r^2 \frac{\partial}{\partial r} \right) + \frac{1}{\sin\theta}\frac{\partial}{\partial \theta}\left(\sin\theta \frac{\partial}{\partial \theta} \right) + \frac{1}{\sin^2\theta}\frac{\partial^2}{\partial \varphi^2} \right] f. \qquad \text{(B.2)}$$

The continuity equation written in terms of the physical components of velocity takes the form

$$\frac{\partial \rho}{\partial t} + \frac{1}{r^2}\frac{\partial(\rho u_r r^2)}{\partial r} + \frac{1}{r\sin\theta}\frac{\partial(\rho u_\theta \sin\theta)}{\partial \theta} + \frac{1}{r\sin\theta}\frac{\partial(\rho u_\varphi)}{\partial \varphi} = 0.$$

The pressure gradient in terms of its physical components is given by

$$\nabla p = \frac{\partial p}{\partial r}\hat{\mathbf{e}}_r + \frac{1}{r}\frac{\partial p}{\partial \theta}\hat{\mathbf{e}}_\theta + \frac{1}{r\sin\theta}\frac{\partial p}{\partial \varphi}\hat{\mathbf{e}}_\varphi.$$

The physical components of acceleration have the form

$$a_r = \frac{\partial u_r}{\partial t} + u_r \frac{\partial u_r}{\partial r} + \frac{u_\theta}{r} \frac{\partial u_r}{\partial \theta} + \frac{u_\varphi}{r \sin \theta} \frac{\partial u_r}{\partial \varphi} - \frac{u_\theta^2 + u_\varphi^2}{r},$$

$$a_\theta = \frac{\partial u_\theta}{\partial t} + u_r \frac{\partial u_\theta}{\partial r} + \frac{u_\theta}{r} \frac{\partial u_\theta}{\partial \theta} + \frac{u_\varphi}{r \sin \theta} \frac{\partial u_\theta}{\partial \varphi} + \frac{u_r u_\theta - u_\varphi^2 \cot \theta}{r},$$

$$a_\varphi = \frac{\partial u_\varphi}{\partial t} + u_r \frac{\partial u_\varphi}{\partial r} + \frac{u_\theta}{r} \frac{\partial u_\varphi}{\partial \theta} + \frac{u_\varphi}{r \sin \theta} \frac{\partial u_\varphi}{\partial \varphi} + \frac{u_r u_\varphi + u_\theta u_\varphi \cot \theta}{r}.$$

The physical components of the rate-of-strain tensor are expressed via the physical components of velocity in the following way

$$e_{rr} = \frac{\partial u_r}{\partial r}, \quad e_{\theta\theta} = \frac{1}{r} \frac{\partial u_\theta}{\partial \theta} + \frac{u_r}{r}, \quad e_{\varphi\varphi} = \frac{1}{r \sin \theta} \frac{\partial u_\varphi}{\partial \varphi} + \frac{u_r}{r} + \frac{u_\theta \cot \theta}{r},$$

$$2e_{r\theta} = \frac{1}{r} \frac{\partial u_r}{\partial \theta} + \frac{\partial u_\theta}{\partial r} - \frac{u_\theta}{r}, \quad 2e_{r\varphi} = \frac{\partial u_\varphi}{\partial r} + \frac{1}{r \sin \theta} \frac{\partial u_r}{\partial \varphi} - \frac{u_\varphi}{r},$$

$$2e_{\varphi\theta} = \frac{1}{r \sin \theta} \frac{\partial u_\theta}{\partial \varphi} + \frac{1}{r} \frac{\partial u_\varphi}{\partial \theta} - \frac{u_\varphi \cot \theta}{r}.$$

The Navier-Stokes equations for an incompressible fluid with constant viscosity in the absence of body forces are given by

$$\rho a_r = -\frac{\partial p}{\partial r} + \mu \left(\Delta u_r - \frac{2u_r}{r^2} - \frac{2}{r^2} \frac{\partial u_\theta}{\partial \theta} - \frac{2u_\theta}{r^2} \cot \theta - \frac{2}{r^2 \sin \theta} \frac{\partial u_\varphi}{\partial \varphi} \right),$$

$$\rho a_\theta = -\frac{1}{r} \frac{\partial p}{\partial \theta} + \mu \left(\Delta u_\theta - \frac{u_\theta}{r^2 \sin^2 \theta} - \frac{2 \cos \theta}{r^2 \sin^2 \theta} \frac{\partial u_\varphi}{\partial \varphi} + \frac{2}{r^2} \frac{\partial u_r}{\partial \theta} \right),$$

$$\rho a_\varphi = -\frac{1}{r \sin \theta} \frac{\partial p}{\partial \varphi} + \mu \left(\Delta u_\varphi - \frac{u_\varphi}{r^2 \sin^2 \theta} + \frac{2 \cos \theta}{r^2 \sin^2 \theta} \frac{\partial u_\theta}{\partial \varphi} + \frac{2}{r^2 \sin \theta} \frac{\partial u_r}{\partial \varphi} \right),$$

where Δ is defined by (B.2).

Appendix C

Complex representation of biharmonic functions

Consider how an arbitrary biharmonic function $W(x, y)$ of Cartesian coordinates x, y in a domain D can be represented in terms of two functions of the complex variable $z = x + iy$.

Let $P = \Delta W$, where $\Delta = \partial^2/\partial x^2 + \partial^2/\partial y^2$. Given that W is biharmonic, $\Delta^2 W = 0$, the function P is harmonic, $\Delta P = 0$, and hence the Cauchy-Riemann equations

$$\frac{\partial P}{\partial x} = \frac{\partial Q}{\partial y}, \qquad \frac{\partial P}{\partial y} = -\frac{\partial Q}{\partial x},$$

allow one to determine (up to an additive constant) a harmonic function Q conjugate to P. Then, the function

$$f(z) = P + iQ$$

and hence the function

$$\phi(z) \equiv p + iq = \frac{1}{4} \int f(z)\, dz$$

are both analytic functions of the complex variable z. Since obviously

$$\phi'(z) = \frac{\partial p}{\partial x} + i\frac{\partial q}{\partial x} = \frac{1}{4}(P + iQ)$$

and functions p and q satisfy the Cauchy-Riemann equations, one has

$$\frac{\partial p}{\partial x} = \frac{\partial q}{\partial y} = \frac{1}{4}P, \qquad \frac{\partial p}{\partial y} = -\frac{\partial q}{\partial x} = -\frac{1}{4}Q.$$

The first of these equations allows one to verify that

$$\Delta(W - px - qy) = 0,$$

which means that

$$p_1 = W - px - qy \tag{C.1}$$

is a harmonic function in D.

Now, using p_1 one can determine a conjugate funciton q_1 from the Cauchy-Riemann equations

$$\frac{\partial p_1}{\partial x} = \frac{\partial q_1}{\partial y}, \qquad \frac{\partial p_1}{\partial y} = -\frac{\partial q_1}{\partial x}$$

so that p_1 becomes the real part of an analytic function

$$\chi(z) = p_1 + i q_1. \tag{C.2}$$

Thus, combining (C.1) and (C.2) we get

$$W = px + qy + p_1 = \operatorname{Re}\{\bar{z}\phi(z) + \chi(z)\}. \tag{C.3}$$

Alternatively, by considering $Q = \Delta W$ and $q_1 = W - qx + py$, one can represent W as the imaginary part of the same expression

$$W = \operatorname{Im}\{\bar{z}\phi(z) + \chi(z)\}. \tag{C.4}$$

Representations (C.3) and (C.4) play a fundamental role in the theory of biharmonic functions and their applications, in particular, in the linear theory of elasticity. They were first derived by Goursat (1898) and in a different form by Muskhelishvili (1919), whose derivation is reproduced here following Muskhelishvili (1963, p. 109). A comprehensive review of two-dimensional biharmonic problems can be found in Meleshko (2003).

Appendix D

Physical properties of some fluids

TABLE D.1: Densities and viscosities of some common fluids

Fluid	$T, °C$	ρ, g/cm^3	μ, cP	ν, cSt
Water	5	1.00	1.514	1.514
	10	1.00	1.304	1.304
	15	0.999	1.137	1.138
	20	0.998	1.002	1.004
	50	0.998	0.548	0.554
Ethyl alcohol	15	0.79	1.34	1.70
Glycerine	20	1.26	8.2	6.5
Mercury	15	13.6	1.58	0.116
Air ($p = 1$ atm)	0	1.293×10^{-3}	1.71×10^{-2}	13.2
	10	1.247×10^{-3}	1.76×10^{-2}	14.1
	20	1.205×10^{-3}	1.81×10^{-2}	15.0

TABLE D.2: Surface tensions (mN/m) of some common liquids against air at different temperatures

Liquid	10°C	25°C	50°C	75°C	100°C
Water	74.23	71.99	67.94	63.57	58.91
Mercury	488.55	488.48	480.36	475.23	470.11
Tetrachloromethane		26.43	23.37	20.31	17.25
Ethanol	23.22	21.97	19.89		
Ethylene glycol		47.99	45.76	43.54	41.31
Acetone		23.46	20.66		
Pentane	17.15	15.49			
Benzene		28.22	25.00	21.77	
Hexane	19.42	17.89	15.33		

The data are for $p = 1$ atm or, if the temperature is above the boiling point, the pressure is that of the saturated vapor. [Data from Lide (2004)].

TABLE D.3: Surface tensions (mN/m) between water and some pure liquids

Liquid	20°C	25°C
Mercury[1]	415	416
n-Hexane[2]	51.0	
Carbon tetrachloride[2]	45.1	43.7
Benzene[2]	35.0	34.71
Nitrobenzene[2]	26.0	
Ethyl ether[2]	10.7	
n-Hexanol[2]		6.8
Ethyl acetate[1]	6.8	
Aniline[2]	5.85	
n-Pentanol[2]		4.4
Isobutanol[2]	2.1	
n-Butanol[2]	1.6	1.8

1: Adamson & Gast (1997), 2: Davies & Rideal (1961).

TABLE D.4: Surface tension (mN/m) for some liquid-liquid systems at 20°

	Ethanol	n-Hexane	Water
Mercury	389	378	415
	n-Heptane	Benzene	Carbon tetrachloride
Diethylene glycol	10.6	35.0	45.0

Data from Adamson & Gast (1997).

Conversion factors: 1 cP $= 10^{-2}$ g/(cm s) $= 1$ mPa s, 1 mN/m $=$ 1 dyn/cm, 1 cSt $= 10^{-2}$ cm^2/s.

References

ABLETT, R. 1923 An investigation of the angle of contact between paraffin wax and water. *Philos. Mag.* **46,** 244–256.

ABRASHKIN, A. A. & YAKUBOVICH, E. I. 2006 *Vortex Dynamics in Lagrangian Description.* Fizmatlit, Moscow.

ABRIKOSOVA, I. I. & DERJAGUIN, B. V. 1953 On the law of intermolecular interaction at large distances. *Dokl. Akad. Nauk SSSR* **90,** 1055–1058.

ABRIKOSOVA, I. I. & DERJAGUIN, B. V. 1958 Direct measurement of the molecular attraction of solid bodies. II. Method for measuring the gap. Results of experiments. *JETP* **4,** 2–10.

ADAMSON, A. W. & GAST, A. P. 1967 *Physical Chemistry of Surfaces.* Wiley, New York.

AGARWAL, R. K., YUN, K.-Y. & BLAKRISHNAN, R. 2001 Beyond Navier-Stokes: Burnett equations for flows in the continuum-transition regime. *Phys. Fluids* **13,** 3061–3085.

AGARWAL, R. K., YUN, K.-Y. & BLAKRISHNAN, R. 2001 Erratum: "Beyond Navier-Stokes: Burnett equations for flows in the continuum-transition regime" [Phys. Fluids 13, 3061 (2001)]. *Phys. Fluids* **14,** 1818.

AIRY, G. B. 1862 On the strains in the interior of beams. *Proc. Roy. Soc.* (London) **12,** 304–306.

AIRY, G. B. 1863 On the strains in the interior of beams. *Philos. Trans. R. Soc. London, Ser. A* **153,** 49–79.

ALAKOÇ, U., MEGARIDIS, C. M., MCNALLAN, M. & WALLACE, D. B. 2004 Dynamic surface tension measurements with submillisecond resolution using a capillary-jet instability technique. *J. Colloid & Interf. Sci.* **276,** 379–391.

ALBANO, A. M., BEDEAUX, D. & VLIEGER, J. 1979 On the description of interfacial properties using singular densities and currents at a dividing surface. *Physica A* **99,** 293–304.

ALLAIN, C., AUSSERRE, D. & RONDELEZ, F. 1985 A new method for contact-angle measurements of sessile drops. *J. Colloid & Interf. Sci.* **107,** 5–13.

AMBRAVANESWARAN, B., WILKES, E. D. & BASARAN, O. A. 2002 Drop formation from a capillary tube: Comparison of one-dimensional and two-dimensional analyses and occurrence of satellite drops. *Phys. Fluids* **14**, 2602–2621.

AMIRFAZLI, A., KWOK, D. Y., GAYDOS, J. & NEUMANN, A. W. 1998 Line tension measurements through drop size dependence of contact angle. *J. Colloid & Interf. Sci.* **205**, 1–11.

ANILKUMAR, A. V., LEE, C. P. & WANG, T. G. 1991 Surface-tension-induced mixing following coalescence of initially stationary drops. *Phys. Fluids* **A3**, 2587–2591.

ANTANOVSKII, L. K. 1994 Influence of surfactants on a creeping free-boundary flow induced by 2 counter-rotating horizontal thin cylinders. *Euro. J. Mech., B/Fluids* **13**, 73–92.

ASHGRIZ, N. & MASHAYEK, F. 1995 Temporal analysis of capillary jet breakup. *J. Fluid Mech.* **291**, 163–190.

BAER, T. A., CAIRNCROSS, R. A., SCHUNK, R., RAO, R. R. & SACKINGER, P. A. 2000 A finite element method for free surface flows of incompressible fluids in three dimensions. Part II. Dynamic wetting lines. *Int. J. Numer. Meth. Fluids* **33**, 405–427.

BAKKER, G. 1928 Kapillarität und Oberflächenspnnung. In: *Handbuch der Experimentalphysik* (W. Wien, F. Harms & H. Lenz, eds.) vol. 6, ch. 10, Akad. Verlags, Leipzig.

BARENBLATT, G. I. & CHERNYI, G. G. 1963 On moment relations on surfaces of discontinuity in dissipative media. *PMM — Sov. Appl. Maths & Mech.* **25**, 784–793.

BARRERO, A. & LOSCERTALES, I. G. 2007 Micro- and nanoparticles via capillary flows. *Annu. Rev. Fluid Mech.* **39**, 89–106.

BASCOM, W. D., COTTINGTON, R. L. & SINGLETERRY, C. R. 1964 Dynamic surface phenomena in the spontaneous spreading of oils on solids. In *Contact Angle, Wettability and Adhesion*, Advances in Chemistry Series 43 (R.F. Gould, ed.), ACS, Washington, DC, pp. 355–379.

BATCHELOR, G. K. 1967 *An Introduction to Fluid Dynamics*. Cambridge Univ. Press.

BATCHELOR, G. K. 1971 The stress generated in a non-dilute suspension of elongated particles by pure straining motion. *J. Fluid Mech.* **46**, 813–829.

BAYER, I. S. & MEGARIDIS, C. M. 2006 Contact angle dynamics in droplets impacting on flat surfaces with different wetting characteristics. *J. Fluid Mech.* **558**, 415–449.

BEAGLEHOLE, D. 1989 Profiles of the precursor of spreading drops of siloxane oil on glass, fused silica and mica. *J. Phys. Chem.* **93**, 893–899.

BEDEAUX, D. 2004 Nonequilibrium thermodynamic description of the three-phase contact line. *J. Chem. Phys.* **120**, 3744–3748.

BEDEAUX, D., ALBANO, A. M. & MAZUR, P. 1976 Boundary conditions and non-equilibrium thermodynamics. *Physica A* **82**, 438–462.

BENKREIRA, H., KHAN, M. & PATEL, R. 2004 Substrate characteristics and their effects on air entrainment. In *Proc. of 12th Intl Symp. Coating Sci. & Technol.*, September 20–22, 2004, Rochester, USA, pp. 123–128.

BIKERMAN, J. J. 1977/78 Capillarity before Laplace: Segner, Monge, Young. *Arch. History Exact Sci.* **18**, No. 2, 103–122.

BIRKHOFF, G. 1950 *Hydrodynamics. A Study in Logic, Fact, and Similitude.* Princeton Univ. Press.

BLACK, W., DE JONGH, J. G. V., OVERBEEK, J. TH. G. & SPARNAAY, M. J. 1960 Measurement of retarded van der Waals forces. *Trans. Farad. Soc.* **56**, 1597–1608.

BLAKE, T. D. 1993 Dynamic contact angles and wetting kinetics. In *Wettability* (ed. J. C. Berg), pp. 251–309. Marcel Dekker.

BLAKE, T. D. 1995 Personal communication.

BLAKE, T. D., BRACKE, M. & SHIKHMURZAEV, Y. D. 1999 Experimental evidence of nonlocal hydrodynamic influence on the dynamic contact angle. *Phys. Fluids* **11**, 1995–2007.

BLAKE, T. D., CLARKE, A. & RUSCHAK, K. J. 1994 Hydrodynamic assist of dynamic wetting. *AIChE J.* **40**, 229–242.

BLAKE, T. D., DOBSON, R. A. & RUSCHAK, K. J. 2004 Wetting at high capillary numbers. *J. Colloid & Interf. Sci.* **279**, 198–205.

BLAKE, T. D. & HAYNES, J. M. 1969 Kinetics of liquid/liquid displacement. *J. Colloid & Interf. Sci.* **30**, 421–423.

BLAKE, T. D. & RUSCHAK, K. J. 1979 A maximum speed of wetting. *Nature* (London) **282**, 489–491.

BLAKE, T. D. & SHIKHMURZAEV, Y. D. 2002 Dynamic wetting by liquids of different viscosity. *J. Colloid & Interf. Sci.* **253**, 196–202.

BLODGETT, K. B. 1935 Films built by depositing successive monomolecular layers on a solid surface. *J. Amer. Chem. Soc.* **57**, 1007–1022.

BOHR, N. 1909 Determination of the surface-tension of water by the method of jet vibration. *Philos. Trans. Roy. Soc.* (London) **A209**, 281–317.

BOHR, N. 1910 Determination of the tension of a recently formed water-surface. *Proc. Roy. Soc. A* (London) **84,** 395–403.

BOLTON, B. & MIDDLEMAN, S. 1980 Air entrainment in a roll coating system. *Chem. Eng. Sci.* **35,** 597–601.

BONGIORNO, V. & DAVIS, H. T. 1975 Modified Van der Waals thoery of fluid interfaces. *Phys. Rev. A* **12,** 2213–2224.

BOURGIN, P. & SAINTLOS, S. 2001 High-velocity coating flows and other related problems. *Intl J. Non-Linear Mech.* **36,** 585–596.

BOUSSINESQ, M. J. 1913 Sur l'existence d'une viscosité superficielle, dans la mince couche de transition séparant un liquide d'une antre fluide contigu. *Ann. Chim. Phys.* **29,** 349–357.

BRACKE, M., DE VOEGHT, F. & JOOS, P. 1989 The kinetics of wetting: the dynamic contact angle. *Progr. Colloid. & Polym. Sci.* **79,** 142–149.

BRADLEY, R. S. 1932 The cohesive force between solid surfaces and the surface energy of solids. *Philos. Mag.* **13,** 853–862.

BRAUN, R. J., MURRAY, B. T., BOETTINGER, W. J. & McFADDEN, G. B. 1995 Lubrication theory for reactive spreading of a thin drop. *Phys. Fluids* **7,** 1797–1810.

BRENNER, H. 1979 A micromechanical derivation of the differential equation of interfacial statics. *J. Colloid & Interf. Sci.* **68,** 422–439.

BRENNER, M. P., LISTER, J. R. & STONE, H. A. 1996 Pinching threads, singularities and the number 0.0304.... *Phys. Fluids* **8,** 2827–2836.

BROWN, C. E., JONES, T. D. & NEUSTADTER, E. L. 1980 Interfacial flow during immiscible displacement. *J. Colloid & Interf. Sci.* **76,** 582–586.

BUONOPANE, R. A., GUTOFF, E. B. & RIMORE, M. M. T. 1986 Effect of plungering tape surface properties on air entrainment velocity. *AIChE J.* **32,** 682–683.

BURLEY, R. & JOLLY, R. P. S. 1984 Entrainment of air into liquids by high speed continuos solid surface. *Chem. Eng. Sci.* **39,** 1357–1372.

BURLEY, R. & KENNEDY, B. S. 1976a A study of the dynamic wetting behavior of polyester tapes. *Br. Polym. J.* **8,** 140–143.

BURLEY, R. & KENNEDY, B. S. 1976b An experimental study of air entrainment at a solid/liquid/gas interface. *Chem. Eng. Sci.* **31,** 901–911.

BURNETT, D. 1935a The distribution of velocities in a slightly non-uniform gas. *Proc. London Math. Soc.* **39,** 385–430.

BURNETT, D. 1935b The distribution of molecular velocities and the mean motion in a non-uniform gas. *Proc. London Math. Soc.* **40,** 382–435.

BURTON, J. C., RUTLEDGE, J. E. & TABOREK, P. 2004 Fluid pinch-off dynamics at nanometer length scale. *Phys. Rev. Lett.* **92,** 244505.

BUYEVICH, YU. A. & WEBBON, B. W. 1996 Bubble formation at a submerged orifice in reduced gravity. *Chem. Eng. Sci.* **51,** 4843–4857.

CAHN, J. W. 1965 Phase separation by spinodal decomposition in isotropic systems. *J. Chem. Phys.* **42,** 93–99.

CAHN, J. W. 1977 Critical point wetting. *J. Chem. Phys.* **66,** 3667–3672.

CAHN, J. W. & HILLIARD, J. E. 1958 Free energy of a nonuniform system. I. Interfacial free energy. *J. Chem. Phys.* **28,** 258–267.

CAIN, J. B., FRANCIS, D. W., VENTER, R. D. & NEUMANN, A. W. 1983 Dynamic contact angles on smooth and rough surfaces. *J. Colloid & Interf. Sci.* **94,** 123–130.

CAMERON, A. 1966 *The Principles of Lubrication.* Longmans, Green & Co., New York.

CARRE, A. & WOEHL, P. 2002 Hydrodynamic behavior at the triple line of spreading liquids and the divergence problem. *Langmuir* **18,** 3600–3603.

CASIMIR, H. B. C. & POLDER, D. 1946 Influence of retardation on London-van der Waals forces. *Nature* (London) **158,** 787–788.

CASIMIR, H. B. C. & POLDER, D. 1948 Influence of retardation on London-van der Waals forces. *Phys. Rev.* **73,** 360–372.

CASSIE, A. B. D. & BAXTER, S. 1994 Wettability of porous surfaces. *Trans. Farad. Soc.* **40,** 546–551.

CATTANEO, C. 1948 Sulla conduzione del calore. *Atti. Sem. Mat. Fis. Univ. Modena* **3,** 83–101.

CATTANEO, C. 1958 A form of heat conduction equation which eliminates the paradox of instantaneous propagation. *C. R. Acad. Sci.* (Paris) **247,** 431–433.

CHAUDHURY, M. K. & WHITESIDES, G. M. 1992 How to make water run uphill. *Science* **256,** No. 5063, 1539–1541.

CHEBBI, R. 2000 Dynamics of wetting. *J. Colloid & Interf. Sci.* **229,** 155–164.

CHECCO, A. & GUENOUN, P. 2003 Nonlinear dependence of the contact angle of nanodroplets on contact line curvature. *Phys. Rev. Lett.* **91,** 186101.

CHEN, A. U., NOTZ, P. K. & BASARAN, O. A. 2002 Computational and experimental analysis of pinch-off and scaling. *Phys. Rev. Lett.* **88,** 174501.

420 *References*

CHEN, H.-Y., JASNOW, D. & VIÑALS, J. 2000 Interface and contact line motion in a two phase fluid under shear flow. *Phys. Rev. Lett.* **85,** 1686–1689.

CHEN, J.-D. 1988 Experiments on a spreading drop and its contact angle on a solid. *J. Colloid & Interf. Sci.* **122,** 60–72.

CHEN, Q., RAMÉ, E. & GAROFF, S. 1996 Experimental studies on parametrization of liquid spreading and dynamic contact angles. *Colloid Surf.* **116,** 115–124.

CHEN, Q., RAMÉ, E. & GAROFF, S. 1997 The velocity field near moving contact lines. *J. Fluid Mech.* **337,** 49–66.

CHEN, Y.-J. & STEEN, P. H. 1997 Dynamics of inviscid capillary breakup: collapse and pinchoff of a film bridge. *J. Fluid Mech.* **341,** 245–267.

CHERRY, B. W. & HOLMES, C. M. 1969 Kinetics of wetting of surfaces by polymers. *J. Colloid & Interf. Sci.* **29,** 174–176.

CHOU, K.-S. & KWONG, Y.-C. 2007 Finite time rupture for thin films under van der Waals forces. *Noninearity* **20,** 299–317.

CHOW, T. S. 1998 Wetting of rough surfaces. *J. Phys.: Condens. Matter* **10,** L445–L451.

CHURAEV, N. V. 2003 Surface forces in wetting films. *Colloid J.* (USSR) **65,** 263–274.

CHURAEV, N. V., SOBOLEV, V. D. & SOMOV, A. N. 1984 Slippage of liquids over lyophobic solid surfaces. *J. Colloid & Interf. Sci.* **97,** 574–581.

CLARKE, A. 1995 The application of particle tracking velocimetry and flow visualisation to curtain coating. *Chem. Eng. Sci.* **50,** 2397–3407.

CLARKE, A. & STATTERSFIELD, E. 2006 Direct evidence supporting nonlocal hydrodynamic influence on the dynamic contact angle. *Phys. Fluids* **18,** 048109.

COONS, J. E., HALLEY, P. J., McGLASHAN, S. A. & TRAN-CONG, T. 2005 Scaling laws for the critical rupture thickness of common thin films. *Colloid Surf. A* **263,** 258–266.

COX, R. G. 1986 The dynamics of the spreading of liquids on a solid surface. Part 1. Viscous flow. *J. Fluid Mech.* **168,** 169–194.

CRISTINI, V., BLAWZDZIEWICZ, J. & LOEWENBERG, M. 1998 Drop breakup in three-dimensional viscous flows. *Phys. Fluids* **10,** 1781–1783.

CUVELIER, C., SEGAL, A. & VAN STEENHOVEN, A. A. 1988 *Finite Element Methods and Navier-Stokes Equations.* D. Reidel Publ. Co.

DAVIES, J. T. & RIDEAL, E. K. 1961 *Interfacial Phenomena.* Academic Press, New York–London.

DAVIS, S. H. 1980 Moving contact lines and rivulet instabilities. Part 1. The static rivulet. *J. Fluid Mech.* **98,** 225–242.

DAVIS, S. H. & HOCKING, L. M. 1999 Spreading and imbibition of viscous liquid on a porous base. *Phys. Fluids* **11,** 48–57.

DAVIS, S. H. & HOCKING, L. M. 2000 Spreading and imbibition of viscous liquid on a porous base. II. *Phys. Fluids* **12,** 1646–1655.

DAY, R. F., HINCH, E. J. & LISTER, J. R. 1998 Self-similar capillary pinchoff of an inviscid fluid. *Phys. Rev. Lett.* **80,** 704–707.

DE GENNES, P. G. 1985 Wetting: statics and dynamics. *Rev. Mod. Phys.* **57,** 827–863.

DE GENNES, P.-G., BROCHARD-WYART, F. & QUÉRÉ, D. 2003 *Capillarity and Wetting Phenomena. Drops, Bubbles, Pearls, Waves.* Springer.

DE GROOT, S. R. & MAZUR, P. 1962 *Non-equilibrium Thermodynamics.* North-Holland, Amsterdam.

DEFAY, R., PRIGOGINE, I. & BELLEMANS, A. 1966 *Surface Tension and Adsorption.* Longmans, Green & Co., London.

DE RUIJTER, M. J., BLAKE, T. D. & DE CONINCK, J. 1999 Dynamic wetting studied by molecular modeling simulations of droplet spreading. *Langmuir* **15,** 7836–7847.

DE RUIJTER, M., BLAKE, T. D., CLARKE, A. & DE CONINCK, J. 1999 Droplet spreading: a tool to characterize surfaces at the microscopic scale. *J. Petrol. Sci. Eng.* **24,** 189–198.

DE RUIJTER, M. J., DE CONINCK, J., BLAKE, T. D., CLARKE, A. & RANKIN, A. 1997 Contact angle relaxation during the spreading of partially wetting drops. *Langmuir* **13,** 7293–7298.

DE RUIJTER, M. J., DE CONINCK, J. & OSHANIN, G. 1999 Droplet spreading: partial wetting regime revisited. *Langmuir* **15,** 2209–2216.

DERJAGUIN, B. V. & ABRIKOSOVA, I. I. 1951 Direct measurement of the molecular attraction as a function of the distance between spheres. *Zhurn. Eksp. Teor. Fiz. (USSR)* **21,** 945–946.

DERJAGUIN, B. V. & ABRIKOSOVA, I. I. 1957 Direct measurement of the molecular attraction of solid bodies. I. Statement of the problem and method of measuring forces by using negative feedback. *JETP* **3,** 819–829.

DERJAGUIN, B. V., CHURAEV, N. V. & MULLER, V. M. 1987 *Surface Forces.* Plenum Press, New York (Russian edn.: Nauka, Moscow, 1985).

DIEZ, J. A., GRATTON, R., THOMAS, L. P. & MARINO, B. Laplace pressure driven drop spreading. *Phys. Fluids* **6,** 24–33.

DING, H. & SPELT, P. D. M. 2007 Inertial effects in droplet spreading: a comparison between diffuse-interface and level-set simulations. *J. Fluid Mech.* **576,** 287–296.

DODGE, F. T. 1988 The spreading of liquid droplets on solid surfaces. *J. Colloid & Interf. Sci.* **121,** 154–169.

DOOLEY, B. S., WARNCKE, A. E., GHARIB, M. & TRYGGVASON, G. 1997 Vortex ring generation due to the coalescence of a water drop at a free surface. *Experim. Fluids* **22,** 369–374.

DOWSON, D. 1979 *History of Tribology.* Longmans, Green & Co., London/New York.

DRAZIN, P. & RILEY, N. 2006 *The Navier-Stokes equations: a classification of flows and exact solutions.* Cambridge Univ. Press.

DURBIN, P. A. 1988 Consideration on the moving contact line singularity with application to frictional drag on a slender drop. *J. Fluid Mech.* **197,** 157–169.

DURBIN, P. A. 1989 Stokes flow near a moving contact line with yield-stress boundary condition. *Quart. J. Mech. App. Maths* **42,** 99–113.

DUSSAN V., E. B. 1976 The moving contact line: the slip boundary condition. *J. Fluid Mech.* **77,** Pt 4, 665–684.

DUSSAN V., E. B. 1977 Immiscible liquid displacement in a capillary tube. The moving contact line. *AIChE J.* **23,** 131–133.

DUSSAN V., E. B. 1979 On the spreading of liquids on solid surfaces: static and dynamic contact lines. *Annu. Rev. Fluid Mech.* **11,** 371–400.

DUSSAN V., E. B. & DAVIS, S. H. 1974 On the motion of a fluid-fluid interface along a solid surface. *J. Fluid Mech.* **65,** 71–95.

DZYALOSHINKII, I. E., LIFSHITZ, E. M. & PITAEVSKII, L. P. 1960 Van der Waals forces in liquid films. *JETP* **10,** 161–170.

EDGERTON, H. E., HAUSER, E. A. & TUCKER, W. B. 1937 Studies in drop formation as revealed by the high-speed motion camera. *J. Phys. Chem.* **41,** 1017–1028.

EGGERS, J. 1993 Universal pinching of 3D axisymmetric free-surface flow. *Phys. Rev. Lett.* **71,** 3458–3460.

EGGERS, J 1995 Theory of drop formation. *Phys. Fluids* **7,** 941–953.

EGGERS, J. 1997 Nonlinear dynamics and breakup of free-surface flows. *Rev. Mod. Phys.* **69,** 865–929.

EGGERS, J. 2001 Air entrainment through free-surface cusps. *Phys. Rev. Lett.* **86,** 4290–4293.

EGGERS, J. 2004 Toward a description of contact line motion at higher capillary numbers. *Phys. Fluids* **16**, 3491–3494.

EGGERS, J. & DUPONT, T. F. 1994 Drop formation in a one-dimensional approximation. *J. Fluid Mech.* **262**, 205–221.

EGGERS, J. & EVANS, R. 2004 Comment on "Dynamic wetting by liquids of different viscosity," by T.D. Blake and Y.D. Shikhmurzaev. *J. Colloid & Interf. Sci.* **280**, 537–538.

EGGERS, J., LISTER, J. R. & STONE, H. A. 1999 Coalescence of liquid drops. *J. Fluid Mech.* **401**, 293–310.

EHRHARD, P. & DAVIS, S. H. 1991 Non-isothermal spreading of liquid drops on horizontal plates. *J. Fluid Mech.* **229**, 365–388.

EINSTEIN, A. 1956 *Investigations of the Theory of Brownnian Movement* (R. Fürth, Ed.) Dover, New York.

ELLIOTT, G. E. P. & RIDDIFORD A. C. 1967 Dynamic contact angles. I. The effect of impressed motion. *J. Colloid & Interf. Sci.* **23**, 389–398.

ERNEUX, T. & DAVIS, S. H. 1993 Nonlinear rupture of free films. *Phys. Fluids* **A5**, 1117–1122.

EXTRAND, C. W. & KUMAGAI, Y. 1996 Contact angles and hysteresis on soft surfaces. *J. Colloid & Interf. Sci.* **1984**, 1991–200.

FERMIGIER, M. & JENFFER, P. 1988 Dynamics of a liquid-liquid interface in a capillary. *Annal. Physique* **13**, Coll. 2, suppl. 3, 37–42.

FERMIGIER, M. & JENFFER, P. 1991 An experimental investigation of the dynamic contact angle in liquid-liquid systems. *J. Colloid & Interf. Sci.* **146**, 226–241.

FINLOW, D. E., KOTA, P. R. & BOSE, A. 1996 Investigations of wetting hydrodynamics using numerical simulations. *Phys. Fluids* **8**, 302–309.

FINN, R. 2006 The contact angle in capillarity. *Phys. Fluids* **18**, 047102.

FIX, G. 1969 Higher-order Rayleigh-Ritz approximations. *J. Math. Mech.* **18**, 645–657.

FLEKKOY, E. G., WAGNER, G. & FEDER, J. 2000 Hybrid model for combined particle and continuum dyanmics. *Europhys. Lett.* **52**, 271–276.

FOISTER, R. T. 1990 The kinetics of displacement wetting in liquid/liquid/solid systems. *J. Colloid & Interf. Sci.* **136**, 266–282.

FOX, H. W., HARE, E. F. & ZISMAN, W. A. 1955 Wetting properties of organic liquids on high energy surfaces. *J. Phys. Chem.* **59**, 1097–1106.

FRENKEL, J. 1945 Viscous flow of crystalline bodies under the action of surface tension. *J. Phys. (USSR)* **9**, 385–391.

GAUDET, S., MCKINLEY, G. H. & STONE, H. A. 1996 Extensional deformation of Newtonian liquid bridges. *Phys. Fluids* **8,** 2568–2579.

GIBBS, J. W. 1928 *The Collected Works of J. Willard Gibbs.* Vol. 1, Longmans, Green & Co., New York.

GLASNER, K. B. 2003 Spreading of droplets under the influence of intermolecular forces. *Phys. Fluids* **15,** 1837–1842.

GLASSTONE, S., LAIDLER, K. J. & EYRING, H. 1941 *The Theory of Rate Processes.* McGraw-Hill, New York.

GOGOSOV, V. V., NALETOVA, V. A. & SHAPOSHNIKOVA, G. A. 1981 Hydrodynamics of magnetic fluids. *Itogi Nauki i Tekhn.: Mekhanika Zhidkosti i Gaza* **16,** 76–210.

GOLDSHTIK, M. A. 1990 Viscous-flow paradoxes. *Annu. Rev. Fluid Mech.* **22,** 441–472.

GOLDSTEIN, S. (Ed.) 1965 *Modern Developments in Fluid Dynamics.* vol. 2, Dover, New York.

GOODWIN, R. & HOMSY, G. M. 1991 Viscous flow down a slope in the vicinity of contact line. *Phys. Fluids* **A3,** 515–528.

GOUIN, H. 2001 The wetting problem of fluids on solid surfaces: Dynamics of lines and contact angle hysteresis. *J. Physique* **11,** 261-268.

GOURSAT, É. 1898 Sur l'équation $\Delta\Delta u = 0$. *Bull. Soc. Math. France* **26,** 236.

GRAD, H. 1949 On the kinetic theory of rarefied gases. *Commun. Pure Appl. Math.* **2,** 331-407.

GREENSPAN, H. P. 1978 On the motion of a small viscous droplet that wets a surface. *J. Fluid Mech.* **84,** Pt 1, 125–143.

GREENSPAN, H. P. & MCCAY, B. M. 1981 On the wetting of a surface by a very viscous fluid. *Stud. Appl. Maths* **64,** 95–112.

GRESHO, P. M. & SANI, R. L. 2000 *Incompressible Flow and the Finite Element Method.* Wiley.

GUGGENHEIM, E. A. 1940 The thermodynamics of interfaces in systems of several components. *Trans. Farad. Soc.* **36,** 397–412.

GUMERMAN, R. J. & HOMSY, G. M. 1975 The stability of radially bounded thin films. *Chem. Eng. Commum.* **2,** 27–36.

GUTOFF, E. B. & KENDRICK, C. E. 1982 Dynamic contact angles. *AIChE J.* **28,** 459–466.

HADJICONSTANTINOU, N. 1999a Combining atomistic and continuum simulations of contact line motion. *Phys. Rev. E* **59,** 2475–2478.

HADJICONSTANTINOU, N. G. 1999 Hybrid atomistic-continuum formulations and the moving contact-line problem. *J. Comp. Phys.* **154**, 245–265.

HALEY, P. J. & MIKSIS, M. J. 1991 The effect of the contact line on droplet spreading. *J. Fluid Mech.* **223**, 57–81.

HAMAKER, H. C. 1937 The London-van der Waals attraction between spherical particles. *Physica* **4**, 1058–1072.

HANSEN, R. J. & TOONG, T. Y. 1971 Interface behavior as one fluid completely displaces another from a small-diameter tube. *J. Colloid & Interf. Sci.* **36**, 410–413.

HANSEN, R. S. & MIOTTO, M. 1957 Relaxation phenomena and contact angle hysteresis. *J. Amer. Chem. Soc.* **79**, 1765.

HAPPEL, J. & BRENNER, H. 1965 *Low Reynolds Number Hydrodynamics.* Prentice-Hall.

HARDY, W. B. 1919 The spreading of fluids on glass. *Philos. Mag.* **38**, 49–55.

HARKINS, W. H. 1952 *The Physical Chemistry of Surface Films.*, Reinhold, New York.

HAUSER, E. A., EDGERTON, H. E. & TUCKER, W. B. 1936 The application of the high-speed motion picture camera to the research on the surface tension of liquids. *J. Phys. Chem.* **40**, 973–988.

HAYES, R. A. & RALSTON, J. 1993 Forced liquid movement on low energy surfaces. *J. Colloid & Interf. Sci.* **159**, 429–438.

HEINE, D. R., GREST, G. S. & WEBB, E. B. 2003 Spreading dynamics of polymer nanodroplets. *Phys. Rev. E* **68**, 061603.

HE, G. & HADJICONSTANTINOU, N. G. 2003 A molecular view of Tanner's law: molecular dynamics simulations of droplet spreading. *J. Fluid Mech.* **497**, 123–132.

HEINE, D. R., GREST, G. S. & WEBB, E. B. 2004 Spreading dynamics of polymer nanodroplets in cylindrical geometries. *Phys. Rev. E* **70**, 011606.

HENDERSON, D. M., PRITCHARD, W. G. & SMOLKA, L. B. 1997 On the pinch-off of a pendant drop of viscous fluid. *Phys. Fluids* **9**, 3188–3200.

HESLOT, F., CAZABAT, A.-M. & LEVINSON, P. 1989 Dynamics of wetting of tiny drops: ellipsometric study of the late stages of spreading. *Phys. Rev. Lett.* **62**, 1286–1288.

HESLOT, F., FRAYSSE, N. & CAZABAT, A. M. 1989 Molecular layering in the spreading of wetting liquid drops. *Nature* (London) **338**, 640–642.

HIRAM, Y. & NIR, A. 1983 A simulation of surface tension driven coalescence. *J. Colloid & Interf. Sci.* **95**, 462–470.

HOCKING, L. M. 1976 A moving fluid interface on a rough surface. *J. Fluid Mech.* **76,** Pt 4, 801–817.

HOCKING, L. M. 1977 A moving fluid interface. Part 2. The removal of the force singularity by a slip flow. *J. Fluid Mech.* **79,** Pt 2, 209–229.

HOCKING, L. M. 1981 Sliding and spreading of two-dimensional drops. *Quart. J. M App. Maths* **34,** Pt 1, 37–55.

HOCKING, L. M. 1987 The damping of capillary-gravity waves at a rigid boundary. *J. Fluid Mech.* **179,** 253–266.

HOCKING, L. M. 1990 Spreading and instability of a viscous fluid sheet. *J. Fluid Mech.* **211,** 373–392.

HOCKING, L. M. 1992 Rival contact-angle models and the spreading of drops. *J. Fluid Mech.* **239,** 671–681.

HOCKING, L. M. 1993 The influence of intermolecular forces on thin fluid layers. *Phys. Fluids* **A5,** 793–799.

HOCKING, L. M. & RIVERS, A. D. 1982 The spreading of a drop by capillary action. *J. Fluid Mech.* **121,** 425–442.

HOFFMAN R. 1975 A study of the advancing interface. I. Interface shape in liquid-gas systems. *J. Colloid & Interf. Sci.* **50,** 228–241.

HOPPER, R. W. 1984 Coalescence of two equal cylinders — exact results for creeping viscous plane flow driven by capillarity. *J. Amer. Ceram. Soc. (Commun.)* **67,** C262–264. Errata, *ibid.* **68,** C138 (1985).

HOPPER, R. W. 1990 Plane Stokes flow driven by capillarity on a free surface. *J. Fluid Mech.* **213,** 349–375.

HOPPER, R. W. 1991 Plane Stokes flow driven by capillarity on a free surface. Part 2. Further developments. *J. Fluid Mech.* **230,** 355–364.

HOPPER, R. W. 1992 Stokes flow of a cylinder and half-space driven by capillarity. *J. Fluid Mech.* **243,** 171–181.

HOPPER, R. W. 1993a Coalescence of 2 viscous cylinders by capillarity: Part I. Theory. *J. Amer. Ceram. Soc.* **76,** 2947–2952.

HOPPER, R. W. 1993b Coalescence of 2 viscous cylinders by capillarity: Part II. Shape evolution. *J. Amer. Ceram. Soc.* **76,** 2953–2960.

HUGHES, T. J. R. 1987 *The Finite Element Method: Linear Static and Dynamic Finite Element Analysis.* Prentice-Hall.

HUH, C. & MASON, S. G. 1977a The steady movement of a liquid meniscus in a capillary tube. *J. Fluid Mech.* **81,** 401–419.

HUH, C. & MASON, S. G. 1977b Effects of surface roughness on wetting (theoretical). *J. Colloid & Interf. Sci.* **60,** 1138.

HUH, C. & SCRIVEN, L. E. 1971 Hydrodynamic model of steady movement of a solid/liquid/fluid contact line. *J. Colloid & Interf. Sci.* **35**, 85–101.

INVERARITY, G. 1969 Dynamic wetting of glass fibre and polymer fibre. *Br. Polym. J.* **1**, 245–251.

ISHIMI, K., HIKITA, H. & ESMAIL, M. N. 1986 Dynamic contact angles on moving plates. *AIChE J.* **32**, 486–500.

ISRAELACHVILI, J. N. & TABOR, D. 1972–1973 The measurement of van der Waals dispersion forces in the range 1.5 to 130 nm. *Proc. Roy. Soc. A* (London) **331**, 19–38.

IWAMOTO, C. & TANAKA, S. 2002 Atomic morphology and chemical reactions of the reactive wetting front. *Acta Mater.* **50**, 749–755.

JACQMIN, D. 2000 Contact-line dynamics of a diffuse fluid interface. *J. Fluid Mech.* **402**, 57–88.

JANSONS, K. M. 1985 Moving contact lines on a two-dimensional rough surface. *J. Fluid Mech.* **154**, 1–28.

JEONG, J.-T. 1999 Formation of cusp on the free surface at low Reynolds number flow. *Phys. Fluids* **11**, 521–526.

JEONG, J.-T. & MOFFATT, H. K. 1992 Free-surface cusps associated with flow at low Reynolds number. *J. Fluid Mech.* **241**, 1–22.

JIANG, T. S., OH, S. G. & SLATTERY, J. C. 1979 Correlation for dynamic contact angle. *J. Colloid & Interf. Sci.* **69**, 74–77.

JOHNSON, R. E., DETTRE, R. H. & BRANDRETH, D. A. 1977 Dynamic contact angles and contact angle hysteresis. *J. Colloid & Interf. Sci.* **62**, 205–212.

JONES, A. F. & WILSON, S. D. R. 1978 The film drainage problem in droplet coalescence. *J. Fluid Mech.* **87**, 263–288.

JOSEPH, D. D. 1992 Understanding cusped interfaces. *J. Non-Newton. Fluid Mech.* **44**, 127–148.

JOSEPH, D. D., NELSON, J., RENARDY, M. & RENARDY, Y. 1991 Two-dimensional cusped interfaces. *J. Fluid Mech.* **223**, 383–409.

KALLMANN, H. & WILLSTAETTER, M. 1932 Zur Theorie des Aufbaues kolloidaler Systeme. *Naturwiss.* **20**, 952–953.

KELLER, J. B. 1983 Breaking of liquid films and threads. *Phys. Fluids* **26**, 3451–3453.

KHATAVKAR, V. V., ANDERSON, P. D. & MEIJER, H. E. H. 2007 Capillary spreading of a droplet in the partially wetting regime using a diffuse-interface model. *J. Fluid Mech.* **572**, 367–387.

KHESHGI, H. S. & SCRIVEN, L. E. 1991 Dewetting: Nucleation and growth of dry regions. *Chem. Eng. Sci.* **46,** 519–526.

KHOROMIN, N. YA. (Ed.) 1918 *Encyclopedia of Thought.* Pantheon, Kiev.

KISS, E. & GÖLANDER, C.-G. 1991 Static and dynamic wetting behaviour of poly(ethyleneoxide) layer formed on mica substrate. *Colloid Surf.* **58,** 263–270.

KISTLER, S. F. & SCHWEIZER, P. M. (Eds.) 1997 *Liquid Film Coating.* Chapman & Hall.

KITCHENER, J. A. & PROSSER, A. P. 1957 Direct measurement of the long-range van der Waals forces. *Proc. Roy. Soc. A* (London) **242,** 403–409.

KOCHIN, N. E., KIBEL, I. A. & ROZE, N. V. 1964 *Theoretical Hydromechanics.* Interscience.

KOCHUROVA, N. N. & RUSANOV, A. I. 1981 Dynamic surface properties of water: surface tension and surface potential. *J. Colloid & Interf. Sci.* **81,** 297–303.

KOENIG, F. O. 1950 On the thermodynamic relation between surface tension and curvature. *J. Chem. Phys.* **18,** 449–459.

KOPLIK, J. & BANAVAR, J. R. 1993 Molecular dynamics of interface rupture. *Phys. Fluids* **A5,** 521–536.

KOPLIK, J., BANAVAR, J. R. & WILLEMSEN, J. F. 1988 Molecular dynamics of Poiseuille flow and moving contact lines. *Phys. Rev. Lett.* **60,** 1282–1285.

KOPLIK, J., BANAVAR, J. R. & WILLEMSEN, J. F. 1989 Molecular dynamics of a fluid flow at solid surfaces. *Phys. Fluids* **A1,** 781–794.

KOVAC, J. 1977 Non-equilibrium thermodynamics of interfacial systems. *Physica A* **86,** 1–24.

KOWALEWSKI, T. A. 1996 On the separation of droplets from a liquid jet. *Fluid Dyn. Res.* **17,** 121–145.

KRÖNER, D. 1987 The flow of a fluid with a free boundary and dynamic contact angle. *ZAMM* **67,** 304–306.

KUZNETSOV, B. G. 1993 Hyperbolic modification of the Navier-Stokes equations. *J. Appl. Mech. Techn. Phys.* **34,** 133–141.

LADYZHENSKAYA, O. A. 1969 *The Mathematical Theory of Viscous Incompressible Flows.* (2nd edition), Gordon & Breach.

LADYZHENSKAYA, O. A. 1975 Mathematical analysis of Navier-Stokes equations for incompressible liquids. *Annu. Rev. Fluid Mech.* **7,** 249–272.

LAFAURIE, B., NARDONE, C., SCARDOVELLI, R., ZALESKI, S. & ZANETTI, G. 1994 Modelling merging and fragmentation in multiphase flows with SURFER. *J. Comp. Phys.* **113**, 134–147.

LAM, C. N. C., KIM, N., HUI, D., KWOK, D. Y., HAIR, M. L. & NEUMANN, A. W. 2001 The effect of liquid properties to contact angle hysteresis. *Colloid Surf. A* **189**, 265–278.

LAM, C. N. C., WU, R., LI, D., HAIR, M. L., & NEUMANN, A. W. 2002 Study of the advancing and receding contact angles: liquid sorption as a cause of contact angle hysteresis. *Adv. Colloid & Interf. Sci.* **96**, 169–191.

LAMB, H. 1932 *Hydrodynamics.* Cambridge Univ. Press.

LANDAU, L. D. & LIFSHITZ, E. M. 1987 *Fluid Mechanics.* Pergamon.

LANGLOIS, W. E. 1964 *Slow Viscous Flow.* Macmillan, New York.

LAPLACE, P.-S. 1806 Supplément au dixiéme livre du Traité de mécanique céleste. Sur l'action capillaire. Paris (see Laplace, P.-S. *Oeuvres complète,* Paris, 1878–1912, v. 4, pp. 349–417.)

LAPLACE, P.-S. 1807 Supplément à la théorie de l'action capillaire. Paris (see Laplace, P.-S. *Oeuvres complète,* Paris, 1878–1912, v. 4, pp. 419–498.)

LE, H. P. 1998 Progress and trends in ink-jet printing technology. *J. Imaging Sci. & Technol.* **42**, 49–62.

LEE, Y.-N. & CHIAO, S.-M. 1996 Visualization of dynamic contact angles on cylinder and fiber. *J. Colloid & Interf. Sci.* **181**, 378–394.

LENARD, P. 1887 Ueber die Schwingungen fallender Tropfen. *Ann. Physik* **30**, 209-243.

LEPPINEN, D. & LISTER, J. R. 2003 Capillary pinch-off in inviscid fluids. *Phys. Fluids* **15**, 568–578.

LESSER, M. B. 1983 The impact of compressible liquids. *Annu. Rev. Fluid Mech.* **15**, 97–122.

LEVINE, S., LOWNDES, J., WATSON, E. J. & NEALE, G. 1980 A theory of capillary rise of a liquid in a vertical cylindrical tube and in a parallel-plate channel. Washburn equation modified to account for the meniscus with slippage at the contact line. *J. Colloid & Interf. Sci.* **73**, 136–151.

LEWIS, P. E. & WARD, J. P. 1991 *The Finite Element Method. Principles and Applications.* Addison-Wesley.

LIDE, D. R. (ED.) 2004 *CRC Handbook of Chemistry and Physics.* 85th edn. CRC Press.

LIN, J. N., BANERJI, S. K. & YASUDA, H. 1994 Role of interfacial tension in the formation and the detachment of air bubbles. I. A single hole on a horizontal plane immersed in water. *Langmuir* **10**, 936–942.

LIN, S. P. 2003 *Breakup of liquid sheets and jets.* Cambridge Univ. Press.

LIN, S. P. & REITZ, R. D. 1998 Drop and spray formation from a liquid jet. *Annu. Rev. Fluid Mech.* **30,** 85–105.

LISTER, J. R., BRENNER, M. P., DAY, R. F., HINCH, E. J. & STONE, H. A. 1997 Singularities and similarity solutions in capillary breakup. In *Proc. IUTAM Symposium on Non-linear Singularities in Deformation and Flow,* Haifa, Israel, 17–21 March 1997 (D. Durban & J.R.A. Pearson, eds.) Kluwer Acad. Publ., pp. 257–269.

LISTER, J. R. & STONE, H. A. 1998 Capillary breakup of a viscous thread surrounded by another viscous fluid. *Phys. Fluids* **10,** 2758–2764.

LONDON, F. 1930 Über das Verhältnis der van der Waalsschen Kräfte zu den homöopolaren Bindungskräften. *Zeit. f. Phys.* **60,** 491–527.

LOPEZ, J., MILLER, C. A. & RUCKENSTEIN, E. 1976 Spreading kinetics of liquid drops on solids. *J. Colloid & Interf. Sci.* **56,** 460–468.

LOPEZ, P. G., BANKOFF, S. G. & MIKSIS, M. J. 1996 Non-isothermal spreading of a thin liquid film. *J. Fluid Mech.* **324,** 261–286.

LORENCEAU, É., RESTAGNO, F. & QUÉRÉ, D. 2003 Fracture of a viscous liquid. *Phys. Rev. Lett.* **90,** 184501.

LOVETT, R., DeHAVEN, P. W., VIECELI, JR., J. J. & BUFF, F. P. 1973 Generalized van der Waals theories for surface tension and interfacial width. *J. Chem. Phys.* **58,** 1880–1885.

LOWNDES, J. 1980 The numerical simulation of the steady movement of a fluid meniscus in a capillary tube. *J. Fluid Mech.* **101,** 631–645.

LUCAS, R. VON 1918 Über das Zeitgesetz des kapillaren Aufstiegs von Flüssigkeiten. *Koll.-Zeitschr.* **23,** 15–22.

LUDVIKSSON, V. & LIGHTFOOT, E. N. 1968 Deformation of advancing menisci. *AIChE J.* **14,** 674–677.

LUKYANOV, A. & SHIKHMURZAEV, Y. D. 2006 Curtain coating in microfluidics and the phenomenon of nonlocality in dynamic wetting. *Phys. Lett. A* **358,** 426–430.

LUKYANOV, A. V. & SHIKHMURZAEV, Y. D. 2007a The effect of flow field and geometry on the dynamic contact angle. *Phys. Rev. E* **75,** 051604.

LUKYANOV, A. V. & SHIKHMURZAEV, Y. D. 2007b A combined BIE-FE method for the Stokes equations. *IMA J. Appl. Maths* (in press).

LYKOV, A. V. 1965 Application of methods of thermodynamics of irreversible processes to the investigation of heat exchange. *Sov. J. Eng. Phys.* **9,** 287–304.

MAJDA, A. J. & BERTOZZI, A. L. 2001 *Vorticity and Incompressible Flow.* Cambridge Univ. Press

MARANGONI, C. & STEFANELLI, P. 1872 Monografia sulle bolle liquide. *Nuovo Cimento* Ser. 2, **7–8**, 301–356.

MARANGONI, C. & STEFANELLI, P. 1873 Monografia sulle bolle liquide. *Nuovo Cimento* Ser. 2, **9**, 236–256.

MARIOTTE, E. 1686 *Traité du Mouvement des Eaux et des Autres Corps Fluides.* E. Michallet, Paris.

MARKOVA, M. P. & SHKADOV, V. YA. 1972 On nonlinear development of capillary waves in a liquid jet. *Sov. Fluid Dynam.* **3**, 30–37.

MARMUR, A. 1998 Line tension effects on contact angles: Axisymmetric and cylindrical systems with rough or heterogeneous surfaces. *Colloid Surf. A* **136**, 81–88.

MARMUR, A. 2006 Soft contact: measurement and interpretation of contact angles. *Soft Matter* **2**, 12–17.

MARMUR, A. & LELAH, M. D. 1980 The dependence of drop spreading on the size of the solid surface. *J. Colloid & Interf. Sci.* **78**, 262–265.

MARTIC, G., GENTNER, F., SEVANO, D., COULON, D., DE CONINCK, J. & BLAKE, T. D. 2002 A molecular dynamics simulations of capillary imbibition. *Lagmuir* **18**, 7971–7976.

MARTIC, G., GENTNER, F., SEVANO, D., DE CONINCK, J. & BLAKE, T. D. 2004 The possibility of different time-scales in the dynamics of pores imbibition. *J. Colloid & Interf. Sci.* **270**, 171–179.

MARTÍNEZ-HERRERA, J. I. & DERBY, J. J. 1995 Viscous sintering of spherical particles via finite element analysis. *J. Amer. Ceram. Soc.* **78**, 645–649.

MAXWELL, J. C. 1875 Capillary action. In: *Encyclopedia Britannica.*

MELESHKO, V. V. 2003 Selected topics in the history of the two-dimensional biharmonic problem. *Appl. Mech. Rev.* **56**, 33–85.

MENCHACA-ROCHA, A., MARTÍNEZ-DÁVALOS, A., NÚÑEZ, R., POPINET, S. & ZALESKI, S. 2001 Coalescence of liquid drops by surface tension. *Phys. Rev. E* **63**, 046309.

MERCHANT, G. J. & KELLER, J. B. 1992 Contact angles. *Phys. Fluids* **A4**, 477–485.

MILLER, C. A. & RUCKENSTEIN, E. 1974 The origin of flow during wetting of solids. *J. Colloid & Interf. Sci.* **48**, 368–373.

MILNE-THOMSON, L. M. 1968 *Theoretical Hydrodynamics.* (5th edn.), Macmillan, London.

MOFFATT, H. K. 1964 Viscous and resistive eddies near a sharp corner. *J. Fluid Mech.* **18,** Pt 1, 1–18.

MONIN, A. S. 1962 On the Lagrangian equations of hydrodynamics of incompressible viscous fluid. *PMM — Sov. Appl. Maths & Mech.* **26,** 320–327.

MOSELER, M. & LANDMAN, U. 2000 Formation, stability, and breakup of nanojets. *Science* **289,** 1165–1169.

MÜLLER, I. & RUGGERI, T. 1998 *Rational Extended Thermodynamics.* (2nd end.), Springer.

MUMLEY, T. E., RADKE, C. J. & WILLIAMS, M. C. 1986a Kinetics of liquid/liquid capillary rise. I. Experimental observations. *J. Colloid & Interf. Sci.* **109,** 398–412.

MUMLEY, T. E., RADKE, C. J. & WILLIAMS, M. C. 1986b Kinetics of liquid/liquid capillary rise. II. Development and test of theory. *J. Colloid & Interf. Sci.* **109,** 413–425.

MUSKHELISHVILI, N. I. 1919 Sur l'intégration de l'équation biharmonique. *Izv. AN SSSR* 663–686.

MUSKHELISHVILI, N. I. 1963 *Some Basic Problems of the Mathematical Theory of Elasticity.* P. Noordhoff Ltd.

NAPOLITANO, L. G. 1978 Thermodynamics and dynamics of pure interfaces. *Acta Astronautica* **5,** 655–670.

NAPOLITANO, L. G. 1979 Thermodynamics and dynamics of surface phases. *Acta Astronautica* **6,** 1093–1112.

NAPOLITANO, L. G. 1986 Recent development of Marangoni flows theory and experimental results. *Adv. Space Res.* **6,** No. 5, 19–34.

NAVIER, M. 1823 Mémoire sur les lois du mouvement des fluides. *Mém. de l'Acad. des Sciences l'Inst de France* **6,** 389–440.

NEOGI, P. & MILLER, C. A. 1983 Spreading kinetics of a drop on a rough solid surface. *J. Colloid & Interf. Sci.* **92,** 338–349.

NEOGI, P. & MILLER, C. A. 1982 Spreading kinetics of a drop on a smooth solid surface. *J. Colloid & Interf. Sci.* **86,** 525–538.

NETO, C., EVANS, D. R., BONACCURSO, E., BUTT, H.-J. & CRAIG, V. S. J. 2005 Boundary slip in Newtonian liquids: a review of experimental studies. *Rep. Prog. Phys.* **68,** 2859–2897.

NEUMANN, A. W. & GOOD, R. J. 1979 Methods of measuring contact angles. In: *Colloind and Surface Science*, R. J. Good and R. R. Stromberg (Eds.), vol. 11, pp. 31–91, Plenum Press, New York.

NGAN, C. G. & DUSSAN V., E. B. 1982 On the nature of the dynamic contact angle: an experimental study. *J. Fluid Mech.* **118,** 27–40.

NIE, X. B., CHEN, S. Y., E, W. N. & ROBBINS, M. O 2004 A continuum and molecular dynamics by hybrid method for micro- and nano-fluid flow. *J. Fluid Mech.* **500**, 55–64.

NIGMATULIN, R. I. 1991 *Dynamics of Multiphase Media.* Vols. 1 & 2, Hemisphere Publ., NY–London.

NORTHRUP, M. A., JENSEN, K. F. & HARRISON, D. J. (EDS.) 2003 *μTAS, 7th Intl Conf. on Micro Total Analysis Systems.* Squaw Valley, California, USA, 2003, Transducers Res. Foundation, San Diego. CA.

NOTZ, P. K. & BASARAN, O. A. 2004 Dynamics and breakup of a contracting liquid filament. *J. Fluid Mech.* **512**, 223–256.

NOTZ, P. K., CHEN, A. U. & BASARAN, O. A. 2001 Satellite drops: Unexpected dynamics and change of scaling during pinch-off. *Phys. Fluids* **13**, 549–551.

OĞUZ, H. N. & PROSPERETTI, A. 1989 Surface-tension effects in the contact of liquid surfaces. *J. Fluid Mech.* **203**, 149–171.

OLIVER, J. F. & MASON, S. G. 1977 Microspreading studies on rough surfaces by scanning electron microscopy. *J. Colloid & Interf. Sci.* **60**, 480–487.

ONSAGER, L. 1931a Reciprocal relations in irreversible processes. I. *Phys. Rev.* **37**, 405–426.

ONSAGER, L. 1931b Reciprocal relations in irreversible processes. II. *Phys. Rev.* **38**, 2265–2279.

ORON, A., DAVIS, S. H. & BANKOFF, S. G. 1997 Long-scale evolution of thin liquid films. *Rev. Mod. Phys.* **69**, 931–980.

PANTON, R. L. 2005 *Incompressible Flow.* (3rd edn.) Wiley, NJ.

PAPAGEORGIOU, D. T. 1995a On the breakup of viscous liquid threads. *Phys. Fluids* **7**, 1529–1544.

PAPAGEORGIOU, D. T. 1995b Analytical description of the breakup of liquid jets. *J. Fluid Mech.* **301**, 109–132.

PETROV, J. G. & PETROV, P. G. 1992a Forced advancement and retraction of polar liquids on a low energy surface. *Colloid Surf.* **64**, 143–149.

PETROV, J. G. & RADOEV, B. P. 1981 Steady motion of the three phase contact line in model Langmuir-Blodgett systems. *Colloid Polym. Sci.* **259**, 753–760.

PETROV, J. G., RALSTON, J., SCHNEEMILCH, M. & HAYES, R. A. 2003a Dynamics of partial wetting and dewetting of an amorphous fluoropolymer by pure liquids. *Langmuir* **19**, 2795–2801.

PETROV, J. G., RALSTON, J., SCHNEEMILCH, M. & HAYES, R. A. 2003b Dynamics of partial wetting and dewetting in well-defined systems. *J. Phys. Chem. B* **107**, 1634–1645.

PETROV, J. P. & SEDEV, R. V. 1985 On the existence of a maximum speed of wetting. *Colloid Surf.* **13**, 313–322.

PETROV, P. G. & PETROV, J. G. 1992b A combined molecular-hydrodynamic approach to wetting kinetics. *Langmuir* **8**, 1762–1767.

PISMEN, L. M. 2001 Nonlocal diffuse interface theory of thin films and the moving contact line. *Phys. Rev. E* **64**, 021603.

PISMEN, L. M. 2002 Mesoscopic hydrodynamics of contact line motion. *Colloid Surf. A* **206**, 11–30.

PISMEN, L. M. & POMEAU, Y. 2000 Disjoining potential and spreading of thin liquid layers in the diffuse-interface model coupled to hydrodynamics. *Phys. Rev. E* **62**, 2480–2492.

PISMEN, L. M. & RUBINSTEIN, B. Y. 2001 Kinetic slip condition, van der Waals forces, and dynamic contact angle. *Langmuir* **17**, 5265–5270.

PISMEN, L. M., RUBINSTEIN, B. Y. & BAZHLEKOV, I. 2000 Spreading of a wetting film under the action of van der Waals forces. *Phys. Fluids* **12**, 480–483.

PLATEAU, J. 1843 Mémoire sur les phénomènes que présente une masse liquide libre et soustraite de l'action de la pesanteur. *Mém. de l'Acad. Roy. Belgique, nuvelle s'er.* **16**, 1–35. English translation: Experimental and theoretical researches on the figures of equilibrium of a liquid mass withdrawn from the action of gravity. The Annu. Rept of the Board of Regents of the Smithsonian Inst., 1863, Government Printing Office, Washington D.C. 207–225.

PLATEAU, J. 1849 Sur les figures d'équilibre d'une masse liquide sans pésanteur. *Mém. de l'Acad. Roy. Belgique, nuvelle s'er.* **23**, 1–159.

PLATEAU, J. 1873 *Statique Expérimentale et Théoretique des Liquides soumis aux seules Forces Moléculaires.* Gautier-Villars, Paris.

POISSON, S. D. 1831 Mémoire sur les equations générales de l'équilibre et du mouvement des curps solides élastiques et des fluides. *J. de l'École Polytechnique* **13**, cahier 20, 1–174.

POZRIKIDIS, C. 1999 Capillary instability and breakup of a viscous thread. *J. Eng. Maths* **36**, 255–275.

PRIGOGINE, I. 1967 *Introduction to Thermodynamics of Irreversible Processes.* Interscience.

QIAN, T., WANG, X.-P. & SHENG, P. 2003 Molecular scale contact line hydrodynamics of immiscible flows. *Phys. Rev. E* **68**, 016306.

QIAN, T., WANG, X.-P. & SHENG, P. 2006 A variational approach to moving contact line hydrodynamics. *J. Fluid Mech.* **564,** 333–360.

RAMÉ, E., GAROFF, S. & WILLSON, K. R. 2004 Characterizing the microscopic physics near moving contact lines using dynamic contact angle data. *Phys. Rev. E* **70,** 031608.

RAMOS, S. M. M., CHARLAIX, E., BENYAGOUB, A. & TOULEMONDE, M. 2003 Wetting on nanorough surfaces. *Phys. Rev. E* **67,** 031604.

RAYLEIGH, LORD 1879a On the instability of jets. *Proc. London Math. Soc.* **10,** 4–13.

RAYLEIGH, LORD 1879b On the capillary phenomena of jets. *Proc. Roy. Soc.* (London) **29,** 71–97.

RAYLEIGH, LORD 1891 Some applications of photography. *Nature* (London) **44,** 249–254.

RAYLEIGH, LORD 1892 On the instability of cylindrical fluid surfaces. *Philos. Mag.* **34,** 177–180.

REDDY, J. N. 1985 *An Introduction to the Finite Element Method.* McGraw-Hill.

REINER, M. 1964 The Deborah number. *Phys. Today* No. 1, 62.

REMOORTERE, P. VAN & JOOS, P. 1993 About the kinetics of partial wetting. *J. Colloid & Interf. Sci.* **160,** 387–396.

RENARDY, M. 2005 A comment on self-similar breakup for inertialess Newtonian liquid jets. *IMA J. Appl. Maths* **70,** 353–358.

RENARDY, M., RENARDY, Y. & LI, J. 2001 Numerical simulation of moving contact line problem using a volume-of-fluid method. *J. Comp. Phys.* **171,** 243–263.

REYNOLDS, O. 1886 On the theory of lubrication and its application to Mr. Beauchamp Tower's experiments, including an experimental determination of the viscosity of olive oil. *Philos. Trans. R. Soc.* (London) **177,** 157–234.

RICHARDSON, S. 1968 Two-dimensional bubbles in slow viscous flows. *J. Fluid Mech.* **33,** 476–493.

RICHARDSON, S. 1992 Two-dimensional slow viscous flows with time-dependent free boundaries driven by surface tension. *Euro. J. Appl. Maths* **3,** 193–207.

RICHARDSON, S. 1997 Two-dimensional Stokes flows with time-dependent free boundaries driven by surface tension. *Euro. J. Appl. Maths* **8,** 311–329.

RIDEAL, E. K. 1922 On the flow of liquids under capillary pressure. *Philos. Mag.* **44,** 1152–1159.

RILLAERTS, E. & JOOS, P. 1980 The dynamic contact angle. *Chem. Eng. Sci.* **35**, 883–887.

RIO, E., DAERR, A., ANDREOTTI, B. & LIMAT, L. 2005 Boundary conditions in the vicinity of a dynamic contact line: Experimental investigation of viscous drops sliding down an inclined plane. *Phys. Rev. Lett.* **94**, 024503.

ROGERS, W. B. 1858 On the formation of rotating rings by air and liquids under certain conditions of discharge. *Amer. J. Sci. Arts* (Ser. 2), **26**, 246–258.

RONIS, D., BEDEAUX, D. & OPPENHEIM, I. 1978 On the derivation of dynamical equations for a system with an interface: I. General theory. *Physica A* **90**, 487–506.

RONIS, D. & OPPENHEIM, I. 1983 On the derivation of dynamical equations for a system with an interface: II. The gas-liquid interface. *Physica A* **117**, 317–354.

ROSE, W. & HEINS, R. W. 1962 Moving interfaces and contact angle rate dependence. *J. Colloid & Interf. Sci.* **17**, 39–48.

ROSS, J. W., MILLER, W. A. & WEATHERLY, G. C. 1981 Dynamic computer simulation of viscous flow sintering kinetics. *J. Appl. Phys.* **52**, 3884-3888.

ROTHER, M. A., ZINCHENKO, A. Z. & DAVIS, R. H. 1997 Buoyancy-driven coalescence of slightly deformable drops. *J. Fluid Mech.* **346**, 117–148.

ROTHERT, A., RICHTER, R. & REHBERG, I. 2001 Transition from symmetric to asymmetric scaling function before drop pinch-off. *Phys. Rev. Lett.* **87**, 084501.

ROTT, N. 1990 Note on the history of the Reynolds number. *Annu. Rev. Fluid Mech.* **22**, 1–11.

ROWLINSON, J. S. 2002 *Cohesion. A Scientific History of Intermolecular Forces.* Cambridge Univ. Press.

ROWLINSON, J. S. & WIDOM, B. 1982 *Molecular Theory of Capillarity.* Clarendon, Oxford.

RUCKENSTEIN, E. 1992 Dynamics of partial wetting. *Langmuir* **8**, 3038–3039.

RUCKENSTEIN, E. & DUNN, C. S. 1977 Slip velocity during wetting of solids. *J. Colloid & Interf. Sci.* **59**, 135–138.

RUCKENSTEIN, E. & JAIN, R. K. 1974 Spontaneous rupture of thin liquid films. *Faraday Trans. 2* **70**, 132–147.

RUGGERI, T. 1983 Symmetric-hyperbolic system of conservative equations for a viscous heat conducting fluid. *Acta Mech.* **47**, 167–183.

RUGGERI, T. 1989 Galilean invariance and entropy principle for system of balance laws. The structure of extended thermodynamics. *Continuum Mech. Thermodyn.* **1**, 3–20.

RUSANOV, A. I. 2005 Surface thermodynamics revisited. *Surf. Sci. Rep.* **58**, No. 5–8, 111–239.

SAINT-VENANT, B. DE 1843 Note á joindre au mémoire sur la dynamique des fluides. *C. R. Acad. Sci.* (Paris) **17**, 1240–1243.

SAVART, F. 1833 Mémoire sur la constitution des veines liquides lancées par des orifices circulaires en mince paroi. *Annal. Chim.* **53**, 337–374, plates in vol. **54**.

SAVELSKI, M. J., SHETTY, S. A., KOLB, W. B. & CERRO, R. L. 1995 Flow patterns associated with steady movement of a solid/liquid/fluid contact line. *J. Colloid & Interf. Sci.* **176**, 117–127.

SCHENA, M., HELLER, R. A., THERIAULT, T. P., KONRAD, K., LACHEN-MEIER E. & DAVIS, R. W. 1998 Microarrays: Biotechnology's discovery platform for functional genomics. *Trends Biotechnol.* **16**, 301–306.

SCHNEEMILCH, M., HAYES, R. A., PETROV, J. G. & RALSTON, J. 1998 Dynamic wetting and dewetting of a low-energy surface by pure liquids. *Langmuir* **14**, 7047–7051.

SCHOFIELD, P. & HENDERSON, J. R. 1982 Statistical mechanics of inhomogeneous fluids. *Proc. Roy. Soc. A* (London) **379**, 231–249.

SCHWARTZ, A. M., RADER, C. A. & HUEY, E. 1964 Resistance to flow in capillary systems of positive contact angle. In *Contact Angle, Wettability and Adhesion*, (R. F. Gould, Ed.), Adv. in Chem. No. 43, Washington DC, Amer. Chem. Soc., pp. 250–267

SCHWARTZ, A. M. & TEJADA, S. B. 1970 Studies of dynamic contact angles on solids. *NASA Contract Rep. CR-72728.*

SCHWARTZ, A. M. & TEJADA, S. B. 1972 Studies of dynamic contact angles on solids. *J. Colloid & Interf. Sci.* **38**, 359–375.

SCRIVEN, L. E. 1960 Dynamics of a fluid interface. Equation of motion for Newtonian surface fluids. *Chem. Eng. Sci.* **12**, 98–108.

SCRIVEN, L. E. & STERNING, C. V. 1960 The Marangoni effects. *Nature* (London) **187**, 186–188.

SEDEV, R. V., BUDZIAK, C. J., PETROV, J. G. & NEUMANN, A. W. 1993 Dynamic contact angles at low velocities. *J. Colloid & Interf. Sci.* **159**, 392–399.

SEDEV, R. V. & PETROV, J. G. 1991 The critical condition for transition from steady wetting to film entrainment. *Colloid Surf.* **53**, 147–156.

SEDEV, R. V. & PETROV, J. G. 1988 Influence of geometry of the three-phase system on the maximum speed of wetting. *Colloid Surf.* **34**, 197–201.

SEDEV, R. V. & PETROV, J. G. 1992 Influence of geometry on steady dewetting kinetics. *Colloid Surf.* **62**, 141–151.

SEDOV, L. I. 1965a *Introduction to the Mechanics of a Continuous Medium.* Addison-Wesley.

SEDOV, L. I. 1965b *Two-Dimensional Problems of Hydrodynamics and Aerodynamics.* Interscience.

SEDOV, L. I. 1997 *Mechanics of Continuous Media.* Vol. 1 & 2, World Scientific, Singapore.

SEPPECHER, P. 1996 Moving contact lines in the Cahn-Hilliard theory. *Int. J. Eng. Sci.* **34**, 977–992.

SHANKAR, P. N. & DESHPANDE, M. D. 2000 Fluid mechanics in the driven cavity. *Annu. Rev. Fluid Mech.* **32**, 93–136.

SHARMA, A. & RUCKENSTEIN, E. 1986 An analytical nonlinear theory of thin film rupture and its application to wetting films. *J. Colloid & Interf. Sci.* **113**, 456–479.

SHELUDKO, A. 1967 Thin liquid films. *Adv. Colloid & Interf. Sci.* **1**, 391–464.

SHEN, C. & RUTH, D. W. 1998 Experimental and numerical investigations of the interface profile close to a moving contact line. *Phys. Fluids* **10**, 789–799.

SHENG, P. & ZHOU, M. 1992 Immiscible-fluid displacement: Contact-line dynamics and the velocity-dependent capillary pressure. *Phys. Rev. A* **45**, 5694–5708.

SHI, X. D., BRENNER, M. P. & NAGEL, S. R. 1994 A cascade of structure in a drop falling from a faucet. *Science* **265**, 219–222.

SHIKHMURZAEV, Y. D. 1993 The moving contact line on a smooth solid surface. *Int. J. Multiphase Flow* **19**, 589–610.

SHIKHMURZAEV, Y. D. 1994 Mathematical modeling of wetting hydrodynamics. *Fluid Dynam. Res.* **13**, 45–64.

SHIKHMURZAEV, Y. D. 1996 Dynamic contact angles in gas/liquid/solid systems and flow in vicinity of moving contact line. *AIChE J.* **42**, 601–612.

SHIKHMURZAEV, Y. D. 1997a Moving contact lines in liquid/liquid/solid systems. *J. Fluid Mech.* **334**, 211–249.

SHIKHMURZAEV, Y. D. 1997b Spreading of drops on solid surfaces in a quasi-static regime. *Phys. Fluids* **9**, 266–275.

SHIKHMURZAEV, Y. D. 1998 On cusped interfaces. *J. Fluid Mech.* **359,** 313–328.

SHIKHMURZAEV, Y. D. 1999 On the initial stage of the bubble growth on a plate. In *Proc. of the Second Internat. Symposium on Two-Phase Modelling and Experimentation*, May 23–26, 1999, Pisa, Italy), (G.P. Celata, P. Di Marco and R.K. Shah, Eds.), Edizioni ETS, v. 2, 907–914.

SHIKHMURZAEV, Y. D. 2002 On metastable regimes of dynamic wetting. *J. Phys.: Condens. Matter* **14,** 319–330.

SHIKHMURZAEV, Y. D. 2005a Singularity of free-surface curvature in convergent flow: cusp or corner? *Phys. Lett. A* **245,** 378–385.

SHIKHMURZAEV, Y. D. 2005b Capillary breakup of liquid threads: A singularity-free solution. *IMA J. Appl. Maths* **70,** 880–907.

SHIKHMURZAEV, Y. D. 2005c Macroscopic mechanism of rupture of free liquid films. *C.R. Mecanique* **333,** 205–210.

SHIKHMURZAEV, Y. D. 2006 Singularities at the moving contact line. Mathematical, physical and computational aspects. *Physica D* **217,** 121–133.

SHIKHMURZAEV, Y. D. & BLAKE, T. D. 2004 Response to the comment on [J. Colloid Interface Sci. 253 (2002) 196] by J. Eggers and R. Evans. *J. Colloid & Interf. Sci.* **280,** 539–541.

SHLIOMIS, M. I. 1971 Effective viscosity of magnetic suspensions. *JETP* **61,** 2411–2418.

SIEROU, A. & LISTER, J. R. 2003 Self-similar solutions for viscous capillary pinch-off. *J. Fluid Mech.* **497,** 381–403.

ŠIKALO, Š., WILHELM, H.-D., ROISMAN, I. V., JAKIRLIĆ, S. & TROPEA, C. 2005 Dynamic contact line of spreading droplets: Experiments and simulations. *Phys. Fluids* **17,** 062103.

SMITH, M. K. 1995 Thermocapillary migration of a two-dimensional liquid droplet on a solid surface. *J. Fluid Mech.* **294,** 209–230.

SOBOLEV, V. D., CHURAEV, N. V., VELARDE, M. G. & ZORIN, Z. M. 2001 Dynamic contact angles of water in ultrathin capillaries. *Colloid J.* (USSR) **63,** 119–123.

SOMALINGA, S. & BOSE, A. 2000 Numerical investigation of boundary conditions for moving contact line problems. *Phys. Fluids* **12,** 499-510.

SPRITTLES, J. E. & SHIKHMURZAEV, Y. D. 2007 Viscous flow over a chemically patterned surface. *Phys. Rev. E* **76,** in press.

STAROV, V. M. 1983 Spreading of droplets of nonvolatile liquids over a flat surface. *Colloid J.* (USSR) **45,** 1009–1015.

STOKES, G. G. 1845 On the theories of the internal friction of fluids in motion. *Trans. Camb. Phil. Soc.* **8**, 287–305. (Reprinted in: Mathematical and Physical Papers (1880) **1**: 75–129, Cambridge Univ. Press)

STRANG, G. & FIX, G. J. 1973 *An Analysis of the Finite Element Method.* Prentice-Hall.

STRÖM, G., FREDRIKSSON, M., STENIUS, P. & RADOEV, B. 1990 Kinetics of steady-state wetting. *J. Colloid & Interf. Sci.* **134**, 107–115.

SUMNER, C. G. 1937 An apparatus for the measurement of contact angles by the plate method. In *Wetting and Detergency,* (W. Clayton, Ed.), pp. 41–52, Chem. Publ. Co., New York.

TABOR, D. & WINTERTON, R. H. S. 1968 Surface forces: Direct measurement of normal and retarded van der Waals forces. *Nature* **219**, 1120–1121.

TABOR, D. & WINTERTON, R. H. S. 1969 The direct measurement of normal and retarded van der Waals forces. *Proc. Roy. Soc. A* (London) **312**, 435–450.

TANNER, L. 1979 The spreading of silicone oil drops on horizontal surfaces. *J. Phys. D* **12**, 1473–1484.

TELETZKE, G. F., SCRIVEN, L. E. & DAVIS, H. T. 1982 Gradien theory of wetting transitions. *J. Colloid & Interf. Sci.* **87**, 550–571.

TEMPLETON, C. C. 1956 Oil-water displacements in microscopic capillaries. *Trans. AIME* **207**, 211–214.

THIELE, U. 2003 Open questions and promising new fields in dewetting. *Euro. Phys. J., Ser. E* **12**, 409–416.

THOMPSON, P. A., BRINCKERHOFF, W. B. & ROBBINS, M. O. 1993 Microscopic studies of static and dynamic contact angles. *J. Adhesion* **7**, 535–554.

THOMPSON, P. A. & ROBBINS, M. O. 1989 Simulations of contact line motion: slip and the dynamic contact angle. *Phys. Rev. Lett.* **63**, 766-769.

THOMPSON, P. A. & ROBBINS, M. O. 1990 Shear flow near solids: Epitaxial order and flow boundary conditions. *Phys. Rev. A* **41**, 6830–6837.

THOMSON, J. J. & NEWALL, H. F. 1885 On the formation of vortex rings by drops falling into liquids and some allied phenomena. *Proc. Roy. Soc.* (London) **39**, 417–436.

THOMSON, W. 1869 On vortex motion. *Trans. Roy. Soc. Edinburgh* **15**, 217–260.

THORODDSEN, S. T. & TAKEHARA, K. 2000 The coalescence cascade of a drop. *Phys. Fluids* **12**, 1265–1267.

THORODDSEN, S. T., TAKEHARA, K. & ETOH, T. G. 2005 The coalescence speed of a pendent and sessile drop. *J. Fluid Mech.* **527**, 85–114.

TILTON, J. N. 1988 The steady motion of an interface between two viscous liquids in a capillary tube. *Chem. Eng. Sci.* **43**, 1371–1384.

TOLMAN, R. C. 1949 The effect of droplet size on surface tension. *J. Chem. Phys.* **17**, 333–337.

TREVIÑO, C., FERRO-FONTÁN, C. & MÉNDEZ, F. 1998 Asymptotic analysis of axisymmetric drop spreading. *Phys. Rev. E* **56**, 4478–4484.

TRIEZENBERG, D. G. & ZWANZIG, R. 1972 Fluctuation theory of surface tension. *Phys. Rev. Lett.* **28**, 1183–1185.

TROIAN, S. M., HERBOLZHEIMER, E., SAFRAN, S. A. & JOANNY, J. F. 1989 Fingering instabilities of driven spreading films. *Europhys. Lett.* **10**, 25–30.

TSAI, W.-T. & YUE, D. K. P. 1996 Computation of nonlinear free-surface flows. *Annu. Rev. Fluid Mech.* **28**, 249–278.

VAILLANT, M. P. 1913 Sur un procédé de mesure des grandes résistances polarisables et son application a la mesure de la résistance de liquides. *C. R. Hebd. Sean. Acad. Sci.* Paris,

VAN MOURIK, S., VELDMAN, A. E. P. & DREYER, M. E. 2005 Simulation of capillary flow with a dynamic contact angle. *Microgravity Sci. Technol.* **17**, 87–94.

VAN DER WAALS, J. D. 1893 The thermodynamic theory of capillarity under the hypothesis of a continuous variation of density. *Verhandel. Konink. Akad. Weten. Amsterdam* (Sect. 1), **1**, No. 8 [English transl. by J. S. Rowlinson in *J. Statist. Phys.* **20** (1979) 197–244].

VERNOTTE, P. 1958 Les paradoxes de la théorie continue de l'équation de la chaleur. *C. R. Acad. Sci.* (Paris) **246**, 3154–3155.

VIGNES-ADLER, M. & BRENNER, H. 1985 A micromechanical derivation of the differential equations of interfacial statics. III. Line tension. *J. Colloid & Interf. Sci.* **103**, 11–44.

VILLANUEVA, W. & AMBERG, G. 2006 Some generic capillary-driven flows. *Int. J. Multiphase Flow* **32**, 1072–1086.

VILLERMAUX, E. 2007 Fragmentation. *Annu. Rev. Fluid Mech.* **39**, 419–446.

VISWANATH, D. S., GHOSH, T. K., PRASAD, D. H. L., DUTT, N. V. K. & RANI, K. Y. 2007 *Viscosity of Liquids: Theory, Estimation, Experiment, and Data.* Springer.

VOINOV, O. V. 1976 Hydrodynamics of wetting. *Fluid Dynam.* **11**, 714–721.

VOINOV, O. V. 1978 Asymptote to the free surface of a viscous liquid creeping on a surface and the velocity dependence of the angle of contact. *Sov. Phys. — Doklady* **23**, 891–893.

VRIJ, A. 1966 Possible mechanism for the spontaneous rupture of thin, free liquid films. *Discuss. Faraday Soc.* **42**, 23–33.

WALDMANN, L. 1967 Non-equilibrium thermodynamics of boundary conditions. *Z. Naturforschung* **22a**, 1269–1279.

WANG, C. Y. 1989 Exact solutions of the unsteady Navier-Stokes equations. *Appl. Mech. Rev.* **42**, No. 11, Pt 2, S269–S282.

WANG, C. Y. 1991 Exact solutions of the steady-state Navier-Stokes equations. *Annu. Rev. Fluid Mech.* **23**, 159–177.

WANG, X. D., PENG, X. F. & LEE, D. Z. 2003 Dynamic wetting and stress singularity on contact line. *Sci. China, Ser. E: Technol. Sci.* **46**, 407–417.

WASHBURN, E. W. 1921 The dynamics of capillary flow. *Phys. Rev.* **17**, 273–283.

WEINSTEIN, S. J. & RUSCHAK, K. J. 2004 Coating flows. *Annu. Rev. Fluid Mech.* **36**, 29–53.

WENZEL, R. N. 1936 Resistance of solid surfaces to wetting by water. *Ind. Eng. Chem.* **28**, 988–994.

WENZEL, R. N. 1949 Surface roughness and contact angle. *J. Phys. Colloid Chem.* **53**, 1466.

WEST, G. D. 1911 On the resistance to the motion of a thread of mercury in a glass tube. *Proc. Roy. Soc. A* (London) **86**, 20–25.

WEYMANN, H. D. 1967 Finite speed of propagation in heat conduction, diffusion and viscous shear motion. *Amer. J. Phys.* **35**, 488-496.

WHETHAM, W. C. D. 1890 On the alleged slipping at the boundary of a liquid in motion. *Philos. Trans. A* **181**, 559–582.

WHITEMAN, J. R. (ED.) 1973 *The Mathematics of Finite Elements and Applications.* (Proc. Brunel Univ. Conf. of the IMA), Academic Press.

WIDOM, B. 1978 Structure of the $\alpha\gamma$ interface. *J. Chem. Phys.* **68**, 3878–3883.

WILKES, E. D., PHILLIPS, S. D. & BASARAN, O. A. 1999 Computational and experimental analysis of dynamics of drop formation. *Phys. Fluids* **11**, 3577–3597.

WILKINSON, W. 1975 Entrainment of air by a solid surface entering a liquid/air interface. *Chem. Eng. Sci.* **30**, 1227–1230.

WILLIAMS, M. B. & DAVIS, S. H. 1982 Nonlinear theory of film rupture. *J. Colloid & Interf. Sci.* **90**, 220–228.

WILLIAMS, R. 1977 The advancing front of a spreading liquid. *Nature* (London) **266,** 153–154.

WILSON, M. C. T., SUMMERS, J. L., SHIKHMURZAEV, Y. D., CLARKE, A. & BLAKE, T. D. 2006 Nonlocal hydrodynamic influence on the dynamic contact angle: Slip models versus experiment. *Phys. Rev. E* **73,** 041606.

WU, M., CUBAUD, T. & HO, C.-M. 2004 Scaling law in liquid drop coalescence driven by surface tension. *Phys. Fluids* **16,** L51–L54.

WU, Q. & WONG, H. 2004 A slope-dependent disjoining pressure for nonzero contact angles. *J. Fluid Mech.* **506,** 157–185.

YARNOLD, G. D. 1938 The motion of a mercury index in a capillary tube. *Proc. Phys. Soc.* (London) **50,** 540–552.

YARNOLD, G. D. & MASON, B. J. 1949 The angle of contact between water and wax. *Proc. Phys. Soc.* (London) **B62,** 125–128.

YOUNG, T. 1805 An essay on the cohesion of fluids. *Philos. Trans. R. Soc.* (London) **95,** 65–87.

ZHANG, N. & CHAO, D. F. 2002 A new laser shadowgraphy method for measurements of dynamic contact angle and simultaneous flow visualization in a sessile drop. *Optics & Laser Technol.* **34,** 243–248.

ZHANG, W. W. & LISTER, J. R. 1999 Similarity solutions for capillary pinch-off in fluids of different viscosity. *Phys. Rev. Lett.* **83,** 1151–1154.

ZHELEZNYI, B. V. & KORNEVA, T. V. Dynamic wetting angle of a dry lyophilic surface. *Sov. Phys. — Doklady* **249,** 569–572.

ZHOU, M. Y. & SHENG, P. 1990 Dynamics of immiscible fluid displacement in a capillary tube. *Phys. Rev. Lett.* **64,** 882–885.

ZOSEL, A. 1993 Studies of the wetting kinetics of liquid drops on solid surfaces. *Colloid Polym. Sci.* **271,** 680–687.

Index of Authors

Subject Index